# Numerical Approximation Methods for Elliptic Boundary Value Problems

Olaf Steinbach

# Numerical Approximation Methods for Elliptic Boundary Value Problems

## Finite and Boundary Elements

Springer

Olaf Steinbach
Institute of Computational Mathematics
Graz University of Technology
Austria

Originally published in the German language by B.G. Teubner Verlag as “Olaf Steinbach: Numerische Näherungsverfahren für elliptische Randwertprobleme. 1. Auflage (1st ed.)”.


English version published by Springer Science+Business Media, LLC

ISBN 978-1-4419-2173-4 e-ISBN 978-0-387-68805-3

Mathematics Subject Classification (2000): 65N30, 65N38



Printed on acid-free paper.

9 8 7 6 5 4 3 2 1

springer.com

# Preface

Finite and boundary element methods belong to the most used numerical discretization methods for the approximate solution of elliptic boundary value problems. Finite element methods (FEM) are based on a variational formulation of the partial differential equation to be solved. The definition of a conforming finite dimensional trial space requires an appropriate decomposition of the computational domain into finite elements. The advantage of using finite element methods is their almost universal applicability, e.g. when considering nonlinear partial differential equations. Contrary, the use of boundary element methods (BEM) requires the explicit knowledge of a fundamental solution, which allows the transformation of the partial differential equation to a boundary integral equation to be solved. The approximate solution then only requires a decomposition of the boundary into boundary elements. Boundary element methods are often used to solve partial differential equations with (piecewise) constant coefficients, and to find solutions of boundary value problems in exterior unbounded domains. In addition, direct boundary element methods provide a direct computation of the complete Cauchy data which are the real target functions in many applications. In finite element methods, the Cauchy data can be computed by using Lagrange multipliers and by solving related saddle point problems. By combining both discretization methods it is possible to profit from the advantages of both methods.

Although the aim of this book is to give a unified introduction into finite and boundary element methods, the main focus of the presentation is on the numerical analysis of boundary integral equation methods. Therefore, we only consider some linear model problems such as the potential equation, the system of linear elasticity, the Stokes system, and the Helmholtz equation. When considering the above mentioned elliptic boundary value problems it is possible to describe and to analyze finite and boundary element methods in a unified manner. After the description of the model problems, we introduce the function spaces which are needed later. Then we discuss variational methods for the solution of operator equations with and without side conditions. In particular, this also includes the formulation of saddle point problems by using

Lagrange multipliers. The variational formulation of boundary value problems is the basis of finite element methods, but on the other hand, domain variational methods are also needed in the analysis of boundary integral operators. After the computation of fundamental solutions we define certain boundary integral operators, and analyze their mapping properties such as boundedness and ellipticity. For the solution of different boundary value problems we then describe and analyze different boundary integral equations to find the complete Cauchy data. Numerical discretization methods are first formulated and investigated in a more abstract setting. Afterwards, appropriate finite dimensional trial spaces are constructed, and corresponding approximation properties are given. For the solution of mixed boundary value problems we then discuss different finite and boundary element methods. In particular, we investigate the properties of the associated linear systems of algebraic equations. For this, we also describe appropriate preconditioned iterative solution strategies where the proposed preconditioning techniques involves both the preconditioning with integral operators of the opposite order, and a hierarchical multilevel preconditioner. Since the Galerkin discretization of boundary integral operators leads to dense stiffness matrices, fast boundary element methods are used to obtain an almost optimal complexity of storage and of matrix by vector multiplications. Finally, we describe domain decomposition methods to handle partial differential equations with jumping coefficients, and to couple and parallelize different discretization techniques such as finite and boundary element methods.

Boundary element methods are a well established numerical method for elliptic boundary value problems as discussed in this textbook. For the sake of simplicity in the presentation, we only consider the case of linear and self–adjoint partial differential equations. For more general partial differential operators one has to consider the fundamental solution of the formally adjoint operator. While the existence of fundamental solutions can be ensured for a large class of partial differential operators their explicit knowledge is mandatory for a numerical realization. For nonlinear partial differential equations there also exist different approaches to formulate boundary integral equation methods which are often based on the use of volume potentials to cover the nonlinear terms.

While there exists a rather huge number of textbooks on finite element methods, e.g. [5, 21, 31, 41, 57] just to mention a few of them, much less is available on boundary integral and boundary element methods. For the analysis of boundary integral operators related to elliptic partial differential equations we refer to [3, 60, 81, 88, 102, 103] while for the numerical analysis of boundary element methods we mention [39, 124, 126, 135]. In addition, the references [9, 15, 19, 30, 61, 77] are on practical aspects of the use of boundary element methods in engineering. In [74, 98, 125, 158] one may find recent results on the use of advanced boundary element algorithms. For a detailed description of fast boundary element methods starting from the basic ideas and proceeding to their practical realization see [117] where also numerous examples are given.

Since the aim of this textbook is to give a unified introduction into finite and boundary element methods, not all topics of interest can be discussed. For a further reading we refer, e.g., to [51, 130] for *hp* finite element methods, to [2, 8, 10, 153] for a posteriori error estimators and adaptive finite element methods, and to [24, 69] for multigrid methods. In the case of boundary element methods we refer, e.g., to [17, 99, 100, 131, 146] for *hp* methods, to [37, 38, 56, 128, 129] for a posteriori error estimators and adaptive methods, and to [96, 111, 147] for multigrid and multilevel methods. Within this textbook we also do not discuss the matter of numerical integration, for this we refer to [55, 67, 87, 93, 115, 123, 132, 148] and the references given therein.

This textbook is based on lectures on the numerical solution of elliptic partial differential equations which I taught at the University of Stuttgart, at the Technical University of Chemnitz, at the Johannes Kepler University of Linz, and at Graz University of Technology. Chapters 1–4, 8, 9, 11, and 13 can be used for an introductory lecture on finite element methods, while chapters 1–8, 10, 12, and 13 are on the basics of the boundary element method. Chapter 14 gives an overview on fast boundary element methods. Besides the use as a complementary textbook it is also recommended for self–study for students and researchers, both in applied mathematics, in scientific computing, and in computational engineering.

It is my great pleasure to thank W. L. Wendland for his encouragement and support over the years. Many results of our joint work influenced this book. Special thanks go to J. Breuer and G. Of, who read the original German manuscript and made valuable comments and corrections. This text book was originally published in a German edition [140]. Once again I would like to thank J. Weiss and B. G. Teubner for the fruitful cooperation.

When preparing the English translation I got many responses, suggestions and hints on the German edition. I would like to thank all who helped to improve the book. In particular I thank G. Of, S. Engleder and D. Copeland who read the English manuscript. Finally I thank Springer New York for the cooperation and the patience when preparing this book.

Graz, August 2007 Olaf Steinbach

# Contents

# 1 Boundary Value Problems

In this chapter we describe some stationary boundary value problems with self–adjoint partial differential operators of second order. As simple model problems we consider the scalar potential equation and the Helmholtz equation, while for a system we consider the equations of linear elasticity and, as for incompressible materials, the Stokes system.

## 1.1 Potential Equation

Let $\Omega \subset \mathbb{R}^d$ $(d = 2, 3)$ be a bounded and simply connected domain with sufficiently smooth boundary $\Gamma = \partial\Omega$, and $\underline{n}(x)$ is the exterior unit normal vector which is defined almost everywhere for $x \in \Gamma$. For $x \in \Omega$ we consider a self–adjoint linear partial differential operator of second order which is applied to a scalar real valued function $u$,

$$(Lu)(x) := -\sum_{i,j=1}^{d} \frac{\partial}{\partial x_j}\left[a_{ji}(x)\frac{\partial}{\partial x_i}u(x)\right] + a_0(x)u(x)\,. \qquad (1.1)$$

The coefficient functions $a_{ji}(x)$ are assumed to be sufficient smooth satisfying $a_{ij}(x) = a_{ji}(x)$ for all $i, j = 1, \ldots, d$, $x \in \Omega$. Partial differential operators of the form (1.1) are used to model, for example, the static heat transfer, electrostatic potentials, or ideal fluids.

For a classification [79] of scalar partial differential operators $L$ we consider the real eigenvalues $\lambda_k(x)$ of the symmetric coefficient matrix

$$A(x) = \left(a_{ij}(x)\right)_{i,j=1}^{d}, \quad x \in \Omega.$$

The partial differential operator $L$ is called elliptic at $x \in \Omega$ iff $\lambda_k(x) > 0$ is satisfied for all $k = 1, \ldots, d$. If this condition is satisfied for all $x \in \Omega$, then $L$ is elliptic in $\Omega$. If there exists a uniform lower bound $\lambda_0 > 0$ satisfying

$$\lambda_k(x) \geq \lambda_0 \quad \text{for } k = 1, \ldots, d \text{ and for all } x \in \Omega, \qquad (1.2)$$

the partial differential operator $L$ is called uniformly elliptic in $\Omega$.

The starting point for what follows is the well known theorem of Gauss and Ostrogradski, i.e.

$$\int_\Omega \frac{\partial}{\partial x_i} f(x)dx = \int_\Gamma \gamma_0^{\text{int}} f(x) n_i(x) ds_x, \quad i = 1, \ldots, d,$$

where

$$\gamma_0^{\text{int}} f(x) := \lim_{\Omega \ni \widetilde{x} \to x \in \Gamma} f(\widetilde{x}) \quad \text{for } x \in \Gamma = \partial\Omega \tag{1.3}$$

is the interior boundary trace of a given function $f(x)$, $x \in \Omega$.

For sufficiently smooth functions $u$, $v$ we consider $f(x) = u(x)v(x)$ to obtain the formula of integration by parts,

$$\int_\Omega v(x) \frac{\partial}{\partial x_i} u(x) dx = \int_\Gamma \gamma_0^{\text{int}} u(x) \gamma_0^{\text{int}} v(x) n_i(x) ds_x - \int_\Omega u(x) \frac{\partial}{\partial x_i} v(x) dx. \tag{1.4}$$

When multiplying the partial differential operator (1.1) with a sufficiently smooth test function $v$, and integrating the result over $\Omega$, this gives

$$\int_\Omega (Lu)(x)v(x)dx = - \sum_{i,j=1}^d \int_\Omega \frac{\partial}{\partial x_j} \left[ a_{ji}(x) \frac{\partial}{\partial x_i} u(x) \right] v(x) dx \,.$$

Applying integration by parts, see (1.4), we obtain

$$\begin{aligned}\int_\Omega (Lu)(x)v(x)dx = &\sum_{i,j=1}^d \int_\Omega a_{ji}(x) \frac{\partial}{\partial x_i} u(x) \frac{\partial}{\partial x_j} v(x) dx \\ &- \sum_{i,j=1}^d \int_\Gamma n_j(x) \gamma_0^{\text{int}} \left[ a_{ji}(x) \frac{\partial}{\partial x_i} u(x) \right] \gamma_0^{\text{int}} v(x) ds_x,\end{aligned}$$

and therefore Green's first formula

$$a(u,v) = \int_\Omega (Lu)(x)v(x)dx + \int_\Gamma \gamma_1^{\text{int}} u(x) \gamma_0^{\text{int}} v(x) ds_x \tag{1.5}$$

by using the symmetric bilinear form

$$a(u,v) := \sum_{i,j=1}^d \int_\Omega a_{ji}(x) \frac{\partial}{\partial x_i} u(x) \frac{\partial}{\partial x_j} v(x) dx \tag{1.6}$$

as well as the interior conormal derivative

$$\gamma_1^{\text{int}} u(x) := \lim_{\Omega \ni \widetilde{x} \to x \in \Gamma} \left[ \sum_{i,j=1}^d n_j(x) a_{ji}(\widetilde{x}) \frac{\partial}{\partial \widetilde{x}_i} u(\widetilde{x}) \right] \quad \text{for } x \in \Gamma. \tag{1.7}$$

As in (1.5) we find, by exchanging the role of $u$ and $v$, the analogue Green's formula

$$a(u,v) = a(v,u) = \int_\Omega (Lv)(x)u(x)dx + \int_\Gamma \gamma_1^{\text{int}} v(x)\gamma_0^{\text{int}} u(x)ds_x .$$

Combining this with (1.5) we therefore obtain Green's second formula

$$\int_\Omega (Lu)(x)v(x)dx + \int_\Gamma \gamma_1^{\text{int}} u(x)\gamma_0^{\text{int}} v(x)ds_x \tag{1.8}$$

$$= \int_\Omega (Lv)(x)u(x)dx + \int_\Gamma \gamma_1^{\text{int}} v(x)\gamma_0^{\text{int}} u(x)ds_x$$

which holds for arbitrary but sufficiently smooth functions $u$ and $v$.

*Example 1.1.* For the special choice $a_{ij}(x) = \delta_{ij}$ where $\delta_{ij}$ is the Kronecker delta with $\delta_{ij} = 1$ for $i = j$ and $\delta_{ij} = 0$ for $i \neq j$, the partial differential operator (1.1) is the Laplace operator

$$(Lu)(x) = -\Delta u(x) := -\sum_{i=1}^d \frac{\partial^2}{\partial x_i^2} u(x) \quad \text{for } x \in \mathbb{R}^d. \tag{1.9}$$

The associated conormal derivative (1.7) coincides with the normal derivative

$$\gamma_1^{\text{int}} u(x) = \frac{\partial}{\partial n_x} u(x) := \underline{n}(x) \cdot \nabla u(x) \quad \text{for } x \in \Gamma.$$

Let $\Gamma = \overline{\Gamma}_D \cup \overline{\Gamma}_N \cup \overline{\Gamma}_R$ be a disjoint decomposition of the boundary $\Gamma = \partial\Omega$. The boundary value problem is to find a scalar function satisfying the partial differential equation

$$(Lu)(x) = f(x) \quad \text{for } x \in \Omega, \tag{1.10}$$

the Dirichlet boundary conditions

$$\gamma_0^{\text{int}} u(x) = g_D(x) \quad \text{for } x \in \Gamma_D, \tag{1.11}$$

the Neumann boundary conditions

$$\gamma_1^{\text{int}} u(x) = g_N(x) \quad \text{for } x \in \Gamma_N, \tag{1.12}$$

and the Robin boundary conditions

$$\gamma_1^{\text{int}} u(x) + \kappa(x)\gamma_0^{\text{int}} u(x) = g_R(x) \quad \text{for } x \in \Gamma_R \tag{1.13}$$

where $f$, $g_D$, $g_N$, $g_R$, and $\kappa$ are some given functions. The boundary value problem (1.10) and (1.11) with $\Gamma = \Gamma_D$ is called a Dirichlet boundary value problem, while the boundary value problem (1.10) and (1.12) with $\Gamma = \Gamma_N$ is called a Neumann boundary value problem. In the case $\Gamma = \Gamma_R$ the problem (1.10) and (1.13) is said to be a Robin boundary value problem. In all other cases we have to solve boundary value problems with boundary conditions of mixed type. Note that one may also consider nonlinear Robin type boundary conditions [119]

$$\gamma_1^{\text{int}} u(x) + G(\gamma_0^{\text{int}} u, x) \;=\; g_R(x) \quad \text{for } x \in \Gamma_R$$

where $G(v, \cdot)$ is some given function, e.g. $G(v, \cdot) = v^3$ or $G(v, \cdot) = v^4$.

In the classical approach, the solution of the boundary value problem (1.10)–(1.13) has to be sufficiently differentiable, in particular we require

$$u \in C^2(\Omega) \cap C^1(\Omega \cup \Gamma_N \cup \Gamma_R) \cap C(\Omega \cup \Gamma_D)$$

where we have to assume that the given data are sufficiently smooth. For results on the unique solvability of boundary value problems in the classical sense we refer, for example, to [92].

In the case of a Neumann boundary value problem (1.10) and (1.12) additional considerations are needed to investigate the solvability of the boundary value problem. Obviously, $v_1(x) = 1$ for $x \in \Omega$ is a solution of the homogeneous Neumann boundary value problem

$$(Lv_1)(x) = 0 \quad \text{for } x \in \Omega, \quad \gamma_1^{\text{int}} v_1(x) \;=\; 0 \quad \text{for } x \in \Gamma. \tag{1.14}$$

Applying Green's second formula (1.8) we then obtain the orthogonality

$$\int_\Omega (Lu)(x)dx + \int_\Gamma \gamma_1^{\text{int}} u(x) ds_x = 0\,. \tag{1.15}$$

When considering the Neumann boundary value problem (1.10) and (1.12),

$$(Lu)(x) \;=\; f(x) \quad \text{for } x \in \Omega, \quad \gamma_1^{\text{int}} u(x) \;=\; g_N(x) \quad \text{for } x \in \Gamma, \tag{1.16}$$

and using the orthogonality (1.15) for the given data $f$ and $g_N$, we have to assume the solvability condition

$$\int_\Omega f(x)dx + \int_\Gamma g_N(x) ds_x = 0. \tag{1.17}$$

Since there exists a non–trivial solution $v_1(x) = 1$ for $x \in \Omega$ of the homogeneous Neumann boundary value problem (1.14), we conclude that the solution of the Neumann boundary value problem (1.16) is only unique up to an additive constant. Let $u$ be a solution of (1.16). Then, for any $\alpha \in \mathbb{R}$ we can define

$$\widetilde{u}(x) \;=\; u(x) + \alpha \quad \text{for } x \in \Omega$$

to be also a solution of the Neumann boundary value problem (1.16). The constant $\alpha \in \mathbb{R}$ is uniquely determined when requiring an additional scaling condition on the solution $u$ of (1.16), e.g.

$$\int_\Omega u(x)\,dx = 0, \quad \text{or} \quad \int_\Gamma \gamma_0^{\text{int}} u(x)\,ds_x = 0.$$

## 1.2 Linear Elasticity

As an example for a system of partial differential equations we consider the system of linear elasticity. For any $x \in \Omega$ we have to find a vector valued function $\underline{u}(x)$ with components $u_i(x)$, $i = 1,2,3$, describing the displacements of an elastic body where we assume a reversible, isotropic and homogeneous material behavior. In particular, we consider the equilibrium equations

$$-\sum_{j=1}^{3} \frac{\partial}{\partial x_j} \sigma_{ij}(\underline{u}, x) = f_i(x) \quad \text{for } x \in \Omega,\ i = 1,2,3. \tag{1.18}$$

In (1.18), $\sigma_{ij}(\underline{u}, x)$ are the components of the stress tensor which is linked to the strain tensor $e_{ij}(\underline{u}, x)$ by Hooke's law,

$$\sigma_{ij}(\underline{u}, x) = \frac{E\nu}{(1+\nu)(1-2\nu)} \delta_{ij} \sum_{k=1}^{3} e_{kk}(\underline{u}, x) + \frac{E}{1+\nu} e_{ij}(\underline{u}, x) \tag{1.19}$$

for $x \in \Omega$, $i, j = 1,2,3$, and with the Young modulus $E > 0$ and with the Poisson ratio $\nu \in (0, \frac{1}{2})$. Moreover, when assuming small deformations the linearized strain tensor is given by

$$e_{ij}(\underline{u}, x) = \frac{1}{2}\left[\frac{\partial}{\partial x_i} u_j(x) + \frac{\partial}{\partial x_j} u_i(x)\right] \quad \text{for } x \in \Omega,\ i, j = 1,2,3. \tag{1.20}$$

Multiplying the equilibrium equations (1.18) with a test function $v_i$, integrating over $\Omega$, and applying integration by parts, this gives for $i = 1,2,3$

$$\begin{aligned} \int_\Omega f_i(x) v_i(x) dx &= -\int_\Omega \sum_{j=1}^{3} \frac{\partial}{\partial x_j} \sigma_{ij}(\underline{u}, x) v_i(x) dx \\ &= \int_\Omega \sum_{j=1}^{3} \sigma_{ij}(\underline{u}, x) \frac{\partial}{\partial x_j} v_i(x) dx - \int_\Gamma \sum_{j=1}^{3} n_j(x) \sigma_{ij}(\underline{u}, x) v_i(x) ds_x. \end{aligned}$$

Taking the sum for $i = 1,2,3$ we obtain Betti's first formula

$$-\int_\Omega \sum_{i,j=1}^{3} \frac{\partial}{\partial x_j} \sigma_{ij}(\underline{u}, x) v_i(x) dx = a(\underline{u}, \underline{v}) - \int_\Gamma \gamma_0^{\text{int}} \underline{v}(x)^\top \gamma_1^{\text{int}} \underline{u}(x) ds_x \tag{1.21}$$

with the bilinear form

$$
\begin{aligned}
a(\underline{u},\underline{v}) &:= \int_\Omega \sum_{i,j=1}^3 \sigma_{ij}(\underline{u},x)\frac{\partial}{\partial x_j} v_i(x)dx \\
&= \int_\Omega \sum_{i,j=1}^3 \sigma_{ij}(\underline{u},x)\frac{1}{2}\left[\frac{\partial}{\partial x_j}v_i(x) + \frac{\partial}{\partial x_i}v_j(x)\right] dx \\
&= \int_\Omega \sum_{i,j=1}^3 \sigma_{ij}(\underline{u},x) e_{ij}(\underline{v},x)dx,
\end{aligned}
\tag{1.22}
$$

and with the conormal derivative

$$
(\gamma_1^{\text{int}}\underline{u})_i(x) := \sum_{j=1}^3 \sigma_{ij}(\underline{u},x)n_j(x) \quad \text{for } x\in\Gamma, i = 1,2,3. \tag{1.23}
$$

Inserting the strain tensor (1.20) as well as Hooke's law (1.19) into the equilibrium equations (1.18) we obtain a system of partial differential equations where the unknown function is the displacement field $\underline{u}$. First we have

$$
\sum_{k=1}^3 e_{kk}(\underline{u},x) = \sum_{k=1}^3 \frac{\partial}{\partial x_k}u_k(x) =: \operatorname{div}\underline{u}(x)
$$

and therefore

$$
\begin{aligned}
\sigma_{ii}(\underline{u},x) &= \frac{E\nu}{(1+\nu)(1-2\nu)}\operatorname{div}\underline{u}(x) + \frac{E}{1+\nu}\frac{\partial}{\partial x_i}u_i(x), \\
\sigma_{ij}(\underline{u},x) &= \frac{E}{2(1+\nu)}\left[\frac{\partial}{\partial x_i}u_j(x) + \frac{\partial}{\partial x_j}u_i(x)\right] \quad \text{for } i\neq j.
\end{aligned}
$$

From this we obtain

$$
-\frac{E}{2(1+\nu)}\Delta u_i(x) - \left[\frac{E\nu}{(1+\nu)(1-2\nu)} + \frac{E}{2(1+\nu)}\right]\frac{\partial}{\partial x_i}\operatorname{div}\underline{u}(x) = f_i(x)
$$

for $x\in\Omega$, $i = 1,2,3$. By introducing the Lamé constants

$$
\lambda = \frac{E\nu}{(1+\nu)(1-2\nu)}, \quad \mu = \frac{E}{2(1+\nu)} \tag{1.24}
$$

we finally conclude the Navier system

$$
-\mu\Delta\underline{u}(x) - (\lambda+\mu)\operatorname{grad}\operatorname{div}\underline{u}(x) = \underline{f}(x) \quad \text{for } x\in\Omega. \tag{1.25}
$$

The bilinear form (1.22) can be written as

$$\begin{aligned} a(\underline{u},\underline{v}) &= \sum_{i,j=1}^{3} \int_\Omega \sigma_{ij}(\underline{u},x)e_{ij}(\underline{v},x)dx \\ &= 2\mu \int_\Omega \sum_{i,j=1}^{3} e_{ij}(\underline{u},x)e_{ij}(\underline{v},x)dx + \lambda \int_\Omega \operatorname{div}\underline{u}(x)\operatorname{div}\underline{v}(x)\,dx \end{aligned} \tag{1.26}$$

implying the symmetry of the bilinear form $a(\cdot,\cdot)$.

The first component of the conormal derivative (1.23) is

$$\begin{aligned} (\gamma_1^{\text{int}}\underline{u})_1(x) &= \sum_{j=1}^{3} \sigma_{1j}(\underline{u},x)n_j(x) \\ &= \left[\lambda \operatorname{div}\underline{u}(x) + 2\mu\frac{\partial}{\partial x_1}u_1(x)\right] n_1(x) + \mu\left[\frac{\partial}{\partial x_1}u_2(x) + \frac{\partial}{\partial x_2}u_1(x)\right] n_2(x) \\ &\quad +\mu\left[\frac{\partial}{\partial x_1}u_3(x) + \frac{\partial}{\partial x_3}u_1(x)\right] n_3(x) \\ &= \lambda \operatorname{div}\underline{u}(x)\, n_1(x) + 2\mu\,\frac{\partial}{\partial n_x}u_1(x) + \mu\left[\frac{\partial}{\partial x_1}u_2(x) - \frac{\partial}{\partial x_2}u_1(x)\right] n_2(x) \\ &\quad +\mu\left[\frac{\partial}{\partial x_1}u_3(x) - \frac{\partial}{\partial x_3}u_1(x)\right] n_3(x). \end{aligned}$$

From this we obtain the following representation of the boundary stress operator for $x \in \Gamma$,

$$\gamma_1^{\text{int}}\underline{u}(x) = \lambda \operatorname{div}\underline{u}(x)\,\underline{n}(x) + 2\mu\,\frac{\partial}{\partial n_x}\underline{u}(x) + \mu\,\underline{n}(x) \times \operatorname{curl}\underline{u}(x). \tag{1.27}$$

In many applications of solid mechanics the boundary conditions (1.11) and (1.12) are given within their components, i.e.

$$\begin{aligned} \gamma_0^{\text{int}}u_i(x) &= g_{D,i}(x) \qquad \text{for } x \in \Gamma_{D,i}, \\ (\gamma_1^{\text{int}}\underline{u})_i(x) &= g_{N,i}(x) \qquad \text{for } x \in \Gamma_{N,i}, \end{aligned} \tag{1.28}$$

where $\Gamma = \overline{\Gamma}_{D,i} \cup \overline{\Gamma}_{N,i}$ for $i = 1, 2, 3$.

The non–trivial solutions of the homogeneous Neumann boundary value problem

$$-\mu\Delta\underline{u}(x) - (\lambda+\mu)\operatorname{grad}\operatorname{div}\underline{u}(x) = \underline{0} \quad \text{for } x \in \Omega, \quad \gamma_1^{\text{int}}\underline{u}(x) = \underline{0} \quad \text{for } x \in \Gamma$$

are given by the rigid body motions $\underline{v}_k \in \mathcal{R}$ where

$$\mathcal{R} = \operatorname{span}\left\{ \begin{pmatrix}1\\0\\0\end{pmatrix}, \begin{pmatrix}0\\1\\0\end{pmatrix}, \begin{pmatrix}0\\0\\1\end{pmatrix}, \begin{pmatrix}-x_2\\x_1\\0\end{pmatrix}, \begin{pmatrix}0\\-x_3\\x_2\end{pmatrix}, \begin{pmatrix}x_3\\0\\-x_1\end{pmatrix} \right\}. \tag{1.29}$$

Note that Betti's first formula (1.21) reads

$$a(\underline{u},\underline{v}) = -\int_\Omega \sum_{i,j=1}^3 \frac{\partial}{\partial x_j}\sigma_{ij}(\underline{u},x)v_i(x)dx + \int_\Gamma \gamma_0^{\text{int}}\underline{v}(x)^\top \gamma_1^{\text{int}}\underline{u}(x)ds_x,$$

$$a(\underline{v},\underline{u}) = -\int_\Omega \sum_{i,j=1}^3 \frac{\partial}{\partial x_j}\sigma_{ij}(\underline{v},x)u_i(x)dx + \int_\Gamma \gamma_0^{\text{int}}\underline{u}(x)^\top \gamma_1^{\text{int}}\underline{v}(x)ds_x$$

when exchanging the role of $\underline{u}$ and $\underline{v}$. From the symmetry of the bilinear form $a(\cdot,\cdot)$ we then obtain Betti's second formula

$$-\int_\Omega \sum_{i,j=1}^3 \frac{\partial}{\partial x_j}\sigma_{ij}(\underline{u},x)v_i(x)dx + \int_\Gamma \gamma_0^{\text{int}}\underline{v}(x)^\top \gamma_1^{\text{int}}\underline{u}(x)ds_x \tag{1.30}$$

$$= -\int_\Omega \sum_{i,j=1}^3 \frac{\partial}{\partial x_j}\sigma_{ij}(\underline{v},x)u_i(x)dx + \int_\Gamma \gamma_0^{\text{int}}\underline{u}(x)^\top \gamma_1^{\text{int}}\underline{v}(x)ds_x.$$

Inserting the rigid body motions $\underline{v}_k \in \mathcal{R}$ into Betti's second formula (1.30) this gives the orthogonality

$$-\int_\Omega \sum_{i,j=1}^3 \frac{\partial}{\partial x_j}\sigma_{ij}(\underline{u},x)v_{k,i}(x)dx + \int_\Gamma \gamma_0^{\text{int}}\underline{v}_k(x)^\top \gamma_1^{\text{int}}\underline{u}(x)ds_x = 0$$

for all $\underline{v}_k \in \mathcal{R}$. Hence, for the solvability of the Neumann boundary value problem

$$-\mu\Delta\underline{u}(x) - (\lambda+\mu)\operatorname{grad}\operatorname{div}\underline{u}(x) = \underline{f}(x) \quad \text{for } x\in\Omega,$$
$$\gamma_1^{\text{int}}\underline{u}(x) = \underline{g}_N(x) \quad \text{for } x \in \Gamma$$

we have to assume the solvability conditions

$$\int_\Omega \underline{v}_k(x)^\top \underline{f}(x)dx + \int_\Gamma \gamma_0^{\text{int}}\underline{v}_k(x)^\top \underline{g}_N(x)ds_x = 0 \quad \text{for all } \underline{v}_k \in \mathcal{R}. \tag{1.31}$$

Note that the solution of the Neumann boundary value problem is only unique up to the rigid body motions (1.29). A unique solution can be defined when considering either nodal or scaling conditions in addition.

The second order system (1.25) of linear elasticity can also be written as a scalar partial differential equation of fourth order. By setting

$$\underline{u}(x) := \Delta\underline{w}(x) - \frac{\lambda+\mu}{\lambda+2\mu}\operatorname{grad}\operatorname{div}\underline{w}(x) \quad \text{for } x\in\Omega \tag{1.32}$$

from the equilibrium equations (1.25) we obtain the scalar Bi–Laplace equation

$$-\mu\Delta^2\underline{w}(x) \;=\; \underline{f}(x) \quad \text{for } x \in \Omega. \tag{1.33}$$

When considering a homogeneous partial differential equation with $\underline{f} \equiv \underline{0}$ the solution of (1.25) can be described by setting $w_2 \equiv w_3 \equiv 0$ where the remaining component $w_1 = \psi$ is the solution of the Bi–Laplace equation

$$-\Delta^2\psi(x) \;=\; 0 \quad \text{for } x \in \Omega.$$

Then,

$$u_1(x) := \Delta\psi(x) - \frac{\lambda+\mu}{\lambda+2\mu}\frac{\partial^2}{\partial x_1^2}\psi(x),$$
$$u_2(x) := -\frac{\lambda+\mu}{\lambda+2\mu}\frac{\partial^2}{\partial x_1 \partial x_2}\psi(x),$$
$$u_3(x) := -\frac{\lambda+\mu}{\lambda+2\mu}\frac{\partial^2}{\partial x_1 \partial x_3}\psi(x)$$

is a solution of the homogeneous system (1.25). The function $\psi$ is known as Airy's stress function.

### 1.2.1 Plane Elasticity

To describe problems of linear elasticity in two space dimensions one may consider two different approaches. In plain stress we assume that the stress tensor depends on two space coordinates $(x_1, x_2)$ only, and that the $x_3$–coordinates of the stress tensor disappear:

$$\sigma_{ij}(x_1,x_2,x_3) = \sigma_{ij}(x_1,x_2) \quad \text{for } i,j = 1,2;$$
$$\sigma_{3i}(x) = \sigma_{i3}(x) = 0 \quad \text{for } i = 1,2,3.$$

Applying Hooke's law (1.19) we then obtain

$$e_{3i}(\underline{u},x) \;=\; e_{i3}(\underline{u},x) \;=\; 0 \quad \text{for } i = 1,2 \tag{1.34}$$

and

$$e_{33}(\underline{u},x) \;=\; -\frac{\nu}{1-\nu}[e_{11}(\underline{u},x) + e_{22}(\underline{u},x)]. \tag{1.35}$$

The resulting stress–strain relation reads

$$\sigma_{ij}(\underline{u},x) \;=\; \frac{E\nu}{(1+\nu)(1-\nu)}\delta_{ij}\sum_{k=1}^{2} e_{kk}(\underline{u},x) + \frac{E}{1+\nu}e_{ij}(\underline{u},x)$$

for $x \in \Omega$ and $i,j = 1,2$. With

$$u_i(x_1,x_2,x_3) \;=\; u_i(x_1,x_2) \quad \text{for } i = 1,2$$

we further obtain from (1.34)

$$u_3(x_1,x_2,x_3) \;=\; u_3(x_3)\,.$$

In addition, (1.35) gives

$$\frac{\partial}{\partial x_3}u_3(x_3) \;=\; -\frac{\nu}{1-\nu}\left[\frac{\partial}{\partial x_1}u_1(x_1,x_2)+\frac{\partial}{\partial x_2}u_2(x_1,x_2)\right],$$

and therefore

$$u_3(x) \;=\; -\frac{\nu}{1-\nu}\left[\frac{\partial}{\partial x_1}u_1(x)+\frac{\partial}{\partial x_2}u_2(x)\right]\,x_3.$$

To ensure compatibility with (1.34) we have to neglect terms of order $\mathcal{O}(x_3)$ in the definition of the strain tensor $e_{ij}(\underline{u},x)$. By introducing the modified Lamé constants

$$\widetilde{\lambda} \;=\; \frac{E\nu}{(1+\nu)(1-\nu)},\quad \widetilde{\mu} \;=\; \frac{E}{2(1+\nu)}$$

we then obtain a system of partial differential equations to find the displacement field $(u_1,u_2)$ such that

$$-\widetilde{\mu}\Delta\underline{u}(x) - (\widetilde{\lambda}+\widetilde{\mu})\operatorname{grad}\operatorname{div}\underline{u}(x) \;=\; \underline{f}(x) \quad \text{for } x\in\Omega\subset\mathbb{R}^2.$$

In plain strain we assume that all components $e_{ij}(\underline{u},x)$ of the strain tensor depend only on the space coordinates $(x_1,x_2)$ and that the $x_3$–coordinates vanish:

$$e_{ij}(\underline{u},x_1,x_2,x_3) = e_{ij}(\underline{u},x_1,x_2) \quad \text{for } i,j=1,2;$$
$$e_{3i}(\underline{u},x) = e_{i3}(\underline{u},x) = 0 \quad \text{for } i=1,2,3.$$

For the associated displacements we then obtain

$$u_i(x_1,x_2,x_3) = u_i(x_1,x_2) \quad \text{for } i=1,2, \quad u_3(x) \;=\; \text{constant},$$

and the stress–strain relation reads

$$\sigma_{ij}(\underline{u},x) \;=\; \frac{E\nu}{(1+\nu)(1-2\nu)}\delta_{ij}\sum_{k=1}^{2}e_{kk}(\underline{u},x)+\frac{E}{1+\nu}e_{ij}(\underline{u},x)$$

for $i,j=1,2$ yielding the equilibrium equations (1.25) to find the displacement field $(u_1,u_2)$. Obviously,

$$\sigma_{3i}(\underline{u},x) = \sigma_{i3}(\underline{u},x) = 0 \quad \text{for } i=1,2,$$

and

$$\sigma_{33}(\underline{u},x) \;=\; \frac{E\nu}{(1+\nu)(1-2\nu)}\left[e_{11}(\underline{u},x)+e_{22}(\underline{u},x)\right].$$

With

$$\sigma_{11}(\underline{u},x)+\sigma_{22}(\underline{u},x) \;=\; \frac{E}{(1+\nu)(1-2\nu)}\left[e_{11}(\underline{u},x)+e_{22}(\underline{u},x)\right]$$

we finally get

$$\sigma_{33}(\underline{u},x) \;=\; \nu\left[\sigma_{11}(\underline{u},x)+\sigma_{22}(\underline{u},x)\right].$$

In both cases of two–dimensional plane stress and plane strain linear elasticity models the rigid body motions are given as

$$\mathcal{R} \;=\; \operatorname{span}\left\{\begin{pmatrix}1\\0\end{pmatrix},\begin{pmatrix}0\\1\end{pmatrix},\begin{pmatrix}-x_2\\x_1\end{pmatrix}\right\}. \tag{1.36}$$

For the first component of the boundary stress we obtain for $x\in\Gamma$

$$\begin{aligned}(\gamma_1^{\text{int}}\underline{u})_1(x) &= \sum_{j=1}^{2}\sigma_{1j}(\underline{u},x)n_j(x)\\ &= \left[\lambda\operatorname{div}\underline{u}(x)+2\mu\frac{\partial}{\partial x_1}u_1(x)\right]n_1(x)+\mu\left[\frac{\partial}{\partial x_1}u_2(x)+\frac{\partial}{\partial x_2}u_1(x)\right]n_2(x)\\ &= \lambda\operatorname{div}\underline{u}(x)\,n_1(x)+2\mu\frac{\partial}{\partial n_x}u_1(x)+\mu\left[\frac{\partial}{\partial x_1}u_2(x)-\frac{\partial}{\partial x_2}u_1(x)\right]n_2(x),\end{aligned}$$

and for the second component

$$\begin{aligned}(\gamma_1^{\text{int}}\underline{u})_2(x) &= \lambda\operatorname{div}\underline{u}(x)\,n_2(x)+2\mu\frac{\partial}{\partial n_x}u_2(x)\\ &\quad+\mu\left[\frac{\partial}{\partial x_2}u_1(x)-\frac{\partial}{\partial x_1}u_2(x)\right]n_1(x).\end{aligned}$$

If we define for a two–dimensional vector field $\underline{v}$ the rotation as

$$\operatorname{curl}\underline{v} = \frac{\partial}{\partial x_1}v_2(x)-\frac{\partial}{\partial x_2}v_1(x),$$

and if we declare

$$\underline{a}\times\alpha := \alpha\begin{pmatrix}a_2\\-a_1\end{pmatrix},\quad \underline{a}\in\mathbb{R}^2,\alpha\in\mathbb{R},$$

we can write the boundary stress as in the representation (1.27),

$$\gamma_1^{\text{int}}\underline{u}(x) \;=\; [\lambda\operatorname{div}\underline{u}(x)]\underline{n}(x)+2\mu\frac{\partial}{\partial n_x}\underline{u}(x)+\mu\,\underline{n}(x)\times\operatorname{curl}\underline{v}(x),\quad x\in\Gamma.$$

### 1.2.2 Incompressible Elasticity

For $d = 2, 3$ we consider the system (1.25) describing the partial differential equations of linear elasticity. Here we are interested in almost incompressible materials, i.e. for $\nu \to \frac{1}{2}$ we conclude $\lambda \to \infty$. Hence we introduce

$$p(x) \;:=\; -(\lambda + \mu) \operatorname{div} \underline{u}(x) \quad \text{for } x \in \Omega$$

to obtain from (1.25)

$$-\mu \Delta \underline{u}(x) + \nabla p(x) = \underline{f}(x), \quad \operatorname{div} \underline{u}(x) = -\frac{1}{\lambda + \mu} p(x) \quad \text{for } x \in \Omega.$$

In the incompressible case $\nu = \frac{1}{2}$ this is equivalent to

$$-\mu \Delta \underline{u}(x) + \nabla p(x) = \underline{f}(x), \quad \operatorname{div} \underline{u}(x) = 0 \quad \text{for } x \in \Omega \tag{1.37}$$

which coincides with the Stokes system which plays an important role in fluid mechanics.

## 1.3 Stokes System

When considering the Stokes system [91] we have to find a velocity field $\underline{u}$ and the pressure $p$ satisfying the system of partial differential equations

$$-\mu \, \Delta \underline{u}(x) + \nabla p(x) = \underline{f}(x), \quad \operatorname{div} \underline{u}(x) = 0 \quad \text{for } x \in \Omega \subset \mathbb{R}^d \tag{1.38}$$

where $\mu$ is the viscosity constant. If we assume Dirichlet boundary conditions $\underline{u}(x) = \underline{g}(x)$ for $x \in \Gamma$, integration by parts of the second equation gives

$$0 = \int_\Omega \operatorname{div} \underline{u}(x) \, dx = \int_\Gamma [\underline{n}(x)]^\top \underline{u}(x) ds_x = \int_\Gamma [\underline{n}(x)]^\top \underline{g}(x) ds_x. \tag{1.39}$$

Therefore, the given Dirichlet data $\underline{g}(x)$ have to satisfy the solvability condition (1.39.) Moreover, the pressure $p$ is only unique up some additive constant.

When multiplying the components of the first partial differential equation in (1.38) with some test function $v_i$, integrating over $\Omega$, and applying integration by parts this gives

$$\int_\Omega f_i(x) v_i(x) dx = -\mu \int_\Omega \Delta u_i(x) v_i(x) dx + \int_\Omega \frac{\partial}{\partial x_i} p(x) v_i(x) dx \tag{1.40}$$

$$= -\mu \int_\Omega \Delta u_i(x) v_i(x) dx - \int_\Omega p(x) \frac{\partial}{\partial x_i} v_i(x) dx + \int_\Gamma p(x) n_i(x) v_i(x) ds_x.$$

Moreover we have

$$\frac{\partial}{\partial x_j}\left[e_{ij}(\underline{u},x)v_i(x)\right] = v_i(x)\frac{\partial}{\partial x_j}e_{ij}(\underline{u},x) + e_{ij}(\underline{u},x)\frac{\partial}{\partial x_j}v_i(x),$$

as well as

$$\sum_{i,j=1}^{d} e_{ij}(\underline{u},x)\frac{\partial}{\partial x_j}v_i(x) = \sum_{i,j=1}^{d} e_{ij}(\underline{u},x)e_{ij}(\underline{v},x).$$

For $d = 3$ and $i = 1$ we compute

$$\begin{aligned}\sum_{j=1}^{3}\frac{\partial}{\partial x_j}e_{1j}(\underline{u},x) &= \frac{\partial}{\partial x_1}e_{11}(\underline{u},x) + \frac{\partial}{\partial x_2}e_{12}(\underline{u},x) + \frac{\partial}{\partial x_3}e_{13}(\underline{u},x)\\ &= \frac{\partial^2}{\partial x_1^2}u_1(x) + \frac{1}{2}\frac{\partial}{\partial x_2}\left[\frac{\partial}{\partial x_1}u_2(x) + \frac{\partial}{\partial x_2}u_1(x)\right]\\ &\quad + \frac{1}{2}\frac{\partial}{\partial x_3}\left[\frac{\partial}{\partial x_1}u_3(x) + \frac{\partial}{\partial x_3}u_1(x)\right]\\ &= \frac{1}{2}\Delta u_1(x) + \frac{1}{2}\frac{\partial}{\partial x_1}\left[\frac{\partial}{\partial x_1}u_1(x) + \frac{\partial}{\partial x_2}u_2(x) + \frac{\partial}{\partial x_3}u_3(x)\right]\\ &= \frac{1}{2}\Delta u_1(x) + \frac{1}{2}\frac{\partial}{\partial x_1}\operatorname{div}\underline{u}(x).\end{aligned}$$

Corresponding results hold for $i = 2, 3$ and $d = 2$, respectively. Then we obtain

$$\begin{aligned}\sum_{i,j=1}^{d}\frac{\partial}{\partial x_j}\left[e_{ij}(\underline{u},x)v_i(x)\right] &= \sum_{i,j=1}^{d}\frac{\partial}{\partial x_j}e_{ij}(\underline{u},x)v_i(x) + \sum_{i,j=1}^{d}e_{ij}(\underline{u},x)\frac{\partial}{\partial x_j}v_i(x)\\ &= \frac{1}{2}\sum_{i=1}^{d}\left[\Delta u_i(x) + \frac{\partial}{\partial x_i}\operatorname{div}\underline{u}(x)\right]v_i(x) + \sum_{i,j=1}^{d}e_{ij}(\underline{u},x)e_{ij}(\underline{v},x).\end{aligned}$$

This can be rewritten as

$$\begin{aligned}-\sum_{i=1}^{d}\Delta u_i(x)v_i(x) &= 2\sum_{i,j=1}^{d}e_{ij}(\underline{u},x)e_{ij}(\underline{v},x) - 2\sum_{i,j=1}^{d}\frac{\partial}{\partial x_j}\left[e_{ij}(\underline{u},x)v_i(x)\right]\\ &\quad + \sum_{i=1}^{d}v_i(x)\frac{\partial}{\partial x_i}\operatorname{div}\underline{u}(x).\end{aligned}$$

Taking the sum of (1.40) for $i = 1, \ldots, d$, substituting the above results, and applying integration by parts this gives

$$\begin{aligned}
\int_\Omega \underline{v}(x)^\top \underline{f}(x)dx &= -\mu \int_\Omega \sum_{i=1}^d v_i(x)\Delta u_i(x)dx + \int_\Omega \sum_{i=1}^d v_i(x)\frac{\partial}{\partial x_i}p(x)\,dx \\
&= 2\mu \int_\Omega \sum_{i,j=1}^d e_{ij}(\underline{u},x)e_{ij}(\underline{v},x)dx - 2\mu \int_\Omega \sum_{i,j=1}^d \frac{\partial}{\partial x_j}\left[e_{ij}(\underline{u},x)v_i(x)\right]dx \\
&\qquad + \mu \int_\Omega \sum_{i=1}^d v_i(x)\frac{\partial}{\partial x_i}\operatorname{div}\underline{u}(x)dx + \int_\Omega \sum_{i=1}^d v_i(x)\frac{\partial}{\partial x_i}p(x)\,dx \\
&= 2\mu \int_\Omega \sum_{i,j=1}^d e_{ij}(\underline{u},x)e_{ij}(\underline{v},x)dx - 2\mu \int_\Gamma \sum_{i,j=1}^d n_j(x)e_{ij}(\underline{u},x)\gamma_0^{\text{int}}v_i(x)ds_x \\
&\qquad -\mu \int_\Omega \operatorname{div}\underline{u}(x)\operatorname{div}\underline{v}(x)\,dx + \mu \int_\Gamma \operatorname{div}\underline{u}(x)\underline{n}(x)^\top \gamma_0^{\text{int}}\underline{v}(x)ds_x \\
&\qquad - \int_\Omega p(x)\operatorname{div}\underline{v}(x)dx + \int_\Gamma p(x)\underline{n}(x)^\top \gamma_0^{\text{int}}\underline{v}(x)ds_x.
\end{aligned}$$

Hence we obtain Green's first formula for the Stokes system

$$\begin{aligned}
a(\underline{u},\underline{v}) = \int_\Omega \sum_{i=1}^d \left[-\mu\Delta u_i(x) + \frac{\partial}{\partial x_i}p(x)\right] v_i(x)dx \qquad\qquad (1.41) \\
+ \int_\Omega p(x)\operatorname{div}\underline{v}(x)dx + \int_\Gamma \sum_{i=1}^d t_i(\underline{u},p)v_i(x)ds_x
\end{aligned}$$

with the symmetric bilinear form

$$a(\underline{u},\underline{v}) := 2\mu \int_\Omega \sum_{i,j=1}^d e_{ij}(\underline{u},x)e_{ij}(\underline{v},x)dx - \mu \int_\Omega \operatorname{div}\underline{u}(x)\operatorname{div}\underline{v}(x)\,dx, \qquad (1.42)$$

and with the conormal derivative

$$t_i(\underline{u},p) := -[p(x) + \mu\operatorname{div}\underline{u}(x)]n_i(x) + 2\mu\sum_{j=1}^d e_{ij}(\underline{u},x)n_j(x)$$

defined for $x \in \Gamma$ and $i = 1,\ldots,d$. For divergence–free functions $\underline{u}$ satisfying $\operatorname{div}\underline{u} = 0$ we obtain, as for the system of linear elastostatics, the representation

$$\underline{t}(\underline{u},p) = -p(x)\,\underline{n}(x) + 2\mu\frac{\partial}{\partial n_x}\underline{u}(x) + \mu\,\underline{n}(x)\times\operatorname{curl}\underline{u}(x), \quad x \in \Gamma. \qquad (1.43)$$

Besides the standard boundary conditions (1.28) sliding boundary conditions are often considered in fluid mechanics. In particular, for $x \in \Gamma_S$ we describe non–penetration in normal direction and no adherence in tangential direction by the boundary conditions

$$\underline{n}(x)^\top \underline{u}(x) = 0, \quad \underline{t}_T(x) := \underline{t}(x) - [\underline{n}(x)^\top \underline{t}(x)]\underline{n}(x) = \underline{0}.$$

## 1.4 Helmholtz Equation

The wave equation

$$\frac{1}{c^2}\frac{\partial^2}{\partial t^2}U(x,t) = \Delta U(x,t) \quad \text{for } x \in \mathbb{R}^d \ \ (d=2,3)$$

describes the wave propagation in a homogeneous, isotropic and friction–free medium having the constant speed of sound $c$. Examples are acoustic scattering and sound radiation problems.

For time harmonic acoustic waves

$$U(x,t) = \mathrm{Re}\left(u(x)e^{-i\omega t}\right)$$

with a frequency $\omega$ we obtain a reduced wave equation or the Helmholtz equation

$$-\Delta u(x) - k^2 u(x) = 0 \quad \text{for } x \in \mathbb{R}^d \tag{1.44}$$

where $u$ is a scalar valued complex function and $k = \omega/c > 0$ is the wave number.

Let us first consider the Helmholtz equation (1.44) in a bounded domain $\Omega \subset \mathbb{R}^d$,

$$-\Delta u(x) - k^2 u(x) = 0 \quad \text{for } x \in \Omega.$$

Multiplying this with a test function $v$, integrating over $\Omega$, and applying integration by parts, this gives Green's first formula

$$a(u,v) = \int_\Omega [-\Delta u(x) - k^2 u(x)]v(x)dx + \int_\Gamma \gamma_1^{\mathrm{int}} u(x)\gamma_0^{\mathrm{int}} v(x) ds_x \tag{1.45}$$

with the symmetric bilinear form

$$a(u,v) = \int_\Omega \nabla u(x)\nabla v(x)dx - k^2 \int_\Omega u(x)v(x)dx.$$

Note that $\gamma_1^{\mathrm{int}} u = \underline{n}(x)\cdot\nabla u(x)$ is the normal derivative of $u$ in $x \in \Gamma$. Exchanging the role of $u$ and $v$ we obtain in the same way

$$a(v,u) = \int_\Omega [-\Delta v(x) - k^2 v(x)]u(x)dx + \int_\Gamma \gamma_1^{\mathrm{int}} v(x)\gamma_0^{\mathrm{int}} u(x) ds_x,$$

and by using the symmetry of the bilinear form $a(\cdot,\cdot)$ we conclude Green's second formula

$$\int_\Omega [-\Delta u(x) - k^2 u(x)]v(x)dx + \int_\Gamma \gamma_1^{\text{int}} u(x)\gamma_0^{\text{int}} v(x) ds_x \tag{1.46}$$
$$= \int_\Omega [-\Delta v(x) - k^2 v(x)]u(x)dx + \int_\Gamma \gamma_1^{\text{int}} v(x)\gamma_0^{\text{int}} u(x) ds_x.$$

For any solution $u$ of the Helmholtz equation (1.44) we find from (1.45) by setting $v = \overline{u}$

$$\int_\Omega |\nabla u(x)|^2 dx - k^2 \int_\Omega |u(x)|^2 dx = \int_\Gamma \gamma_1^{\text{int}} u(x)\gamma_0^{\text{int}} \overline{u(x)} ds_x. \tag{1.47}$$

Next we consider the Helmholtz equation (1.44) in an unbounded domain,

$$-\Delta u(x) - k^2 u(x) = 0 \quad \text{for } x \in \Omega^c = \mathbb{R}^d \backslash \overline{\Omega}$$

where we have to add the Sommerfeld radiation condition

$$\left| \frac{x}{|x|} \cdot \nabla u(x) - iku(x) \right| = \mathcal{O}\left( \frac{1}{|x|^2} \right) \quad \text{as } |x| \to \infty. \tag{1.48}$$

For $x_0 \in \Omega$ let $B_R(x_0)$ be a ball with center $x_0$ and radius $R$ such that $\Omega \subset B_R(x_0)$ is satisfied. Then, $\Omega_R = B_R(x_0) \backslash \overline{\Omega}$ is a bounded domain for which we can write (1.47) as

$$\int_{\Omega_R} |\nabla u(x)|^2 dx - k^2 \int_{\Omega_R} |u(x)|^2 dx$$
$$= -\int_\Gamma \gamma_1^{\text{ext}} u(x)\gamma_0^{\text{ext}} \overline{u(x)} ds_x + \int_{\partial B_R(x_0)} \gamma_1^{\text{int}} u(x)\gamma_0^{\text{int}} \overline{u(x)} ds_x$$

taking into account the opposite direction of the normal vector on $\Gamma$. This clearly implies

$$\text{Im} \int_{\partial B_R(x_0)} \gamma_1^{\text{int}} u(x)\gamma_0^{\text{int}} \overline{u(x)} ds_x = \text{Im} \int_\Gamma \gamma_1^{\text{ext}} u(x)\gamma_0^{\text{ext}} \overline{u(x)} ds_x = \mathcal{O}(1).$$

On the other hand, from the Sommerfeld radiation condition (1.48) we also conclude the weaker condition due to Rellich,

$$\lim_{R\to\infty} \int_{\partial B_R(x_0)} \left| \gamma_1^{\text{int}} u(x) - ik\gamma_0^{\text{int}} u(x) \right|^2 ds_x = 0. \tag{1.49}$$

From this we find

$$\begin{aligned}
0 &= \lim_{R\to\infty} \int\limits_{\partial B_R(x_0)} \left|\gamma_1^{\text{int}} u(x) - ik\gamma_0^{\text{int}} u(x)\right|^2 ds_x \\
&= \lim_{R\to\infty} \left[ \int\limits_{\partial B_R(x_0)} \left|\gamma_1^{\text{int}} u(x)\right|^2 ds_x + k^2 \int\limits_{\partial B_R(x_0)} \left|\gamma_0^{\text{int}} u(x)\right|^2 ds_x \right. \\
&\qquad \left. -2k \operatorname{Im} \int\limits_{\partial B_R(x_0)} \gamma_1^{\text{int}} u(x) \gamma_0^{\text{int}} \overline{u(x)} ds_x \right] \\
&= \lim_{R\to\infty} \left[ \int\limits_{\partial B_R(x_0)} \left|\gamma_1^{\text{int}} u(x)\right|^2 ds_x + k^2 \int\limits_{\partial B_R(x_0)} \left|\gamma_0^{\text{int}} u(x)\right|^2 ds_x \right. \\
&\qquad \left. -2k \operatorname{Im} \int\limits_{\Gamma} \gamma_1^{\text{ext}} u(x) \gamma_0^{\text{ext}} \overline{u(x)} ds_x \right]
\end{aligned}$$

and therefore

$$\begin{aligned}
2k \operatorname{Im} \int\limits_{\Gamma} \gamma_1^{\text{ext}} u(x) \gamma_0^{\text{ext}} \overline{u(x)} ds_x \\
= \lim_{R\to\infty} \left[ \int\limits_{\partial B_R(x_0)} \left|\gamma_1^{\text{int}} u(x)\right|^2 ds_x + k^2 \int\limits_{\partial B_R(x_0)} \left|\gamma_0^{\text{int}} u(x)\right|^2 ds_x \right] \geq 0
\end{aligned}$$

implying

$$\lim_{R\to\infty} \int\limits_{\partial B_R(x_0)} \left|\gamma_0^{\text{int}} u(x)\right|^2 ds_x = \mathcal{O}(1)$$

as well as

$$|u(x)| = \mathcal{O}\left(\frac{1}{|x|}\right) \quad \text{as } |x| \to \infty. \tag{1.50}$$

## 1.5 Exercises

**1.1** For $x \in \mathbb{R}^2$ we consider polar coordinates

$$x_1 = x_1(r,\varphi) = r\cos\varphi, \quad x_2 = x_2(r,\varphi) = r\sin\varphi \quad \text{for } r > 0,\ \varphi \in [0, 2\pi).$$

Then, a given function $u(x)$ can be written as

$$u(x_1,x_2) = u(x_1(r,\varphi), x_2(r,\varphi)) = \widetilde{u}(r,\varphi).$$

Express the gradient of $\widetilde{u}(r,\varphi)$ in terms of the gradient of $u(x_1,x_2)$, i.e. find a matrix $J$ such that

$$\nabla_{(r,\varphi)}\widetilde{u}(r,\varphi) \;=\; J\,\nabla_x u(x).$$

Derive a representation of $\nabla_x u(x)$ in terms of $\nabla_{(r,\varphi)}\widetilde{u}(r,\varphi)$.

**1.2** Rewrite the two–dimensional Laplace operator

$$\Delta u(x) \;=\; \frac{\partial^2}{\partial x_1^2}u(x) + \frac{\partial^2}{\partial x_2^2}u(x)$$

when using polar coordinates.

**1.3** Prove that for any $\alpha \in \mathbb{R}_+$ and $x \neq 0$

$$u(x) \;=\; \widetilde{u}(r,\varphi) \;=\; r^\alpha \sin(\alpha\varphi)$$

is a solution of the two–dimensional Laplace equation.

**1.4** Rewrite the three–dimensional Laplace operator

$$\Delta u(x) \;=\; \frac{\partial^2}{\partial x_1^2}u(x) + \frac{\partial^2}{\partial x_2^2}u(x) + \frac{\partial^2}{\partial x_3^2}u(x) \quad \text{for } x \in \mathbb{R}^3$$

when using spherical coordinates

$$x_1 = r\cos\varphi\sin\vartheta, \quad x_2 = r\sin\varphi\sin\vartheta, \quad x_3 = r\cos\vartheta$$

where $r > 0$, $\varphi \in [0,2\pi)$, $\vartheta \in [0,\pi]$.

**1.5** Consider the Navier system

$$-\mu\Delta\underline{u}(x) - (\lambda+\mu)\text{grad div}\underline{u}(x) \;=\; \underline{f}(x) \quad \text{for } x \in \mathbb{R}^d.$$

Determine the constant $\alpha \in \mathbb{R}$ such that the solution of the Navier system

$$\underline{u}(x) \;=\; \Delta\underline{w}(x) + \alpha\,\text{grad div}\underline{w}(x)$$

can be found via the solution of a Bi–Laplace equation.

**1.6** Compute all eigenvalues $\lambda_k$ and associated eigenvectors $u_k$ of the Dirichlet eigenvalue problem

$$-u_k''(x) \;=\; \lambda u_k(x) \quad \text{for } x \in (0,1), \quad u_k(0) = u_k(1) = 0.$$

# 2 Function Spaces

In this chapter we introduce the most important function spaces as needed for the weak formulation of boundary value problems. For a further reading we refer to [1, 103, 106].

## 2.1 The Spaces $C^k(\Omega)$, $C^{k,\kappa}(\Omega)$ and $L_p(\Omega)$

For $d \in \mathbb{N}$ we call a vector $\alpha = (\alpha_1, \ldots, \alpha_d)$, $\alpha_i \in \mathbb{N}_0$, multi index with the absolute value $|\alpha| = \alpha_1 + \cdots + \alpha_d$ and with the factorial $\alpha! = \alpha_1! \ldots \alpha_d!$. For $x \in \mathbb{R}^d$ we can therefore write

$$x^\alpha = x_1^{\alpha_1} \cdots x_d^{\alpha_d}.$$

If $u$ is a sufficient smooth real valued function, then we can write partial derivatives as

$$D^\alpha u(x) := \left(\frac{\partial}{\partial x_1}\right)^{\alpha_1} \cdots \left(\frac{\partial}{\partial x_d}\right)^{\alpha_d} u(x_1, \ldots, x_d).$$

Let $\Omega \subseteq \mathbb{R}^d$ be some open subset and assume $k \in \mathbb{N}_0$. $C^k(\Omega)$ is the space of functions which are bounded and $k$ times continuously differentiable in $\Omega$. In particular, for $u \in C^k(\Omega)$ the norm

$$\|u\|_{C^k(\Omega)} := \sum_{|\alpha| \le k} \sup_{x \in \Omega} |D^\alpha u(x)|$$

is finite. Correspondingly, $C^\infty(\Omega)$ is the space of functions which are bounded and infinitely often continuously differentiable. For a function $u(x)$ defined for $x \in \Omega$ we denote

$$\operatorname{supp} u := \overline{\{x \in \Omega : u(x) \neq 0\}}$$

to be the support of the function $u$. Then,

$$C_0^\infty(\Omega) := \{u \in C^\infty(\Omega) \, : \, \operatorname{supp} u \subset \Omega\}$$

is the space of $C^\infty(\Omega)$ functions with compact support.

For $k \in \mathbb{N}_0$ and $\kappa \in (0,1)$ we define $C^{k,\kappa}(\Omega)$ to be the space of Hölder continuous functions equipped with the norm

$$\|u\|_{C^{k,\kappa}(\Omega)} := \|u\|_{C^k(\Omega)} + \sum_{|\alpha|=k} \sup_{x,y\in\Omega, x\neq y} \frac{|D^\alpha u(x) - D^\alpha u(y)|}{|x-y|^\kappa} \, .$$

In particular for $\kappa = 1$ we have $C^{k,1}(\Omega)$ to be the space of functions $u \in C^k(\Omega)$ where the derivatives $D^\alpha u$ of order $|\alpha| = k$ are Lipschitz continuous.

The boundary of an open set $\Omega \subset \mathbb{R}^d$ is defined as

$$\Gamma := \partial\Omega = \overline{\Omega} \cap (\mathbb{R}^d \backslash \Omega) \, .$$

We require that for $d \geq 2$ the boundary $\Gamma = \partial\Omega$ can be represented locally as the graph of a Lipschitz function using different systems of Cartesian coordinates for different parts of $\Gamma$, as necessary. The simplest case occurs when there is a function $\gamma : \mathbb{R}^{d-1} \to \mathbb{R}$ such that

$$\Omega := \left\{x \in \mathbb{R}^d \, : \, x_d < \gamma(\widetilde{x}) \text{ for all } \widetilde{x} = (x_1, \ldots, x_{d-1}) \in \mathbb{R}^{d-1}\right\} .$$

If $\gamma(\cdot)$ is Lipschitz,

$$|\gamma(\widetilde{x}) - \gamma(\widetilde{y})| \leq L \, |\widetilde{x} - \widetilde{y}| \quad \text{for all } \widetilde{x}, \widetilde{y} \in \mathbb{R}^{d-1}$$

then $\Omega$ is said to be a Lipschitz hypograph with boundary

$$\Gamma = \left\{x \in \mathbb{R}^d \, : \, x_n = \gamma(\widetilde{x}) \quad \text{for all } \widetilde{x} \in \mathbb{R}^{d-1}\right\} .$$

**Definition 2.1.** *The open set $\Omega \subset \mathbb{R}^d$, $d \geq 2$, is a Lipschitz domain if its boundary $\Gamma = \partial\Omega$ is compact and if there exist finite families $\{W_i\}$ and $\{\Omega_j\}$ having the following properties:*

*i. The family $\{W_j\}$ is a finite open cover of $\Gamma$, i.e. $W_j \subset \mathbb{R}^d$ is an open subset and $\Gamma \subseteq \cup_j W_j$.*
*ii. Each $\Omega_j$ can be transformed to a Lipschitz hypograph by a rigid motion, i.e. by rotations and translations.*
*iii. For all $j$ the equality $W_j \cap \Omega = W_j \cap \Omega_j$ is satisfied.*

The local representation of a Lipschitz boundary $\Gamma = \partial\Omega$, i.e. the choice of families $W_j$ and $\Omega_j$, is in general not unique. Examples for non–Lipschitz domains are given in Fig. 2.1, see also [103].

If the parametrizations satisfy $\gamma \in C^k(\mathbb{R}^{d-1})$ or $\gamma \in C^{k,\kappa}(\mathbb{R}^{d-1})$ we call the boundary $k$ times differentiable or Hölder continuous, respectively. If this holds only locally, we call the boundary piecewise smooth.

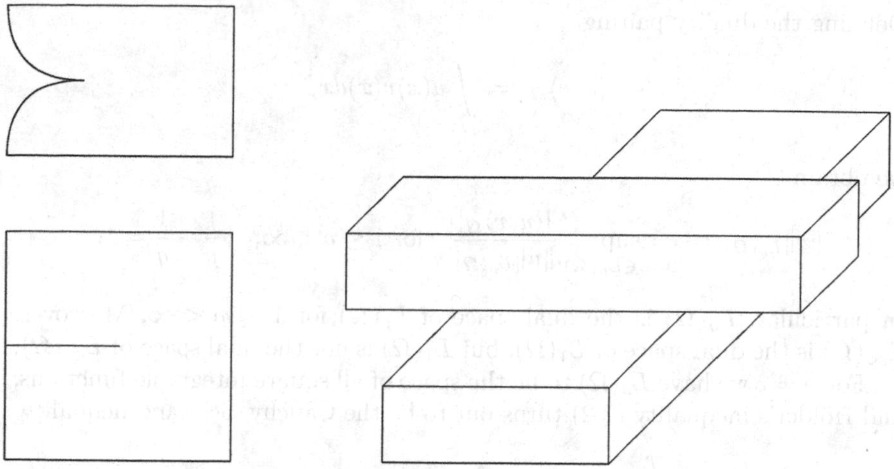

**Fig. 2.1.** Examples for non–Lipschitz domains.

By $L_p(\Omega)$ we denote the space of all equivalence classes of measurable functions on $\Omega$ whose powers of order $p$ are integrable. The associated norm is

$$\|u\|_{L_p(\Omega)} := \left\{ \int_\Omega |u(x)|^p dx \right\}^{1/p} \quad \text{for } 1 \leq p < \infty.$$

Two elements $u, v \in L_p(\Omega)$ are identified with each other if they are different only on a set $K$ of zero measure $\mu(K) = 0$. In what follows we always consider one represent $u \in L_p(\Omega)$. In addition, $L_\infty(\Omega)$ is the space of functions $u$ which are measurable and bounded almost everywhere with the norm

$$\|u\|_{L_\infty(\Omega)} := \operatorname*{ess\,sup}_{x\in\Omega} \{|u(x)|\} := \inf_{K\subset\Omega, \mu(K)=0} \sup_{x\in\Omega\setminus K} |u(x)|.$$

The spaces $L_p(\Omega)$ are Banach spaces with respect to the norm $\|\cdot\|_{L_p(\Omega)}$. There holds the Minkowski inequality

$$\|u+v\|_{L_p(\Omega)} \leq \|u\|_{L_p(\Omega)} + \|v\|_{L_p(\Omega)} \quad \text{for all } u, v \in L_p(\Omega). \tag{2.1}$$

For $u \in L_p(\Omega)$ and $v \in L_q(\Omega)$ with adjoint parameters $p$ and $q$, i.e.

$$\frac{1}{p} + \frac{1}{q} = 1,$$

we further have Hölder's inequality

$$\int_\Omega |u(x)v(x)|dx \leq \|u\|_{L_p(\Omega)} \|v\|_{L_q(\Omega)}. \tag{2.2}$$

Defining the duality pairing

$$\langle u, v\rangle_\Omega := \int_\Omega u(x)v(x)dx,$$

we obtain

$$\|v\|_{L_q(\Omega)} = \sup_{0\neq u\in L_p(\Omega)} \frac{|\langle u, v\rangle_\Omega|}{\|u\|_{L_p(\Omega)}} \quad \text{for } 1\le p<\infty, \quad \frac{1}{p}+\frac{1}{q}=1.$$

In particular, $L_q(\Omega)$ is the dual space of $L_p(\Omega)$ for $1 \le p < \infty$. Moreover, $L_\infty(\Omega)$ is the dual space of $L_1(\Omega)$, but $L_1(\Omega)$ is not the dual space of $L_\infty(\Omega)$.

For $p = 2$ we have $L_2(\Omega)$ to be the space of all square integrable functions, and Hölder's inequality (2.2) turns out to be the Cauchy–Schwarz inequality

$$\int_\Omega |u(x)v(x)|dx \le \|u\|_{L_2(\Omega)}\|v\|_{L_2(\Omega)}. \tag{2.3}$$

Moreover, for $u, v \in L_2(\Omega)$we can define the inner product

$$\langle u, v\rangle_{L_2(\Omega)} := \int_\Omega u(x)v(x)dx$$

and with

$$\langle u, u\rangle_{L_2(\Omega)} = \|u\|^2_{L_2(\Omega)} \quad \text{for all } u \in L_2(\Omega)$$

we conclude that $L_2(\Omega)$ is a Hilbert space.

## 2.2 Generalized Derivatives and Sobolev Spaces

By $L_1^{\text{loc}}(\Omega)$ we denote the space of locally integrable functions, i.e. $u \in L_1^{\text{loc}}(\Omega)$ is integrable with respect to any closed bounded subset $K \subset \Omega$ .

*Example 2.2.* Let $\Omega = (0, 1)$ and let $u(x) = 1/x$. Due to

$$\int_0^1 u(x)dx = \lim_{\varepsilon\to 0}\int_\varepsilon^1 \frac{1}{x}\,dx = \lim_{\varepsilon\to 0} \ln\frac{1}{\varepsilon} = \infty$$

we find $u \notin L_1(\Omega)$. For an arbitrary closed interval $K := [a, b] \subset (0, 1) = \Omega$ with $0 < a < b < 1$ we obtain

$$\int_K u(x)dx = \int_a^b \frac{1}{x}\,dx = \ln\frac{b}{a} < \infty$$

and therefore $u \in L_1^{\text{loc}}(\Omega)$.

For functions $\varphi, \psi \in C_0^\infty(\Omega)$ we may apply integration by parts,

$$\int_\Omega \frac{\partial}{\partial x_i}\varphi(x)\,\psi(x)\,dx = -\int_\Omega \varphi(x)\,\frac{\partial}{\partial x_i}\psi(x)\,dx.$$

Note that all integrals may be defined even for non–smooth functions. This motivates the following definition of a generalized derivative.

**Definition 2.3.** *A function $u \in L_1^{\text{loc}}(\Omega)$ has a generalized partial derivative with respect to $x_i$, if there exists a function $v \in L_1^{\text{loc}}(\Omega)$ satisfying*

$$\int_\Omega v(x)\varphi(x)dx = -\int_\Omega u(x)\frac{\partial}{\partial x_i}\varphi(x)dx \quad \textit{for all } \varphi \in C_0^\infty(\Omega). \tag{2.4}$$

*Again we denote the generalized derivative by* $\dfrac{\partial}{\partial x_i}u(x) := v(x)$.

The recursive application of (2.4) enables us to define a generalized partial derivative $D^\alpha u(x) \in L_1^{\text{loc}}(\Omega)$ by

$$\int_\Omega [D^\alpha u(x)]\varphi(x)dx = (-1)^{|\alpha|}\int_\Omega u(x)D^\alpha\varphi(x)dx \quad \text{for all } \varphi \in C_0^\infty(\Omega). \tag{2.5}$$

*Example 2.4.* Let $u(x) = |x|$ for $x \in \Omega = (-1,1)$. For an arbitrary $\varphi \in C_0^\infty(\Omega)$ we have

$$\begin{aligned}\int_{-1}^{1} u(x)\frac{\partial}{\partial x}\varphi(x)dx &= -\int_{-1}^{0} x\,\frac{\partial}{\partial x}\varphi(x)dx + \int_{0}^{1} x\,\frac{\partial}{\partial x}\varphi(x)dx\\ &= -\left[x\,\varphi(x)\right]_{-1}^{0} + \int_{-1}^{0}\varphi(x)dx + \left[x\,\varphi(x)\right]_{0}^{1} - \int_{0}^{1}\varphi(x)dx\\ &= \int_{-1}^{0}\varphi(x)dx - \int_{0}^{1}\varphi(x)dx = -\int_{-1}^{1}\text{sign}(x)\,\varphi(x)dx\end{aligned}$$

with

$$\text{sign}(x) := \begin{cases} 1 & \text{for } x > 0,\\ -1 & \text{for } x < 0.\end{cases}$$

The generalized derivative of $u(x) = |x|$ is therefore given by

$$\frac{\partial}{\partial x}u(x) = \text{sign}(x) \in L_1^{\text{loc}}(\Omega).$$

To compute the second derivative of $u(x) = |x|$ we obtain

$$\int_{-1}^{1} \operatorname{sign}(x) \frac{\partial}{\partial x}\varphi(x)dx = -\int_{-1}^{0} \frac{\partial}{\partial x}\varphi(x)dx + \int_{0}^{1} \frac{\partial}{\partial x}\varphi(x)dx = -2\varphi(0).$$

However, there exists no locally integrable function $v \in L_1^{\mathrm{loc}}(\Omega)$ satisfying

$$\int_{-1}^{1} v(x)\varphi(x)dx = 2\varphi(0)$$

for all $\varphi \in C_0^\infty(\Omega)$. Later we will find the generalized derivative of $\operatorname{sign}(x)$ in the distributional sense.

For $k \in \mathbb{N}_0$ we define norms

$$\|u\|_{W_p^k(\Omega)} := \begin{cases} \left\{ \sum\limits_{|\alpha|\le k} \|D^\alpha u\|^p_{L_p(\Omega)} \right\}^{1/p} & \text{for } 1 \le p < \infty, \\ \max\limits_{|\alpha|\le k} \|D^\alpha u\|_{L_\infty(\Omega)} & \text{for } p = \infty. \end{cases} \tag{2.6}$$

By taking the closure of $C^\infty(\Omega)$ with respect to the norm $\|\cdot\|_{W_p^k(\Omega)}$ we define the Sobolev space

$$W_p^k(\Omega) := \overline{C^\infty(\Omega)}^{\|\cdot\|_{W_p^k(\Omega)}}. \tag{2.7}$$

In particular, for any $u \in W_p^k(\Omega)$ there exists a sequence $\{\varphi_j\}_{j\in\mathbb{N}} \subset C^\infty(\Omega)$ such that

$$\lim_{j\to\infty} \|u - \varphi_j\|_{W_p^k(\Omega)} = 0.$$

Correspondingly, the closure of $C_0^\infty(\Omega)$ with respect to $\|\cdot\|_{W_p^k(\Omega)}$ defines the Sobolev space

$$\overset{\circ}{W}{}_p^k(\Omega) := \overline{C_0^\infty(\Omega)}^{\|\cdot\|_{W_p^k(\Omega)}}. \tag{2.8}$$

The definition of Sobolev norms $\|\cdot\|_{W_p^k(\Omega)}$ and therefore of the Sobolev spaces (2.7) and (2.8) can be extended for any arbitrary $s \in \mathbb{R}$. We first consider $0 < s \in \mathbb{R}$ with $s = k + \kappa$ and $k \in \mathbb{N}_0$, $\kappa \in (0,1)$. Then,

$$\|u\|_{W_p^s(\Omega)} := \left\{ \|u\|^p_{W_p^k(\Omega)} + |u|^p_{W_p^s(\Omega)} \right\}^{1/p}$$

is the Sobolev–Slobodeckii norm, and

$$|u|^p_{W_p^s(\Omega)} = \sum_{|\alpha|=k} \int_\Omega \int_\Omega \frac{|D^\alpha u(x) - D^\alpha u(y)|^p}{|x-y|^{d+p\kappa}} dxdy$$

is the associated semi–norm. In particular for $p = 2$ we have $W_2^s(\Omega)$ to be a Hilbert space with inner product

$$\langle u, v\rangle_{W_2^k(\Omega)} := \sum_{|\alpha|\le k} \int_\Omega D^\alpha u(x)\, D^\alpha v(x)\, dx$$

for $s = k \in \mathbb{N}_0$ and

$$\langle u, v\rangle_{W_2^s(\Omega)} := \langle u, v\rangle_{W_2^k(\Omega)} \tag{2.9}$$

$$+ \sum_{|\alpha|=k} \int_\Omega \int_\Omega \frac{(D^\alpha u(x) - D^\alpha u(y))(D^\alpha v(x) - D^\alpha v(y))}{|x-y|^{d+2\kappa}} dxdy.$$

for $s = k+\kappa, k \in \mathbb{N}_0, \kappa \in (0,1)$.

For $s < 0$ and $1 < p < \infty$ the Sobolev space $W_p^s(\Omega)$ is defined as the dual space of $\mathring{W}_q^{-s}(\Omega)$. Hereby we have $1/q + 1/p = 1$, and the associated norm is

$$\|u\|_{W_p^s(\Omega)} := \sup_{0\ne v\in \mathring{W}_q^{-s}(\Omega)} \frac{|\langle u, v\rangle_\Omega|}{\|v\|_{W_q^{-s}(\Omega)}}.$$

Correspondingly, $\mathring{W}_p^s(\Omega)$ is the dual space of $W_q^{-s}(\Omega)$.

## 2.3 Properties of Sobolev Spaces

In this section we state some properties of Sobolev spaces $W_p^s(\Omega)$ which are needed later in the numerical analysis of finite and boundary element methods.

Assuming a certain relation for the indices $s \in \mathbb{R}$ and $p \in \mathbb{N}$ a function $u \in W_p^s(\Omega)$ turns out to be bounded and continuous.

**Theorem 2.5 (Imbedding Theorem of Sobolev).** *Let $\Omega \subset \mathbb{R}^d$ be a bounded domain with Lipschitz boundary $\partial\Omega$ and let*

$$d \le s \quad for\, p = 1, \qquad d/p < s \quad for\, p > 1.$$

*For $u \in W_p^s(\Omega)$ we obtain $u \in C(\Omega)$ satisfying*

$$\|u\|_{L_\infty(\Omega)} \le c\,\|u\|_{W_p^s(\Omega)} \quad for\ all\ u \in W_p^s(\Omega).$$

For a proof of Theorem 2.5 see, for example, [31, Theorem 1.4.6], [103, Theorem 3.26].

The norm (2.6) of the Sobolev space $W_2^1(\Omega)$ is

$$\|v\|_{W_2^1(\Omega)} = \left\{ \|v\|^2_{L_2(\Omega)} + \|\nabla v\|^2_{L_2(\Omega)} \right\}^{1/2}$$

where

$$|v|_{W_2^1(\Omega)} = \|\nabla v\|_{L_2(\Omega)}$$

is a semi–norm. Applying the following theorem we may deduce equivalent norms in $W_2^1(\Omega)$.

**Theorem 2.6 (Norm Equivalence Theorem of Sobolev).**
*Let $f : W_2^1(\Omega) \to \mathbb{R}$ be a bounded linear functional satisfying*

$$0 \leq |f(v)| \leq c_f \, \|v\|_{W_2^1(\Omega)} \quad \textit{for all } v \in W_2^1(\Omega).$$

*If $f(constant) = 0$ is only satisfied for $constant = 0$, then*

$$\|v\|_{W_2^1(\Omega),f} := \Big\{ |f(v)|^2 + \|\nabla v\|_{L_2(\Omega)}^2 \Big\}^{1/2} \tag{2.10}$$

*defines an equivalent norm in $W_2^1(\Omega)$.*

*Proof.* Since the linear functional $f$ is bounded we conclude

$$\begin{aligned} \|v\|_{W_2^1(\Omega),f}^2 &= |f(v)|^2 + \|\nabla v\|_{L_2(\Omega)}^2 \\ &\leq c_f^2 \|v\|_{W_2^1(\Omega)}^2 + \|\nabla v\|_{L_2(\Omega)}^2 \;\leq\; (1 + c_f^2) \|v\|_{W_2^1(\Omega)}^2 . \end{aligned}$$

The proof of the opposite direction is indirect. Assume that there is no constant $c_0 > 0$ such that

$$\|v\|_{W_2^1(\Omega)} \;\leq\; c_0 \, \|v\|_{W_2^1(\Omega),f} \quad \text{for all } v \in W_2^1(\Omega).$$

Then there would exist a sequence $\{v_n\}_{n \in \mathbb{N}} \subset W_2^1(\Omega)$ with

$$n \leq \frac{\|v_n\|_{W_2^1(\Omega)}}{\|v_n\|_{W_2^1(\Omega),f}} \quad \text{for } n \in \mathbb{N}.$$

For the normalized sequence $\{\bar{v}_n\}_{n \in \mathbb{N}}$ with

$$\bar{v}_n := \frac{v_n}{\|v_n\|_{W_2^1(\Omega)}}$$

we therefore have

$$\|\bar{v}_n\|_{W_2^1(\Omega)} = 1, \quad \|\bar{v}_n\|_{W_2^1(\Omega),f} = \frac{\|v_n\|_{W_2^1(\Omega),f}}{\|v_n\|_{W_2^1(\Omega)}} \leq \frac{1}{n} \to 0 \quad \text{as } n \to \infty.$$

From this and (2.10) we conclude

$$\lim_{n \to \infty} |f(\bar{v}_n)| = 0, \quad \lim_{n \to \infty} \|\nabla \bar{v}_n\|_{L_2(\Omega)} = 0.$$

Since the sequence $\{\bar{v}_n\}_{n \in \mathbb{N}}$ is bounded in $W_2^1(\Omega)$ and since the imbedding $W_2^1(\Omega) \hookrightarrow L_2(\Omega)$ is compact [160], there exists a subsequence $\{\bar{v}_{n'}\}_{n' \in \mathbb{N}} \subset \{\bar{v}_n\}_{n \in \mathbb{N}}$ which converges in $L_2(\Omega)$. In particular, $\bar{v} := \lim_{n' \to \infty} \bar{v}_{n'} \in L_2(\Omega)$. From

$$\lim_{n' \to \infty} \|\nabla \bar{v}_{n'}\|_{L_2(\Omega)} = 0$$

we obtain $\bar{v} \in W_2^1(\Omega)$ with $\|\nabla \bar{v}\|_{L_2(\Omega)} = 0$, i.e. $\bar{v} = \text{constant}$. With

$$0 \le |f(\bar{v})| = \lim_{n' \to \infty} |f(\bar{v}_{n'})| = 0$$

we then conclude $f(\bar{v}) = 0$ and therefore $\bar{v} = 0$. However, this is a contradiction to

$$\|\bar{v}\|_{W_2^1(\Omega)} = \lim_{n' \to \infty} \|\bar{v}_{n'}\|_{W_2^1(\Omega)} = 1. \qquad \Box$$

*Example 2.7.* Equivalent norms in $W_2^1(\Omega)$ are given by

$$\|v\|_{W_2^1(\Omega),\Omega} := \left\{ \left[ \int_\Omega v(x)\,dx \right]^2 + \|\nabla v\|^2_{L_2(\Omega)} \right\}^{1/2} \tag{2.11}$$

and

$$\|v\|_{W_2^1(\Omega),\Gamma} := \left\{ \left[ \int_\Gamma v(x)\,ds_x \right]^2 + \|\nabla v\|^2_{L_2(\Omega)} \right\}^{1/2} .$$

We therefore obtain $\|\nabla \cdot \|_{L_2(\Omega)}$ to be an equivalent norm in $\mathring{W}_2^1(\Omega)$

Using the equivalent norm (2.11) as given in Example 2.7 the Poincaré inequality

$$\int_\Omega |v(x)|^2 dx \le c_P \left\{ \left[ \int_\Omega v(x)\,dx \right]^2 + \int_\Omega |\nabla v(x)|^2 dx \right\} \tag{2.12}$$

for all $v \in W_2^1(\Omega)$ follows.

To derive some approximation properties of (piecewise) polynomial trial spaces the following result is needed.

**Theorem 2.8 (Bramble–Hilbert Lemma).**
*For $k \in \mathbb{N}_0$ let $f : W_2^{k+1}(\Omega) \to \mathbb{R}$ be a bounded linear functional satisfying*

$$|f(v)| \le c_f \, \|v\|_{W_2^{k+1}(\Omega)} \quad \text{for all } v \in W_2^{k+1}(\Omega).$$

*By $P_k(\Omega)$ we denote the space of all polynomials of degree $k$ defined in $\Omega$. If*

$$f(q) = 0$$

*is satisfied for all $q \in P_k(\Omega)$ then we also have*

$$|f(v)| \le c(c_p)\, c_f \, |v|_{W_2^{k+1}(\Omega)} \quad \text{for all } v \in W_2^{k+1}(\Omega)$$

*where the constant $c(c_p)$ depends only on the constant $c_p$ of the Poincaré inequality* (2.12).

*Proof.* We give only the proof for the case $k = 1$, i.e. $P_1(\Omega)$ is the space of linear functions defined on $\Omega$. For $v \in W_2^2(\Omega)$ and $q \in P_1(\Omega)$ we have, due to the assumptions,

$$|f(v)| = |f(v) + f(q)| = |f(v+q)| \le c_f \, \|v+q\|_{W_2^2(\Omega)}.$$

Moreover,

$$\begin{aligned}\|v+q\|^2_{W_2^2(\Omega)} &= \|v+q\|^2_{L_2(\Omega)} + |v+q|^2_{W_2^1(\Omega)} + |v+q|^2_{W_2^2(\Omega)} \\ &= \|v+q\|^2_{L_2(\Omega)} + \|\nabla(v+q)\|^2_{L_2(\Omega)} + |v|^2_{W_2^2(\Omega)},\end{aligned}$$

since the second derivatives of a linear function disappear. Applying the Poincaré inequality (2.12) this gives

$$\|v+q\|^2_{W_2^2(\Omega)} \le c_P \left[\int_\Omega [v(x)+q(x)]dx\right]^2 + (1+c_P)\,\|\nabla(v+q)\|^2_{L_2(\Omega)} + |v|^2_{W_2^2(\Omega)}.$$

For the second term we apply the Poincaré inequality (2.12) once again to obtain

$$\begin{aligned}\|\nabla(v+q)\|^2_{L_2(\Omega)} &= \sum_{i=1}^d \int_\Omega \left|\frac{\partial}{\partial x_i}[v(x)+q(x)]\right|^2 dx \\ &\le c_P \sum_{i=1}^d \left\{\left[\int_\Omega \frac{\partial}{\partial x_i}[v(x)+q(x)]dx\right]^2 + \sum_{j=1}^d \int_\Omega \left[\frac{\partial^2}{\partial x_i \partial x_j}[v(x)+q(x)]\right]^2 dx\right\} \\ &= c_P \sum_{i=1}^d \left[\int_\Omega \frac{\partial}{\partial x_i}[v(x)+q(x)]dx\right]^2 + c_P\,|v|^2_{W_2^2(\Omega)}\end{aligned}$$

and hence

$$\begin{aligned}\|v+q\|^2_{W_2^2(\Omega)} &\le c_P \left[\int_\Omega [v(x)+q(x)]dx\right]^2 \\ &\quad + (1+c_P)c_P \sum_{i=1}^d \left[\int_\Omega \frac{\partial}{\partial x_i}[v(x)+q(x)]dx\right]^2 + [1+(1+c_P)c_P]\,|v|^2_{W_2^2(\Omega)}.\end{aligned}$$

The assertion is proved if we can choose $q \in P_1(\Omega)$ so that the first two terms are zero, i.e.

$$\int_\Omega [v(x)+q(x)]dx = 0, \quad \int_\Omega \frac{\partial}{\partial x_i}[v(x)+q(x)]dx = 0 \quad \text{for } i = 1, \ldots, d.$$

With

$$q(x) = a_0 + \sum_{i=1}^{d} a_i x_i$$

we get

$$a_i = -\frac{1}{|\Omega|}\int_\Omega \frac{\partial}{\partial x_i} v(x)dx \quad \text{for } i = 1, \ldots, d$$

and therefore

$$a_0 = -\frac{1}{|\Omega|}\int_\Omega \left[ v(x) + \sum_{i=1}^{d} a_i x_i \right] dx.$$

The proof for general $k \in \mathbb{N}$ is almost the same. □

## 2.4 Distributions and Sobolev Spaces

As it was observed in Example 2.4 not every function in $L_1^{\text{loc}}(\Omega)$ has a generalized derivative in $L_1^{\text{loc}}(\Omega)$. Hence we also introduce derivatives in the sense of distributions, see also [103, 150, 156, 161].

For $\Omega \subseteq \mathbb{R}^d$ we first define $\mathcal{D}(\Omega) := C_0^\infty(\Omega)$ to be the space of all test functions.

**Definition 2.9.** *A complex valued continuous linear form $T$ acting on $\mathcal{D}(\Omega)$ is called a distribution. $T$ is continuous on $\mathcal{D}(\Omega)$, if $\varphi_k \to \varphi$ in $\mathcal{D}(\Omega)$ always implies $T(\varphi_k) \to T(\varphi)$. The set of all distributions on $\mathcal{D}(\Omega)$ is denoted by $\mathcal{D}'(\Omega)$.*

For $u \in L_1^{\text{loc}}(\Omega)$ we define the distribution

$$T_u(\varphi) := \int_\Omega u(x)\varphi(x)dx \quad \text{for } \varphi \in \mathcal{D}(\Omega). \tag{2.13}$$

Distributions of the type (2.13) are called regular. Local integrable functions $u \in L_1^{\text{loc}}(\Omega)$ can be identified with a subset of $\mathcal{D}'(\Omega)$. Hence, instead of $T_u \in \mathcal{D}'(\Omega)$ we simply write $u \in \mathcal{D}'(\Omega)$. Nonregular distributions are called singular. For example, the Dirac distribution for $x_0 \in \Omega$,

$$\delta_{x_0}(\varphi) = \varphi(x_0) \quad \text{for } \varphi \in \mathcal{D}(\Omega),$$

can not be represented as in (2.13).

For the computation of the derivative of the function $v(x) = \text{sign}(x)$ as considered in Example 2.4 we now obtain:

*Example 2.10.* Using integration by parts we have for $v(x) = \mathrm{sign}(x)$

$$\int_{-1}^{1} \mathrm{sign}(x) \frac{\partial}{\partial x}\varphi(x)dx = -2\varphi(0) = -\int_{-1}^{1} \frac{\partial}{\partial x}v(x)\,\varphi(x)dx \quad \text{for all } \varphi \in \mathcal{D}(\Omega).$$

Hence the derivative of $v$ in the distributional sense is given by

$$\frac{\partial}{\partial x}v = 2\,\delta_0 \in \mathcal{D}'(\Omega).$$

As for the generalized derivative (2.5) we can define higher order derivatives $D^\alpha T_u \in \mathcal{D}'(\Omega)$ of a distribution $T_u \in \mathcal{D}(\Omega)$ by

$$(D^\alpha T_u)(\varphi) = (-1)^{|\alpha|} T_u(D^\alpha \varphi) \quad \text{for } \varphi \in \mathcal{D}(\Omega).$$

In what follows we introduce Sobolev spaces $H^s(\Omega)$ which may be equivalent to the previously introduced Sobolev spaces $W_2^s(\Omega)$ when some regularity assumptions on $\Omega$ are satisfied. The definition of Sobolev spaces $H^s(\Omega)$ is based on the Fourier transform of distributions. Hence we need to introduce first the space $\mathcal{S}(\mathbb{R}^d)$ of rapidly decreasing functions.

**Definition 2.11.** *$\mathcal{S}(\mathbb{R}^d)$ is the space of functions $\varphi \in C^\infty(\mathbb{R}^d)$ satisfying*

$$\|\varphi\|_{k,\ell} := \sup_{x\in\mathbb{R}^d} \left(|x|^k + 1\right) \sum_{|\alpha|\le \ell} |D^\alpha \varphi(x)| < \infty \quad \textit{for all } k,\ell \in \mathbb{N}_0.$$

*In particular, the function $\varphi$ and all of their derivatives decreases faster than any polynomial.*

*Example 2.12.* For the function $\varphi(x) := e^{-|x|^2}$ we have $\varphi \in \mathcal{S}(\mathbb{R}^d)$, but $\varphi \notin \mathcal{D}(\Omega) = C_0^\infty(\mathbb{R}^d)$.

As in Definition 2.9 we can introduce the space $\mathcal{S}'(\mathbb{R}^d)$ of tempered distributions as the space of all complex valued linear forms $T$ over $\mathcal{S}(\mathbb{R}^d)$.

For a function $\varphi \in \mathcal{S}(\mathbb{R}^d)$ we can define the Fourier transform $\widehat{\varphi} \in \mathcal{S}(\mathbb{R}^d)$,

$$\widehat{\varphi}(\xi) := (\mathcal{F}\varphi)(\xi) = (2\pi)^{-\frac{d}{2}} \int_{\mathbb{R}^d} e^{-i\langle x,\xi\rangle}\varphi(x)dx \quad \text{for } \xi \in \mathbb{R}^d. \tag{2.14}$$

The mapping $\mathcal{F} : \mathcal{S}(\mathbb{R}^d) \to \mathcal{S}(\mathbb{R}^d)$ is invertible and the inverse Fourier transform is given by

$$(\mathcal{F}^{-1}\widehat{\varphi})(x) = (2\pi)^{-\frac{d}{2}} \int_{\mathbb{R}^d} e^{i\langle x,\xi\rangle}\widehat{\varphi}(\xi)d\xi \quad \text{for } x \in \mathbb{R}^d. \tag{2.15}$$

In general, $\varphi \in \mathcal{D}(\mathbb{R}^d)$ does not imply $\widehat{\varphi} \in \mathcal{D}(\mathbb{R}^d)$.

For $\varphi \in \mathcal{S}(\mathbb{R}^d)$ we have

$$D^\alpha(\mathcal{F}\varphi)(\xi) = (-i)^{|\alpha|}\mathcal{F}(x^\alpha\varphi)(\xi) \tag{2.16}$$

as well as

$$\xi^\alpha(\mathcal{F}\varphi)(\xi) = (-i)^{|\alpha|}\mathcal{F}(D^\alpha\varphi)(\xi). \tag{2.17}$$

**Lemma 2.13.** *The Fourier transform maintains rotational symmetries, i.e. for $u \in \mathcal{S}(\mathbb{R}^d)$ we have $\widehat{u}(\xi) = \widehat{u}(|\xi|)$ for all $\xi \in \mathbb{R}^d$ iff $u(x) = u(|x|)$ for all $x \in \mathbb{R}^d$.*

*Proof.* Let us first consider the two–dimensional case $d = 2$. Using polar coordinates,

$$\xi = \begin{pmatrix} |\xi|\cos\psi \\ |\xi|\sin\psi \end{pmatrix}, \quad x = \begin{pmatrix} r\cos\phi \\ r\sin\phi \end{pmatrix},$$

we obtain

$$\begin{aligned}
\widehat{u}(\xi) = \widehat{u}(|\xi|, \psi) &= \frac{1}{2\pi}\int_0^\infty\int_0^{2\pi} e^{-ir\,|\xi|[\cos\phi\cos\psi + \sin\phi\sin\psi]}u(r)r\,d\phi\,dr \\
&= \frac{1}{2\pi}\int_0^\infty\int_0^{2\pi} e^{-ir|\xi|\cos(\phi-\psi)}u(r)r\,d\phi\,dr.
\end{aligned}$$

With $\psi_0 \in [0, 2\pi)$ and substituting $\widetilde{\phi} := \phi - \psi_0$ it follows that

$$\begin{aligned}
\widehat{u}(|\xi|, \psi + \psi_0) &= \frac{1}{2\pi}\int_0^\infty\int_0^{2\pi} e^{-ir|\xi|\cos(\phi-\psi-\psi_0)}u(r)r\,d\phi\,dr \\
&= \frac{1}{2\pi}\int_0^\infty\int_{-\psi_0}^{2\pi-\psi_0} e^{-ir|\xi|\cos(\widetilde{\phi}-\psi)}u(r)r\,d\widetilde{\phi}\,dr.
\end{aligned}$$

By using

$$\int_{-\psi_0}^{0} e^{-ir|\xi|\cos(\widetilde{\phi}-\psi)}d\widetilde{\phi} = \int_{2\pi-\psi_0}^{2\pi} e^{-ir|\xi|\cos(\widetilde{\phi}-\psi)}d\widetilde{\phi}$$

we then obtain

$$\widehat{u}(|\xi|, \psi) = \widehat{u}(|\xi|, \psi + \psi_0) \quad \text{for all } \psi_0 \in [0, 2\pi)$$

and therefore the assertion $\widehat{u}(\xi) = \widehat{u}(|\xi|)$.

For $d = 3$ we use spherical coordinates

$$\xi = \begin{pmatrix} |\xi| \cos\psi \sin\vartheta \\ |\xi| \sin\psi \sin\vartheta \\ |\xi| \cos\vartheta \end{pmatrix}, \quad x = \begin{pmatrix} r\cos\phi \sin\theta \\ r\sin\phi\sin\theta \\ r\cos\theta \end{pmatrix}$$

to obtain

$$\widehat{u}(|\xi|, \psi, \vartheta) =$$
$$= \frac{1}{(2\pi)^{3/2}} \int_0^\infty \int_0^{2\pi} \int_0^\pi e^{-ir|\xi|[\cos(\phi-\psi)\sin\theta\sin\vartheta + \cos\theta\cos\vartheta]} u(r) r^2 \sin\theta d\theta d\phi dr.$$

As for the two–dimensional case $d = 2$ we conclude

$$\widehat{u}(|\xi|, \psi + \psi_0, \vartheta) = \widehat{u}(|\xi|, \psi, \vartheta) \quad \text{for all } \psi_0 \in [0, 2\pi).$$

For a fixed $\vartheta \in [0, \pi]$ and for a given radius $\varrho$ we also have $\widehat{u}(\xi) = \widehat{u}(|\xi|) = \widehat{u}(\varrho)$ along the circular lines

$$\xi_1^2 + \xi_2^2 = \varrho^2 \sin^2\vartheta, \quad \xi_3 = \varrho\cos\vartheta.$$

Using permutated spherical coordinates we also find $\widehat{u}(\xi) = \widehat{u}(\varrho)$ along the circular lines

$$\xi_1^2 + \xi_3^2 = \varrho^2 \sin^2\vartheta, \quad \xi_2 = \varrho\cos\vartheta. \qquad \square$$

For a distribution $T \in \mathcal{S}'(\mathbb{R}^d)$ we can define the Fourier transform $\widehat{T} \in \mathcal{S}'(\mathbb{R}^d)$

$$\widehat{T}(\varphi) := T(\widehat{\varphi}) \quad \text{for } \varphi \in \mathcal{S}(\mathbb{R}^d).$$

The mapping $\mathcal{F} : \mathcal{S}'(\mathbb{R}^d) \to \mathcal{S}'(\mathbb{R}^d)$ is invertible and the inverse Fourier transform is given by

$$(\mathcal{F}^{-1}T)(\varphi) := T(\mathcal{F}^{-1}\varphi) \quad \text{for } \varphi \in \mathcal{S}(\mathbb{R}^d).$$

The rules (2.16) and (2.17) remain valid for distributions $T \in \mathcal{S}'(\mathbb{R}^d)$.

For $s \in \mathbb{R}$ and $u \in \mathcal{S}(\mathbb{R}^d)$ we define the Bessel potential operator

$$\mathcal{J}^s u(x) := (2\pi)^{-d/2} \int_{\mathbb{R}^d} (1 + |\xi|^2)^{s/2} \widehat{u}(\xi) e^{i\langle x, \xi\rangle} d\xi, \quad x \in \mathbb{R}^d,$$

which is a bounded linear operator $\mathcal{J}^s : \mathcal{S}(\mathbb{R}^d) \to \mathcal{S}(\mathbb{R}^d)$. The application of the Fourier transform gives

$$(\mathcal{F}\mathcal{J}^s u)(\xi) = (1 + |\xi|^2)^{s/2} (\mathcal{F}u)(\xi).$$

From this we conclude that the application of $\mathcal{J}^s$ corresponds in the Fourier space to a multiplication with a function of order $\mathcal{O}(|\xi|^s)$. Therefore, using (2.17) we can see $\mathcal{J}^s$ as a differential operator of order $s$.

For $T \in \mathcal{S}'(\mathbb{R}^d)$ we define a bounded and linear operator $\mathcal{J}^s : \mathcal{S}'(\mathbb{R}^d) \to \mathcal{S}'(\mathbb{R}^d)$ acting on the space of tempered distributions,

$$(\mathcal{J}^s T)(\varphi) := T(\mathcal{J}^s \varphi) \quad \text{for all } \varphi \in \mathcal{S}(\mathbb{R}^d).$$

The Sobolev space $H^s(\mathbb{R}^d)$ is the space of all distributions $v \in \mathcal{S}'(\mathbb{R}^d)$ with $\mathcal{J}^s v \in L_2(\mathbb{R}^d)$ where the associated inner product

$$\langle u, v\rangle_{H^s(\mathbb{R}^d)} := \langle \mathcal{J}^s u, \mathcal{J}^s v\rangle_{L_2(\mathbb{R}^d)}$$

implies the norm

$$\|u\|^2_{H^s(\mathbb{R}^d)} := \|\mathcal{J}^s u\|^2_{L_2(\mathbb{R}^d)} = \int\limits_{\mathbb{R}^d} (1+|\xi|^2)^s |\widehat{u}(\xi)|^2 d\xi .$$

The connection with the Sobolev spaces $W_2^s(\mathbb{R}^d)$ can be seen from the following theorem, see for example [103, 160].

**Theorem 2.14.** *For all $s \in \mathbb{R}$ there holds*

$$H^s(\mathbb{R}^d) = W_2^s(\mathbb{R}^d) .$$

For a bounded domain $\Omega \subset \mathbb{R}^d$ we define the Sobolev space $H^s(\Omega)$ by restriction,

$$H^s(\Omega) := \left\{ v = \widetilde{v}_{|\Omega} : \widetilde{v} \in H^s(\mathbb{R}^d) \right\},$$

with the norm

$$\|v\|_{H^s(\Omega)} := \inf_{\widetilde{v} \in H^s(\mathbb{R}^d), \widetilde{v}_{|\Omega} = v} \|\widetilde{v}\|_{H^s(\mathbb{R}^d)} .$$

In addition we introduce Sobolev spaces

$$\widetilde{H}^s(\Omega) := \overline{C_0^\infty(\Omega)}^{\|\cdot\|_{H^s(\mathbb{R}^d)}}, \quad H_0^s(\Omega) := \overline{C_0^\infty(\Omega)}^{\|\cdot\|_{H^s(\Omega)}}$$

which will coincide for almost all $s \in \mathbb{R}_+$, see for example [103, Theorem 3.33].

**Theorem 2.15.** *Let $\Omega \subset \mathbb{R}^d$ be a Lipschitz domain. For $s \geq 0$ we have*

$$\widetilde{H}^s(\Omega) \subset H_0^s(\Omega).$$

*In particular,*

$$\widetilde{H}^s(\Omega) = H_0^s(\Omega) \quad \text{for } s \notin \left\{ \frac{1}{2}, \frac{3}{2}, \frac{5}{2}, \ldots \right\}.$$

*Moreover,*

$$\widetilde{H}^s(\Omega) = [H^{-s}(\Omega)]', \quad H^s(\Omega) = [\widetilde{H}^{-s}(\Omega)]' \quad \text{for all } s \in \mathbb{R}.$$

The equivalence of Sobolev spaces $W_2^s(\Omega)$ and $H^s(\Omega)$ holds only when certain assumptions on $\Omega$ are satisfied. Sufficient for the norm equivalence is the existence of a linear bounded extension operator

$$E_\Omega : W_2^s(\Omega) \to W_2^s(\mathbb{R}^d).$$

This condition is ensured for a bounded domain $\Omega \subset \mathbb{R}^d$, if a uniform cone condition is satisfied, see for example [161, Theorem 5.4].

**Theorem 2.16.** *For a Lipschitz domain $\Omega \subset \mathbb{R}^d$ we have*

$$W_2^s(\Omega) \;=\; H^s(\Omega) \quad \textit{for all } s > 0.$$

For the analysis of the Stokes system we need to have some mapping properties of the gradient $\nabla$.

**Theorem 2.17.** [53, Theorem 3.2, p. 111] *Let the Lipschitz domain $\Omega \subset \mathbb{R}^d$ be bounded and connected. Then there holds*

$$\|q\|_{L_2(\Omega)} \;\leq\; c_1 \left\{ \|q\|_{H^{-1}(\Omega)} + \|\nabla q\|_{[H^{-1}(\Omega)]^d} \right\} \quad \textit{for all } q \in L_2(\Omega)$$

*as well as*

$$\|q\|_{L_2(\Omega)} \;\leq\; c_2 \, \|\nabla q\|_{[H^{-1}(\Omega)]^d} \quad \textit{for all } q \in L_2(\Omega) \textit{ with } \int_\Omega q(x)dx = 0. \quad (2.18)$$

Bounded linear operators can be seen as maps between different Sobolev spaces inducing different operator norms. Then one can extend the boundedness properties to Sobolev spaces between. For a general overview on interpolation spaces we refer to [16, 103] . Here we will use only the following result.

**Theorem 2.18 (Interpolation Theorem).** *Let $A : H^{\alpha_1}(\Omega) \to H^\beta(\Omega)$ be some bounded and linear operator with norm*

$$\|A\|_{\alpha_1,\beta} \;:=\; \sup_{0 \neq v \in H^{\alpha_1}(\Omega)} \frac{\|Av\|_{H^\beta(\Omega)}}{\|v\|_{H^{\alpha_1}(\Omega)}} \,.$$

*For $\alpha_2 > \alpha_1$ let $A : H^{\alpha_2}(\Omega) \to H^\beta(\Omega)$ be bounded with norm $\|A\|_{\alpha_2,\beta}$. Then the operator $A : H^\alpha(\Omega) \to H^\beta(\Omega)$ is bounded for all $\alpha \in [\alpha_1, \alpha_2]$ and the corresponding operator norm is given by*

$$\|A\|_{\alpha,\beta} \leq \left(\|A\|_{\alpha_1,\beta}\right)^{\frac{\alpha-\alpha_2}{\alpha_1-\alpha_2}} \left(\|A\|_{\alpha_2,\beta}\right)^{\frac{\alpha-\alpha_1}{\alpha_2-\alpha_1}} .$$

*Let the operator $A : H^\alpha(\Omega) \to H^{\beta_1}(\Omega)$ be bounded with norm $\|A\|_{\alpha,\beta_1}$ and let $A : H^\alpha(\Omega) \to H^{\beta_2}(\Omega)$ be bounded with norm $\|A\|_{\alpha,\beta_2}$ assuming $\beta_1 < \beta_2$. Then the operator $A : H^\alpha(\Omega) \to H^\beta(\Omega)$ is bounded for all $\beta \in [\beta_1, \beta_2]$ and the corresponding operator norm is given by*

$$\|A\|_{\alpha,\beta} \leq \left(\|A\|_{\alpha,\beta_1}\right)^{\frac{\beta-\beta_2}{\beta_1-\beta_2}} \left(\|A\|_{\alpha,\beta_2}\right)^{\frac{\beta-\beta_1}{\beta_2-\beta_1}} .$$

## 2.5 Sobolev Spaces on Manifolds

Let $\Omega \subset \mathbb{R}^d$ be a bounded domain ($d = 2, 3$) and let the boundary $\Gamma = \partial\Omega$ be given by some arbitrary overlapping piecewise parametrization

$$\Gamma = \bigcup_{i=1}^{J} \Gamma_i, \quad \Gamma_i := \left\{ x \in \mathbb{R}^d \,:\, x = \chi_i(\xi) \text{ for } \xi \in \mathcal{T}_i \subset \mathbb{R}^{d-1} \right\}. \tag{2.19}$$

With respect to (2.19) we also consider a partition of unity, $\{\varphi_i\}_{i=1}^{p}$, of non–negative cut off functions $\varphi_i \in C_0^\infty(\mathbb{R}^d)$ satisfying

$$\sum_{i=1}^{J} \varphi_i(x) = 1 \quad \text{for } x \in \Gamma, \quad \varphi_i(x) = 0 \quad \text{for } x \in \Gamma \backslash \Gamma_i.$$

For any function $v$ defined on the boundary $\Gamma$ we can write

$$v(x) = \sum_{i=1}^{J} \varphi_i(x) v(x) = \sum_{i=1}^{J} v_i(x) \quad \text{for } x \in \Gamma$$

with $v_i(x) := \varphi_i(x)v(x)$. Inserting the local parametrizations (2.19) we obtain for $i = 1, \ldots, J$

$$v_i(x) = \varphi_i(x)v(x) = \varphi_i(\chi_i(\xi))v(\chi_i(\xi)) =: \widetilde{v}_i(\xi) \quad \text{for } \xi \in \mathcal{T}_i \subset \mathbb{R}^{d-1}.$$

The functions $\widetilde{v}_i$ are defined with respect to the parameter domains $\mathcal{T}_i \subset \mathbb{R}^{d-1}$ for which we can introduce appropriate Sobolev spaces. Taking into account the chain rule we have to ensure the existence of all corresponding derivatives of the local parametrization $\chi_i(\xi)$. For the definition of derivatives of order $|s| \leq k$ we therefore have to assume $\chi_i \in C^{k-1,1}(\mathcal{T}_i)$. In particular for a Lipschitz domain with a local parametrization $\chi_i \in C^{0,1}(\mathcal{T}_i)$ we can only introduce Sobolev spaces $H^s(\mathcal{T}_i)$ for $|s| \leq 1$.

In general we can define the Sobolev norm

$$\|v\|_{H^s_\chi(\Gamma)} := \left\{ \sum_{i=1}^{J} \|\widetilde{v}_i\|^2_{H^s(\mathcal{T}_i)} \right\}^{1/2} \tag{2.20}$$

for $0 \leq s \leq k$ and therefore the corresponding Sobolev space $H^s(\Gamma)$.

**Lemma 2.19.** *For $s = 0$ an equivalent norm in $H^0_\chi(\Gamma)$ is given by*

$$\|v\|_{L_2(\Gamma)} := \left\{ \int_\Gamma |v(x)|^2 ds_x \right\}^{1/2}.$$

*Proof.* First we note that

$$\|v\|^2_{H^0_\chi(\Gamma)} = \sum_{i=1}^{J} \int_{T_i} [\varphi_i(\chi_i(\xi))v(\chi_i(\xi))]^2 d\xi$$

and

$$\|v\|^2_{L_2(\Gamma)} = \int_\Gamma [v(x)]^2 ds_x = \sum_{i=1}^{J} \int_{\Gamma_i} \varphi_i(x)[v(x)]^2 ds_x .$$

Inserting the local parametrization this gives

$$\|v\|^2_{L_2(\Gamma)} = \sum_{i=1}^{J} \int_{T_i} \varphi_i(\chi_i(\xi))[v(\chi_i(\xi))]^2 \det\chi_i(\xi) d\xi .$$

From this the assertion follows, where the constants depend on both the chosen parametrization (2.19) and on the particular definition of the cut off functions $\varphi_i$. □

For $s \in (0,1)$ we find in the same way that the Sobolev–Slobodeckii norm

$$\|v\|_{H^s(\Gamma)} := \left\{ \|v\|^2_{L_2(\Gamma)} + \int_\Gamma \int_\Gamma \frac{[v(x)-v(y)]^2}{|x-y|^{d-1+2s}} ds_x ds_y \right\}^{1/2}$$

is an equivalent norm in $H^s_\chi(\Gamma)$.

As in the Equivalence Theorem of Sobolev (Theorem 2.6) we may also define other equivalent norms in $H^s(\Gamma)$. For example,

$$\|v\|_{H^{1/2}(\Gamma),\Gamma} := \left\{ \left[ \int_\Gamma v(x) ds_x \right]^2 + \int_\Gamma \int_\Gamma \frac{[v(x)-v(y)]^2}{|x-y|^d} ds_x ds_y \right\}^{1/2}$$

defines an equivalent norm in $H^{1/2}(\Gamma)$.

Up to now we only considered Sobolev spaces $H^s(\Gamma)$ for $s \geq 0$. For $s < 0$ $H^s(\Gamma)$ is defined as the dual space of $H^{-s}(\Gamma)$,

$$H^s(\Gamma) := [H^{-s}(\Gamma)]',$$

where the associated norm is

$$\|w\|_{H^s(\Gamma)} := \sup_{0 \neq v \in H^{-s}(\Gamma)} \frac{\langle w, v\rangle_\Gamma}{\|v\|_{H^{-s}(\Gamma)}}$$

with respect to the duality pairing

$$\langle w, v\rangle_\Gamma := \int_\Gamma w(x)v(x) ds_x .$$

Let $\Gamma_0 \subset \Gamma$ be some open part of a sufficient smooth boundary $\Gamma = \partial\Omega$. For $s \geq 0$ we introduce the Sobolev space

$$H^s(\Gamma_0) := \{v = \widetilde{v}_{|\Gamma_0} : \widetilde{v} \in H^s(\Gamma)\}$$

with the norm

$$\|v\|_{H^s(\Gamma_0)} := \inf_{\widetilde{v} \in H^s(\Gamma) : \widetilde{v}_{|\Gamma_0} = v} \|\widetilde{v}\|_{H^s(\Gamma)}$$

as well as the Sobolev space

$$\widetilde{H}^s(\Gamma_0) := \left\{v = \widetilde{v}_{|\Gamma_0} : \widetilde{v} \in H^s(\Gamma),\ \operatorname{supp} \widetilde{v} \subset \Gamma_0\right\}.$$

For $s < 0$ we define the appropriate Sobolev spaces by duality,

$$H^s(\Gamma_0) := [\widetilde{H}^{-s}(\Gamma_0)]', \quad \widetilde{H}^s(\Gamma_0) := [H^{-s}(\Gamma_0)]'.$$

Finally we consider a closed boundary $\Gamma = \partial\Omega$ which is piecewise smooth,

$$\Gamma = \bigcup_{i=1}^{J} \overline{\Gamma}_i, \quad \Gamma_i \cap \Gamma_j = \emptyset \quad \text{for } i \neq j.$$

For $s > 0$ we define by

$$H^s_{\text{pw}}(\Gamma) := \{v \in L_2(\Gamma) : v_{|\Gamma_i} \in H^s(\Gamma_i), i = 1, \ldots, J\}$$

the space of piecewise smooth functions with the norm

$$\|v\|_{H^s_{\text{pw}}(\Gamma)} := \left\{\sum_{i=1}^{J} \|v_{|\Gamma_i}\|^2_{H^s(\Gamma_i)}\right\}^{1/2}$$

while for $s < 0$ we have

$$H^s_{\text{pw}}(\Gamma) := \prod_{j=1}^{J} \widetilde{H}^s(\Gamma_j) \tag{2.21}$$

with the norm

$$\|w\|_{H^s_{\text{pw}}(\Gamma)} := \sum_{j=1}^{J} \|w_{|\Gamma_j}\|_{\widetilde{H}^s(\Gamma_j)}. \tag{2.22}$$

**Lemma 2.20.** *For $w \in H^s_{pw}(\Gamma)$ and $s < 0$ we have*

$$\|w\|_{H^s(\Gamma)} \leq \|w\|_{H^s_{pw}(\Gamma)}.$$

*Proof.* By duality we conclude

$$\begin{aligned}
\|w\|_{H^s(\Gamma)} &= \sup_{0\neq v\in H^{-s}(\Gamma)} \frac{|\langle w, v\rangle_\Gamma|}{\|v\|_{H^{-s}(\Gamma)}} \leq \sup_{0\neq v\in H^{-s}(\Gamma)} \sum_{j=1}^J \frac{|\langle w, v\rangle_{\Gamma_j}|}{\|v\|_{H^{-s}(\Gamma)}} \\
&\leq \sup_{0\neq v\in H^{-s}(\Gamma)} \sum_{j=1}^J \frac{|\langle w_{|\Gamma_j}, v_{|\Gamma_j}\rangle_{\Gamma_j}|}{\|v_{|\Gamma_j}\|_{H^{-s}(\Gamma_j)}} \\
&\leq \sum_{j=1}^J \sup_{0\neq v_j\in H^{-s}(\Gamma_j)} \frac{|\langle w_{|\Gamma_j}, v_j\rangle_{\Gamma_j}|}{\|v_j\|_{H^{-s}(\Gamma_j)}} = \|w\|_{H^s_{\text{pw}}(\Gamma)}. \qquad \square
\end{aligned}$$

If $\Gamma = \partial\Omega$ is the boundary of a Lipschitz domain $\Omega \subset \mathbb{R}^d$, then we have to assume $|s| \leq 1$ to ensure the above statements.

For a function $u$ given in a bounded domain $\Omega \subset \mathbb{R}^d$ the application of the interior trace (1.3) gives $\gamma_0^{\text{int}} u$ as a function on the boundary $\Gamma = \partial\Omega$. The relations between the corresponding function spaces are stated in the next two theorems, see, for example, [1, 103, 160].

**Theorem 2.21 (Trace Theorem).** *Let $\Omega \subset \mathbb{R}^d$ be a $C^{k-1,1}$–domain. For $\frac{1}{2} < s \leq k$ the interior trace operator*

$$\gamma_0^{int} : H^s(\Omega) \to H^{s-1/2}(\Gamma)$$

*is bounded satisfying*

$$\|\gamma_0^{int} v\|_{H^{s-1/2}(\Gamma)} \leq c_T \, \|v\|_{H^s(\Omega)} \quad \text{for all } v \in H^s(\Omega).$$

For a Lipschitz domain $\Omega$ we can apply Theorem 2.21 with $k = 1$ to obtain the boundedness of the trace operator $\gamma_0^{\text{int}} : H^s(\Omega) \to H^{s-1/2}(\Gamma)$ for $s \in (\frac{1}{2}, 1]$. This remains true for $s \in (\frac{1}{2}, \frac{3}{2})$, see [44] and [103, Theorem 3.38].

**Theorem 2.22 (Inverse Trace Theorem).** *Let $\Omega$ be a $C^{k-1,1}$–domain. For $\frac{1}{2} < s \leq k$ the interior trace operator $\gamma_0^{int} : H^s(\Omega) \to H^{s-1/2}(\Gamma)$ has a continuous right inverse operator*

$$\mathcal{E} : H^{s-1/2}(\Gamma) \to H^s(\Omega)$$

*satisfying $\gamma_0^{int}\mathcal{E}w = w$ for all $w \in H^{s-1/2}(\Gamma)$ as well as*

$$\|\mathcal{E}w\|_{H^s(\Omega)} \leq c_{IT} \, \|w\|_{H^{s-1/2}(\Gamma)} \quad \text{for all } w \in H^{s-1/2}(\Gamma).$$

Therefore, for $s > 0$ we can define Sobolev spaces $H^s(\Gamma)$ also as trace spaces of $H^{s+1/2}(\Omega)$. The corresponding norm is given by

$$\|v\|_{H^s(\Gamma),\gamma_0} := \inf_{V\in H^{s+1/2}(\Omega),\gamma_0^{\text{int}}V=v} ||V||_{H^{1/2+s}(\Omega)}.$$

However, for a Lipschitz domain $\Omega \subset \mathbb{R}^d$ the norms $\|v\|_{H^s(\Gamma),\gamma_0}$ and $\|v\|_{H^s(\Gamma)}$ are only equivalent for $|s| \leq 1$.

*Remark 2.23.* The interpolation theorem (Theorem 2.18) holds also for appropriate Sobolev spaces $H^s(\Gamma)$.

As in Theorem 2.8 we also have the Bramble–Hilbert lemma:

**Theorem 2.24.** *Let $\Gamma = \partial\Omega$ the boundary of a $C^{k-1,1}$–domain $\Omega \subset \mathbb{R}^d$ and let $f : H^{k+1}(\Gamma) \to \mathbb{R}$ be a bounded linear functional satisfying*

$$|f(v)| \leq c_f \, \|v\|_{H^{k+1}(\Gamma)} \quad \textit{for all } v \in H^{k+1}(\Gamma).$$

*If*

$$f(q) = 0$$

*is satisfied for all $q \in P_k(\Gamma)$ then we also have*

$$|f(v)| \leq c \, c_f \, |v|_{H^{k+1}(\Gamma)} \quad \textit{for all } v \in H^{k+1}(\Gamma).$$

## 2.6 Exercises

**2.1** Let $u(x)$, $x \in (0,1)$, be a continuously differentiable function satisfying $u(0) = u(1) = 0$. Prove

$$\int_0^1 [u(x)]^2 dx \leq c \int_0^1 [u'(x)]^2 dx$$

where $c$ should be as small as possible.

**2.2** Consider the function

$$u(x) = \begin{cases} 0 & \text{for } x \in [0, \frac{1}{2}], \\ 1 & \text{for } x \in (\frac{1}{2}, 1]. \end{cases}$$

Determine those values of $s \in (0,1)$ such that

$$\int_0^1 \int_0^1 \frac{[u(x) - u(y)]^2}{|x-y|^{1+2s}} \, dx \, dy < \infty$$

is finite.

# 3

# Variational Methods

The weak formulation of boundary value problems leads to variational problems and associated operator equations. In particular, the representation of solutions of partial differential equations by using surface and volume potentials requires the solution of boundary integral operator equations to find the complete Cauchy data. In this chapter we describe the basic tools from functional analysis which are needed to investigate the unique solvability of operator equations.

## 3.1 Operator Equations

Let $X$ be a Hilbert space with the inner product $\langle \cdot, \cdot \rangle_X$ and with the induced norm $\| \cdot \|_X = \sqrt{\langle \cdot, \cdot \rangle_X}$. Let $X'$ be the dual space of $X$ with respect to the duality pairing $\langle \cdot, \cdot \rangle$. Then it holds that

$$\|f\|_{X'} = \sup_{0 \neq v \in X} \frac{|\langle f, v \rangle|}{\|v\|_X} \quad \text{for all } f \in X'. \tag{3.1}$$

Let $A : X \to X'$ be a bounded linear operator satisfying

$$\|Av\|_{X'} \leq c_2^A \, \|v\|_X \quad \text{for all } v \in X. \tag{3.2}$$

We assume that $A$ is self–adjoint, i.e., we have

$$\langle Au, v \rangle = \langle u, Av \rangle \quad \text{for all } u, v \in X.$$

For a given $f \in X'$ we want to find the solution $u \in X$ of the operator equation

$$Au = f\,. \tag{3.3}$$

Instead of the operator equation (3.3) we may consider an equivalent variational problem to find $u \in X$ such that

$$\langle Au, v\rangle = \langle f, v\rangle \quad \text{for all } v \in X. \tag{3.4}$$

Obviously, any solution $u \in X$ of the operator equation (3.3) is also a solution of the variational problem (3.4). To show the reverse direction we now consider $u \in X$ to be a solution of the variational problem (3.4). Using the norm definition (3.1) we then obtain

$$\|Au - f\|_{X'} = \sup_{0 \neq v \in X} \frac{|\langle Au - f, v\rangle|}{\|v\|_X} = 0$$

and therefore $0 = Au - f \in X'$, i.e., $u \in X$ is a solution of the operator equation (3.3).

The operator $A : X \to X'$ induces a bilinear form

$$a(u, v) := \langle Au, v\rangle \quad \text{for all } u, v \in X$$

with the mapping property

$$a(\cdot, \cdot) : X \times X \to \mathbb{R}. \tag{3.5}$$

In the reverse case, any bilinear form (3.5) defines an operator $A : X \to X'$.

**Lemma 3.1.** *Let $a(\cdot, \cdot) : X \times X \to \mathbb{R}$ be a bounded bilinear form satisfying*

$$|a(u, v)| \leq c_2^A \, \|u\|_X \|v\|_X \quad \textit{for all } u, v \in X.$$

*For any $u \in X$ there exists an element $Au \in X'$ such that*

$$\langle Au, v\rangle = a(u, v) \quad \textit{for all } v \in X.$$

*The operator $A : X \to X'$ is linear and bounded satisfying*

$$\|Au\|_{X'} \leq c_2^A \, \|u\|_X \quad \textit{for all } u \in X.$$

*Proof.* For a given $u \in X$ we define $\langle f_u, v\rangle := a(u, v)$ which is a bounded linear form in $X$, i.e., we have $f_u \in X'$. The map $u \in X \to f_u \in X'$ defines a linear operator $A : X \to X'$ with $Au = f_u \in X'$ and satisfying

$$\|Au\|_{X'} = \|f_u\|_{X'} = \sup_{0 \neq v \in X} \frac{|\langle f_u, v\rangle|}{\|v\|_X} = \sup_{0 \neq v \in X} \frac{|a(u, v)|}{\|v\|_X} \leq c_2^A \, \|u\|_X. \quad \square$$

If $A : X \to X'$ is a self–adjoint and positive semi–definite operator we can derive a minimization problem which is equivalent to the variational formulation (3.4).

**Lemma 3.2.** *Let $A : X \to X'$ be self–adjoint and positive semi–definite, i.e.,*

$$\langle Av, v\rangle \geq 0 \quad \textit{for all } v \in X.$$

*Let $F$ be the functional*

$$F(v) := \frac{1}{2}\langle Av, v\rangle - \langle f, v\rangle \quad \textit{for}\, v \in X\,.$$

*The solution of the variational formulation (3.4) is then equivalent to the solution of the minimization problem*

$$F(u) = \min_{v\in X} F(v)\,. \tag{3.6}$$

*Proof.* For $u, v \in X$ we choose an arbitrary $t \in \mathbb{R}$. Then we have

$$\begin{aligned} F(u+tv) &= \frac{1}{2}\langle A(u+tv), u+tv\rangle - \langle f, u+tv\rangle \\ &= F(u) + t\,[\langle Au, v\rangle - \langle f, v\rangle] + \frac{1}{2}t^2\,\langle Av, v\rangle. \end{aligned}$$

If $u \in X$ is a solution of the variational problem (3.4) we then obtain

$$F(u) \le F(u) + \frac{1}{2}t^2\,\langle Av, v\rangle = F(u+tv)$$

for all $v \in X$ and $t \in \mathbb{R}$. Therefore, $u \in X$ is also a solution of the minimization problem (3.6).

Let $u \in X$ be now a solution of (3.6). Then, as a necessary condition,

$$\frac{d}{dt}F(u+tv)_{|t=0} = 0 \quad \text{for all } v \in X.$$

From this we obtain

$$\langle Au, v\rangle = \langle f, v\rangle \quad \text{for all } v \in X$$

and therefore the equivalence of both the variational and the minimization problem. □

To investigate the unique solvability of the operator equation (3.3) we now consider a fixed point iteration. For this we need to formulate the following Riesz representation theorem.

**Theorem 3.3 (Riesz Representation Theorem).** *Any linear and bounded functional $f \in X'$ can be written as*

$$\langle f, v\rangle = \langle u, v\rangle_X$$

*where $u \in X$ is uniquely determined by $f \in X'$, and*

$$\|u\|_X = \|f\|_{X'}. \tag{3.7}$$

*Proof.* Let $f \in X'$ be arbitrary but fixed. Then we can find $u \in X$ as the solution of the variational problem

$$\langle u, v\rangle_X = \langle f, v\rangle \quad \text{for all } v \in X. \tag{3.8}$$

Using Lemma 3.2 this variational problem is equivalent to the minimization problem

$$F(u) = \min_{v\in X} F(v) \tag{3.9}$$

where the functional is given by

$$F(v) = \frac{1}{2}\,\langle v, v\rangle_X - \langle f, v\rangle \quad \text{for } v \in X.$$

Hence we have to investigate the unique solvability of the minimization problem (3.9). From

$$\begin{aligned} F(v) &= \frac{1}{2}\,\langle v, v\rangle_X - \langle f, v\rangle \geq \frac{1}{2}\,\|v\|_X^2 - \|f\|_{X'}\|v\|_X \\ &= \frac{1}{2}\,[\|v\|_X - \|f\|_{X'}]^2 - \frac{1}{2}\,\|f\|_{X'}^2 \geq -\frac{1}{2}\,\|f\|_{X'}^2 \end{aligned}$$

we find that $F(v)$ is bounded below for all $v \in X$. Hence there exists the infimum

$$\alpha := \inf_{v\in X} F(v) \in \mathbb{R}.$$

Let $\{u_k\}_{k\in\mathbb{N}} \subset X$ be a sequence approaching the minimum, i.e., $F(u_k) \to \alpha$ as $k \to \infty$. With the identity

$$\|u_k - u_\ell\|_X^2 + \|u_k + u_\ell\|_X^2 = 2\,\left\{\|u_k\|_X^2 + \|u_\ell\|_X^2\right\}$$

we then obtain

$$\begin{aligned} 0 \leq \|u_k - u_\ell\|_X^2 &= 2\,\|u_k\|_X^2 + 2\,\|u_\ell\|_X^2 - \|u_k + u_\ell\|_X^2 \\ &= 4\left\{\frac{1}{2}\,\|u_k\|_X^2 - \langle f, u_k\rangle\right\} + 4\left\{\frac{1}{2}\,\|u_\ell\|_X^2 - \langle f, u_\ell\rangle\right\} \\ &\quad +4\,\langle f, u_k + u_\ell\rangle - \|u_k + u_\ell\|_X^2 \\ &= 4\,F(u_k) + 4\,F(u_\ell) - 8\,F\left(\frac{1}{2}(u_k + u_\ell)\right) \\ &\leq 4\,F(u_k) + 4\,F(u_\ell) - 8\alpha \to 0 \quad \text{as } k, \ell \to \infty. \end{aligned}$$

Therefore, $\{u_k\}_{k\in\mathbb{N}}$ is a Cauchy sequence, and since $X$ is a Hilbert space, we find the limit

$$u = \lim_{k\to\infty} u_k \in X.$$

Moreover,

$$\begin{aligned}|F(u_k) - F(u)| &\leq \frac{1}{2}\,|\langle u_k, u_k\rangle_X - \langle u, u\rangle_X| + |\langle f, u_k - u\rangle| \\ &= \frac{1}{2}\,|\langle u_k, u_k - u\rangle_X + \langle u, u_k - u\rangle_X| + |\langle f, u_k - u\rangle| \\ &\leq \left\{\frac{1}{2}\,\|u_k\|_X + \frac{1}{2}\,\|u\|_X + \|f\|_{X'}\right\} \|u_k - u\|_X,\end{aligned}$$

and hence

$$F(u) = \lim_{k\to\infty} F(u_k) = \alpha.$$

In particular, $u \in X$ is a solution of the minimization problem (3.9) and therefore also a solution of the variational problem (3.8).

It remains to prove the uniqueness. Let $\widetilde{u} \in X$ be another solution of (3.9) and (3.8), respectively. Then,

$$\langle \widetilde{u}, v\rangle_X = \langle f, v\rangle \quad \text{for all } v \in X.$$

Subtracting this from (3.8) this gives

$$\langle u - \widetilde{u}, v\rangle_X = 0 \quad \text{for all } v \in X.$$

Choosing $v = u - \widetilde{u}$ we now obtain

$$\|u - \widetilde{u}\|_X^2 = 0$$

and therefore $u = \widetilde{u}$, i.e., $u \in X$ is the unique solution of (3.8) and (3.9), respectively.

Finally,

$$\|u\|_X^2 = \langle u, u\rangle_X = \langle f, u\rangle \leq \|f\|_{X'}\|u\|_X$$

and

$$\|f\|_{X'} = \sup_{0\neq v\in X} \frac{|\langle f, v\rangle|}{\|v\|_X} = \sup_{0\neq v\in X} \frac{|\langle u, v\rangle_X|}{\|v\|_X} \leq \|u\|_X,$$

imply the norm equality (3.7). □

The map $J : X' \to X$ as introduced in Theorem 3.3 is called the Riesz map $u = Jf$ and satisfies the variational problem

$$\langle Jf, v\rangle_X = \langle f, v\rangle \quad \text{for all } v \in X. \tag{3.10}$$

Moreover,

$$\|Jf\|_X = \|f\|_{X'}. \tag{3.11}$$

## 3.2 Elliptic Operators

To ensure the unique solvability of the operator equation (3.3) and of the variational problem (3.4) we need to have a further assumption for the operator $A$ and for the bilinear form $a(\cdot,\cdot)$, respectively. The operator $A : X \to X'$ is called $X$–elliptic if

$$\langle Av, v\rangle \geq c_1^A \, \|v\|_X^2 \quad \text{for all } v \in X \tag{3.12}$$

is satisfied with some positive constant $c_1^A$.

**Theorem 3.4 (Lax–Milgram Lemma).** *Let the operator $A : X \to X'$ be bounded and $X$–elliptic. For any $f \in X'$ there exists a unique solution of the operator equation* (3.3) *satisfying the estimate*

$$\|u\|_X \leq \frac{1}{c_1^A} \, \|f\|_{X'}.$$

*Proof.* Let $J : X' \to X$ be the Riesz operator as defined by (3.10). The operator equation (3.3) is then equivalent to the fixed point equation

$$u = u - \varrho J(Au - f) = T_\varrho u + \varrho Jf$$

with the operator

$$T_\varrho := I - \varrho JA \, : \, X \to X$$

and with a suitable chosen parameter $0 < \varrho \in \mathbb{R}$. From the boundedness estimate (3.2) and from the ellipticity assumption (3.12) of $A$ as well as from the properties (3.10) and (3.11) of the Riesz map $J$ we conclude

$$\langle JAv, v\rangle_X = \langle Av, v\rangle \geq c_1^A \, \|v\|_X^2, \quad \|JAv\|_X = \|Av\|_{X'} \leq c_2^A \, \|v\|_X$$

and therefore

$$\begin{aligned}
\|T_\varrho v\|_X^2 &= \|(I - \varrho JA)v\|_X^2 \\
&= \|v\|_X^2 - 2\varrho\langle JAv, v\rangle_X + \varrho^2 \|JAv\|_X^2 \\
&\leq [1 - 2\varrho c_1^A + \varrho^2 (c_2^A)^2] \, \|v\|_X^2.
\end{aligned}$$

Hence we obtain that for $\varrho \in (0, 2c_1^A/(c_2^A)^2)$ the operator $T_\varrho$ is a contraction in $X$, and the unique solvability of (3.3) follows from Banach's fixed point theorem [163]. Let $u \in X$ be the unique solution of the operator equation (3.3). Then,

$$c_1^A \, \|u\|_X^2 \leq \langle Au, u\rangle = \langle f, u\rangle \leq \|f\|_{X'} \|u\|_X,$$

which is equivalent to the remaining bound. □

Applying Theorem 3.4 this gives the inverse operator $A^{-1} : X' \to X$ and we obtain

$$\|A^{-1}f\|_X \leq \frac{1}{c_1^A} \|f\|_{X'} \quad \text{for all } f \in X'. \tag{3.13}$$

From the boundedness of the self–adjoint and invertible operator $A$ we also conclude an ellipticity estimate for the inverse operator $A^{-1}$.

**Lemma 3.5.** *Let $A : X \to X'$ be bounded, self–adjoint and $X$–elliptic. In particular we assume* (3.2), *i.e.,*

$$\|Av\|_{X'} \leq c_2^A \|v\|_X \quad \textit{for all } v \in X.$$

*Then,*

$$\langle A^{-1}f, f\rangle \geq \frac{1}{c_2^A} \|f\|_{X'}^2 \quad \textit{for all } f \in X'.$$

*Proof.* Let us consider the operator $B := JA : X \to X$ satisfying

$$\|Bv\|_X = \|JAv\|_X = \|Av\|_{X'} \leq c_2^A \|v\|_X \quad \text{for all } v \in X.$$

Since

$$\langle Bu, v\rangle_X = \langle JAu, v\rangle_X = \langle Au, v\rangle = \langle u, Av\rangle = \langle u, Bv\rangle_X$$

holds for all $u, v \in X$ the operator $B$ is self–adjoint satisfying the ellipticity estimate

$$\langle Bv, v\rangle_X = \langle Av, v\rangle \geq c_1^A \|v\|_X^2 \quad \text{for all } v \in X.$$

Hence there exists a self–adjoint and invertible operator $B^{1/2}$ satisfying $B = B^{1/2}B^{1/2}$, see, e.g., [118]. In addition we define $B^{-1/2} := (B^{1/2})^{-1}$. Then we obtain

$$\|B^{1/2}v\|_X^2 = \langle Bv, v\rangle_X \leq \|Bv\|_X \|v\|_X \leq c_2^A \|v\|_X^2 \quad \text{for all } v \in X$$

and further

$$\|B^{1/2}v\|_X \leq \sqrt{c_2^A} \|v\|_X \quad \text{for all } v \in X.$$

For an arbitrary $f \in X'$ we then conclude

$$\begin{aligned} \|f\|_{X'} &= \sup_{0 \neq v \in X} \frac{|\langle f, v\rangle|}{\|v\|_X} = \sup_{0 \neq v \in X} \frac{|\langle Jf, v\rangle_X|}{\|v\|_X} = \sup_{0 \neq v \in X} \frac{|\langle B^{-1/2}Jf, B^{1/2}v\rangle_X|}{\|v\|_X} \\ &\leq \sup_{0 \neq v \in X} \frac{\|B^{-1/2}Jf\|_X \|B^{1/2}v\|_X}{\|v\|_X} \leq \sqrt{c_2^A} \|B^{-1/2}Jf\|_X, \end{aligned}$$

and therefore

$$\|f\|_{X'}^2 \leq c_2^A \|B^{-1/2}Jf\|_X^2 = c_2^A \langle B^{-1}Jf, Jf\rangle_X = c_2^A \langle A^{-1}f, f\rangle. \quad \square$$

## 3.3 Operators and Stability Conditions

Let $\Pi$ be a Banach space and let $B : X \to \Pi'$ be a bounded linear operator satisfying

$$\|Bv\|_{\Pi'} \leq c_2^B \, \|v\|_X \quad \text{for all } v \in X. \tag{3.14}$$

The operator $B$ implies a bounded bilinear form $b(\cdot,\cdot) : X \times \Pi \to \mathbb{R}$,

$$b(v,q) := \langle Bv, q\rangle \quad \text{for } (v,q) \in X \times \Pi.$$

The null space of the operator $B$ is

$$\ker B := \{v \in X : Bv = 0\}.$$

The orthogonal complement of $\ker B$ in $X$ is given as

$$(\ker B)^\perp := \{w \in X : \langle w, v\rangle_X = 0 \quad \text{for all } v \in \ker B\} \subset X.$$

Finally,

$$(\ker B)^0 := \{f \in X' : \langle f, v\rangle = 0 \quad \text{for all } v \in \ker B\} \subset X' \tag{3.15}$$

is the polar space which is induced by $\ker B$.

For a given $g \in \Pi'$ we want to find solutions $u \in X$ of the operator equation

$$Bu = g. \tag{3.16}$$

Obviously we have to require the solvability condition

$$g \in \mathrm{Im}_X B := \{Bv \in \Pi' \quad \text{for all } v \in X\}. \tag{3.17}$$

Let $B' : \Pi \to X'$ the adjoint of $B : X \to \Pi'$, i.e.

$$\langle v, B'q\rangle := \langle Bv, q\rangle \quad \text{for all } (v,q) \in X \times \Pi.$$

Then we have

$$\begin{aligned}
\ker B' &:= \{q \in \Pi : \langle Bv, q\rangle = 0 \quad \text{for all } v \in X\}, \\
(\ker B')^\perp &:= \{p \in \Pi : \langle p, q\rangle_\Pi = 0 \quad \text{for all } q \in \ker B'\}, \\
(\ker B')^0 &:= \{g \in \Pi' : \langle g, q\rangle = 0 \quad \text{for all } q \in \ker B'\}.
\end{aligned}$$

To characterize the image $\mathrm{Im}_X B$ we will use the following result, see, for example, [163].

**Theorem 3.6 (Closed range theorem).** *Let $X$ and $\Pi$ be Banach spaces, and let $B : X \to \Pi'$ be a bounded linear operator. Then the following properties are all equivalent:*

*i.* $\mathit{Im}_X B$ *is closed in* $\Pi'$.
*ii.* $\mathit{Im}_\Pi B'$ *is closed in* $X'$.
*iii.* $\mathit{Im}_X B = (\ker B')^0$.
*iv.* $\mathit{Im}_\Pi B' = (\ker B)^0$.

*Proof.* Here we only prove that *iii.* follows from *i.*, see also [163]. From the definition of the polar space with respect to $B'$ we find that

$$\begin{aligned}(\ker B')^0 &= \{g \in \Pi' : \langle g, q\rangle = 0 \quad \text{for all } q \in \ker B'\} \\ &= \{g \in \Pi' : \langle g, q\rangle = 0 \quad \text{for all } q \in \Pi : \langle Bv, q\rangle = 0 \quad \text{for all } v \in X\}\end{aligned}$$

and therefore

$$\mathrm{Im}_X B \subset (\ker B')^0 .$$

Let $g \in (\ker B')^0$ with $g \notin \mathrm{Im}_X B$. Applying the separation theorem for closed convex sets there exists a $\bar{q} \in \Pi$ and a constant $\alpha \in \mathbb{R}$ such that

$$\langle g, \bar{q}\rangle > \alpha > \langle f, \bar{q}\rangle \quad \text{for all } f \in \mathrm{Im}_X(B) \subset \Pi'.$$

Since $B$ is linear we obtain for an arbitrary given $f \in \mathrm{Im}_X B$ also $-f \in \mathrm{Im}_X B$ and therefore

$$\alpha > -\langle f, \bar{q}\rangle.$$

From this we obtain $\alpha > 0$ as well as $|\langle f, \bar{q}\rangle| < \alpha$. For $f \in \mathrm{Im}_\Pi B$ and for any arbitrary $n \in \mathbb{N}$ we also conclude $nf \in \mathrm{Im}_\Pi B$ and therefore

$$|\langle f, \bar{q}\rangle| < \frac{\alpha}{n} \quad \text{for all } n \in \mathbb{N}$$

which is equivalent to

$$\langle f, \bar{q}\rangle = 0 \quad \text{for all } f \in \mathrm{Im}_X(B).$$

For any $f \in \mathrm{Im}_X(B)$ there exists at least one $u \in X$ with $f = Bu$. Hence,

$$0 = \langle f, \bar{q}\rangle = \langle Bu, \bar{q}\rangle = \langle u, B'\bar{q}\rangle \quad \text{for all } u \in X,$$

and therefore $\bar{q} \in \ker B'$. On the other hand, for $g \in (\ker B)^0$ we have

$$\langle g, q\rangle = 0 \quad \text{for all } q \in \ker B'$$

and therefore $\langle g, \bar{q}\rangle = 0$ which is a contradiction to $\langle g, \bar{q}\rangle > \alpha > 0$. □

The solvability condition (3.17) is equivalent to

$$\langle g, q\rangle = 0 \quad \text{for all } q \in \ker B' \subset \Pi. \tag{3.18}$$

If the equivalent solvability conditions (3.17) and (3.18) are satisfied, then there exists at least one solution $u \in X$ satisfying $Bu = g$. When the null space $\ker B$ is non–trivial, we can add an arbitrary $u_0 \in \ker B$, in particular, $u + u_0$ is still a solution of $B(u + u_0) = g$. In this case, the solution is not unique in general. Instead we consider only solutions $u \in (\ker B)^\perp$. To ensure unique solvability in this case, we have to formulate additional assumptions.

**Theorem 3.7.** *Let $X$ and $\Pi$ be Hilbert spaces and let $B : X \to \Pi'$ be a bounded linear operator. Further we assume the stability condition*

$$c_S \|v\|_X \le \sup_{0 \neq q \in \Pi} \frac{\langle Bv, q\rangle}{\|q\|_\Pi} \quad \text{for all } v \in (\ker B)^\perp. \tag{3.19}$$

*For a given $g \in Im_X(B)$ there exists a unique solution $u \in (\ker B)^\perp$ of the operator equation $Bu = g$ satisfying*

$$\|u\|_X \le \frac{1}{c_S} \|g\|_{\Pi'}.$$

*Proof.* Since we assume $g \in \mathrm{Im}_X B$ there exists at least one solution $u \in (\ker B)^\perp$ of the operator equation $Bu = g$ satisfying

$$\langle Bu, q\rangle = \langle g, q\rangle \quad \text{for all } q \in \Pi.$$

Let $\bar{u} \in (\ker B)^\perp$ be a second solution satisfying

$$\langle B\bar{u}, q\rangle = \langle g, q\rangle \quad \text{for all } q \in \Pi.$$

Then,

$$\langle B(u - \bar{u}), q\rangle = 0 \quad \text{for all } q \in \Pi.$$

Obviously, $u - \bar{u} \in (\ker B)^\perp$. From the stability condition (3.19) we then conclude

$$0 \le c_S \|u - \bar{u}\|_X \le \sup_{0 \neq q \in \Pi} \frac{\langle B(u - \bar{u}), q\rangle}{\|q\|_\Pi} = 0$$

and therefore uniqueness, $u = \bar{u}$. Applying (3.19) for the solution $u$ this gives

$$c_S \|u\|_X \le \sup_{0 \neq q \in \Pi} \frac{\langle Bu, q\rangle}{\|q\|_\Pi} = \sup_{0 \neq q \in \Pi} \frac{\langle g, q\rangle}{\|q\|_\Pi} \le \|g\|_{\Pi'}. \qquad \square$$

## 3.4 Operator Equations with Constraints

In many applications we have to solve an operator equation $Au = f$ where the solution $u$ has to satisfy an additional constraint $Bu = g$. In this case we have to assume first the solvability condition (3.17). For a given $g \in \Pi'$ we then define the manifold

$$V_g := \{v \in X : Bv = g\}.$$

In particular, $V_0 = \ker B$. Further, the given $f \in X'$ has to satisfy the solvability condition

$$f \in \mathrm{Im}_{V_g} A := \{Av \in X' \quad \text{for all } v \in V_g\}.$$

Then we have to find $u \in V_g$ satisfying the variational problem

$$\langle Au, v\rangle \;=\; \langle f, v\rangle \quad \text{for all } v \in V_0. \tag{3.20}$$

The unique solvability of (3.20) now follows from the following result.

**Theorem 3.8.** *Let $A : X \to X'$ be bounded and $V_0$–elliptic, i.e.*

$$\langle Av, v\rangle \ge c_1^A \, \|v\|_X^2 \quad \textit{for all } v \in V_0 := \ker B,$$

*where $B : X \to \Pi'$. For $f \in \mathrm{Im}_{V_g} A$ and $g \in \mathrm{Im}_X B$ there exists a unique solution $u \in X$ of the operator equation $Au = f$ satisfying the constraint $Bu = g$.*

*Proof.* Since $g \in \mathrm{Im}_X B$ is satisfied there exists at least one $u_g \in X$ with $Bu_g = g$. It remains to find $u_0 = u - u_g \in V_0$ satisfying the operator equation

$$Au_0 \;=\; f - Au_g$$

which is equivalent to the variational problem

$$\langle Au_0, v\rangle \;=\; \langle f - Au_g, v\rangle \quad \text{for all } v \in V_0.$$

From the assumption $f \in \mathrm{Im}_{V_g} A$ we conclude $f - Au_g \in \mathrm{Im}_{V_0} A$. Then there exists at least one $u_0 \in V_0$ with $Au_0 = f - Au_g$. It remains to show the uniqueness of $u_0 \in V_0$. Let $\bar{u}_0 \in V_0$ be another solution with $A\bar{u}_0 = f - Au_g$. From the $V_0$–ellipticity of $A$ we then obtain

$$0 \le c_1^A \, \|u_0 - \bar{u}_0\|_X^2 \le \langle A(u_0 - \bar{u}_0), u_0 - \bar{u}_0\rangle = \langle Au_0 - A\bar{u}_0, u_0 - \bar{u}_0\rangle = 0$$

and therefore $u_0 = \bar{u}_0$ in $X$.

Note that $u_g \in V_g$ is in general not unique. However, the final solution $u = u_0 + u_g$ is unique independent of the chosen $u_g \in V_g$: For $\hat{u}_g \in X$ with $B\hat{u}_g = g$ there exists a unique $\hat{u}_0 \in V_0$ satisfying $A(\hat{u}_0 + \hat{u}_g) = f$. Due to

$$B(u_g - \hat{u}_g) = Bu_g - B\hat{u}_g = g - g = 0 \quad \text{in } \Pi'$$

we have $u_g - \hat{u}_g \in \ker B = V_0$ . Using

$$A(u_0 + u_g) = f, \quad A(\hat{u}_0 + \hat{u}_g) = f$$

we obtain

$$A(u_0 + u_g - \hat{u}_0 - \hat{u}_g) = 0.$$

Obviously, $u_0 - \hat{u}_0 + (u_g - \hat{u}_g) \in V_0$, and from the $V_0$–ellipticity of $A$ we conclude

$$u_0 - \hat{u}_0 + (u_g - \hat{u}_g) \;=\; 0$$

and therefore uniqueness, $u \;=\; u_0 + u_g \;=\; \hat{u}_0 + \hat{u}_g$. $\square$

For what follows we assume that for $g \in \mathrm{Im}_X B$ there exists a $u_g \in V_g$ satisfying

$$\|u_g\|_X \leq c_B \, \|g\|_{\Pi'} \tag{3.21}$$

with some positive constant $c_B$. Then we can bound the norm of the unique solution $u \in V_g$ satisfying the variational problem (3.20) by the norms of the given data $f \in X'$ and $g \in \Pi'$.

**Corollary 3.9.** *Let us assume the assumptions of Theorem* 3.8 *as well as assumption* (3.21). *The solution $u \in V_g$ of $Au = f$ satisfies the estimate*

$$\|u\|_X \leq \frac{1}{c_1^A} \, \|f\|_{X'} + \left(1 + \frac{c_2^A}{c_1^A}\right) c_B \, \|g\|_{\Pi'} .$$

*Proof.* Applying Theorem 3.8, the solution $u$ of $Au = f$ admits the representation $u = u_0 + u_g$ where $u_0 \in V_0$ is the unique solution of the variational problem

$$\langle Au_0, v\rangle = \langle f - Au_g, v\rangle \quad \text{for all } v \in V_0 .$$

From the $V_0$–ellipticity of $A$ we obtain

$$c_1^A \, \|u_0\|_X^2 \leq \langle Au_0, u_0\rangle = \langle f - Au_g, u_0\rangle \leq \|f - Au_g\|_{X'} \|u_0\|_X$$

and therefore

$$\|u_0\|_X \leq \frac{1}{c_1^A} \left[\|f\|_{X'} + c_2^A \, \|u_g\|_X\right] .$$

Now the assertion follows from the triangle inequality and using assumption (3.21). □

## 3.5 Mixed Formulations

Instead of the operator equation $Au = f$ with the constraint $Bu = g$ we may introduce a Lagrange multiplier $p \in \Pi$ to formulate an extended variational problem: Find $(u, p) \in X \times \Pi$ such that

$$\begin{aligned} \langle Au, v\rangle + \langle Bv, p\rangle &= \langle f, v\rangle \\ \langle Bu, q\rangle \qquad\qquad &= \langle g, q\rangle \end{aligned} \tag{3.22}$$

is satisfied for all $(v, q) \in X \times \Pi$. Note that for any solution $(u, p) \in X \times \Pi$ of the extended variational problem (3.22) we conclude that $u \in V_g$ is a solution of $Au = f$. The second equation in (3.22) describes just the constraint $u \in V_g$ while the first equation in (3.22) coincides with the variational formulation to find $u_0 \in V_0$ when choosing as test function $v \in V_0$. It remains to ensure the existence of the Lagrange multiplier $p \in \Pi$ such that the first equation in (3.22) is satisfied for all $v \in X$, see Theorem 3.11.

For the Lagrange functional

$$\mathcal{L}(v,q) := \frac{1}{2}\langle Av, v\rangle - \langle f, v\rangle + \langle Bv, q\rangle - \langle g, q\rangle,$$

which is defined for $(v,q) \in X \times \Pi$, we first find the following characterization.

**Theorem 3.10.** *Let $A : X \to X'$ be a self-adjoint bounded and positive semi-definite operator, i.e. $\langle Av, v\rangle \geq 0$ for all $v \in X$. Further, let $B : X \to \Pi'$ be bounded. $(u,p) \in X \times \Pi$ is a solution of the variational problem (3.22) iff*

$$\mathcal{L}(u,q) \leq \mathcal{L}(u,p) \leq \mathcal{L}(v,p) \quad \textit{for all } (v,q) \in X \times \Pi. \tag{3.23}$$

*Proof.* Let $(u,p)$ be a solution of the variational problem (3.22). From the first equation in (3.22) we then obtain

$$\begin{aligned}\mathcal{L}(v,p) - \mathcal{L}(u,p) &= \frac{1}{2}\langle Av, v\rangle - \langle f, v\rangle + \langle Bv, p\rangle - \langle g, p\rangle \\ &\quad -\frac{1}{2}\langle Au, u\rangle + \langle f, u\rangle - \langle Bu, p\rangle + \langle g, p\rangle \\ &= \frac{1}{2}\langle A(u-v), u-v\rangle + \langle Au, v-u\rangle + \langle B(v-u), p\rangle - \langle f, v-u\rangle \\ &= \frac{1}{2}\langle A(u-v), u-v\rangle \geq 0,\end{aligned}$$

and therefore

$$\mathcal{L}(u,p) \leq \mathcal{L}(v,p) \quad \text{for all } v \in X.$$

Using the second equation of (3.22) this gives

$$\begin{aligned}\mathcal{L}(u,p) - \mathcal{L}(u,q) &= \frac{1}{2}\langle Au, u\rangle - \langle f, u\rangle + \langle Bu, p\rangle - \langle g, p\rangle \\ &\quad -\frac{1}{2}\langle Au, u\rangle + \langle f, u\rangle - \langle Bu, q\rangle + \langle g, q\rangle \\ &= \langle Bu, p-q\rangle - \langle g, p-q\rangle = 0\end{aligned}$$

and therefore

$$\mathcal{L}(u,q) \leq \mathcal{L}(u,p) \quad \text{for all } q \in \Pi.$$

For a fixed $p \in \Pi$ we consider $u \in X$ as the solution of the minimization problem

$$\mathcal{L}(u,p) \leq \mathcal{L}(v,p) \quad \text{for all } v \in X.$$

Then we have for any arbitrary $w \in X$

$$\frac{d}{dt}\mathcal{L}(u+tw,p)_{|t=0} = 0. \tag{3.24}$$

From

$$\mathcal{L}(u+tw,p) = \frac{1}{2}\langle Au,u\rangle - \langle f,u\rangle + \langle Bu,p\rangle - \langle g,p\rangle + \frac{1}{2}t^2\langle Aw,w\rangle$$
$$+\,t\,[\langle Au,w\rangle + \langle Bw,p\rangle - \langle f,w\rangle]\,,$$

and using (3.24) we obtain the first equation of (3.22),

$$\langle Au,w\rangle + \langle Bw,p\rangle - \langle f,w\rangle \;=\; 0 \quad \text{for all } w \in X.$$

Now, let $p \in \Pi$ satisfy

$$\mathcal{L}(u,\widetilde{q}) \;\leq\; \mathcal{L}(u,p) \quad \text{for all } \widetilde{q} \in \Pi.$$

For an arbitrary $q \in \Pi$ we define $\widetilde{q} := p + q$. Then,

$$\begin{aligned} 0 &\leq \mathcal{L}(u,p) - \mathcal{L}(u,p+q) \\ &= \frac{1}{2}\langle Au,u\rangle - \langle f,u\rangle + \langle Bu,p\rangle - \langle g,p\rangle \\ &\quad -\frac{1}{2}\langle Au,u\rangle + \langle f,u\rangle - \langle Bu,p+q\rangle + \langle g,p+q\rangle \\ &= -\langle Bu,q\rangle + \langle g,q\rangle\,. \end{aligned}$$

For $\widetilde{q} := p - q$ we obtain in the same way

$$0 \leq \mathcal{L}(u,p) - \mathcal{L}(u,p-q) \;=\; \langle Bu,q\rangle - \langle g,q\rangle,$$

and therefore,

$$\langle Bu,q\rangle \;=\; \langle g,q\rangle \quad \text{for all } q \in \Pi,$$

which is the second equation of (3.22). □

Any solution $(u,p) \in X \times \Pi$ of the variational problem (3.22) is hence a saddle point of the Lagrange functional $\mathcal{L}(\cdot,\cdot)$. This is why the variational problem (3.22) is often called a saddle point problem. The unique solvability of (3.22) now follows from the following result.

**Theorem 3.11.** *Let $X$ and $\Pi$ be Banach spaces and let $A : X \to X'$ and $B : X \to \Pi'$ be bounded operators. Further, we assume that $A$ is $V_0$–elliptic,*

$$\langle Av,v\rangle \;\geq\; c_1^A\,\|v\|_X^2 \quad \text{for all } v \in V_0 = \ker B\,,$$

*and that the stability condition*

$$c_S\,\|q\|_\Pi \;\leq\; \sup_{0\neq v\in X} \frac{\langle Bv,q\rangle}{\|v\|_X} \quad \text{for all } q \in \Pi \tag{3.25}$$

*is satisfied.*

*For $g \in \mathrm{Im}_X B$ and $f \in \mathrm{Im}_{V_g} A$ there exists a unique solution $(u,p) \in X \times \Pi$ of the variational problem (3.22) satisfying*

$$\|u\|_X \leq \frac{1}{c_1^A} \|f\|_{X'} + \left(1 + \frac{c_2^A}{c_1^A}\right) c_B \|g\|_{\Pi'}, \tag{3.26}$$

*and*

$$\|p\|_\Pi \leq \frac{1}{c_S} \left(1 + \frac{c_2^A}{c_1^A}\right) \left\{\|f\|_{X'} + c_B\, c_2^A \|g\|_{\Pi'}\right\}. \tag{3.27}$$

*Proof.* Applying Theorem 3.8 we first find a unique $u \in X$ satisfying

$$\langle Au, v\rangle = \langle f, v\rangle \quad \text{for all } v \in V_0$$

and

$$\langle Bu, q\rangle = \langle g, q\rangle \quad \text{for all } q \in \Pi.$$

The estimate (3.26) is just the estimate of Corollary 3.9.

It remains to find $p \in \Pi$ as the solution of the variational problem

$$\langle Bv, p\rangle = \langle f - Au, v\rangle \quad \text{for all } v \in X.$$

First we have $f - Au \in (\ker B)^0$, and using Theorem 3.6 we obtain $f - Au \in \mathrm{Im}_\Pi(B')$ and therefore the solvability of the variational problem.

To prove the uniqueness of $p \in \Pi$ we assume that there are given two arbitrary solutions $p, \hat{p} \in \Pi$ satisfying

$$\langle Bv, p\rangle = \langle f - Au, v\rangle \quad \text{for all } v \in X$$

and

$$\langle Bv, \hat{p}\rangle = \langle f - Au, v\rangle \quad \text{for all } v \in X.$$

Then,

$$\langle Bv, p - \hat{p}\rangle = 0 \quad \text{for all } v \in X.$$

Using the stability condition (3.25) we obtain

$$0 \leq c_S \|p - \hat{p}\|_\Pi \leq \sup_{0 \neq v \in X} \frac{\langle Bv, p - \hat{p}\rangle}{\|v\|_X} = 0$$

and therefore $p = \hat{p}$ in $\Pi$.

Using again (3.25) for the unique solution $p \in \Pi$ this gives

$$c_S \|p\|_\Pi \leq \sup_{0 \neq v \in X} \frac{\langle Bv, p\rangle}{\|v\|_X} = \sup_{0 \neq v \in X} \frac{\langle f - Au, v\rangle}{\|v\|_X} \leq \|f\|_{X'} + c_2^A \|u\|_X$$

and applying (3.26) we finally obtain (3.27). □

The statement of Theorem 3.11 remains valid when we assume that $A : X \to X'$ is $X$–elliptic, i.e.

$$\langle Av, v\rangle \;\geq\; c_1^A \, \|v\|_X^2 \quad \text{for all } v \in X.$$

For an arbitrary $p \in \Pi$ there exists a unique solution $u \;=\; A^{-1}[f - B'p] \in X$ of the first equation of (3.22). Inserting this into the second equation of (3.22) we obtain a variational problem to find $p \in \Pi$ such that

$$\langle BA^{-1}B'p, q\rangle \;=\; \langle BA^{-1}f - g, q\rangle \tag{3.28}$$

is satisfied for all $q \in \Pi$. To investigate the unique solvability of the variational problem (3.28) we have to check the assumptions of Theorem 3.4 (Lax–Milgram theorem).

**Lemma 3.12.** *Let the assumptions of Theorem 3.11 be satisfied. The operator* $S := BA^{-1}B' : \Pi \to \Pi'$ *is then bounded and from the stability condition* (3.25) *it follows that* $S$ *is* $\Pi$*–elliptic,*

$$\langle Sq, q\rangle \geq c_1^S \, \|q\|_\Pi^2 \quad \textit{for all } q \in \Pi. \tag{3.29}$$

*Proof.* For $q \in \Pi$ we have $u := A^{-1}B'q$ as unique solution of the variational problem

$$\langle Au, v\rangle \;=\; \langle Bv, q\rangle \quad \text{for all } v \in X.$$

Using the $X$–ellipticity of $A : X \to X'$ and applying Theorem 3.4 we conclude the existence of the unique solution $u \in X$ satisfying

$$\|u\|_X \;=\; \|A^{-1}B'q\|_X \;\leq\; \frac{1}{c_1^A}\, \|B'q\|_{X'} \;\leq\; \frac{c_2^B}{c_1^A}\, \|q\|_\Pi .$$

From this we obtain

$$\|Sq\|_{\Pi'} \;=\; \|BA^{-1}B'q\|_{\Pi'} \;=\; \|Bu\|_{\Pi'} \;\leq\; c_2^B \, \|u\|_X \;\leq\; \frac{[c_2^B]^2}{c_1^A}\, \|q\|_\Pi$$

for all $q \in \Pi$ and therefore the boundedness of $S : \Pi \to \Pi'$. Further,

$$\langle Sq, q\rangle \;=\; \langle BA^{-1}B'q, q\rangle \;=\; \langle Bu, q\rangle \;=\; \langle Au, u\rangle \;\geq\; c_1^A \, \|u\|_X^2 \, .$$

On the other hand, the stability condition (3.25) gives

$$c_S \, \|q\|_\Pi \;\leq\; \sup_{0 \neq v \in X} \frac{\langle Bv, q\rangle}{\|v\|_X} \;=\; \sup_{0 \neq v \in X} \frac{\langle Au, v\rangle}{\|v\|_X} \;\leq\; c_2^A \, \|u\|_X$$

and therefore the ellipticity estimate (3.29) with $c_1^S = c_1^A [c_S / c_2^A]^2$. □

From Lemma 3.12 we see that (3.28) is an elliptic variational problem to find $p \in \Pi$. Hence we obtain the unique solvability of (3.28) when applying Theorem 3.4. Moreover, for the solution of the variational problem (3.22) we obtain the following result.

**Theorem 3.13.** *Let $X$ and $\Pi$ be Banach spaces and let $A : X \to X'$ and $B : X \to \Pi'$ be bounded operators. We assume that $A$ is $X$–elliptic, and that the stability condition (3.25) is satisfied. For $f \in X'$ and $g \in \Pi'$ there exists the unique solution $(u,p) \in X \times \Pi$ of the variational problem (3.22) satisfying*

$$\|p\|_\Pi \leq \frac{1}{c_1^S} \|BA^{-1}f - g\|_{\Pi'} \leq \frac{1}{c_1^S} \left[ \frac{c_2^B}{c_1^A} \|f\|_{X'} + \|g\|_{\Pi'} \right] \tag{3.30}$$

*and*

$$\|u\|_X \leq \frac{1}{c_1^A} \left( 1 + \frac{[c_2^B]^2}{c_1^A c_1^S} \right) \|f\|_{X'} + \frac{c_2^B}{c_1^A c_1^S} \|g\|_{\Pi'}. \tag{3.31}$$

*Proof.* The application of Theorem 3.4 (Lax–Milgram lemma) gives the unique solvability of the variational problem (3.28) as well as the estimate (3.30). For a known $p \in X$ we find $u \in X$ as the unique solution of the variational problem

$$\langle Au, v \rangle = \langle f - B'p, v \rangle \quad \text{for all } v \in X.$$

From the $X$–ellipticity of $A$ we obtain

$$c_1^A \|u\|_X^2 \leq \langle Au, u \rangle = \langle f - B'p, u \rangle \leq \|f - B'p\|_{X'} \|u\|_X$$

and therefore

$$\|u\|_X \leq \frac{1}{c_1^A} \|f\|_{X'} + \frac{c_2^B}{c_1^A} \|p\|_\Pi.$$

Applying (3.30) this gives the estimate (3.31). □

## 3.6 Coercive Operators

Since the ellipticity assumption (3.12) is too restrictive for some applications we now consider the more general case of coercive operators. An operator $A : X \to X'$ is called coercive if there exists a compact operator $C : X \to X'$ such that there holds a Gårdings inequality, i.e.

$$\langle (A + C)v, v \rangle \geq c_1^A \|v\|_X^2 \quad \text{for all } v \in X. \tag{3.32}$$

An operator $C : X \to Y$ is said to be compact if the image of the unit sphere of $X$ is relatively compact in $Y$. Note that the product of a compact operator with a bounded linear operator is compact. Applying the Riesz–Schauder theory, see for example [163], we can state the following result.

**Theorem 3.14 (Fredholm alternative).** *Let $K : X \to X$ be a compact operator. Either the homogeneous equation*

$$(I - K)u = 0$$

*has a non–trivial solution $u \in X$ or the inhomogeneous equation*

$$(I - K)u = g$$

*has, for every given $g \in X$, a uniquely determined solution $u \in X$ satisfying*

$$\|u\|_X \leq c\,\|g\|_X.$$

Based on Fredholm's alternative we can derive a result on the solvability of operator equations $Au = f$ when $A$ is assumed to be coercive.

**Theorem 3.15.** *Let $A : X \to X'$ be a bounded coercive linear operator and let $A$ be injective, i.e., from $Au = 0$ it follows that $u = 0$. Then there exists the unique solution $u \in X$ of the operator equation $Au = f$ satisfying*

$$\|u\|_X \leq c\,\|f\|_{X'}.$$

*Proof.* The linear operator $D = A + C : X \to X'$ is bounded and, due to assumption (3.32), $X$–elliptic. Applying the Lax–Milgram lemma (Theorem 3.4) this gives the inverse operator $D^{-1} : X' \to X$. Hence, instead of the operator equation $Au = f$ we consider the equivalent equation

$$Bu = D^{-1}Au = D^{-1}f \tag{3.33}$$

with the bounded operator

$$B = D^{-1}A = D^{-1}(D - C) = I - D^{-1}C : X \to X.$$

Since the operator $D^{-1}C : X \to X$ is compact we can apply Theorem 3.14 to investigate the unique solvability of the operator equation (3.33). Since $A$ is assumed to be injective the homogeneous equation $D^{-1}Au = 0$ has only the trivial solution. Hence there exists a unique solution $u \in X$ of the inhomogeneous equation $Bu = D^{-1}f$ satisfying

$$\|u\|_X \leq c\,\|D^{-1}f\|_X \leq \widetilde{c}\,\|f\|_{X'}. \qquad \square$$

# 4

# Variational Formulations of Boundary Value Problems

In this chapter we describe and analyze variational methods for second order elliptic boundary value problems as given in Chapter 1. To establish the unique solvability of the associated variational formulations we will use the methods which were given in the previous Chapter 3. The weak formulation of boundary value problem is the basis to introduce finite element methods. Moreover, from these results we can also derive mapping properties of boundary integral operators (cf. Chapter 6) as used in boundary element methods.

## 4.1 Potential Equation

Let us consider the scalar partial differential operator (1.1),

$$(Lu)(x) = -\sum_{i,j=1}^{d} \frac{\partial}{\partial x_j}\left[a_{ji}(x)\frac{\partial}{\partial x_i}u(x)\right] \quad \text{for } x \in \Omega \subset \mathbb{R}^d, \tag{4.1}$$

the trace operator (1.3),

$$\gamma_0^{\text{int}} u(x) = \lim_{\Omega \ni \widetilde{x} \to x \in \Gamma} u(\widetilde{x}) \quad \text{for } x \in \Gamma = \partial\Omega,$$

and the associated conormal derivative (1.7),

$$\gamma_1^{\text{int}} u(x) = \lim_{\Omega \ni \widetilde{x} \to x \in \Gamma} \sum_{i,j=1}^{d} n_j(x) a_{ji}(\widetilde{x}) \frac{\partial}{\partial \widetilde{x}_i} u(\widetilde{x}) \quad \text{for } x \in \Gamma = \partial\Omega. \tag{4.2}$$

Note that Green's first formula (1.5),

$$a(u,v) = \int_\Omega (Lu)(x)v(x)dx + \int_\Gamma \gamma_1^{\text{int}} u(x) \gamma_0^{\text{int}} v(x) ds_x,$$

remains valid for $u \in H^1(\Omega)$ with $Lu \in \widetilde{H}^{-1}(\Omega)$ and $v \in H^1(\Omega)$, i.e. we have

$$a(u,v) \;=\; \langle Lu, v\rangle_\Omega + \langle \gamma_1^{\text{int}} u, \gamma_0^{\text{int}} v\rangle_\Gamma \tag{4.3}$$

where $a(\cdot,\cdot)$ is the symmetric bilinear form as defined in (1.6),

$$a(u,v) = \sum_{i,j=1}^{d} \int_\Omega a_{ji}(x) \frac{\partial}{\partial x_i} u(x) \frac{\partial}{\partial x_j} v(x)\, dx. \tag{4.4}$$

**Lemma 4.1.** *Assume that $a_{ij} \in L_\infty(\Omega)$ for $i,j = 1,\ldots,d$ with*

$$\|a\|_{L_\infty(\Omega)} \;:=\; \max_{i,j=1,\ldots,d} \sup_{x\in\Omega} |a_{ij}(x)|. \tag{4.5}$$

*The bilinear form $a(\cdot,\cdot) \,:\, H^1(\Omega) \times H^1(\Omega) \to \mathbb{R}$ is bounded satisfying*

$$|a(u,v)| \;\le c_2^A \,|u|_{H^1(\Omega)} |v|_{H^1(\Omega)} \quad \textit{for all } u, v \in H^1(\Omega) \tag{4.6}$$

*with $c_2^A := d\,\|a\|_{L_\infty(\Omega)}$.*

*Proof.* Using (4.5) we first have

$$\begin{aligned}|a(u,v)| &= \left| \sum_{i,j=1}^{d} \int_\Omega a_{ji}(x) \frac{\partial}{\partial x_i} u(x) \frac{\partial}{\partial x_j} v(x) dx \right| \\ &\le \|a\|_{L_\infty(\Omega)} \int_\Omega \sum_{i=1}^{d} \left| \frac{\partial}{\partial x_i} u(x) \right| \sum_{j=1}^{d} \left| \frac{\partial}{\partial x_j} v(x) \right| dx.\end{aligned}$$

Applying the Cauchy–Schwarz inequality twice we then obtain

$$\begin{aligned}|a(u,v)| &\le \|a\|_{L_\infty(\Omega)} \left( \int_\Omega \left[ \sum_{i=1}^{d} \left| \frac{\partial}{\partial x_i} u(x) \right| \right]^2 dx \right)^{1/2} \left( \int_\Omega \left[ \sum_{j=1}^{d} \left| \frac{\partial}{\partial x_j} v(x) \right| \right]^2 dx \right)^{1/2} \\ &\le \|a\|_{L_\infty(\Omega)} \left( \int_\Omega d \sum_{i=1}^{d} \left| \frac{\partial}{\partial x_i} u(x) \right|^2 dx \right)^{1/2} \left( \int_\Omega d \sum_{j=1}^{d} \left| \frac{\partial}{\partial x_j} v(x) \right|^2 dx \right)^{1/2} \\ &= d\,\|a\|_{L_\infty(\Omega)}\, \|\nabla u\|_{L_2(\Omega)} \|\nabla v\|_{L_2(\Omega)}. \qquad \square\end{aligned}$$

From (4.6) we further get the estimate

$$|a(u,v)| \le c_2^A \, \|u\|_{H^1(\Omega)} \|v\|_{H^1(\Omega)} \quad \text{for all } u, v \in H^1(\Omega). \tag{4.7}$$

**Lemma 4.2.** *Let $L$ be a uniform elliptic partial differential operator as given in* (4.1). *For the bilinear form* (4.4) *we then have*

$$a(v,v) \ge \lambda_0 \, |v|^2_{H^1(\Omega)} \quad \textit{for all } v \in H^1(\Omega) \tag{4.8}$$

*where $\lambda_0$ is the positive constant of the uniform ellipticity estimate* (1.2).

*Proof.* By using $w_i(x) := \frac{\partial}{\partial x_i} v(x)$ for $i = 1, \ldots, d$ we have

$$\begin{aligned} a(v,v) &= \int_\Omega (A(x)\underline{w}(x), \underline{w}(x))\, dx \\ &\ge \lambda_0 \int_\Omega (\underline{w}(x), \underline{w}(x))\, dx = \lambda_0 \, \|\nabla v\|^2_{L_2(\Omega)}. \end{aligned}$$ □

## 4.1.1 Dirichlet Boundary Value Problem

We start to consider the Dirichlet boundary value problem (1.10) and (1.11),

$$(Lu)(x) = f(x) \quad \text{for } x \in \Omega, \quad \gamma_0^{\text{int}} u(x) = g(x) \quad \text{for } x \in \Gamma. \tag{4.9}$$

The manifold to be used in the weak formulation is defined as

$$V_g := \left\{ v \in H^1(\Omega) : \gamma_0^{\text{int}} v(x) = g(x) \quad \text{for } x \in \Gamma \right\}, \quad V_0 = H_0^1(\Omega).$$

The variational formulation of the Dirichlet boundary value problem (4.9) then follows from Green's first formula (4.3): Find $u \in V_g$ such that

$$a(u,v) = \langle f, v \rangle_\Omega \tag{4.10}$$

is satisfied for all $v \in V_0$. Since the Dirichlet boundary condition is explicitly incorporated as a side condition in the manifold $V_g$, we call boundary conditions of Dirichlet type also essential boundary conditions.

The variational problem (4.10) corresponds to the abstract formulation (3.20). Hence we can apply Theorem 3.8 and Corollary 3.9 to establish the unique solvability of the variational problem (4.10).

**Theorem 4.3.** *For $f \in H^{-1}(\Omega)$ and $g \in H^{1/2}(\Gamma)$ there exists a unique solution $u \in H^1(\Omega)$ of the variational problem* (4.10) *satisfying*

$$\|u\|_{H^1(\Omega)} \le \frac{1}{c_1^A} \, \|f\|_{H^{-1}(\Omega)} + \left(1 + \frac{c_2^A}{c_1^A}\right) c_{IT} \, \|g\|_{H^{1/2}(\Gamma)}. \tag{4.11}$$

*Proof.* For any given Dirichlet datum $g \in H^{1/2}(\Gamma)$ we find, by applying the inverse trace theorem (Theorem 2.22), a bounded extension $u_g \in H^1(\Omega)$ satisfying $\gamma_0^{\text{int}} u_g = g$ and

$$\|u_g\|_{H^1(\Omega)} \leq c_{IT} \, \|g\|_{H^{1/2}(\Gamma)}.$$

It remains to find $u_0 := u - u_g \in V_0$ as the solution of the variational problem

$$a(u_0, v) = \langle f, v\rangle_\Omega - a(u_g, v) \quad \text{for all } v \in V_0 \, . \tag{4.12}$$

Recall that

$$\|v\|_{W_2^1(\Omega),\Gamma} := \left\{ \left[ \int_\Gamma \gamma_0^{\text{int}} v(x) \, ds_x \right]^2 + \|\nabla v\|^2_{L_2(\Omega)} \right\}^{1/2}$$

defines an equivalent norm in $H^1(\Omega)$ (cf. Example 2.7). For $v \in V_0 = H_0^1(\Omega)$ we then find from Lemma 4.2

$$a(v, v) \geq \lambda_0 \, |v|^2_{H^1(\Omega)} = \lambda_0 \, \|v\|^2_{W_2^1(\Omega),\Gamma} \geq c_1^A \, \|v\|^2_{H^1(\Omega)}. \tag{4.13}$$

Therefore, all assumptions of Theorem 3.4 (Lax–Milgram lemma) are satisfied. Hence we conclude the unique solvability of the variational problem (4.12).

For the unique solution $u_0 \in V_0$ of the variational problem (4.12) we have, since the bilinear form $a(\cdot,\cdot)$ is $V_0$–elliptic and bounded,

$$\begin{aligned} c_1^A \, \|u_0\|^2_{H^1(\Omega)} &\leq a(u_0, u_0) = \langle f, u_0\rangle_\Omega - a(u_g, u_0) \\ &\leq \left[ \|f\|_{H^{-1}(\Omega)} + c_2^A \, \|u_g\|_{H^1(\Omega)} \right] \|u_0\|_{H^1(\Omega)}, \end{aligned}$$

from which we finally get the estimate (4.11). □

The unique solution $u \in V_g$ of the variational problem (4.10) is also denoted as weak solution of the Dirichlet boundary value problem (4.9). For $f \in \widetilde{H}^{-1}(\Omega)$ we can determine the associated conormal derivative $\gamma_1^{\text{int}} u \in H^{-1/2}(\Gamma)$ as the solution of the variational problem

$$\langle \gamma_1^{\text{int}} u, z\rangle_\Gamma = a(u, \mathcal{E}z) - \langle f, \mathcal{E}z\rangle_\Omega \tag{4.14}$$

for all $z \in H^{1/2}(\Gamma)$. In (4.14), $\mathcal{E} : H^{1/2}(\Gamma) \to H^1(\Omega)$ is the bounded extension operator as defined by the inverse trace theorem (Theorem 2.22). The unique solvability of the variational formulation (4.14) follows when applying Theorem 3.7. Hence we need to assume the stability condition

$$\|w\|_{H^{-1/2}(\Gamma)} = \sup_{0 \neq z \in H^{1/2}(\Gamma)} \frac{\langle w, z\rangle_\Gamma}{\|z\|_{H^{1/2}(\Gamma)}} \quad \text{for all } w \in H^{-1/2}(\Gamma). \tag{4.15}$$

**Lemma 4.4.** *Let $u \in H^1(\Omega)$ be the unique solution of the Dirichlet boundary value problem* (4.10) *when assuming $g \in H^{1/2}(\Gamma)$ and $f \in \widetilde{H}^{-1}(\Omega)$. For the associated conormal derivative $\gamma_1^{int} u \in H^{-1/2}(\Gamma)$ we then have*

$$\|\gamma_1^{int} u\|_{H^{-1/2}(\Gamma)} \le c_{IT} \left\{ \|f\|_{\widetilde{H}^{-1}(\Omega)} + c_2^A \, |u|_{H^1(\Omega)} \right\}. \tag{4.16}$$

*Proof.* Using the stability condition (4.15) and the variational formulation (4.14) we find from the boundedness of the bilinear form $a(\cdot,\cdot)$ and by applying the inverse trace theorem

$$\begin{aligned}
\|\gamma_1^{int} u\|_{H^{-1/2}(\Gamma)} &= \sup_{0 \ne z \in H^{1/2}(\Gamma)} \frac{|\langle \gamma_1^{int} u, z\rangle_\Gamma|}{\|z\|_{H^{1/2}(\Gamma)}} \\
&= \sup_{0 \ne z \in H^{1/2}(\Gamma)} \frac{|a(u, \mathcal{E}z) - \langle f, \mathcal{E}z\rangle_\Omega|}{\|z\|_{H^{1/2}(\Gamma)}} \\
&\le \left\{ c_2^A \, |u|_{H^1(\Omega)} + \|f\|_{\widetilde{H}^{-1}(\Omega)} \right\} \sup_{0 \ne z \in H^{1/2}(\Gamma)} \frac{\|\mathcal{E}z\|_{H^1(\Omega)}}{\|z\|_{H^{1/2}(\Gamma)}} \\
&\le c_{IT} \left\{ \|f\|_{\widetilde{H}^{-1}(\Omega)} + c_2^A \, |u|_{H^1(\Omega)} \right\}. \qquad \square
\end{aligned}$$

In particular for the solution $u$ of the Dirichlet boundary value problem with a homogeneous partial differential equation, i.e. $f \equiv 0$, we obtain the following result which is essential for the analysis of boundary integral operators.

**Corollary 4.5.** *Let $u \in H^1(\Omega)$ be the weak solution of the Dirichlet boundary value problem*

$$(Lu)(x) = 0 \quad \text{for } x \in \Omega, \quad \gamma_0^{int} u(x) = g(x) \quad \text{for } x \in \Gamma$$

*where $L$ is a uniform elliptic partial differential operator of second order. Then,*

$$a(u,u) \ge c \, \|\gamma_1^{int} u\|^2_{H^{-1/2}(\Gamma)} . \tag{4.17}$$

*Proof.* By setting $f \equiv 0$ the estimate (4.16) first gives

$$\|\gamma_1^{int} u\|^2_{H^{-1/2}(\Gamma)} \le [c_{IT} \, c_2^A]^2 \, |u|^2_{H^1(\Omega)} .$$

The assertion now follows from the semi–ellipticity (4.8) of the bilinear form $a(\cdot,\cdot)$. $\square$

When $\Omega$ is a Lipschitz domain we can formulate stronger assumptions on the given data $f$ and $g$ to establish higher regularity results for the solution $u$ of the Dirichlet boundary value problem and for the associated conormal derivative $\gamma_1^{int} u$.

**Theorem 4.6.** [106, Theorem 1.1, p. 249] *Let $\Omega \subset \mathbb{R}^d$ be a bounded Lipschitz domain with boundary $\Gamma = \partial\Omega$. Let $u \in H^1(\Omega)$ be the weak solution of the Dirichlet boundary value problem*

$$(Lu)(x) = f(x) \quad \text{for } x \in \Omega, \quad \gamma_0^{int} u(x) = g(x) \quad \text{for } x \in \Gamma.$$

*If $f \in L_2(\Omega)$ and $g \in H^1(\Gamma)$ are satisfied, then we have $u \in H^{3/2}(\Omega)$ with*

$$\|u\|_{H^{3/2}(\Omega)} \leq c_1 \left\{ \|f\|_{L_2(\Omega)} + \|g\|_{H^1(\Gamma)} \right\}$$

*as well as $\gamma_1^{int} u \in L_2(\Gamma)$ satisfying*

$$\|\gamma_1^{int} u\|_{L_2(\Gamma)} \leq c_2 \left\{ \|f\|_{L_2(\Omega)} + \|g\|_{H^1(\Gamma)} \right\}.$$

When formulating stronger assumptions both on the domain $\Omega$ and on the given data $f$ and $g$ we can establish even higher regularity results for the solution $u$ of the Dirichlet boundary value problem (4.9). Let the boundary $\Gamma = \partial\Omega$ be either smooth or piecewise smooth, but assume that $\Omega$ is convex, and let $f \in L_2(\Omega)$. If $g = \gamma_0^{\text{int}} u_g$ is the trace of a function $u_g \in H^2(\Omega)$, then we have $u \in H^2(\Omega)$. For more general results on the regularity of solutions of boundary value problems we refer, for example, to [66].

### 4.1.2 Lagrange Multiplier Methods

In what follows we will consider a saddle point variational formulation which is equivalent to the variational problem (4.10). The Dirichlet boundary conditions are now formulated as side conditions, and the associated conormal derivative corresponds to the Lagrange multiplier [7, 24]. Starting from Green's first formula (4.3) we obtain by introducing the Lagrange multiplier $\lambda := \gamma_1^{\text{int}} u \in H^{-1/2}(\Gamma)$ the following saddle point problem: Find $(u, \lambda) \in H^1(\Omega) \times H^{-1/2}(\Gamma)$ such that

$$\begin{aligned} a(u,v) - b(v,\lambda) &= \langle f, v \rangle_\Omega \\ b(u,\mu) &= \langle g, \mu \rangle_\Gamma \end{aligned} \tag{4.18}$$

is satisfied for all $(v, \mu) \in H^1(\Omega) \times H^{-1/2}(\Gamma)$. Here we have used the bilinear form

$$b(v,\mu) := \langle \gamma_0^{\text{int}} v, \mu \rangle_\Gamma \quad \text{for } (v,\mu) \in H^1(\Omega) \times H^{-1/2}(\Gamma).$$

To investigate the unique solvability of the saddle point problem (4.18) we will apply Theorem 3.11. Obviously,

$$\ker B := \left\{ v \in H^1(\Omega) : \langle \gamma_0^{\text{int}} v, \mu \rangle_\Gamma = 0 \quad \text{for all } \mu \in H^{-1/2}(\Gamma) \right\} = H_0^1(\Omega).$$

Hence, due to (4.13), we have the $\ker B$–ellipticity of the bilinear form $a(\cdot,\cdot)$. It remains to establish the stability condition

$$c_S \, \|\mu\|_{H^{-1/2}(\Gamma)} \leq \sup_{0 \neq v \in H^1(\Omega)} \frac{\langle \gamma_0^{\text{int}} v, \mu \rangle_\Gamma}{\|v\|_{H^1(\Omega)}} \quad \text{for all } \mu \in H^{-1/2}(\Gamma). \tag{4.19}$$

**Lemma 4.7.** *The stability condition* (4.19) *is satisfied for all* $\mu \in H^{-1/2}(\Gamma)$.

*Proof.* Let an arbitrary $\mu \in H^{-1/2}(\Gamma)$ be given. Applying Theorem 3.3 (Riesz Representation Theorem) we find a uniquely determined $u_\mu \in H^{1/2}(\Gamma)$ satisfying

$$\langle u_\mu, v\rangle_{H^{1/2}(\Gamma)} = \langle \mu, v\rangle_\Gamma \quad \text{for all } v \in H^{1/2}(\Gamma)$$

and

$$\|u_\mu\|_{H^{1/2}(\Gamma)} = \|\mu\|_{H^{-1/2}(\Gamma)}\,.$$

Using the inverse trace theorem (Theorem 2.22) there exists an extension $\mathcal{E}u_\mu \in H^1(\Omega)$ with

$$\|\mathcal{E}u_\mu\|_{H^1(\Omega)} \le c_{IT}\,\|u_\mu\|_{H^{1/2}(\Gamma)}.$$

For $v = \mathcal{E}u_\mu \in H^1(\Omega)$ we then have

$$\begin{aligned}\frac{\langle v, \mu\rangle_\Gamma}{\|v\|_{H^1(\Omega)}} &= \frac{\langle u_\mu, \mu\rangle_\Gamma}{\|\mathcal{E}u_\mu\|_{H^1(\Omega)}} = \frac{\langle u_\mu, u_\mu\rangle_{H^{1/2}(\Gamma)}}{\|\mathcal{E}u_\mu\|_{H^1(\Omega)}}\\ &\ge \frac{1}{c_{IT}}\|u_\mu\|_{H^{1/2}(\Gamma)} = \frac{1}{c_{IT}}\|\mu\|_{H^{-1/2}(\Gamma)}\end{aligned}$$

and therefore the stability condition (4.19) is satisfied. □

Hence we can conclude the unique solvability of the saddle point problem (4.18) due to Theorem 3.11.

Recall that the bilinear form $a(\cdot,\cdot)$ in the saddle point formulation (4.18) is only $H_0^1(\Omega)$–elliptic. However, the saddle point problem (4.18) can be reformulated to obtain a formulation where the modified bilinear form $\widetilde{a}(\cdot,\cdot)$ is now $H^1(\Omega)$–elliptic. Since the Lagrange multiplier $\lambda := \gamma_1^{\text{int}} u \in H^{-1/2}(\Gamma)$ describes the conormal derivative of the solution $u$, using the orthogonality relation (1.15) we have

$$\int_\Omega f(x)dx + \int_\Gamma \lambda(x)ds_x = 0\,. \tag{4.20}$$

On the other hand, with the Dirichlet boundary condition $\gamma_0^{\text{int}} u = g$ we also have

$$\int_\Gamma \gamma_0^{\text{int}} u(x)ds_x = \int_\Gamma g(x)ds_x\,. \tag{4.21}$$

Hence we can reformulate the saddle point problem (4.18) to find $(u,\lambda) \in H^1(\Omega) \times H^{-1/2}(\Gamma)$ such that

$$\int_\Gamma \gamma_0^{\text{int}} u(x) ds_x \int_\Gamma \gamma_0^{\text{int}} v(x) ds_x + a(u,v) - b(v,\lambda) \tag{4.22}$$
$$= \langle f, v\rangle_\Omega + \int_\Gamma g(x) ds_x \int_\Gamma \gamma_0^{\text{int}} v(x) ds_x$$

$$b(u,\mu) + \int_\Gamma \lambda(x) ds_x \int_\Gamma \mu(x) ds_x = \langle g, \mu\rangle_\Gamma - \int_\Omega f(x) dx \int_\Gamma \mu(x) ds_x \tag{4.23}$$

is satisfied for all $(v,\mu) \in H^1(\Omega) \times H^{-1/2}(\Gamma)$.

The modified saddle point problem (4.22) and (4.23) is uniquely solvable, and the solution is also the unique solution of the original saddle point problem (4.18), i.e. the saddle point formulations (4.22)–(4.23) and (4.18) are equivalent.

**Theorem 4.8.** *The modified saddle point problem* (4.22) *and* (4.23) *has a unique solution* $(u,\lambda) \in H^1(\Omega) \times H^{-1/2}(\Gamma)$, *which is also the unique solution of the saddle point formulation* (4.18).

*Proof.* The extended bilinear form

$$\widetilde{a}(u,v) := \int_\Gamma \gamma_0^{\text{int}} u(x) ds_x \int_\Gamma \gamma_0^{\text{int}} v(x) ds_x + a(u,v)$$

is bounded for all $u, v \in H^1(\Omega)$. Using Lemma 4.2 and Example 2.7 we find

$$\widetilde{a}(v,v) = \left[\int_\Gamma \gamma_0^{\text{int}} v(x) ds_x\right]^2 + a(v,v) \geq \min\{1, \lambda_0\} \|v\|^2_{W_2^1(\Omega),\Gamma} \geq c_1^{\widetilde{A}} \|v\|^2_{H^1(\Omega)}$$

for all $v \in H^1(\Omega)$ and therefore the $H^1(\Omega)$–ellipticity of the extended bilinear form $\widetilde{a}(\cdot,\cdot)$. Applying Theorem 3.11 we obtain as in Theorem 3.13 the unique solvability of the saddle point problem (4.22) and (4.23). In particular for $(v,\mu) \equiv (1,1)$ we have

$$|\Gamma| \int_\Gamma \gamma_0^{\text{int}} u(x) ds_x - \int_\Gamma \lambda(x) ds_x = \int_\Omega f(x) dx + |\Gamma| \int_\Gamma g(x) ds_x,$$
$$\int_\Gamma \gamma_0^{\text{int}} u(x) ds_x + |\Gamma| \int_\Gamma \lambda(x) ds_x = \int_\Gamma g(x) ds_x - |\Gamma| \int_\Omega f(x) dx.$$

Multiplying the first equation with $|\Gamma| > 0$ and adding the result to the second equation this gives

$$(1 + |\Gamma|^2) \int_\Gamma \gamma_0^{\text{int}} u(x) ds_x = (1 + |\Gamma|^2) \int_\Gamma g(x) ds_x$$

and therefore (4.21). Then we immediately get also (4.20), i.e. $(u,\lambda)$ is also a solution of the saddle point problem (4.18). □

### 4.1.3 Neumann Boundary Value Problem

In addition to the Dirichlet boundary value problem (4.9) we now consider the Neumann boundary value problem (1.10) and (1.12),

$$(Lu)(x) = f(x) \quad \text{for } x \in \Omega, \quad \gamma_1^{\text{int}} u(x) = g(x) \quad \text{for } x \in \Gamma. \tag{4.24}$$

Hereby we have to assume the solvability condition (1.17),

$$\int_\Omega f(x)dx + \int_\Gamma g(x)ds_x = 0. \tag{4.25}$$

Moreover, the solution of the Neumann boundary value problem (4.24) is only unique up to an additive constant. To fix this constant, we formulate a suitable scaling condition. For this we define

$$H^1_*(\Omega) := \left\{ v \in H^1(\Omega) : \int_\Omega v(x)dx = 0 \right\}.$$

Using Green's first formula (4.3) we obtain the variational formulation of the Neumann boundary value problem (4.24) to find $u \in H^1_*(\Omega)$ such that

$$a(u,v) = \langle f, v\rangle_\Omega + \langle g, \gamma_0^{\text{int}} v\rangle_\Gamma \tag{4.26}$$

is satisfied for all $v \in H^1_*(\Omega)$.

**Theorem 4.9.** *Let $f \in \widetilde{H}^{-1}(\Omega)$ and $g \in H^{-1/2}(\Gamma)$ be given satisfying the solvability condition* (4.25). *Then there exists a unique solution $u \in H^1_*(\Omega)$ of the variational problem* (4.26) *satisfying*

$$\|u\|_{H^1(\Omega)} \leq \frac{1}{\widetilde{c}_1^A} \left\{ \|f\|_{\widetilde{H}^{-1}(\Omega)} + c_T \|g\|_{H^{-1/2}(\Gamma)} \right\}.$$

*Proof.* Recall that

$$\|v\|_{W_2^1(\Omega),\Omega} := \left\{ \left[ \int_\Omega v(x)\,dx \right]^2 + \|\nabla v\|^2_{L_2(\Omega)} \right\}^{1/2}$$

defines an equivalent norm in $H^1(\Omega)$ (cf. Example 2.7). Using Lemma 4.2 we then have

$$a(v,v) \geq \lambda_0 \|\nabla v\|^2_{L_2(\Omega)} = \lambda_0 \|v\|^2_{W_2^1(\Omega),\Omega} \geq \widetilde{c}_1^A \|v\|^2_{H^1(\Omega)} \tag{4.27}$$

for all $v \in H^1_*(\Omega)$ and therefore the $H^1_*(\Omega)$–ellipticity of the bilinear form $a(\cdot,\cdot)$ follows. The unique solvability of the variational problem (4.26) we now conclude from Theorem 3.4 (Lax–Milgram lemma). Using the $H^1_*(\Omega)$–ellipticity of the bilinear form $a(\cdot,\cdot)$ we further have

$$\begin{aligned}\widetilde{c}_1^A \, \|u\|^2_{H^1(\Omega)} \leq a(u,u) &= \langle f, u\rangle_\Omega + \langle g, \gamma_0^{\text{int}} u\rangle_\Gamma \\ &\leq \|f\|_{\widetilde{H}^{-1}(\Omega)} \|u\|_{H^1(\Omega)} + \|g\|_{H^{-1/2}(\Gamma)} \|\gamma_0^{\text{int}} u\|_{H^{1/2}(\Gamma)}.\end{aligned}$$

Applying the trace theorem (Theorem 2.21) gives the assertion. □

In what follows we will consider a saddle point formulation which is equivalent to the variational problem (4.26). The scaling condition to define the trial space $H^1_*(\Omega)$ is now formulated as a side condition. By using a scalar Lagrange multiplier we obtain the following variational problem (cf. Section 3.5) to find $u \in H^1(\Omega)$ and $\lambda \in \mathbb{R}$ such that

$$\begin{aligned} a(u,v) + \lambda \int_\Omega v(x)dx &= \langle f, v\rangle_\Omega + \langle g, \gamma_0^{\text{int}} v\rangle_\Gamma \\ \int_\Omega u(x)dx \qquad &= 0 \end{aligned} \tag{4.28}$$

is satisfied for all $v \in H^1(\Omega)$. To establish the unique solvability of the saddle point problem (4.28) we have to investigate the assumptions of Theorem 3.11. The bilinear form

$$b(v,\mu) := \mu \int_\Omega v(x)dx \quad \text{for all } v \in H^1(\Omega), \mu \in \mathbb{R}$$

is bounded, and we have $\ker B = H^1_*(\Omega)$. Hence we obtain the $\ker B$–ellipticity of the bilinear form $a(\cdot,\cdot)$ from the ellipticity estimate (4.27). It remains to prove the stability condition

$$c_S \, |\mu| \leq \sup_{0 \neq v \in H^1(\Omega)} \frac{b(v,\mu)}{\|v\|_{H^1(\Omega)}} \quad \text{for all } \mu \in \mathbb{R}. \tag{4.29}$$

For an arbitrary given $\mu \in \mathbb{R}$ we define $v^* := \mu \in H^1(\Omega)$ to obtain the stability estimate (4.29) with $c_S = 1/\sqrt{|\Omega|}$. By applying Theorem 3.11 we now conclude the unique solvability of the saddle point problem (4.28).

Choosing in (4.28) the test function $v \equiv 1$ we obtain for the Lagrange parameter $\lambda$ from the solvability condition (4.25)

$$\lambda = 0.$$

Instead of (4.28) we may now consider an equivalent saddle point formulation to find $(u,\lambda) \in H^1(\Omega) \times \mathbb{R}$ such that

$$\begin{aligned} a(u,v) + \lambda \int_\Omega v(x)dx &= \langle f, v\rangle_\Omega + \langle g, \gamma_0^{\text{int}} v\rangle_\Gamma \\ \int_\Omega u(x)dx - \lambda \qquad &= 0 \end{aligned} \tag{4.30}$$

is satisfied for all $v \in H^1(\Omega)$. Using the second equation we can eliminate the scalar Lagrange multiplier $\lambda \in \mathbb{R}$ to obtain a modified variational problem to find $u \in H^1(\Omega)$ such that

$$a(u,v) + \int_\Omega u(x)dx \int_\Omega v(x)dx = \langle f, v\rangle_\Omega + \langle g, \gamma_0^{\text{int}} v\rangle_\Gamma \tag{4.31}$$

is satisfied for all $v \in H^1(\Omega)$.

**Theorem 4.10.** *For any $f \in \widetilde{H}^{-1}(\Omega)$ and for any $g \in H^{-1/2}(\Gamma)$ there is a unique solution $u \in H^1(\Omega)$ of the modified variational problem* (4.31).
*If $f \in \widetilde{H}^{-1}(\Omega)$ and $g \in H^{-1/2}(\Gamma)$ satisfy the solvability condition* (4.25), *then we have $u \in H^1_*(\Omega)$, i.e. the modified variational problem* (4.31) *and the saddle point formulation* (4.28) *are equivalent.*

*Proof.* The modified bilinear form

$$\widetilde{a}(u,v) := a(u,v) + \int_\Omega u(x)dx \int_\Omega v(x)dx$$

is $H^1(\Omega)$–elliptic, i.e for all $v \in H^1(\Omega)$ we have

$$\widetilde{a}(v,v) \geq \lambda_0 \, \|\nabla v\|^2_{L_2(\Omega)} + \left[ \int_\Omega v(x)dx \right]^2$$

$$\geq \min\{\lambda_0, 1\} \, \|v\|^2_{W_2^1(\Omega),\Omega} \geq \hat{c}_1^A \, \|v\|^2_{H^1(\Omega)}.$$

Hence we conclude the unique solvability of the modified variational problem (4.31) due the Theorem 3.4 (Lax–Milgram lemma) for arbitrary given data $f \in \widetilde{H}^{-1}(\Omega)$ and $g \in H^{-1/2}(\Gamma)$.

Choosing as test function $v \equiv 1$ we get, when assuming the solvability condition (4.25),

$$|\Omega| \int_\Omega u(x)dx = \langle f, 1\rangle_\Omega + \langle g, 1\rangle_\Gamma = 0$$

and therefore $u \in H^1_*(\Omega)$. The solution of the modified variational problem (4.31) is therefore also a solution of the saddle point formulation (4.28), i.e. both formulations are equivalent. □

Since the solution of the Neumann boundary value problem (4.24) is not unique, we can add an arbitrary constant $\alpha \in \mathbb{R}$ to the solution $u \in H^1_*(\Omega)$ to obtain the general solution $\widetilde{u} := u + \alpha \in H^1(\Omega)$.

### 4.1.4 Mixed Boundary Value Problem

We now consider the boundary value problem (1.10)–(1.12) with boundary conditions of mixed type,

$$\begin{aligned} (Lu)(x) &= f(x) && \text{for } x \in \Omega, \\ \gamma_0^{\text{int}} u(x) &= g_D(x) && \text{for } x \in \Gamma_D, \\ \gamma_1^{\text{int}} u(x) &= g_N(x) && \text{for } x \in \Gamma_N. \end{aligned}$$

We assume $\Gamma = \overline{\Gamma}_D \cup \overline{\Gamma}_N$ as well as $\text{meas}(\Gamma_D) > 0$. The associated variational problem again follows from Green's first formula (4.3) to find $u \in H^1(\Omega)$ with $\gamma_0^{\text{int}} u(x) = g_D(x)$ for $x \in \Gamma_D$ such that

$$a(u, v) \;=\; \langle f, v\rangle_\Omega + \langle g_N, \gamma_0^{\text{int}} v\rangle_{\Gamma_N} \tag{4.32}$$

is satisfied for all $v \in H_0^1(\Omega, \Gamma_D)$ where

$$H_0^1(\Omega, \Gamma_D) \;:=\; \Big\{ v \in H^1(\Omega) \,:\, \gamma_0^{\text{int}} v(x) = 0 \quad \text{for } x \in \Gamma_D \Big\}.$$

The unique solvability of the variational problem (4.32) is a consequence of the following theorem.

**Theorem 4.11.** *Let $f \in \widetilde{H}^{-1}(\Omega)$, $g_D \in H^{1/2}(\Gamma_D)$ and $g_N \in H^{-1/2}(\Gamma_N)$ be given. Then there exists a unique solution $u \in H^1(\Omega)$ of the variational problem* (4.32) *satisfying*

$$\|u\|_{H^1(\Omega)} \;\le\; c\left[ \|f\|_{\widetilde{H}^{-1}(\Omega)} + \|g_D\|_{H^{1/2}(\Gamma_D)} + \|g_N\|_{H^{-1/2}(\Gamma_N)} \right]. \tag{4.33}$$

*Proof.* For $g_D \in H^{1/2}(\Gamma_D)$ we first find a bounded extension $\widetilde{g}_D \in H^{1/2}(\Gamma)$ satisfying

$$\|\widetilde{g}_D\|_{H^{1/2}(\Gamma)} \;\le\; c\,\|g_D\|_{H^{1/2}(\Gamma_D)}.$$

Applying the inverse trace theorem (Theorem 2.22) there exists a second extension $u_{\widetilde{g}_D} \in H^1(\Omega)$ with $\gamma_0^{\text{int}} u_{\widetilde{g}_D} = \widetilde{g}_D$ and satisfying

$$\|u_{\widetilde{g}_D}\|_{H^1(\Omega)} \;\le\; c_{IT}\,\|\widetilde{g}_D\|_{H^{1/2}(\Gamma)}.$$

It remains to find $u_0 \in H_0^1(\Omega, \Gamma_D)$ as the unique solution of the variational formulation

$$a(u_0, v) \;=\; \langle f, v\rangle_\Omega + \langle g_N, \gamma_0^{\text{int}} v\rangle_{\Gamma_N} - a(u_{\widetilde{g}_D}, v)$$

for all $v \in H_0^1(\Omega, \Gamma_D)$. As in Example 2.7 we can define

$$\|v\|_{W_2^1(\Omega), \Gamma_D} \;:=\; \left\{ \left[ \int_{\Gamma_D} \gamma_0^{\text{int}} v(x)\, ds_x \right]^2 + \|\nabla v\|^2_{L_2(\Omega)} \right\}^{1/2}$$

which is an equivalent norm in $H^1(\Omega)$. Using Lemma 4.2 we find

$$a(v,v) \geq \lambda_0 \, \|\nabla v\|^2_{L_2(\Omega)} = \lambda_0 \, \|v\|^2_{W_2^1(\Omega),\Gamma_D} \geq c_1^A \, \|v\|^2_{W_2^1(\Omega)}$$

for all $v \in H_0^1(\Omega, \Gamma_D)$. Hence all assumptions of Theorem 3.4 (Lax–Milgram lemma) are satisfied, and therefore, the unique solvability of the variational problem (4.32) follows.

For the unique solution $u_0 \in H_0^1(\Omega, \Gamma_D)$ of the variational problem (4.32) we then obtain

$$\begin{aligned} c_1^A \, \|u_0\|^2_{H^1(\Omega)} &\leq a(u_0, u_0) \\ &= \langle f, u_0\rangle_\Omega + \langle g_N, \gamma_0^{\text{int}} u_0\rangle_{\Gamma_N} - a(u_{\widetilde{g}_D}, u_0) \\ &\leq \Big[\|f\|_{\widetilde{H}^{-1}(\Omega)} + c_2^A \, \|u_{\widetilde{g}_D}\|_{H^1(\Omega)}\Big] \, \|u_0\|_{H^1(\Omega)} \\ &\quad + \|g_N\|_{H^{-1/2}(\Gamma_N)} \|\gamma_0^{\text{int}} u_0\|_{\widetilde{H}^{1/2}(\Gamma_N)}, \end{aligned}$$

from which we conclude the estimate (4.33). □

### 4.1.5 Robin Boundary Value Problems

We finally consider the boundary value problem (1.10) and (1.13) with boundary conditions of Robin type,

$$(Lu)(x) = f(x) \quad \text{for } x \in \Omega, \quad \gamma_1^{\text{int}} u(x) + \kappa(x)\gamma_0^{\text{int}} u(x) = g(x) \quad \text{for } x \in \Gamma.$$

The associated variational formulation is again a direct consequence of Green's first formula (4.3) to find $u \in H^1(\Omega)$ such that

$$a(u,v) + \int_\Gamma \kappa(x)\gamma_0^{\text{int}} u(x)\gamma_0^{\text{int}} v(x) ds_x = \langle f, v\rangle_\Omega + \langle g, \gamma_0^{\text{int}} v\rangle_\Gamma \tag{4.34}$$

is satisfied for all $v \in H^1(\Omega)$.

**Theorem 4.12.** *Let $f \in \widetilde{H}^{-1}(\Omega)$ and $g \in H^{-1/2}(\Gamma)$ be given. Assume that $\kappa(x) \geq \kappa_0 > 0$ holds for all $x \in \Gamma$. Then there exists a unique solution $u \in H^1(\Omega)$ of the variational problem (4.34) satisfying*

$$\|u\|_{H^1(\Omega)} \leq c \left[\|f\|_{\widetilde{H}^{-1}(\Omega)} + \|g\|_{H^{-1/2}(\Gamma)}\right]. \tag{4.35}$$

*Proof.* As in Example 2.7 we can define

$$\|v\|_{W_2^1(\Omega),\Gamma} := \left\{\|\gamma_0^{\text{int}} v\|^2_{L_2(\Gamma)} + \|\nabla v\|^2_{L_2(\Omega)}\right\}^{1/2}$$

which is an equivalent norm in $H^1(\Omega)$. Applying Lemma 4.2 and by using $\kappa(x) \geq \kappa_0 > 0$ for $x \in \Gamma$ we obtain

$$a(v,v) + \int_\Gamma \kappa(x)[\gamma_0^{\text{int}} v(x)]^2 ds_x \geq \lambda_0 \, \|\nabla v\|^2_{L_2(\Omega)} + \kappa_0 \, \|\gamma_0^{\text{int}} v\|^2_{L_2(\Gamma)}$$

$$\geq \min\{\lambda_0, \kappa_0\} \, \|v\|^2_{H^1(\Omega),\Gamma} \geq c_1^A \, \|v\|^2_{H^1(\Omega)}.$$

Hence we can apply Theorem 3.4 (Lax–Milgram lemma) to conclude the unique solvability of the variational problem (4.34). The estimate (4.35) then follows as in the proof of Theorem 4.11. □

## 4.2 Linear Elasticity

Next we consider the system of linear elasticity,

$$L_i \underline{u}(x) = -\sum_{j=1}^d \frac{\partial}{\partial x_j} \sigma_{ij}(\underline{u}, x) \quad \text{for } x \in \Omega \subset \mathbb{R}^d, \; i = 1, \ldots, d.$$

Inserting Hooke's law (1.19) this is equivalent to

$$L\underline{u}(x) = -\mu \Delta \underline{u}(x) - (\lambda + \mu) \operatorname{grad} \operatorname{div} \underline{u}(x) \quad \text{for } x \in \Omega \subset \mathbb{R}^d$$

with the Lamé constants

$$\lambda = \frac{E\nu}{(1+\nu)(1-2\nu)}, \qquad \mu = \frac{E}{2(1+\nu)}$$

where we assume $E > 0$ and $\nu \in (0, 1/2)$. Using the associated conormal derivative (1.23) and the bilinear form (1.26),

$$a(\underline{u}, \underline{v}) = 2\mu \int_\Omega \sum_{i,j=1}^d e_{ij}(\underline{u}, x) e_{ij}(\underline{v}, x) dx + \lambda \int_\Omega \operatorname{div} \underline{u}(x) \operatorname{div} \underline{v}(x) dx$$

$$= \int_\Omega \sum_{i,j=1}^d \sigma_{ij}(\underline{u}, x) e_{ij}(\underline{v}, x) dx,$$

we can write Betti's first formula as (1.21),

$$a(\underline{u}, \underline{v}) = \langle L\underline{u}, \underline{v} \rangle_\Omega + \langle \gamma_1^{\text{int}} \underline{u}, \gamma_0^{\text{int}} \underline{v} \rangle_\Gamma.$$

First we show that the bilinear form $a(\cdot, \cdot)$ is bounded.

**Lemma 4.13.** *The bilinear form* (1.26) *is bounded, i.e.*

$$|a(\underline{u}, \underline{v})| \leq \frac{2E}{1-2\nu} \, |\underline{u}|_{[H^1(\Omega)]^d} |\underline{v}|_{[H^1(\Omega)]^d} \tag{4.36}$$

*for all* $\underline{u}, \underline{v} \in [H^1(\Omega)]^d$.

*Proof.* In the case $d = 3$ we can write Hooke's law (1.19) as

$$\begin{pmatrix} \sigma_{11} \\ \sigma_{22} \\ \sigma_{33} \\ \sigma_{12} \\ \sigma_{13} \\ \sigma_{23} \end{pmatrix} = \frac{E}{(1+\nu)(1-2\nu)} \begin{pmatrix} 1-\nu & \nu & \nu & & & \\ \nu & 1-\nu & \nu & & & \\ \nu & \nu & 1-\nu & & & \\ & & & 1-2\nu & & \\ & & & & 1-2\nu & \\ & & & & & 1-2\nu \end{pmatrix} \begin{pmatrix} e_{11} \\ e_{22} \\ e_{33} \\ e_{12} \\ e_{13} \\ e_{23} \end{pmatrix}$$

or in short,

$$\underline{\sigma} = \frac{E}{(1+\nu)(1-2\nu)} C \underline{e} .$$

Due to the symmetries $\sigma_{ij}(\underline{u}, x) = \sigma_{ji}(\underline{u}, x)$ and $e_{ij}(\underline{v}, x) = e_{ji}(\underline{v}, x)$ we have

$$a(\underline{u}, \underline{v}) = \frac{E}{(1+\nu)(1-2\nu)} \int_\Omega (DC\underline{e}(\underline{u}, x), \underline{e}(\underline{v}, x))\, dx$$

with the diagonal matrix $D = \operatorname{diag}(1,1,1,2,2,2)$. The eigenvalues of the matrix $DC \in \mathbb{R}^{6\times 6}$ are

$$\lambda_1(DC) = 1+\nu, \quad \lambda_{2,3}(DC) = 1-2\nu, \quad \lambda_{4,5,6}(DC) = 2(1-2\nu).$$

Hence, by applying the Cauchy–Schwarz inequality,

$$\begin{aligned} |a(\underline{u}, \underline{v})| &= \left| \frac{E}{(1+\nu)(1-2\nu)} \int_\Omega (DC\underline{e}(\underline{u}, x), \underline{e}(\underline{v}, x))\, dx \right| \\ &\le \frac{E}{(1+\nu)(1-2\nu)} \int_\Omega \|DC\underline{e}(\underline{u}, x)\|_2 \|\underline{e}(\underline{v}, x)\|_2\, dx \\ &\le \frac{E}{(1+\nu)(1-2\nu)} \max\{(1+\nu), 2(1-2\nu)\} \int_\Omega \|\underline{e}(\underline{v}, x)\|_2 \|\underline{e}(\underline{v}, x)\|_2\, dx \\ &\le \frac{2E}{1-2\nu} \left( \int_\Omega \|\underline{e}(\underline{u}, x)\|_2^2\, dx \right)^{1/2} \left( \int_\Omega \|\underline{e}(\underline{v}, x)\|_2^2\, dx \right)^{1/2} . \end{aligned}$$

Using

$$\begin{aligned} \|\underline{e}(\underline{v}, x)\|_2^2 &= \sum_{i=1}^d \sum_{j=i}^d [e_{ij}(\underline{u}, x)]^2 = \frac{1}{4} \sum_{i=1}^d \sum_{j=i}^d \left[ \frac{\partial}{\partial x_i} u_j(x) + \frac{\partial}{\partial x_j} u_i(x) \right]^2 \\ &\le \frac{1}{2} \sum_{i,j=1}^d \left\{ \left[ \frac{\partial}{\partial x_j} u_i(x) \right]^2 + \left[ \frac{\partial}{\partial x_i} u_j(x) \right]^2 \right\} = \sum_{i,j=1}^d \left[ \frac{\partial}{\partial x_j} u_i(x) \right]^2 \end{aligned}$$

we obtain

$$\int_\Omega \|\underline{e}(\underline{u}, x)\|_2^2 \, dx \;\leq\; \int_\Omega \sum_{i,j=1}^d \left[\frac{\partial}{\partial x_j} u_i(x)\right]^2 dx \;=\; |\underline{u}|^2_{[H^1(\Omega)]^d}$$

and therefore the estimate (4.36). In the case $d = 2$ the assertion follows in the same way. □

The proof of the $[H_0^1(\Omega)]^d$–ellipticity of the bilinear form $a(\cdot,\cdot)$ requires several steps.

**Lemma 4.14.** *For $\underline{v} \in [H^1(\Omega)]^d$ we have*

$$a(\underline{v}, \underline{v}) \;\geq\; \frac{E}{1+\nu} \int_\Omega \sum_{i,j=1}^d [e_{ij}(\underline{v}, x)]^2 dx \,. \tag{4.37}$$

*Proof.* The assertion follows from the representation (1.26), i.e.

$$\begin{aligned} a(\underline{v}, \underline{v}) &= 2\mu \int_\Omega \sum_{i,j=1}^d [e_{ij}(\underline{v}, x)]^2 dx + \lambda \int_\Omega [\operatorname{div} \underline{v}(x)]^2 \, dx \\ &\geq 2\,\mu \int_\Omega \sum_{i,j=1}^d [e_{ij}(\underline{v}, x)]^2 \, dx \;=\; \frac{E}{1+\nu} \int_\Omega \sum_{i,j=1}^d [e_{ij}(\underline{v}, x)]^2 \, dx. \end{aligned}$$ □

Next we can formulate Korn's first inequality for $\underline{v} \in [H_0^1(\Omega)]^d$.

**Lemma 4.15 (Korn's First Inequality).** *For $\underline{v} \in [H_0^1(\Omega)]^d$ we have*

$$\int_\Omega \sum_{i,j=1}^d [e_{ij}(\underline{v}, x)]^2 dx \;\geq\; \frac{1}{2}\, |\underline{v}|^2_{[H^1(\Omega)]^d}. \tag{4.38}$$

*Proof.* For $\underline{\varphi} \in C_0^\infty(\Omega)$ we first have

$$\begin{aligned} \int_\Omega \sum_{i,j=1}^d [e_{ij}(\underline{\varphi}, x)]^2 dx &= \frac{1}{4} \int_\Omega \sum_{i,j=1}^d \left[\frac{\partial}{\partial x_j}\varphi_i(x) + \frac{\partial}{\partial x_i}\varphi_j(x)\right]^2 dx \\ &= \frac{1}{2} \sum_{i,j=1}^d \int_\Omega \left[\frac{\partial}{\partial x_j}\varphi_i(x)\right]^2 dx + \frac{1}{2} \sum_{i,j=1}^d \int_\Omega \frac{\partial}{\partial x_j}\varphi_i(x) \frac{\partial}{\partial x_i}\varphi_j(x) dx. \end{aligned}$$

Applying integration by parts twice, this gives

$$\int_\Omega \frac{\partial}{\partial x_j}\varphi_i(x) \frac{\partial}{\partial x_i}\varphi_j(x) dx \;=\; \int_\Omega \frac{\partial}{\partial x_i}\varphi_i(x) \frac{\partial}{\partial x_j}\varphi_j(x) dx$$

and therefore

$$\sum_{i,j=1}^{d} \int_{\Omega} \frac{\partial}{\partial x_j}\varphi_i(x)\frac{\partial}{\partial x_i}\varphi_j(x)dx = \sum_{i,j=1}^{d} \int_{\Omega} \frac{\partial}{\partial x_i}\varphi_i(x)\frac{\partial}{\partial x_j}\varphi_j(x)dx$$
$$= \int_{\Omega} \left[\sum_{i=1}^{d} \frac{\partial}{\partial x_i}\varphi_i(x)\right]^2 dx \geq 0.$$

Hence we have

$$\int_{\Omega} \sum_{i,j=1}^{d} [e_{ij}(\underline{\varphi}, x)]^2 dx \geq \frac{1}{2} \sum_{i,j=1}^{d} \int_{\Omega} \left[\frac{\partial}{\partial x_j}\varphi_i(x)\right]^2 dx = \frac{1}{2}\,|\underline{\varphi}|^2_{[H^1(\Omega)]^d}.$$

Considering the closure of $C_0^\infty(\Omega)$ with respect to the norm $\|\cdot\|_{H^1(\Omega)}$ we conclude the assertion for $\underline{v} \in [H_0^1(\Omega)]^d$. □

Using suitable equivalent norms in $[H^1(\Omega)]^d$ we now conclude the $[H_0^1(\Omega)]^d$–ellipticity of the bilinear form $a(\cdot,\cdot)$.

**Corollary 4.16.** *For* $\underline{v} \in [H_0^1(\Omega)]^d$ *we have*

$$a(\underline{v}, \underline{v}) \geq c\,\|\underline{v}\|^2_{[H^1(\Omega)]^d}. \tag{4.39}$$

*Proof.* For $\underline{v} \in [H_0^1(\Omega)]^d$ we can define by

$$\|\underline{v}\|_{[H^1(\Omega)]^d,\Gamma} := \left\{ \sum_{i=1}^{d} \left[ \int_{\Gamma} v_i(x)\,ds_x \right]^2 + |\underline{v}|^2_{[H^1(\Omega)]^d} \right\}^{1/2} \tag{4.40}$$

an equivalent norm in $[H^1(\Omega)]^d$ (cf. Theorem 2.6). The assertion then follows from Lemma 4.14 and by using Korn's first inequality (4.38). □

The ellipticity estimate (4.39) remains valid for vector functions $\underline{v}$, where only some components $v_i(x)$ are zero for $x \in \Gamma_{D,i} \subset \Gamma$. Let

$$[H_0^1(\Omega, \Gamma_D)]^d = \left\{ \underline{v} \in [H^1(\Omega)]^d \;:\; \gamma_0^{\text{int}} v_i(x) = 0 \text{ for } x \in \Gamma_{D,i}, i = 1, \ldots, d \right\}.$$

Then we have

$$a(\underline{v}, \underline{v}) \geq c\,\|\underline{v}\|^2_{[H^1(\Omega)]^d} \quad \text{for all } \underline{v} \in [H_0^1(\Omega, \Gamma_D)]^d. \tag{4.41}$$

As for the scalar Laplace operator we can extend the bilinear form $a(\cdot,\cdot)$ of the system of linear elasticity by some $L_2$ norm to obtain an equivalent norm in $[H^1(\Omega)]^d$. This is a direct consequence of Korn's second inequality, see [53].

**Theorem 4.17 (Korn's Second Inequality).** *Let $\Omega \subset \mathbb{R}^d$ be a bounded domain with piecewise smooth boundary $\Gamma = \partial\Omega$. Then we have*

$$\int\limits_\Omega \sum_{i,j=1}^d [e_{ij}(\underline{v},x)]^2 dx + \|\underline{v}\|^2_{[L_2(\Omega)]^d} \geq c\,\|\underline{v}\|^2_{[H^1(\Omega)]^d} \quad \textit{for all } \underline{v} \in [H^1(\Omega)]^d.$$

Using Theorem 2.6 we can introduce further equivalent norms in $[H^1(\Omega)]^d$.

**Corollary 4.18.** *Let $\mathcal{R} = span\{\underline{v}_k\}_{k=1}^{dim\,(\mathcal{R})}$ be the space of all rigid body motions as given in* (1.29). *Then we can define*

$$\|\underline{v}\|_{[H^1(\Omega)]^d,\Gamma} := \left\{ \sum_{k=1}^{dim\,(\mathcal{R})} \left[ \int\limits_\Omega \underline{v}_k(x)^\top \underline{v}(x) dx \right]^2 + \int\limits_\Omega \sum_{i,j=1}^d [e_{ij}(\underline{v},x)]^2 dx \right\}^{1/2}$$

*as an equivalent norm in $[H^1(\Omega)]^d$.*

### 4.2.1 Dirichlet Boundary Value Problem

The Dirichlet boundary value problem of linear elasticity reads

$$-\mu\Delta\underline{u}(x) - (\lambda+\mu)\text{grad div}\,\underline{u}(x) = \underline{f}(x) \quad \text{for } x \in \Omega, \ \gamma_0^{\text{int}}\underline{u}(x) = \underline{g}(x) \text{ for } x \in \Gamma.$$

As for the scalar potential equation we define the solution manifold

$$V_g := \left\{ \underline{v} \in [H^1(\Omega)]^d : \gamma_0^{\text{int}} v_i(x) = g_i(x) \quad \text{for } x \in \Gamma,\ i = 1,\ldots,d \right\}$$

where $V_0 = [H_0^1(\Omega)]^d$. Then we have to find $\underline{u} \in V_g$ such that

$$a(\underline{u},\underline{v}) = \langle \underline{f}, \underline{v} \rangle_\Omega \tag{4.42}$$

is satisfied for all $\underline{v} \in V_0$. Since the bilinear form $a(\cdot,\cdot)$ is bounded (cf. (4.36)) and $[H_0^1(\Omega)]^d$–elliptic (cf. (4.39)) we conclude the unique solvability of the variational problem (4.42) by applying Theorem 3.8. Moreover, the unique solution of (4.42) satisfies

$$\|\underline{u}\|_{[H^1(\Omega)]^d} \leq c_1 \|\underline{f}\|_{[H^{-1}(\Omega)]^d} + c_2 \|\underline{g}\|_{[H^{1/2}(\Gamma)]^d}.$$

For the solution $\underline{u} \in [H^1(\Omega)]^d$ of the Dirichlet boundary value problem we now compute the associated boundary stress $\gamma_1^{\text{int}}\underline{u} \in [H^{-1/2}(\Gamma)]^d$ as the solution of the variational problem

$$\langle \gamma_1^{\text{int}}\underline{u}, \underline{w} \rangle_\Gamma = a(\underline{u}, \mathcal{E}\underline{w}) - \langle \underline{f}, \mathcal{E}\underline{w} \rangle_\Omega$$

for all $\underline{w} \in [H^{1/2}(\Gamma)]^d$. Here, $\mathcal{E} : H^{1/2}(\Gamma) \to H^1(\Omega)$ is the extension operator which is applied to the components $w_i \in H^{1/2}(\Omega)$. Note that

$$\|\gamma_1^{\text{int}}\underline{u}\|_{[H^{-1/2}(\Gamma)]^d} \leq c_{IT} \left\{ \|\underline{f}\|_{[\widetilde{H}^{-1}(\Omega)]^d} + \frac{E}{1-2\nu} |\underline{u}|_{[H^1(\Omega)]^d} \right\}.$$

**Lemma 4.19.** *Let* $\underline{u} \in [H^1(\Omega)]^d$ *be the weak solution of the Dirichlet boundary value problem*

$$-\mu\Delta\underline{u}(x) - (\lambda+\mu)\,grad\,div\,\underline{u}(x) = \underline{0} \quad for\; x \in \Omega, \quad \underline{u}(x) = \underline{g}(x) \quad for\; x \in \Gamma.$$

*Then we have*

$$a(\underline{u},\underline{u}) \;\geq\; c\,\|\gamma_1^{int}\underline{u}\|^2_{[H^{-1/2}(\Gamma)]^d}. \tag{4.43}$$

*Proof.* The associated conormal derivative $\gamma_1^{\text{int}}\underline{u} \in [H^{-1/2}(\Gamma)]^d$ is defined as the unique solution of the variational problem

$$\langle \gamma_1^{\text{int}}\underline{u}, \underline{w}\rangle_\Gamma \;=\; a(\underline{u}, \mathcal{E}\underline{w}) \quad \text{for all } \underline{w} \in [H^{1/2}(\Gamma)]^d.$$

As in the proof of Lemma 4.13 we have

$$|a(\underline{u},\mathcal{E}\underline{w})| \leq \frac{2E}{1-2\nu}\left(\int_\Omega \|\underline{e}(\underline{u},x)\|_2^2 dx\right)^{1/2}\left(\int_\Omega \|\underline{e}(\mathcal{E}\underline{w},x)\|_2^2 dx\right)^{1/2}$$

and

$$\int_\Omega \|\underline{e}(\mathcal{E}\underline{w},x)\|_2^2 dx \;\leq\; |\mathcal{E}\underline{w}|^2_{[H^1(\Omega)]^d}.$$

Moreover,

$$\int_\Omega \|\underline{e}(\underline{u},x)\|_2^2 dx \;\leq\; \int_\Omega \sum_{i,j=1}^d [e_{ij}(\underline{u},x)]^2 dx \;\leq\; \frac{1+\nu}{E}\,a(\underline{u},\underline{u}).$$

Applying the inverse trace theorem (Theorem 2.22) this gives

$$\begin{aligned}\|\gamma_1^{\text{int}}\underline{u}\|_{[H^{-1/2}(\Gamma)]^d} &= \sup_{0\neq\underline{w}\in[H^{1/2}(\Gamma)]^d} \frac{\langle\gamma_1^{\text{int}}\underline{u},\underline{w}\rangle_\Gamma}{\|\underline{w}\|_{[H^{1/2}(\Gamma)]^d}} \\ &= \sup_{0\neq\underline{w}\in[H^{1/2}(\Gamma)]^d} \frac{a(\underline{u},\mathcal{E}\underline{w})}{\|\underline{w}\|_{[H^{1/2}(\Gamma)]^d}} \;\leq\; c\sqrt{a(\underline{u},\underline{u})}.\end{aligned}$$

Note that the constant $c$ in the estimate (4.43) tends to zero when $\nu \to \frac{1}{2}$. □

### 4.2.2 Neumann Boundary Value Problem

For the solvability of the Neumann boundary value problem

$$-\mu\Delta\underline{u}(x) - (\lambda+\mu)\,\text{grad div}\,\underline{u}(x) = \underline{f}(x) \quad \text{for } x\in\Omega,\; \gamma_1^{\text{int}}\underline{u}(x) = \underline{g}(x) \text{ for } x\in\Gamma$$

we have to assume the solvability conditions (1.31),

$$\int_\Omega \underline{v}_k(x)^\top \underline{f}(x)dx + \int_\Gamma \gamma_0^{\rm int}\underline{v}_k(x)^\top \underline{g}(x)ds_x = 0 \quad \text{for all } \underline{v}_k \in \mathcal{R}$$

where $\underline{v}_k$ are the rigid body motions (cf. (1.29)). On the other hand, the solution of the Neumann boundary value problem is only unique up to the rigid body motions. To fix the rigid body motions, we formulate appropriate scaling conditions. For this we define

$$[H^1_*(\Omega)]^d = \left\{ \underline{v} \in [H^1(\Omega)]^d : \int_\Omega \underline{v}_k(x)^\top \underline{v}(x)dx = 0 \quad \text{for all } \underline{v}_k \in \mathcal{R} \right\}.$$

The weak formulation of the Neumann boundary value problem is to find $\underline{u} \in [H^1_*(\Omega)]^d$ such that

$$a(\underline{u}, \underline{v}) = \langle \underline{f}, \underline{v} \rangle_\Omega + \langle \underline{g}, \gamma_0^{\rm int}\underline{v} \rangle_\Gamma \tag{4.44}$$

is satisfied for all $\underline{v} \in [H^1_*(\Omega)]^d$. Using Corollary 4.18 we can establish the $[H^1_*(\Omega)]^d$–ellipticity of the bilinear form $a(\cdot,\cdot)$ and therefore we can conclude the unique solvability of the variational problem (4.44) in $[H^1_*(\Omega)]^d$.

By introducing Lagrange multipliers we can formulate the scaling conditions conditions of $[H^1_*(\Omega)]^d$ as side conditions in a saddle point problem. Then we have to find $\underline{u} \in [H^1(\Omega)]^d$ and $\underline{\lambda} \in \mathbb{R}^{\dim(\mathcal{R})}$ such that

$$\begin{aligned} a(\underline{u}, \underline{v}) + \sum_{k=1}^{\dim \mathcal{R}} \lambda_k \int_\Omega \underline{v}_k(x)^\top \underline{v}(x)dx &= \langle \underline{f}, \underline{v} \rangle_\Omega + \langle \underline{g}, \gamma_0^{\rm int}\underline{v} \rangle_\Gamma \\ \int_\Omega \underline{v}_\ell(x)^\top \underline{u}(x)dx &= 0 \end{aligned} \tag{4.45}$$

is satisfied for all $\underline{v} \in [H^1(\Omega)]^d$ and $\ell = 1, \ldots, \dim(\mathcal{R})$. When choosing as test functions $\underline{v}_\ell \in \mathcal{R}$ we then find from the solvability conditions (1.31)

$$\sum_{k=1}^{\dim \mathcal{R}} \lambda_k \int_\Omega \underline{v}_k(x)^\top \underline{v}_\ell(x)dx = 0 \quad \text{for } \ell = 1, \ldots, \dim(\mathcal{R}).$$

Since the rigid body motions are linear independent, we obtain $\underline{\lambda} = \underline{0}$. Inserting this result into the second equation in (4.45), and eliminating the Lagrange multiplier $\underline{\lambda}$ we finally obtain a modified variational problem to find $\underline{u} \in [H^1(\Omega)]^d$ such that

$$a(\underline{u}, \underline{v}) + \sum_{k=1}^{\dim \mathcal{R}} \int_\Omega \underline{v}_k(x)^\top \underline{u}(x)dx \int_\Omega \underline{v}_k(x)^\top \underline{v}(x)dx = \langle \underline{f}, \underline{v} \rangle_\Omega + \langle \underline{g}, \gamma_0^{\rm int}\underline{v} \rangle_\Gamma \tag{4.46}$$

is satisfied for all $\underline{v} \in [H^1(\Omega)]^d$.

The extended bilinear form of the modified variational formulation (4.46) is $[H^1(\Omega)]^d$–elliptic (cf. Corollary 4.18). Hence there exists a unique solution $\underline{u} \in [H^1(\Omega)]^d$ of the variational problem (4.46) for any given $\underline{f} \in [\widetilde{H}^{-1}(\Omega)]^d$ and $\underline{g} \in [H^{-1/2}(\Gamma)]^d$. If the solvability conditions (1.31) are satisfied, then we have $\underline{u} \in [H^1_*(\Omega)]^d$, i.e. the variational problems (4.46) and (4.44) are equivalent.

If $\underline{u} \in [H^1_*(\Omega)]^d$ is a weak solution of the Neumann boundary value problem, then we can define

$$\widetilde{\underline{u}} := \underline{u} + \sum_{k=1}^{\dim \mathcal{R}} \alpha_k \underline{v}_k \in [H^1(\Omega)]^d$$

which is also a solution of the Neumann boundary value problem.

### 4.2.3 Mixed Boundary Value Problems

We now consider a boundary value problem with boundary conditions of mixed type,

$$\begin{aligned} -\mu\Delta\underline{u}(x) - (\lambda+\mu)\operatorname{grad}\operatorname{div}\underline{u}(x) &= \underline{f}(x) && \text{for } x \in \Omega, \\ \gamma_0^{\text{int}} u_i(x) &= g_{D,i}(x) && \text{for } x \in \Gamma_{D,i}, \\ (\gamma_1^{\text{int}}\underline{u})_i(x) &= g_{N,i}(x) && \text{for } x \in \Gamma_{N,i} \end{aligned}$$

where $\Gamma = \overline{\Gamma}_{D,i} \cup \overline{\Gamma}_{N,i}$ and $\operatorname{meas}(\Gamma_{D,i}) > 0$ for $i = 1, \ldots, d$. The associated variational formulation is to find $\underline{u} \in [H^1(\Omega)]^d$ with $\gamma_0^{\text{int}} u_i(x) = g_{D,i}(x)$ for $x \in \Gamma_{D,i}$ such that

$$a(\underline{u}, \underline{v}) \;=\; \langle \underline{f}, \underline{v}\rangle_\Omega + \sum_{i=1}^{d} \langle g_{N,i}, \gamma_0^{\text{int}} v_i\rangle_{\Gamma_{N,i}} \tag{4.47}$$

is satisfied for all $\underline{v} \in [H^1_0(\Omega, \Gamma_D)]^d$.

Using (4.41) we conclude the $[H^1_0(\Omega, \Gamma_D)]^d$–ellipticity of the bilinear form $a(\cdot,\cdot)$ and therefore the unique solvability of the variational problem (4.47).

## 4.3 Stokes Problem

Next we consider the Dirichlet boundary value problem for the Stokes system (1.38),

$$-\mu\Delta\underline{u}(x) + \nabla p(x) = \underline{f}(x), \; \operatorname{div}\underline{u}(x) = 0 \text{ for } x \in \Omega, \; \gamma_0^{\text{int}}\underline{u}(x) = \underline{g}(x) \text{ for } x \in \Gamma.$$

Due to (1.39) we have to assume the solvability condition

$$\int_\Gamma [\underline{n}(x)]^\top \underline{g}(x) ds_x = 0. \tag{4.48}$$

Note that the pressure $p$ is only unique up to an additive constant. However, as for the Neumann boundary value problem for the potential equation we can introduce an appropriate scaling condition to fix this constant. For this we define

$$L_{2,0}(\Omega) = \left\{ q \in L_2(\Omega) : \int_\Omega q(x)dx = 0 \right\}.$$

To derive a variational formulation for the solution $\underline{u}$ of the Dirichlet boundary value problem of the Stokes system, we consider Green's first formula (1.41),

$$a(\underline{u}, \underline{v}) = \int_\Omega p(x) \operatorname{div} \underline{v}(x) dx + \langle \underline{f}, \underline{v} \rangle_\Omega + \langle \underline{t}(\underline{u}, p), \gamma_0^{\text{int}} \underline{v} \rangle_\Gamma .$$

Due to $\operatorname{div} \underline{u}(x) = 0$ for $x \in \Omega$, the bilinear form (1.42) is given as

$$a(\underline{u}, \underline{v}) := 2\mu \int_\Omega \sum_{i,j=1}^d e_{ij}(\underline{u}, x) e_{ij}(\underline{v}, x) dx.$$

Hence we have to find $\underline{u} \in [H^1(\Omega)]^d$ satisfying $\underline{u}(x) = \underline{g}(x)$ for $x \in \Gamma$ and $p \in L_{2,0}(\Omega)$ such that

$$\begin{aligned} a(\underline{u}, \underline{v}) - \int_\Omega p(x) \operatorname{div} \underline{v}(x) dx &= \langle \underline{f}, \underline{v} \rangle_\Omega, \\ \int_\Omega q(x) \operatorname{div} \underline{u}(x)\, dx &= 0 \end{aligned} \tag{4.49}$$

is satisfied for all $\underline{v} \in [H_0^1(\Omega)]^d$ and $q \in L_{2,0}(\Omega)$.

Let $u_g \in [H^1(\Omega)]^d$ be any arbitrary but fixed extension of the given Dirichlet datum $\underline{g} \in [H^{1/2}(\Gamma)]^d$. It remains to find $\underline{u}_0 \in [H_0^1(\Omega)]^d$ and $p \in L_{2,0}(\Omega)$ such that

$$\begin{aligned} a(\underline{u}_0, \underline{v}) - \int_\Omega p(x) \operatorname{div} \underline{v}(x) dx &= \langle \underline{f}, \underline{v} \rangle_\Omega - a(\underline{u}_g, \underline{v}), \\ \int_\Omega q(x) \operatorname{div} \underline{u}_0(x)\, dx &= -\langle \operatorname{div} \underline{u}_g, q \rangle_\Omega \end{aligned} \tag{4.50}$$

is satisfied for all $\underline{v} \in [H_0^1(\Omega)]^d$ and $q \in L_{2,0}(\Omega)$.

To investigate the unique solvability of the saddle point problem (4.50) we have to check the assumptions of Theorem 3.11. The bilinear form $a(\cdot,\cdot) : [H_0^1(\Omega)]^d \times [H_0^1(\Omega)]^d \to \mathbb{R}$ induces an operator $A : [H_0^1(\Omega)]^d \to [H^{-1}(\Omega)]^d$.

The bilinear form

$$b(\underline{v}, q) := \int_\Omega q(x) \operatorname{div} \underline{v}(x)\, dx \quad \text{for } \underline{v} \in [H_0^1(\Omega)]^d, q \in L_2(\Omega)$$

induces an operator $B : [H_0^1(\Omega)]^d \to L_2(\Omega)$. Note that

$$|b(\underline{v}, q)| = \left| \int_\Omega q(x) \operatorname{div} \underline{v}(x)\, dx \right| \le \|q\|_{L_2(\Omega)} \|\operatorname{div} \underline{v}\|_{L_2(\Omega)} \le \|q\|_{L_2(\Omega)} \|\underline{v}\|_{[H^1(\Omega)]^d}.$$

We further have

$$\ker B := \left\{ \underline{v} \in [H_0^1(\Omega)]^d : \operatorname{div} \underline{v} = 0 \right\} \subset [H_0^1(\Omega)]^d.$$

Applying Korn's first inequality (4.38) this gives

$$a(\underline{v}, \underline{v}) = 2\mu \int_\Omega \sum_{i,j=1}^d [e_{ij}(\underline{v}, x)]^2 dx \ge \mu\, |\underline{v}|^2_{[H^1(\Omega)]^d}$$

for all $\underline{v} \in [H_0^1(\Omega)]^d$. Using the equivalent norm (4.40) we then find the $[H_0^1(\Omega)]^d$–ellipticity of the bilinear form $a(\cdot, \cdot)$,

$$a(\underline{v}, \underline{v}) \ge c\, \|\underline{v}\|^2_{[H^1(\Omega)]^d} \quad \text{for all } \underline{v} \in [H_0^1(\Omega)]^d.$$

Due to $\ker B \subset [H_0^1(\Omega)]^d$ we also have the $\ker B$–ellipticity of the bilinear form $a(\cdot, \cdot)$. It remains to prove the stability condition (3.25),

$$c_S\, \|q\|_{L_2(\Omega)} \le \sup_{0 \ne \underline{v} \in [H_0^1(\Omega)]^d} \frac{\int_\Omega q(x) \operatorname{div} \underline{v}(x) dx}{\|\underline{v}\|_{[H^1(\Omega)]^d}} \quad \text{for all } q \in L_{2,0}(\Omega). \tag{4.51}$$

This is a direct consequence of Theorem 2.17:

**Lemma 4.20.** *Let $\Omega \subset \mathbb{R}^d$ be a bounded and connected Lipschitz domain. Then there holds the stability condition* (4.51) .

*Proof.* For $q \in L_{2,0}(\Omega)$ we have $\nabla q \in [H^{-1}(\Omega)]^d$ satisfying, by using Theorem 2.17,

$$\|q\|_{L_2(\Omega)} \le c\, \|\nabla q\|_{[H^{-1}(\Omega)]^d}.$$

Recalling the norm definition in $[H^{-1}(\Omega)]^d$ by duality, this gives

$$\frac{1}{c}\, \|q\|_{L_2(\Omega)} \le \|\nabla q\|_{[H^{-1}(\Omega)]^d} = \sup_{0 \ne \underline{w} \in [H_0^1(\Omega)]^d} \frac{\langle \underline{w}, \nabla q \rangle_\Omega}{\|\underline{w}\|_{[H^1(\Omega)]^d}} = \sup_{0 \ne \underline{w} \in [H_0^1(\Omega)]^d} \frac{-\int_\Omega q(x) \operatorname{div} \underline{w}(x) dx}{\|\underline{w}\|_{[H^1(\Omega)]^d}}.$$

Hence, choosing $\underline{v} := -\underline{w}$ we finally obtain the stability condition (4.51). □

Therefore, all assumptions of Theorem 3.11 are satisfied, and hence, the saddle point problem (4.50) is unique solvable.

The scaling condition to fix the pressure $p \in L_{2,0}(\Omega)$ can now be reformulated as for the Neumann boundary value problem for the potential equation. By introducing a scalar Lagrange multiplier $\lambda \in \mathbb{R}$ we may consider the following extended saddle point problem to find $\underline{u} \in [H^1(\Omega)]^d$ with $\underline{u}(x) = \underline{g}(x)$ for $x \in \Gamma$ as well as $p \in L_2(\Omega)$ and $\lambda \in \mathbb{R}$ such that

$$\begin{aligned} a(\underline{u}, \underline{v}) \quad - \int_\Omega p(x) \operatorname{div} \underline{v}(x) dx &= \langle \underline{f}, \underline{v} \rangle_\Omega, \\ \int_\Omega q(x) \operatorname{div} \underline{u}(x)\, dx + \quad \lambda \int_\Omega q(x) dx &= 0, \\ \int_\Omega p(x) dx &= 0 \end{aligned} \tag{4.52}$$

is satisfied for all $\underline{v} \in [H_0^1(\Omega)]^d$ and $q \in L_2(\Omega)$. Choosing as test function $q \equiv 1$ this gives

$$\lambda\, |\Omega| = - \int_\Omega \operatorname{div} \underline{u}(x)\, dx = - \int_\Gamma \underline{n}(x)^\top \underline{g}(x) ds_x = 0$$

and using the solvability condition (4.48) we get $\lambda = 0$. Hence we can write the third equation in (4.52) as

$$\int_\Omega p(x) dx - \lambda = 0.$$

Eliminating the Lagrange multiplier $\lambda$ we finally obtain a modified saddle point problem which is equivalent to the variational formulation (4.50) to find $\underline{u}_0 \in [H_0^1(\Omega)]^d$ and $p \in L_2(\Omega)$ such that

$$\begin{aligned} a(\underline{u}_0, \underline{v}) \quad - \int_\Omega p(x) \operatorname{div} \underline{v}(x) dx &= \langle \widetilde{\underline{f}}, \underline{v} \rangle_\Omega \\ \int_\Omega q(x) \operatorname{div} \underline{u}_0(x)\, dx + \int_\Omega p(x) dx \int_\Omega q(x) dx &= -\langle \operatorname{div} \underline{u}_g, q \rangle_\Omega \end{aligned} \tag{4.53}$$

is satisfied for all $\underline{v} \in [H_0^1(\Omega)]^d$ and $q \in L_2(\Omega)$ where

$$\langle \widetilde{\underline{f}}, \underline{v} \rangle_\Omega := \langle \underline{f}, \underline{v} \rangle_\Omega - a(\underline{u}_g, \underline{v}) \quad \text{for all } \underline{v} \in [H_0^1(\Omega)]^d$$

induces $\widetilde{\underline{f}} \in [H^{-1}(\Omega)]^d$.

By

$$\begin{aligned}\langle A\underline{u},\underline{v}\rangle_\Omega &:= a(\underline{u},\underline{v}),\\ \langle B\underline{u},q\rangle_{L_2(\Omega)} &:= b(\underline{u},q),\\ \langle \underline{v},B'p\rangle_\Omega &:= \langle B\underline{v},p\rangle_\Omega \,=\, b(\underline{v},p),\\ \langle Dp,q\rangle_{L_2(\Omega)} &:= \int\limits_\Omega p(x)dx \int\limits_\Omega q(x)dx\end{aligned}$$

for all $\underline{u},\underline{v} \in [H_0^1(\Omega)]^d$ and $p,q \in L_2(\Omega)$ we can define bounded operators

$$\begin{aligned}&A : [H_0^1(\Omega)]^d \to [H^{-1}(\Omega)]^d,\\ &B : [H_0^1(\Omega)]^d \to L_2(\Omega),\\ &B' : L_2(\Omega) \to [H^{-1}(\Omega)]^d,\\ &D : L_2(\Omega) \to L_2(\Omega).\end{aligned}$$

Hence we can write the variational problem (4.53) as an operator equation,

$$\begin{pmatrix} A & -B' \\ B & D \end{pmatrix} \begin{pmatrix} \underline{u}_0 \\ p \end{pmatrix} = \begin{pmatrix} \underline{\widetilde{f}} \\ -B\underline{u}_g \end{pmatrix}.$$

Since $A$ is $[H_0^1(\Omega)]^d$–elliptic we find

$$\underline{u}_0 \,=\, A^{-1}\left[\underline{\widetilde{f}} + B'p\right]$$

and inserting this into the second equation we obtain the Schur complement system

$$\left[BA^{-1}B' + D\right]p = -B\underline{u}_g - A^{-1}\underline{\widetilde{f}}. \tag{4.54}$$

**Lemma 4.21.** *The operator $S := BA^{-1}B' + D : L_2(\Omega) \to L_2(\Omega)$ is bounded and $L_2(\Omega)$–elliptic.*

*Proof.* For $p \in L_2(\Omega)$ we have $\underline{u}_p = A^{-1}B'p \in [H_0^1(\Omega)]^d$ which is defined as the unique solution of the variational problem

$$a(\underline{u}_p,\underline{v}) \,=\, b(\underline{v},p) \quad \text{for all } \underline{v} \in [H_0^1(\Omega)]^d.$$

Since $A$ is $[H_0^1(\Omega)]^d$–elliptic, and since $B$ is bounded, we obtain

$$c_1^A \,\|\underline{u}_p\|^2_{[H^1(\Omega)]^d} \,\leq\, a(\underline{u}_p,\underline{u}_p) \,=\, b(\underline{u}_p,p) \,\leq\, c_2^B \,\|\underline{u}_p\|_{[H^1(\Omega)]^d}\|p\|_{L_2(\Omega)}$$

and therefore

$$\|\underline{u}_p\|_{[H^1(\Omega)]^d} \,\leq\, \frac{c_2^B}{c_1^A}\,\|p\|_{L_2(\Omega)}.$$

Applying the Cauchy–Schwarz inequality this gives

$$\begin{aligned}\langle Sp, q\rangle_{L_2(\Omega)} &= \langle [BA^{-1}B + D]p, q\rangle_{L_2(\Omega)} \\ &= \int_\Omega q(x)\operatorname{div}\underline{u}_p(x)dx + \int_\Omega p(x)dx \int_\Omega q(x)dx \\ &\leq c_2^B\, \|q\|_{L_2(\Omega)}\|\underline{u}_p\|_{[H^1(\Omega)]^d} + |\Omega|\, \|p\|_{L_2(\Omega)}\|q\|_{L_2(\Omega)} \\ &\leq c\,\|p\|_{L_2(\Omega)}\|q\|_{L_2(\Omega)}\end{aligned}$$

and hence we conclude the boundedness of $S : L_2(\Omega) \to L_2(\Omega)$.

For an arbitrary given $p \in L_2(\Omega)$ we consider the decomposition

$$p = p_0 + \frac{1}{|\Omega|}\int_\Omega p(x)dx$$

where $p_0 \in L_{2,0}(\Omega)$ and

$$\|p\|^2_{L_2(\Omega)} = \|p_0\|^2_{L_2(\Omega)} + \frac{1}{|\Omega|^2}\left[\int_\Omega p(x)\,dx\right]^2 .$$

For $p_0 \in L_{2,0}(\Omega)$ we can use the stability condition (4.51), the definition of $\underline{u}_{p_0} = A^{-1}B'p_0 \in [H^1_0(\Omega)]^d$ and the boundedness of $A$ to obtain

$$\begin{aligned}c_S\,\|p_0\|_{L_2(\Omega)} &\leq \sup_{0\neq \underline{v}\in[H^1_0(\Omega)]^d} \frac{b(\underline{v}, p_0)}{\|\underline{v}\|_{[H^1(\Omega)]^d}} \\ &= \sup_{0\neq \underline{v}\in[H^1_0(\Omega)]^d} \frac{a(\underline{u}_{p_0}, \underline{v})}{\|\underline{v}\|_{[H^1(\Omega)]^d}} \leq c_2^A\,\|\underline{u}_{p_0}\|_{[H^1(\Omega)]^d}\end{aligned}$$

and therefore

$$\|p_0\|^2_{L_2(\Omega)} \leq \left[\frac{c_2^A}{c_S}\right]^2 \|\underline{u}_{p_0}\|^2_{[H^1(\Omega)]^d} \leq \frac{1}{c_1^A}\left[\frac{c_2^A}{c_S}\right]^2 a(\underline{u}_{p_0}, \underline{u}_{p_0}) = c\,b(\underline{u}_{p_0}, p_0).$$

Inserting the definition of $\underline{u}_{p_0} = A^{-1}B'p_0 \in [H^1_0(\Omega)]^d$ we get

$$\langle BA^{-1}B'p_0, p_0\rangle_{L_2(\Omega)} = \langle B\underline{u}_{p_0}, p_0\rangle_{L_2(\Omega)} \geq \frac{1}{c}\,\|p_0\|^2_{L_2(\Omega)}.$$

For $\underline{v} \in [H^1_0(\Omega)]^d$ we have

$$\langle B'p, \underline{v}\rangle_\Omega = \int_\Omega p(x)\operatorname{div}\underline{v}(x)dx = -\langle \underline{v}, \nabla p\rangle_\Omega = -\langle \underline{v}, \nabla p_0\rangle_\Omega$$

and hence

$$\langle BA^{-1}B'p, p\rangle_{L_2(\Omega)} = \langle BA^{-1}B'p_0, p_0\rangle_{L_2(\Omega)} \geq \frac{1}{c}\,\|p_0\|^2_{L_2(\Omega)}.$$

From this we obtain

$$\begin{aligned}\langle Sp, p\rangle_{L_2(\Omega)} &= \langle BA^{-1}B'p, p\rangle_{L_2(\Omega)} + \langle Dp, p\rangle_{L_2(\Omega)} \\ &= \langle BA^{-1}B'p_0, p_0\rangle_{L_2(\Omega)} + \left[\int_\Omega p(x)\,dx\right]^2 \\ &\geq \frac{1}{c}\,\|p_0\|^2_{L_2(\Omega)} + \left[\int_\Omega p(x)\,dx\right]^2 \geq \min\left\{\frac{1}{c}, |\Omega|^2\right\}\,\|p\|^2_{L_2(\Omega)},\end{aligned}$$

i.e., $S$ is $L_2(\Omega)$–elliptic. □

Applying Theorem 3.4 (Lax–Milgram lemma) we finally obtain the unique solvability of the operator equation (4.54).

## 4.4 Helmholtz Equation

Finally we consider the interior Dirichlet boundary value problem for the Helmholtz equation,

$$-\Delta u(x) - k^2 u(x) = 0 \quad \text{for } x \in \Omega, \quad \gamma_0^{\text{int}} u(x) = g(x) \quad \text{for } x \in \Gamma. \tag{4.55}$$

The related variational formulation is to find $u \in H^1(\Omega)$ with $\gamma_0^{\text{int}} u(x) = g(x)$ for $x \in \Gamma$ such that

$$\int_\Omega \nabla u(x)\nabla v(x)dx - k^2 \int_\Omega u(x)v(x)dx = 0 \tag{4.56}$$

is satisfied for all $v \in H^1_0(\Omega)$. The bilinear form

$$a(u, v) = \int_\Omega \nabla u(x)\nabla v(x)\,dx - k^2 \int_\Omega u(x)v(x)dx$$

can be written as

$$a(u, v) = a_0(u, v) - c(u, v) \tag{4.57}$$

where the symmetric and bounded bilinear form

$$a_0(u, v) = \int_\Omega \nabla u(x)\nabla v(x)\,dx \quad \text{for } u, v \in H^1(\Omega)$$

is $H^1_0(\Omega)$–elliptic, and the bilinear form

$$c(u,v) \,=\, k^2 \int\limits_\Omega u(x)v(x)dx \,=\, \langle Cu, v\rangle_\Omega$$

induces a compact operator $C : H^1(\Omega) \to H^{-1}(\Omega)$. Hence, the bilinear form $a(\cdot,\cdot) : H^1_0(\Omega) \times H^1_0(\Omega) \to \mathbb{R}$ induces a coercive and bounded operator $A : H^1_0(\Omega) \to H^{-1}(\Omega)$. Therefore it remains to investigate the injectivity of $A$, i.e. we have to consider the solvability of the homogeneous Dirichlet boundary value problem

$$-\Delta u(x) - k^2 u(x) \,=\, 0 \quad \text{for } x \in \Omega, \quad \gamma_0^{\text{int}} u(x) \,=\, 0 \quad \text{for } x \in \Gamma. \qquad (4.58)$$

**Proposition 4.22.** *If $k^2 = \lambda$ is an eigenvalue of the interior Dirichlet eigenvalue problem of the Laplace equation,*

$$-\Delta u(x) = \lambda\, u(x) \quad \textit{for } x \in \Omega, \quad \gamma_0^{int}(x) = 0 \quad \textit{for } x \in \Gamma,$$

*then there exist non–trivial solutions of the homogeneous Dirichlet boundary value problem* (4.58).
*If $k^2$ is not an eigenvalue of the Dirichlet problem of the Laplace operator, then the operator $A : H^1_0(\Omega) \to H^{-1}(\Omega)$ which is induced by the bilinear form* (4.57) *is injective.*

Hence we conclude, that if $k^2$ is not an eigenvalue of the Dirichlet eigenvalue problem of the Laplace operator the operator $A : H^1_0(\Omega) \to H^{-1}(\Omega)$ which is induced by the bilinear form (4.57) is coercive and injective, and therefore the variational problem (4.56) admits a unique solution.

## 4.5 Exercises

**4.1** Derive the variational formulation of the following boundary value problem with nonlinear Robin boundary conditions

$$-\Delta u(x) + u(x) = f(x) \quad \text{for } x \in \Omega, \quad \gamma_1^{\text{int}} u(x) + [\gamma_0^{\text{int}} u(x)]^3 \,=\, g(x) \quad \text{for } x \in \Gamma$$

and discuss the unique solvability of the variational problem.

**4.2** Consider the bilinear form

$$a(u,v) = \int\limits_\Omega \sum_{i,j=1}^d a_{ji}(x)\frac{\partial}{\partial x_i}u(x)\frac{\partial}{\partial x_j}v(x)dx + \int\limits_\Omega \sum_{i=1}^d b_i(x)\frac{\partial}{\partial x_i}u(x)v(x)dx + \int\limits_\Omega c(x)u(x)v(x)dx$$

which is related to a uniform elliptic partial differential operator, and where we assume $c(x) \geq c_0 > 0$. Formulate a sufficient condition on the coefficients $b_i(x)$ such that the bilinear form $a(\cdot,\cdot)$ is $H^1(\Omega)$–elliptic.

**4.3** Consider the Dirichlet boundary value problem

$$-\mathrm{div}[\alpha(x)\nabla u(x)] = f(x) \quad \text{for } x \in \Omega, \quad \gamma_0^{\mathrm{int}} u(x) = 0 \quad \text{for } x \in \Gamma$$

where

$$\alpha(x) = \begin{cases} \varepsilon & \text{for } x \in \Omega_0 \subset \Omega, \\ 1 & \text{for } x \in \Omega \backslash \overline{\Omega_0}. \end{cases}$$

Prove the unique solvability of the related variational formulation. Discuss the dependence of the constants on the parameter $\varepsilon \ll 1$. Can these constants be improved when using other norms?

**4.4** Formulate a sufficient condition on the wave number $k$ such that the variational formulation of the interior Neumann boundary value problem of the Helmholtz equation,

$$-\Delta u(x) - k^2 u(x) = 0 \quad \text{for } x \in \Omega, \quad \gamma_1^{\mathrm{int}} u(x) = g(x) \quad \text{for } x \in \Gamma,$$

admits a unique solution.

# 5

# Fundamental Solutions

We now consider the scalar partial differential equation (1.10),

$$(Lu)(x) = f(x) \quad \text{for } x \in \Omega \subset \mathbb{R}^d,$$

with an elliptic partial differential operator of second order,

$$(Lu)(x) \;=\; -\sum_{i,j=1}^{d} \frac{\partial}{\partial x_j}\left[a_{ji}(x)\frac{\partial}{\partial x_i}u(x)\right].$$

The associated conormal derivative (1.7) is

$$\gamma_1^{\text{int}} u(x) \;=\; \sum_{i,j=1}^{d} n_j(x) a_{ji}(x) \frac{\partial}{\partial x_i} u(x) \quad \text{for } x \in \Gamma$$

and Green's second formula (1.8) reads for the solution $u$ of the partial differential equation (1.10) and for an arbitrary test function $v$

$$\int_\Omega (Lv)(y)u(y)dy = \int_\Gamma \gamma_1^{\text{int}} u(y)\gamma_0^{\text{int}} v(y)ds_y - \int_\Gamma \gamma_1^{\text{int}} v(y)\gamma_0^{\text{int}} u(y)ds_y + \int_\Omega f(y)v(y)dy.$$

If there exists for any $x \in \Omega$ a function $v(y) := U^*(x,y)$ satisfying

$$\int_\Omega (L_y U^*)(x,y)u(y)dy \;=\; u(x) \tag{5.1}$$

then the solution of the partial differential equation (1.10) is given by the representation formula for $x \in \Omega$

$$u(x) = \int_\Gamma U^*(x,y)\gamma_1^{\text{int}} u(y) ds_y - \int_\Gamma \gamma_{1,y}^{\text{int}} U^*(x,y)\gamma_0^{\text{int}} u(y) ds_y \qquad (5.2)$$
$$+ \int_\Omega U^*(x,y) f(y) dy.$$

Hence we can describe any solution of the partial differential equation (1.10) just by knowing the Cauchy data $[\gamma_0^{\text{int}} u(x), \gamma_1^{\text{int}} u(x)]$ for $x \in \Gamma$.

Due to

$$u(x) = \int_\Omega \delta_0(y-x) u(y) dy \quad \text{for } x \in \Omega$$

we have to solve a partial differential equation in the distributional sense to find the solution of (5.1),

$$(L_y U^*)(x,y) = \delta_0(y-x) \quad \text{for } x, y \in \mathbb{R}^d. \qquad (5.3)$$

Any solution $U^*(x,y)$ of (5.3) is called a fundamental solution.

The existence of a fundamental solution $U^*(x,y)$ is essential to derive the representation formula (5.2), and therefore to formulate appropriate boundary integral equations to find the complete Cauchy data. For general results on the existence of fundamental solutions for partial differential operators we refer to [79, 90]. In particular for partial differential operators with piecewise constant coefficients the existence of a fundamental solution is ensured. But here we will only consider the explicit computation of fundamental solutions for the Laplace operator, for the system of linear elastostatics, for the Stokes system, and for the Helmholtz operator.

## 5.1 Laplace Operator

Let us first consider the Laplace operator

$$(Lu)(x) := -\Delta u(x) \quad \text{for } x \in \mathbb{R}^d, \quad d = 2, 3.$$

The corresponding fundamental solution $U^*(x,y)$ is the distributional solution of the partial differential equation

$$-\Delta_y U^*(x,y) = \delta_0(y-x) \quad \text{for } x, y \in \mathbb{R}^d.$$

Since the Laplace operator is invariant with respect to translations and rotations, we can find the fundamental solution as $U^*(x,y) = v(z)$ with $z := y - x$. Hence we have to solve

$$-\Delta v(z) = \delta_0(z) \quad \text{for } z \in \mathbb{R}^d. \qquad (5.4)$$

Applying the Fourier transformation (2.14) this gives, when considering the derivation rule (2.17),

$$|\xi|^2 \widehat{v}(\xi) = \frac{1}{(2\pi)^{d/2}}$$

and therefore

$$\widehat{v}(\xi) = \frac{1}{(2\pi)^{d/2}} \frac{1}{|\xi|^2} \in \mathcal{S}'(\mathbb{R}^d).$$

For the Fourier transform $\widehat{v}$ of a tempered distribution $v \in \mathcal{S}'(\mathbb{R}^d)$ we have by definition

$$\langle \widehat{v}, \varphi \rangle_{L_2(\mathbb{R}^d)} = \langle v, \widehat{\varphi} \rangle_{L_2(\mathbb{R}^d)} \quad \text{for all } \varphi \in \mathcal{S}(\mathbb{R}^d).$$

Using

$$\varphi(\xi) = (2\pi)^{-d/2} \int_{\mathbb{R}^d} e^{i\langle z, \xi \rangle} \widehat{\varphi}(z) dz$$

it follows that

$$\langle \widehat{v}, \varphi \rangle_{L_2(\mathbb{R}^d)} = \frac{1}{(2\pi)^d} \int_{\mathbb{R}^d} \frac{1}{|\xi|^2} \int_{\mathbb{R}^d} e^{i\langle z, \xi \rangle} \widehat{\varphi}(z) dz d\xi.$$

Since the integral

$$\int_{\mathbb{R}^d} \frac{1}{|\xi|^2} d\xi$$

does not exist we can not exchange the order of integration. However, using

$$\Delta_z e^{i\langle z, \xi \rangle} = -|\xi|^2 e^{i\langle z, \xi \rangle}$$

we can consider a splitting of the exterior integral, apply integration by parts and exchange the order of integration, and repeat integration by parts to obtain

$$\begin{aligned}
\langle \widehat{v}, \varphi \rangle_{L_2(\mathbb{R}^d)} &= \frac{1}{(2\pi)^d} \int_{\mathbb{R}^d} \frac{1}{|\xi|^2} \int_{\mathbb{R}^d} e^{i\langle z, \xi \rangle} \widehat{\varphi}(z) dz d\xi \\
&= \frac{1}{(2\pi)^d} \int_{|\xi| \le 1} \frac{1}{|\xi|^2} \int_{\mathbb{R}^d} e^{i\langle z, \xi \rangle} \widehat{\varphi}(z) dz d\xi + \frac{1}{(2\pi)^d} \int_{|\xi|>1} \frac{1}{|\xi|^2} \int_{\mathbb{R}^d} \left[ -\Delta_z \frac{e^{i\langle z, \xi \rangle}}{|\xi|^2} \right] \widehat{\varphi}(z) dz d\xi \\
&= \frac{1}{(2\pi)^d} \int_{|\xi| \le 1} \frac{1}{|\xi|^2} \int_{\mathbb{R}^d} e^{i\langle z, \xi \rangle} \widehat{\varphi}(z) dz d\xi + \frac{1}{(2\pi)^d} \int_{|\xi|>1} \frac{1}{|\xi|^4} \int_{\mathbb{R}^d} e^{i\langle z, \xi \rangle} \left[ -\Delta_z \widehat{\varphi}(z) \right] dz d\xi \\
&= \frac{1}{(2\pi)^d} \int_{\mathbb{R}^d} \widehat{\varphi}(z) \int_{|\xi| \le 1} \frac{e^{i\langle z, \xi \rangle}}{|\xi|^2} d\xi dz + \frac{1}{(2\pi)^d} \int_{\mathbb{R}^d} \left[ -\Delta_z \widehat{\varphi}(z) \right] \int_{|\xi|>1} \frac{e^{i\langle z, \xi \rangle}}{|\xi|^4} d\xi dz \\
&= \int_{\mathbb{R}^d} \widehat{\varphi}(z) \frac{1}{(2\pi)^d} \left[ \int_{|\xi| \le 1} \frac{e^{i\langle z, \xi \rangle}}{|\xi|^2} d\xi - \Delta_z \int_{|\xi|>1} \frac{e^{i\langle z, \xi \rangle}}{|\xi|^4} d\xi \right] dz.
\end{aligned}$$

In the three–dimensional case $d = 3$ we use spherical coordinates

$$\xi_1 = r\cos\varphi\sin\theta, \quad \xi_2 = r\sin\varphi\sin\theta, \quad \xi_3 = r\cos\theta$$

for $r \in (0,\infty), \varphi \in (0,2\pi), \theta \in (0,\pi)$ to obtain, by using Lemma 2.13,

$$\begin{aligned} v(z) = v(|z|) &= \frac{1}{(2\pi)^3}\left[\int\limits_{|\xi|\le 1} \frac{e^{i\langle z,\xi\rangle}}{|\xi|^2}d\xi - \Delta_z \int\limits_{|\xi|>1} \frac{e^{i\langle z,\xi\rangle}}{|\xi|^4}d\xi\right] \\ &= \frac{1}{(2\pi)^3}\left[\int\limits_0^{2\pi}\int\limits_0^{\pi}\int\limits_0^{1} e^{i|z|r\cos\theta}\sin\theta\,dr\,d\theta\,d\varphi \right. \\ &\qquad \left. -\Delta_z \int\limits_0^{2\pi}\int\limits_0^{\pi}\int\limits_1^{\infty} \frac{e^{i|z|r\cos\theta}\sin\theta}{r^2}dr\,d\theta\,d\varphi\right] \\ &= \frac{1}{(2\pi)^2}\left[\int\limits_0^{\pi}\int\limits_0^{1} e^{i|z|r\cos\theta}\sin\theta\,dr\,d\theta - \Delta_z \int\limits_0^{\pi}\int\limits_1^{\infty} \frac{e^{i|z|r\cos\theta}\sin\theta}{r^2}dr\,d\theta\right]. \end{aligned}$$

The transformation $u := \cos\theta$ gives

$$\int\limits_0^{\pi} e^{i|z|r\cos\theta}\sin\theta d\theta = \int\limits_{-1}^{1} e^{i|z|ru}du = \frac{1}{i|z|r}\left[e^{i|z|r} - e^{-i|z|r}\right] = \frac{2}{|z|r}\sin|z|r$$

and therefore

$$v(z) = \frac{1}{2\pi^2}\left[\int\limits_0^1 \frac{\sin|z|r}{|z|r}dr - \Delta_z \int\limits_1^{\infty} \frac{\sin|z|r}{|z|r^3}dr\right].$$

Using the transformation $s := |z|r$ we obtain for the first integral

$$I_1 := \int\limits_0^1 \frac{\sin|z|r}{|z|r}dr = \frac{1}{|z|}\int\limits_0^{|z|} \frac{\sin s}{s}ds = \frac{\mathrm{Si}(|z|)}{|z|}.$$

With

$$\begin{aligned} \int \frac{\sin ax}{x^3}dx &= -\frac{1}{2}\frac{\sin ax}{x^2} + \frac{a}{2}\int \frac{\cos ax}{x^2}dx \\ &= -\frac{1}{2}\frac{\sin ax}{x^2} - \frac{a}{2}\frac{\cos ax}{x} - \frac{a^2}{2}\int \frac{\sin ax}{x}dx \end{aligned}$$

the computation of the second integral gives

$$
\begin{aligned}
I_2 := \int_1^\infty \frac{\sin|z|r}{|z|r^3}dr &= \left[-\frac{1}{2}\frac{\sin|z|r}{|z|r^2} - \frac{1}{2}\frac{\cos|z|r}{r}\right]_1^\infty - \frac{|z|}{2}\int_1^\infty \frac{\sin|z|r}{r}dr \\
&= \frac{1}{2}\frac{\sin|z|}{|z|} + \frac{1}{2}\cos|z| - \frac{|z|}{2}\left[\int_0^\infty \frac{\sin|z|r}{r}dr - \int_0^1 \frac{\sin|z|r}{r}dr\right] \\
&= \frac{1}{2}\frac{\sin|z|}{|z|} + \frac{1}{2}\cos|z| - \frac{|z|}{2}\left[\frac{\pi}{2} - \mathrm{Si}(|z|)\right].
\end{aligned}
$$

Inserting this and applying the differentiation we obtain

$$
\begin{aligned}
v(z) &= \frac{1}{2\pi^2}\left\{\frac{\mathrm{Si}(|z|)}{|z|} - \Delta_z\left[\frac{1}{2}\frac{\sin|z|}{|z|} + \frac{1}{2}\cos|z| - \frac{\pi}{4}|z| + \frac{1}{2}|z|\mathrm{Si}(|z|)\right]\right\} \\
&= \frac{1}{8\pi}\Delta_z|z| + \frac{1}{2\pi^2}\underbrace{\left\{\frac{\mathrm{Si}(|z|)}{|z|} - \Delta_z\left[\frac{1}{2}\frac{\sin|z|}{|z|} + \frac{1}{2}\cos|z| + \frac{1}{2}|z|\mathrm{Si}(|z|)\right]\right\}}_{=0} \\
&= \frac{1}{4\pi}\frac{1}{|z|}.
\end{aligned}
$$

Hence the fundamental solution of the Laplace operator in three space dimensions is

$$
U^*(x,y) = \frac{1}{4\pi}\frac{1}{|x-y|} \quad \text{for } x,y \in \mathbb{R}^3.
$$

For the two–dimensional case $d = 2$ the inverse Fourier transform of the fundamental solution has to be regularized in some appropriate way [154]. By

$$
\langle \mathcal{P}\frac{1}{|x|^2}, \varphi\rangle_{L_2(\mathbb{R}^d)} = \int\limits_{x\in\mathbb{R}^2:|x|\le 1} \frac{\varphi(x)-\varphi(0)}{|x|^2}dx + \int\limits_{x\in\mathbb{R}^2:|x|>1} \frac{\varphi(x)}{|x|^2}dx
$$

we first define the tempered distribution $\mathcal{P}\frac{1}{|x|^2} \in \mathcal{S}'(\mathbb{R}^d)$. Then,

$$
2\pi\,\langle v, \widehat{\varphi}\rangle_{L_2(\mathbb{R}^2)} = \langle \mathcal{P}\frac{1}{|\xi|^2}, \varphi\rangle_{L_2(\mathbb{R}^2)} = \int\limits_{\xi\in\mathbb{R}^2:|\xi|\le 1} \frac{\varphi(\xi)-\varphi(0)}{|\xi|^2}d\xi + \int\limits_{\xi\in\mathbb{R}^2:|\xi|>1} \frac{\varphi(\xi)}{|\xi|^2}d\xi
$$

for all $\varphi \in \mathcal{S}(\mathbb{R}^2)$. With

$$
\varphi(\xi) = \frac{1}{2\pi}\int\limits_{\mathbb{R}^2} e^{i\langle z,\xi\rangle}\widehat{\varphi}(z)dz, \quad \varphi(0) = \frac{1}{2\pi}\int\limits_{\mathbb{R}^2}\widehat{\varphi}(z)dz
$$

we then obtain

$$
(2\pi)^2\langle v, \widehat{\varphi}\rangle_{L_2(\mathbb{R}^2)} = \int\limits_{\xi\in\mathbb{R}^2:|\xi|\le 1}\frac{1}{|\xi|^2}\int\limits_{\mathbb{R}^2}[e^{i\langle z,\xi\rangle}-1]\widehat{\varphi}(z)dzd\xi + \int\limits_{\xi\in\mathbb{R}^2:|\xi|>1}\frac{1}{|\xi|^2}\int\limits_{\mathbb{R}^2}e^{i\langle z,\xi\rangle}\widehat{\varphi}(z)dzd\xi.
$$

Again we can not exchange the order of integration in the second term. However, as in three–dimensional case we can write

$$(2\pi)^2\langle v,\widehat{\varphi}\rangle_{L_2(\mathbb{R}^2)} = \int\limits_{\mathbb{R}^2}\widehat{\varphi}(z)\left[\int\limits_{\xi\in\mathbb{R}^2:|\xi|\le 1}\frac{e^{i\langle z,\xi\rangle}-1}{|\xi|^2}d\xi + \int\limits_{\xi\in\mathbb{R}^2:|\xi|>1}\frac{e^{i\langle z,\xi\rangle}}{|\xi|^2}d\xi\right]dz.$$

With Lemma 2.13 we further have

$$v(z) = v(|z|) = \frac{1}{(2\pi)^2}\int\limits_{\xi\in\mathbb{R}^2:|\xi|\le 1}\frac{e^{i\langle z,\xi\rangle}-1}{|\xi|^2}d\xi + \int\limits_{\xi\in\mathbb{R}^2:|\xi|>1}\frac{e^{i\langle z,\xi\rangle}}{|\xi|^2}d\xi$$

and using polar coordinates we obtain

$$\begin{aligned} v(z) &= \frac{1}{(2\pi)^2}\int\limits_0^1\int\limits_0^{2\pi}\frac{1}{r}\left[e^{ir|z|\cos\varphi}-1\right]d\varphi dr + \frac{1}{(2\pi)^2}\int\limits_1^\infty\int\limits_0^{2\pi}\frac{1}{r}e^{ir|z|\cos\varphi}d\varphi dr \\ &= \frac{1}{2\pi}\int\limits_0^1\frac{1}{r}[J_0(r|z|)-1]dr + \frac{1}{2\pi}\int\limits_1^\infty\frac{1}{r}J_0(r|z|)dr \end{aligned}$$

with the first order Bessel function [63, Subsection 8.411],

$$J_0(s) = \frac{1}{2\pi}\int\limits_0^{2\pi}e^{is\cos\varphi}d\varphi.$$

Substituting $r := s/\varrho$ we compute

$$\begin{aligned} v(z) &= \frac{1}{2\pi}\int\limits_0^\varrho\frac{J_0(s)-1}{s}ds + \frac{1}{2\pi}\int\limits_\varrho^\infty\frac{J_0(s)}{s}ds \\ &= \frac{1}{2\pi}\int\limits_0^1\frac{J_0(s)-1}{s}ds + \frac{1}{2\pi}\int\limits_1^\infty\frac{J_0(s)}{s}ds + \frac{1}{2\pi}\int\limits_\varrho^1\frac{1}{s}ds \\ &= -\frac{1}{2\pi}\log|z| - \frac{C_0}{2\pi} \end{aligned} \tag{5.5}$$

with the constant

$$C_0 := \int\limits_0^1\frac{1-J_0(s)}{s}ds - \int\limits_1^\infty\frac{J_0(s)}{s}ds.$$

Since any constant satisfies the homogeneous Laplace equation we can neglect constant terms in the definition of the fundamental solution. Hence the fundamental solution of the Laplace operator in two space dimensions is

$$U^*(x,y) = -\frac{1}{2\pi}\log|x-y| \quad \text{for } x,y\in\mathbb{R}^2.$$

In what follows we will describe an alternative approach to compute the fundamental solution for the Laplace operator in two space dimensions. From Lemma 2.13 we know that the solution $v(z)$ depends only on the absolute value $\varrho := |z|$. For $z \neq 0$ the partial differential equation (5.4) can be rewritten in polar coordinates as

$$\left[\frac{\partial^2}{\partial\varrho^2} + \frac{1}{\varrho}\frac{\partial}{\partial\varrho}\right] v(\varrho) = 0 \quad \text{for } \varrho > 0.$$

The general solution of this ordinary differential equation is given by

$$v(\varrho) = a\log\varrho + b, \quad a,b\in\mathbb{R}.$$

In particular for $b = 0$ we have

$$U^*(x,y) = a\,\log|x-y|\,.$$

For $x\in\Omega$ and for a sufficient small $\varepsilon > 0$ let $B_\varepsilon(x)\subset\Omega$ be a ball with center $x$ and with radius $\varepsilon$. For $y\in\Omega\backslash\overline{B_\varepsilon(x)}$ the fundamental solution $U^*(x,y)$ is a solution of the homogeneous Laplace equation $-\Delta_y U^*(x,y) = 0$. Applying Green's second formula (1.8) with respect to the bounded domain $\Omega\backslash\overline{B_\varepsilon(x)}$ we obtain

$$0 = \int_\Gamma U^*(x,y)\frac{\partial}{\partial n_y}u(y)ds_y - \int_\Gamma \frac{\partial}{\partial n_y}U^*(x,y)u(y)ds_y + \int_{\Omega\backslash\overline{B}_\varepsilon(x)} U^*(x,y)f(y)dy$$
$$+ \int_{\partial B_\varepsilon(x)} U^*(x,y)\frac{\partial}{\partial n_y}u(y)ds_y - \int_{\partial B_\varepsilon(x)} \frac{\partial}{\partial n_y}U^*(x,y)u(y)ds_y.$$

Taking the limit $\varepsilon\to 0$ we first bound

$$\left|\int_{\partial B_\varepsilon(x)} U^*(x,y)\frac{\partial}{\partial n_y}u(y)ds_y\right| = |a|\,|\log\varepsilon|\left|\int_{\partial B_\varepsilon(x)} \frac{\partial}{\partial n_y}u(y)ds_y\right|$$
$$\leq |a|\,2\pi\,\varepsilon\,|\log\varepsilon|\,\|u\|_{C^1(\Omega)}.$$

Using $n_y = \dfrac{1}{\varepsilon}(x-y)$ for $y\in\partial B_\varepsilon(x)$ we have

$$\int_{\partial B_\varepsilon(x)} \frac{\partial}{\partial n_y}U^*(x,y)u(y)ds_y = a\int_{\partial B_\varepsilon(x)} \frac{(n_y, y-x)}{|x-y|^2}u(y)ds_y = -\frac{a}{\varepsilon}\int_{\partial B_\varepsilon(x)} u(y)ds_y.$$

The Taylor expansion

$$u(y) \;=\; u(x) + (y-x)\nabla u(\xi)$$

with a suitable $\xi = x + t(y-x)$, $t \in (0,1)$ yields

$$\int\limits_{\partial B_\varepsilon(x)} \frac{\partial}{\partial n_y} U^*(x,y) u(y) ds_y \;=\; -a\,2\pi\, u(x) - \frac{a}{\varepsilon} \int\limits_{\partial B_\varepsilon(x)} (y-x)\nabla u(\xi) dy$$

where

$$\left| \frac{a}{\varepsilon} \int\limits_{\partial B_\varepsilon(x)} (y-x)\nabla u(\xi) dy \right| \;\leq\; |a|\, 2\pi\, \varepsilon\, \|u\|_{C^1(\Omega)} \,.$$

Taking the limit $\varepsilon \to 0$ gives

$$-a\,2\pi\, u(x) = \int\limits_\Gamma U^*(x,y) \frac{\partial}{\partial n_y} u(y) ds_y - \int\limits_\Gamma \frac{\partial}{\partial n_y} U^*(x,y) u(y) ds_y + \int\limits_\Omega U^*(x,y) f(y) dy$$

and therefore the representation formula when choosing $a = -1/2\pi$.

To summarize, the fundamental solution of the Laplace operator is given by

$$U^*(x,y) \;=\; \begin{cases} -\dfrac{1}{2\pi} \log|x-y| & \text{for } d=2, \\[2ex] \dfrac{1}{4\pi} \dfrac{1}{|x-y|} & \text{for } d=3. \end{cases} \tag{5.6}$$

Any solution of the partial differential equation

$$-\Delta u(x) \;=\; f(x) \quad \text{for } x \in \Omega \subset \mathbb{R}^d$$

is therefore given by the representation formula for $x \in \Omega$

$$u(x) = \int\limits_\Gamma U^*(x,y) \frac{\partial}{\partial n_y} u(y) ds_y - \int\limits_\Gamma \frac{\partial}{\partial n_y} U^*(x,y) u(y) ds_y + \int\limits_\Omega U^*(x,y) f(y) dy. \tag{5.7}$$

## 5.2 Linear Elasticity

Let us now consider the system of linear elastostatics (1.25),

$$-\mu \Delta \underline{u}(x) - (\lambda + \mu) \operatorname{grad} \operatorname{div} \underline{u}(x) \;=\; \underline{f}(x) \quad \text{for } x \in \Omega \subset \mathbb{R}^d,$$

and the associated second Betti formula (1.30),

$$\int_\Omega \sum_{i,j=1}^d \frac{\partial}{\partial y_j}\sigma_{ij}(\underline{v},y)u_i(y)dy = \int_\Gamma \gamma_0^{\rm int}\underline{v}(y)^\top \gamma_1^{\rm int}\underline{u}(y)ds_y - \int_\Gamma \gamma_0^{\rm int}\underline{u}(y)^\top \gamma_1^{\rm int}\underline{v}(y)ds_y + \int_\Omega \underline{f}(y)^\top \underline{v}(y)dy. \tag{5.8}$$

To derive a representation formula for the components $u_k(x)$, $x \in \Omega$, we therefore have to find solutions $\underline{v}^k(x,y)$ satisfying

$$\int_\Omega \sum_{i,j=1}^d \frac{\partial}{\partial y_j}\sigma_{ij}(\underline{v}^k(x,y),y)u_i(y)dy \;=\; u_k(x) \quad \text{for } x\in\Omega,\ k=1,\ldots,d.$$

Let $\underline{e}^k \in \mathbb{R}^d$ be the unit vector with $e^k_\ell = \delta_{k\ell}$ for $k,\ell = 1,\ldots,d$. Using the transformation $z := y - x$ we have to solve the partial differential equations, $k = 1,\ldots,d$,

$$-\mu\Delta_z\underline{v}^k(z) - (\lambda+\mu)\operatorname{grad}_z\operatorname{div}_z\underline{v}^k(z) \;=\; \delta_0(z)\,\underline{e}^k \quad \text{for } z\in\mathbb{R}^d.$$

Using (1.32),

$$\underline{v}^k(z) \;:=\; \Delta[\psi(z)\underline{e}^k] - \frac{\lambda+\mu}{\lambda+2\mu}\operatorname{grad}\operatorname{div}[\psi(z)\underline{e}^k],$$

we have to find the Airy stress function $\psi$ satisfying the Bi–Laplace equation

$$-\mu\Delta^2\psi(z) \;=\; \delta_0(z) \quad \text{for } z\in\mathbb{R}^d$$

or

$$-\mu\Delta\varphi(z) \;=\; \delta_0(z), \quad \Delta\psi(z) \;=\; \varphi(z) \quad \text{for } z\in\mathbb{R}^d.$$

From the fundamental solution of the Laplace operator we find

$$\varphi(z) = \begin{cases} -\dfrac{1}{\mu}\dfrac{1}{2\pi}\log|z|, & \text{for } d=2,\\[2ex] \dfrac{1}{\mu}\dfrac{1}{4\pi}\dfrac{1}{|z|} & \text{for } d=3.\end{cases}$$

For $d = 2$ we have to solve the remaining Poisson equation when using polar coordinates,

$$\left[\frac{\partial^2}{\partial\varrho^2} + \frac{1}{\varrho}\frac{\partial}{\partial\varrho}\right]\widetilde{\psi}(\varrho) \;=\; -\frac{1}{\mu}\frac{1}{2\pi}\log\varrho \quad \text{for } \varrho > 0,$$

with the general solution

$$\widetilde{\psi}(\varrho) = -\frac{1}{\mu}\frac{1}{8\pi}\left[\varrho^2\log\varrho - \varrho^2\right] + a\log\varrho + b \quad \text{for } \varrho > 0,\ a, b \in \mathbb{R}.$$

In particular for $a = b = 0$ we have

$$\widetilde{\psi}(\varrho) = -\frac{1}{\mu}\frac{1}{8\pi}\left[\varrho^2\log\varrho - \varrho^2\right].$$

Due to $\Delta^2|z|^2 = 0$ we then obtain for $\varrho = |z|$

$$\psi(z) = -\frac{1}{\mu}\frac{1}{8\pi}|z|^2\log|z|.$$

For $k = 1$ we find for $\underline{v}^1(z)$

$$v_1^1(z) = \Delta\psi(z) - \frac{\lambda+\mu}{\lambda+2\mu}\frac{\partial^2}{\partial z_1^2}\psi(z), \quad v_2^1(z) = -\frac{\lambda+\mu}{\lambda+2\mu}\frac{\partial^2}{\partial z_1\partial z_2}\psi(z).$$

Using $\dfrac{\partial}{\partial z_i}|z| = \dfrac{z_i}{|z|}$ we obtain

$$\begin{aligned}
\frac{\partial}{\partial z_i}\psi(z) &= -\frac{1}{\mu}\frac{1}{8\pi}\left[2z_i\log|z| + z_i\right] \quad (i = 1, 2)\\
\frac{\partial^2}{\partial z_i^2}\psi(z) &= -\frac{1}{\mu}\frac{1}{8\pi}\left[2\log|z| + 2\frac{z_1^2}{|z|^2} + 1\right] \quad (i = 1, 2)\\
\frac{\partial^2}{\partial z_1\partial z_2}\psi(z) &= -\frac{1}{\mu}\frac{1}{4\pi}\frac{z_1z_2}{|z|^2}
\end{aligned}$$

and therefore

$$\begin{aligned}
v_1^1(z) &= -\frac{1}{4\pi}\frac{\lambda+3\mu}{\mu(\lambda+2\mu)}\log|z| + \frac{1}{4\pi}\frac{\lambda+\mu}{\mu(\lambda+2\mu)}\left[\frac{z_1^2}{|z|^2} - \frac{3}{2}\right],\\
v_2^1(z) &= \frac{1}{4\pi}\frac{\lambda+\mu}{\mu(\lambda+2\mu)}\frac{z_1z_2}{|z|^2}.
\end{aligned}$$

For $k = 2$ the computation is almost the same. Since the constants are solutions of the homogeneous system we can neglect them when defining the fundamental solution. From $\underline{v}^k$ for $k = 1, 2$ we then find the Kelvin solution tensor $U^*(x, y) = (\underline{v}^1, \underline{v}^2)$ with the components

$$U_{k\ell}^*(x, y) = \frac{1}{4\pi}\frac{\lambda+\mu}{\mu(\lambda+2\mu)}\left[-\frac{\lambda+3\mu}{\lambda+\mu}\log|x-y|\,\delta_{k\ell} + \frac{(y_k-x_k)(y_\ell-x_\ell)}{|x-y|^2}\right]$$

for $k, \ell = 1, 2$. Inserting the Lamé constants (1.24) this gives

$$U_{k\ell}^*(x, y) = \frac{1}{4\pi}\frac{1}{E}\frac{1+\nu}{1-\nu}\left[(4\nu-3)\log|x-y|\,\delta_{k\ell} + \frac{(y_k-x_k)(y_\ell-x_\ell)}{|x-y|^2}\right].$$

This is the fundamental solution of linear elastostatics in two space dimensions which even exists for incompressible materials with $\nu = 1/2$.

For $d = 3$ we have to solve the Poisson equation, by using spherical coordinates we obtain

$$\frac{1}{\varrho^2}\frac{\partial}{\partial\varrho}\left[\varrho^2\frac{\partial}{\partial\varrho}\widetilde{\psi}(\varrho)\right] = \frac{1}{\mu}\frac{1}{4\pi}\frac{1}{\varrho} \quad \text{for } \varrho > 0,$$

with the general solution

$$\widetilde{\psi}(\varrho) = \frac{1}{\mu}\frac{1}{4\pi}\left[\frac{1}{2}\varrho + \frac{a}{\varrho} + b\right], \quad \text{for } \varrho > 0,\ a, b \in \mathbb{R}.$$

For $a = b = 0$ we have

$$\psi(z) = \frac{1}{\mu}\frac{1}{8\pi}|z|.$$

For $k = 1$ we obtain $\underline{v}^1(z)$

$$v_1^1(z) = \Delta\psi(z) - \frac{\lambda+\mu}{\lambda+2\mu}\frac{\partial^2}{\partial z_1^2}\psi(z),$$
$$v_2^1(z) = -\frac{\lambda+\mu}{\lambda+2\mu}\frac{\partial^2}{\partial z_1\partial z_2}\psi(z),$$
$$v_3^1(z) = -\frac{\lambda+\mu}{\lambda+2\mu}\frac{\partial}{\partial z_1\partial z_3}\psi(z)$$

and with the derivatives

$$\frac{\partial}{\partial z_i}\psi(z) = \frac{1}{\mu}\frac{1}{8\pi}\frac{z_i}{|z|}, \quad \frac{\partial^2}{\partial z_i^2}\psi(z) = \frac{1}{\mu}\frac{1}{8\pi}\left[\frac{1}{|z|} - \frac{z_i^2}{|z|^3}\right],$$

$$\frac{\partial^2}{\partial z_i\partial z_j}\psi(z) = -\frac{1}{\mu}\frac{1}{8\pi}\frac{z_i z_j}{|z|^3} \quad \text{for } i \neq j$$

we then find for the components of the solution $\underline{v}^1$

$$v_1^1(z) = \frac{1}{8\pi}\frac{\lambda+3\mu}{\mu(\lambda+2\mu)}\frac{1}{|z|} + \frac{1}{8\pi}\frac{\lambda+\mu}{\mu(\lambda+2\mu)}\frac{z_1^2}{|z|^3},$$

$$v_2^1(z) = \frac{1}{8\pi}\frac{\lambda+\mu}{\mu(\lambda+2\mu)}\frac{z_1 z_2}{|z|^3},$$

$$v_3^1(z) = \frac{1}{8\pi}\frac{\lambda+\mu}{\mu(\lambda+2\mu)}\frac{z_1 z_3}{|z|^3}.$$

For $k = 2, 3$ the computations are almost the same. Hence, the Kelvin solution tensor is given by $U^*(x, y) = (\underline{v}^1, \underline{v}^2, \underline{v}^3)$ where

$$U_{k\ell}^*(x,y) = \frac{1}{8\pi}\frac{\lambda+\mu}{\mu(\lambda+2\mu)}\left[\frac{\lambda+3\mu}{\lambda+\mu}\frac{\delta_{k\ell}}{|x-y|} + \frac{(y_k - x_k)(y_\ell - x_\ell)}{|x-y|^3}\right]$$

for $k, \ell = 1, \ldots, 3$. Inserting the Lamé constants (1.24) this gives the fundamental solution of linear elastostatics in three space dimensions

$$U^*_{k\ell}(x,y) = \frac{1}{8\pi}\frac{1}{E}\frac{1+\nu}{1-\nu}\left[(3-4\nu)\frac{\delta_{k\ell}}{|x-y|} + \frac{(y_k-x_k)(y_\ell-x_\ell)}{|x-y|^3}\right].$$

Hence we have the fundamental solution of linear elastostatics

$$U^*_{k\ell}(x,y) = \frac{1}{4(d-1)\pi}\frac{1}{E}\frac{1+\nu}{1-\nu}\left[(3-4\nu)E(x,y)\delta_{k\ell} + \frac{(x_k-y_k)(x_\ell-y_\ell)}{|x-y|^d}\right] \tag{5.9}$$

for $k, \ell = 1, \ldots, d$ with

$$E(x,y) = \begin{cases} -\log|x-y| & \text{for } d=2, \\ \dfrac{1}{|x-y|} & \text{for } d=3. \end{cases}$$

Inserting the solution vectors $\underline{v}(y) = \underline{U}^*_k(x,y)$ into the second Betti formula (5.8) this gives the representation formula

$$u_k(x) = \int_\Gamma \underline{U}^*_k(x,y)^\top \gamma_1^{\text{int}}\underline{u}(y)ds_y - \int_\Gamma \underline{u}(y)^\top \gamma_{1,y}^{\text{int}}\underline{U}^*_k(x,y)ds_y + \int_\Omega \underline{f}(y)^\top \underline{U}^*_k(x,y)dy \tag{5.10}$$

for $x \in \Omega$ and $k = 1, \ldots, d$. Thereby, the boundary stress $\underline{T}^*_k(x,y)$ of the fundamental solution $\underline{U}^*_k(x,y)$ is given for almost all $y \in \Gamma$ by applying (1.27) as

$$\begin{aligned}\underline{T}^*_k(x,y) &:= \gamma_{1,y}^{\text{int}}\underline{U}^*_k(x,y) \\ &= \lambda\,\text{div}_y\underline{U}^*_k(x,y)\,\underline{n}(y) + 2\mu\frac{\partial}{\partial n_y}\underline{U}^*_k(x,y) + \mu\,\underline{n}(y)\times\text{curl}_y\,\underline{U}^*_k(x,y).\end{aligned}$$

Using

$$\text{div}\,\underline{U}^*_k(x,y) = \frac{1}{4(d-1)\pi}\frac{1}{E}\frac{1+\nu}{1-\nu}2(2\nu-1)\frac{y_k-x_k}{|x-y|^d}$$

we then obtain

$$\underline{T}^*_k(x,y) = -\frac{1}{2(d-1)\pi}\frac{\nu}{1-\nu}\frac{y_k-x_k}{|x-y|^d}\underline{n}(y) + \frac{E}{1+\nu}\frac{\partial}{\partial n_y}\underline{U}^*_k(x,y) + \frac{E}{2(1+\nu)}\underline{n}(y)\times\text{curl}_y\,\underline{U}^*_k(x,y). \tag{5.11}$$

Obviously, both the fundamental solutions $\underline{U}^*_k(x,y)$ and the corresponding boundary stress functions $\underline{T}^*_k(x,y)$ exist also for incompressible materials with $\nu = 1/2$.

## 5.3 Stokes Problem

Next we consider the Stokes system (1.38),

$$-\mu\Delta\underline{u}(x) + \nabla p(x) = \underline{f}(x), \quad \operatorname{div}\underline{u}(x) = 0 \quad \text{for } x \in \Omega \subset \mathbb{R}^d.$$

For the solution $\underline{u}$ and for an arbitrary vector field $\underline{v}$ we obtain from Green's first formula (1.41) by using the symmetry $a(\underline{u},\underline{v}) = a(\underline{v},\underline{u})$ Green's second formula

$$\int_\Omega \sum_{i=1}^d \left[-\mu\Delta v_i(y) + \frac{\partial}{\partial y_i} q(y)\right] u_i(y)dy + \int_\Omega p(y)\operatorname{div}\underline{v}(y)dy \tag{5.12}$$
$$= \int_\Gamma \sum_{i=1}^d t_i(\underline{u},p)v_i(y)ds_y - \int_\Gamma \sum_{i=1}^d t_i(\underline{v},q)u_i(y)ds_y + \int_\Omega \underline{f}(y)^\top \underline{v}(y)dy$$

with the conormal derivative $\underline{t}(\underline{u},p)$ as given in (1.43).

To obtain representation formulae for the components $u_k(x)$, $x \in \Omega$, for the velocity field $\underline{u}$ we have to find solutions $\underline{v}^k(x,y)$ and $q^k(x,y)$ such that

$$\int_\Omega \sum_{i=1}^d \left[-\mu\Delta v_i^k(x,y) + \frac{\partial}{\partial y_i} q^k(x,y)\right] u_i(y)dy = u_k(x), \quad \operatorname{div}_y \underline{v}^k(x,y) = 0$$

for $x \in \Omega$, $k = 1,\ldots,d$. With the transformation $z := y - x$ we have to solve for $k = 1,\ldots,d$

$$-\mu\Delta\underline{v}^k(z) + \nabla q^k(z) = \delta_0(z)\underline{e}^k, \quad \operatorname{div}\underline{v}^k(z) = 0 \quad \text{for } z \in \mathbb{R}^d.$$

The application of the Fourier transform (2.14) gives

$$\mu\,|\xi|^2\,\widehat{v}_j^k(\xi) + i\,\xi_j\,\widehat{q}^k(\xi) = \frac{1}{(2\pi)^{d/2}}\,\delta_{jk} \quad (j = 1,\ldots,d), \quad i\sum_{j=1}^d \xi_j\widehat{v}_j^k(\xi) = 0.$$

In particular for $d = 2$ and $k = 1$ we have to solve a linear system,

$$\begin{aligned} \mu\,|\xi|^2\,\widehat{v}_1^1(\xi) + i\,\xi_1\,\widehat{q}^1(\xi) &= \frac{1}{2\pi}, \\ \mu\,|\xi|^2\,\widehat{v}_2^1(\xi) + i\,\xi_2\,\widehat{q}^1(\xi) &= 0, \\ i\xi_1\widehat{v}_1^1(\xi) + i\xi_2\widehat{v}_2^1(\xi) &= 0, \end{aligned}$$

yielding the solution

$$\widehat{v}_1^1(\xi) = \frac{1}{\mu}\frac{1}{2\pi}\left[\frac{1}{|\xi|^2} - \frac{\xi_1^2}{|\xi|^4}\right], \quad \widehat{v}_2^1(\xi) = -\frac{1}{\mu}\frac{1}{2\pi}\frac{\xi_1\xi_2}{|\xi|^4}, \quad \widehat{q}^1(\xi) = -\frac{i}{2\pi}\frac{\xi_1}{|\xi|^2}.$$

As for the scalar Laplace equation we obtain

$$\begin{aligned} v_1^1(z) &= \frac{1}{\mu}\frac{1}{(2\pi)^2}\int_{\mathbb{R}^2} e^{i\langle z,\xi\rangle}\left[\frac{1}{|\xi|^2} - \frac{\xi_1^2}{|\xi|^4}\right] d\xi \\ &= \frac{1}{\mu}\frac{1}{(2\pi)^2}\int_{R^2} e^{i\langle z,\xi\rangle}\frac{1}{|\xi|^2} d\xi + \frac{1}{\mu}\frac{\partial^2}{\partial z_1^2}\left[\frac{1}{(2\pi)^2}\int_{R^2} e^{i\langle z,\xi\rangle}\frac{1}{|\xi|^4}d\xi\right] \end{aligned}$$

and with (5.5) we have

$$\frac{1}{(2\pi)^2}\int_{\mathbb{R}^2} e^{i\langle z,\xi\rangle}\frac{1}{|\xi|^2}d\xi = -\frac{1}{2\pi}\log|z| - \frac{C_0}{2\pi}.$$

On the other hand,

$$\Delta\left(\frac{1}{(2\pi)^2}\int_{\mathbb{R}^2} e^{i\langle z,\xi\rangle}\frac{1}{|\xi|^4}d\xi\right) = -\frac{1}{(2\pi)^2}\int_{\mathbb{R}^2} e^{i\langle z,\xi\rangle}\frac{1}{|\xi|^2}d\xi = \frac{1}{2\pi}\log|z| + \frac{C_0}{2\pi}$$

implies

$$\frac{1}{(2\pi)^2}\int_{\mathbb{R}^2} e^{i\langle z,\xi\rangle}\frac{1}{|\xi|^4}d\xi = \frac{1}{8\pi}\left[|z|^2\log|z| - |z|^2\right] + \frac{C_0}{8\pi}|z|^2 + C_1 + C_2\log|z|$$

with some constants $C_1, C_2 \in \mathbb{R}$. In particular for $C_1 = C_2 = 0$ this gives

$$\begin{aligned} v_1^1(z) &= \frac{1}{\mu}\left[-\frac{1}{2\pi}\log|z| - \frac{C_0}{2\pi}\right] + \frac{1}{\mu}\frac{\partial^2}{\partial z_1^2}\left[\frac{1}{8\pi}\left(|z|^2\log|z| - |z|^2\right) + \frac{C_0}{8\pi}|z|^2\right] \\ &= \frac{1}{\mu}\frac{1}{4\pi}\left[-\log|z| + \frac{z_1^2}{|z|^2} - \frac{2C_0+1}{2}\right]. \end{aligned}$$

Analogous computations yield

$$\begin{aligned} v_2^1(z) &= -\frac{1}{\mu}\frac{1}{(2\pi)^2}\int_{\mathbb{R}^2} e^{i\langle z,\xi\rangle}\frac{\xi_1\xi_2}{|\xi|^4}d\xi = \frac{1}{\mu}\frac{\partial^2}{\partial z_1\partial z_2}\left[\frac{1}{(2\pi)^2}\int_{\mathbb{R}^2} e^{i\langle z,\xi\rangle}\frac{1}{|\xi|^4}d\xi\right] \\ &= \frac{1}{\mu}\frac{\partial^2}{\partial z_1\partial z_2}\left[\frac{1}{8\pi}\left(|z|^2\log|z| - |z|^2\right) + \frac{C_0}{8\pi}|z|^2\right] = \frac{1}{\mu}\frac{1}{4\pi}\frac{z_1 z_2}{|z|^2} \end{aligned}$$

and

$$\begin{aligned} q^1(z) &= -\frac{i}{(2\pi)^2}\int_{\mathbb{R}^2} e^{i\langle z,\xi\rangle}\frac{\xi_1}{|\xi|^2}d\xi = -\frac{\partial}{\partial z_1}\left[\frac{1}{(2\pi)^2}\int_{\mathbb{R}^2} e^{i\langle z,\xi\rangle}\frac{1}{|\xi|^2}d\xi\right] \\ &= -\frac{\partial}{\partial z_1}\left[-\frac{1}{2\pi}\log|z| - \frac{C_0}{2\pi}\right] = \frac{1}{2\pi}\frac{\partial}{\partial z_1}\log|z|. \end{aligned}$$

For $d = 2$ and $k = 2$ the computations are almost the same. Neglecting the constants we finally have the fundamental solution for the Stokes system in two space dimensions,

$$U^*_{k\ell}(x,y) = \frac{1}{4\pi}\frac{1}{\mu}\left[-\log|x-y|\,\delta_{k\ell} + \frac{(y_k-x_k)(y_\ell-x_\ell)}{|x-y|^2}\right] \tag{5.13}$$

$$Q^*_k(x,y) = \frac{1}{2\pi}\frac{y_k-x_k}{|x-y|^2} \tag{5.14}$$

and $k,\ell = 1,2$.

For $d = 3$ we obtain in the same way the fundamental solution for the Stokes system as

$$U^*_{k\ell}(x,y) = \frac{1}{8\pi}\frac{1}{\mu}\left[\frac{\delta_{k\ell}}{|x-y|} + \frac{(y_k-x_k)(y_\ell-x_\ell)}{|x-y|^3}\right]$$

$$Q^*_k(x,y) = \frac{1}{4\pi}\frac{y_k-x_k}{|x-y|^3}$$

and $k,\ell = 1,\ldots,3$.

Comparing the above results with the fundamental solution of the system of linear elasticity we obtain the equality for

$$\frac{1}{E}\frac{1+\nu}{1-\nu} = \frac{1}{\mu}, \quad (3-4\nu) = 1$$

and therefore for

$$\nu = \frac{1}{2}, \quad E = 3\mu\,.$$

The fundamental solution of the linear elasticity system with incompressible material therefore coincides with the fundamental solution of the Stokes system.

Inserting the fundamental solutions $\underline{v}(y) = \underline{U}^*_k(x,y)$ and $q(y) = Q^*_k(x,y)$ into the second Greens formula (5.12) this gives the representation formulae

$$u_k(x) = \int_\Gamma \underline{U}^*_k(x,y)^\top \underline{t}(\underline{u}(y),p(y))ds_y - \int_\Gamma \underline{u}(y)^\top \underline{t}(\underline{U}^*_k(x,y),Q^*_k(x,y))ds_y + \int_\Omega \underline{f}(y)^\top \underline{U}^*_k(x,y)dy \tag{5.15}$$

for $x \in \Omega$ and $k = 1,\ldots,d$. Hereby the conormal derivative $\underline{T}^*_k(x,y)$ is defined via (1.43) for almost all $y \in \Gamma$ by

$$\begin{aligned}\underline{T}^*_k(x,y) &= \underline{t}(\underline{U}^*_k(x,y),Q^*_k(x,y))\\ &= -Q^*_k(x,y)\underline{n}(x) + 2\mu\frac{\partial}{\partial n_y}\underline{U}^*_k(x,y) + \mu\,\underline{n}(x)\times\operatorname{curl}\underline{U}^*_k(x,y)\\ &= -\frac{1}{2(d-1)\pi}\frac{y_k-x_k}{|x-y|^d}\underline{n}(x) + 2\mu\frac{\partial}{\partial n_y}\underline{U}^*_k(x,y) + \mu\,\underline{n}(x)\times\operatorname{curl}\underline{U}^*_k(x,y).\end{aligned}$$

Hence the boundary stress (1.43) of the fundamental solution of the Stokes system also coincides with the boundary stress (1.27) of the fundamental solution of the linear elasticity system when choosing $\nu = \frac{1}{2}$ and $E = 3\mu$.

It remains to find some appropriate representation formulae for the pressure $p$. Let us first consider the case $d = 2$ and the second Green formula (5.12) where we have to find solutions $\underline{v}^3(z)$ and $q^3(z)$ with $z := y - x$ such that

$$-\mu\Delta\underline{v}^3(z) + \nabla q^3(z) = 0, \quad \operatorname{div}\underline{v}^3(z) = \delta_0(z) \quad \text{for } z \in \mathbb{R}^2.$$

By applying the Fourier transformation we obtain the linear system

$$\begin{aligned}\mu\,|\xi|^2\,\widehat{v}_1^3(\xi) + i\,\xi_1\,\widehat{q}^3(\xi) &= 0,\\ \mu\,|\xi|^2\,\widehat{v}_2^3(\xi) + i\,\xi_2\,\widehat{q}^3(\xi) &= 0,\\ i\xi_1\widehat{v}_1^3(\xi) + i\xi_2\widehat{v}_2^3(\xi) &= \frac{1}{2\pi}\end{aligned}$$

with the solution

$$\widehat{v}_1^3(\xi) = -\frac{i}{2\pi}\frac{\xi_1}{|\xi|^2}, \quad \widehat{v}_2^3(\xi) = -\frac{i}{2\pi}\frac{\xi_2}{|\xi|^2}, \quad \widehat{q}^3(\xi) = \frac{\mu}{2\pi}.$$

As before we obtain

$$v_i^3(z) = \frac{1}{2\pi}\frac{\partial}{\partial z_i}\log|z| \quad (i = 1,2), \quad q^3(z) = \mu\delta_0(z).$$

Using $z := y - x$ we conclude for $x \in \Omega$ a representation formula for the pressure

$$\begin{aligned}p(x) = \int_\Gamma \sum_{i=1}^2 t_i(\underline{u},p)v_i^3(x,y)ds_y - \int_\Gamma \sum_{i=1}^2 t_i(\underline{v}^3(x,y),q^3(x,y))u_i(y)ds_y\\ + \int_\Omega \sum_{i=1}^2 \underline{v}_i^3(x,y)f_i(y)dy\end{aligned}$$

where the conormal derivative (1.43) implies

$$\begin{aligned}t_i(\underline{v}^3(x,y),q^3(x,y)) = -[\mu\delta_0(y-x) + \operatorname{div}\underline{v}^3(x,y)]n_i(x)\\ +2\mu\sum_{j=1}^2 e_{ij}(\underline{v}^3(x,y),y)n_j(y)\end{aligned}$$

for $i = 1,2$, $x \in \Omega$ and $y \in \Gamma$. Since $\underline{v}^3$ is divergence–free,

$$\operatorname{div}\underline{v}^3(x,y) = \sum_{i=1}^2 \frac{\partial}{\partial y_i}v_i^3(x,y) = \frac{1}{2\pi}\sum_{i=1}^2 \frac{\partial^2}{\partial y_i^2}\log|x-y| = 0,$$

we obtain for $\Gamma \ni y \neq x \in \Omega$

$$t_i(\underline{v}^3(x,y), q^3(x,y)) = 2\mu \sum_{j=1}^{2} e_{ij}(\underline{v}^3(x,y), y) n_j(y).$$

Moreover,

$$\begin{aligned} e_{ij}(\underline{v}^3(x,y),y) &= \frac{1}{2}\left[\frac{\partial}{\partial y_i} v_j^3(x,y) + \frac{\partial}{\partial y_j} v_i^3(x,y)\right] \\ &= \frac{1}{4\pi}\left[\frac{\partial}{\partial y_i}\frac{\partial}{\partial y_j}\log|x-y| + \frac{\partial}{\partial y_j}\frac{\partial}{\partial y_i}\log|x-y|\right] \\ &= \frac{1}{2\pi}\frac{\partial}{\partial y_i}\frac{\partial}{\partial y_j}\log|x-y| \\ &= -\frac{1}{2\pi}\frac{\partial}{\partial x_j}\frac{\partial}{\partial y_i}\log|x-y| = -\frac{\partial}{\partial x_j} Q_i^*(x,y). \end{aligned}$$

Finally we obtain the representation formula for the pressure $p$, $x \in \Omega$,

$$\begin{aligned} p(x) = \int_\Gamma \sum_{i=1}^{2} t_i(\underline{u}, p) Q_i^*(x,y) dy + 2\mu \int_\Gamma \sum_{i,j=1}^{2} \frac{\partial}{\partial x_j} Q_i^*(x,y) n_j(y) u_i(y) ds_y \\ + \int_\Omega f_i(y) Q_i^*(x,y) dy\,. \end{aligned} \tag{5.16}$$

For $d = 3$ one may obtain a similar formula, we skip the details.

## 5.4 Helmholtz Equation

Finally we consider the Helmholtz

$$-\Delta u(x) - k^2 u(x) = 0 \quad \text{for } x \in \mathbb{R}^d, k \in \mathbb{R} \tag{5.17}$$

where the computation of the fundamental solution can be done as in the alternative approach for the Laplace equation.

For $d = 3$, and by using spherical coordinates we have to solve the partial differential equation to find $v(\varrho) = v(|x|) = u(x)$ such that

$$-\frac{1}{\varrho^2}\frac{\partial}{\partial \varrho}\left[\varrho^2 \frac{\partial}{\partial \varrho} v(\varrho)\right] - k^2 v(\varrho) = 0 \quad \text{for } \varrho > 0.$$

With the transformation

$$v(\varrho) = \frac{1}{\varrho} V(\varrho)$$

this is equivalent to

$$V''(\varrho) + k^2 V(\varrho) = 0 \quad \text{for } \varrho > 0$$

where the general solution is given by

$$V(\varrho) = A_1 \cos k\varrho + A_2 \sin k\varrho$$

and therefore we obtain

$$v(\varrho) = \frac{1}{\varrho} V(\varrho) = A_1 \frac{\cos k\varrho}{\varrho} + A_2 \frac{\sin \varrho}{\varrho}.$$

When considering the behavior as $\varrho \to 0$ we find a fundamental solution of the Helmholtz equation given by

$$U_k^*(x,y) = \frac{1}{4\pi} \frac{\cos k|x-y|}{|x-y|} \quad \text{for } x,y \in \mathbb{R}^3.$$

However, it is more common to use a complex combination of the above fundamental system to define the fundamental solution by

$$U_k^*(x,y) = \frac{1}{4\pi} \frac{e^{ik|x-y|}}{|x-y|} \quad \text{for } x,y \in \mathbb{R}^3. \tag{5.18}$$

For $d = 2$, and by using polar coordinates the Helmholtz equation (5.17) reads

$$-\frac{\partial^2}{\partial \varrho^2} v(\varrho) - \frac{1}{\varrho} \frac{\partial}{\partial \varrho} v(\varrho) - k^2 v(\varrho) = 0 \quad \text{for } \varrho > 0,$$

or

$$\varrho^2 \frac{\partial^2}{\partial \varrho^2} v(\varrho) + \varrho \frac{\partial}{\partial \varrho} v(\varrho) + k^2 \varrho^2 v(\varrho) = 0 \quad \text{for } \varrho > 0.$$

With the substitution

$$s = k\varrho, \quad v(\varrho) = v(\frac{s}{k}) = V(s), \quad V'(s) = \frac{1}{k} \frac{\partial}{\partial \varrho} v(\varrho)$$

we then obtain a Bessel differential equation of order zero,

$$s^2 V''(s) + sV'(s) + s^2 V(s) = 0 \quad \text{for } s > 0. \tag{5.19}$$

To find a fundamental system of the Bessel differential equation (5.19) we first consider the ansatz

$$V_1(s) = \sum_{k=0}^{\infty} v_k s^k, \quad V_1'(s) = \sum_{k=1}^{\infty} v_k k s^{k-1}, \quad V_1''(s) = \sum_{k=2}^{\infty} v_k k(k-1) s^{k-2}.$$

By inserting this into the differential equation (5.19) we obtain

$$\begin{aligned}0 &= s^2V_1''(s) + sV_1'(s) + s^2V_1(s)\\ &= \sum_{k=2}^{\infty} v_k k(k-1)s^k + \sum_{k=1}^{\infty} v_k k s^k + \sum_{k=0}^{\infty} v_k s^{k+2}\\ &= \sum_{k=2}^{\infty}\left[v_{k-2} + k^2 v_k\right] s^k + v_1 s \quad \text{for } s > 0\end{aligned}$$

and thus

$$v_1 = 0, \quad v_k = -\frac{1}{k^2} v_{k-2} \quad \text{for } k \geq 2.$$

Hence we obtain

$$v_{2\ell-1} = 0, \quad v_{2\ell} = -\frac{1}{4\ell^2} v_{2(\ell-1)} \quad \text{for } \ell = 1, 2, \ldots$$

and therefore

$$v_{2\ell} = \frac{(-1)^\ell}{4^\ell(\ell!)^2} v_0 \quad \text{for } \ell = 1, 2, \ldots.$$

In particular for $v_0 = 1$ we have

$$V_1(s) = 1 + \sum_{\ell=1}^{\infty} \frac{(-1)^\ell}{4^\ell(\ell!)^2} s^{2\ell} =: J_0(s)$$

which is the first kind Bessel function of order zero. Note that

$$\lim_{s\to 0} J_0(s) = 1.$$

To find a second solution of the fundamental system including a logarithmic singularity we consider the ansatz

$$V_2(s) = J_0(s)\ln s + W(s) \quad \text{for } s > 0.$$

By using

$$\begin{aligned}V_2'(s) &= J_0'(s)\ln s + \frac{1}{s}J_0(s) + W'(s),\\ V_2''(s) &= J_0''(s)\ln s + \frac{2}{s}J_0'(s) - \frac{1}{s^2}J_0(s) + W''(s)\end{aligned}$$

we obtain

$$\begin{aligned}0 &= s^2V_2''(s) + sV_2'(s) + s^2V_2(s)\\ &= \left[s^2J_0''(s) + sJ_0'(s) + s^2J_0(s)\right]\ln s\\ &\qquad + 2sJ_0'(s) + s^2W''(s) + sW'(s) + s^2W(s)\\ &= 2sJ_0'(s) + s^2W''(s) + sW'(s) + s^2W(s)\end{aligned}$$

since $J_0(s)$ is a solution of the Bessel differential equation (5.19). Hence we have to solve the differential equation

$$s^2W''(s) + sW'(s) + s^2W(s) + 2sI_0'(s) \quad \text{for } s > 0.$$

With

$$W(s) = \sum_{k=0}^{\infty} w_k s^k, \quad J_0'(s) = \sum_{k=1}^{\infty} v_k k s^{k-1}$$

we have to solve

$$\begin{aligned} 0 &= \sum_{k=2}^{\infty} \left[k^2 w_k + w_{k-2}\right] s^k + w_1 s + 2\sum_{k=1}^{\infty} v_k k s^k \\ &= \sum_{k=2}^{\infty} \left[k^2 w_k + w_{k-2} + 2k v_k\right] s^k + [w_1 + 2v_1]\, s \quad \text{for } s > 0. \end{aligned}$$

Hence we find

$$w_1 \;=\; -2v_1 \;=\; 0,$$

and

$$k^2 w_k + w_{k-2} + 2k v_k \;=\; 0 \quad \text{for } k \geq 2,$$

i.e.

$$w_k \;=\; -\frac{1}{k^2}\left[w_{k-2} + 2k v_k\right] \quad \text{for } k \geq 2.$$

By using $v_{2\ell-1} = 0$ for $\ell \in \mathbb{N}$ we then obtain $w_{2\ell-1} = 0$ for $\ell \in \mathbb{N}$ and

$$w_{2\ell} \;=\; -\frac{1}{4\ell^2}\left[w_{2(\ell-1)} + 4\ell v_{2\ell}\right] \;=\; -\frac{1}{4\ell^2} w_{2(\ell-1)} - \frac{1}{\ell}\frac{(-1)^\ell}{4^\ell(\ell!)^2}\,.$$

When choosing $w_0 = 0$ we find by induction

$$w_{2\ell} \;=\; \frac{(-1)^{\ell+1}}{4^\ell(\ell!)^2}\sum_{j=1}^{\ell}\frac{1}{j} \quad \text{for } \ell \in \mathbb{N}.$$

Hence we have

$$\begin{aligned} V_2(s) &= J_0(s)\ln s + W(s) \\ &= \left[1 + \sum_{\ell=1}^{\infty}\frac{(-1)^\ell}{4^\ell(\ell!)^2}s^{2\ell}\right]\ln s - \sum_{\ell=1}^{\infty}\left(\sum_{j=1}^{\ell}\frac{1}{j}\right)\frac{(-1)^\ell}{4^\ell(\ell!)^2}s^{2\ell}\,. \end{aligned}$$

Instead of $V_2(s)$ we will use a linear combination of $V_1(s)$ and $V_2(s)$ to define a second solution of the fundamental system, in particular we introduce the second kind Bessel function of order zero,

$$Y_0(s) \;=\; [\ln 2 - \gamma]\, J_0(s) - V_2(s)$$

where

$$\gamma = \lim_{n\to\infty} \left[ \sum_{j=1}^{n} \frac{1}{j} - \ln\, n \right] \approx 0.57721566490\ldots$$

is the Euler–Mascheroni constant. Note that $Y_0(s)$ behaves like $-\ln s$ as $s \to 0$. The fundamental solution of the Helmholtz equation is then given by

$$U_k^*(x,y) = \frac{1}{2\pi} Y_0(k\,|x-y|) \quad \text{for } x,y \in \mathbb{R}^2. \tag{5.20}$$

## 5.5 Exercises

**5.1** Consider the recursion

$$w_0 = 0, \quad w_{2\ell} = -\frac{1}{4\ell^2} w_{2(\ell-1)} - \frac{1}{\ell} \frac{(-1)^\ell}{4^\ell (\ell!)^2} \quad \text{for } \ell \in \mathbb{N}.$$

Prove by induction that

$$w_{2\ell} = \frac{(-1)^{\ell+1}}{4^\ell (\ell!)^2} \sum_{j=1}^{\ell} \frac{1}{j} \quad \text{for } \ell \in \mathbb{N}.$$

**5.2** Compute the Green function $G(x,y)$ such that

$$u(x) = \int_0^1 G(x,y) f(y) dy \quad \text{for } x \in (0,1)$$

is the unique solution of the Dirichlet boundary value problem

$$-u''(x) = f(x) \quad \text{for } x \in (0,1), \quad u(0) = u(1) = 0.$$

# 6

# Boundary Integral Operators

As a model problem we first consider the Poisson equation for $d = 2, 3$

$$-\Delta u(x) \;=\; f(x) \quad \text{for } x \in \Omega \subset \mathbb{R}^d.$$

The fundamental solution of the Laplace operator is (cf. 5.6)

$$U^*(x,y) \;=\; \begin{cases} -\dfrac{1}{2\pi} \log|x-y| & \text{for } d = 2, \\[2ex] \dfrac{1}{4\pi} \dfrac{1}{|x-y|} & \text{for } d = 3, \end{cases}$$

and the solution of the above Poisson equation is given by the representation formula (5.2)

$$u(x) = \int_\Gamma U^*(x,y)\gamma_1^{\text{int}} u(y) ds_y - \int_\Gamma \gamma_{1,y}^{\text{int}} U^*(x,y) \gamma_0^{\text{int}} u(y) ds_y \tag{6.1}$$
$$+ \int_\Omega U^*(x,y) f(y) dy \quad \text{for } x \in \Omega.$$

To derive appropriate boundary integral equations to find the complete Cauchy data $[\gamma_0^{\text{int}} u(x), \gamma_1^{\text{int}} u(x)]$ for $x \in \Gamma$ we first have to investigate the mapping properties of several surface and volume potentials.

## 6.1 Newton Potential

By

$$(\widetilde{N}_0 f)(x) \;:=\; \int_\Omega U^*(x,y) f(y) dy \quad \text{for } x \in \mathbb{R}^d \tag{6.2}$$

we define the volume or Newton potential of a given function $f(y)$, $y \in \Omega$.

For $\varphi, \psi \in \mathcal{S}(\mathbb{R}^d)$ we have

$$\langle \widetilde{N}_0\varphi, \psi\rangle_\Omega = \int_\Omega \psi(x) \int_\Omega U^*(x,y)\varphi(y)dydx = \langle \varphi, \widetilde{N}_0\psi\rangle_\Omega$$

and, therefore, $\widetilde{N}_0\varphi \in \mathcal{S}(\mathbb{R}^d)$. Then we can define the Newton potential $\widetilde{N}_0 : \mathcal{S}'(\mathbb{R}^d) \to \mathcal{S}'(\mathbb{R}^d)$ by

$$\langle \widetilde{N}_0 f, \psi\rangle_\Omega := \langle f, \widetilde{N}_0\psi\rangle_\Omega \quad \text{for all } \psi \in \mathcal{S}(\mathbb{R}^d).$$

**Theorem 6.1.** *The volume potential $\widetilde{N}_0 : \widetilde{H}^{-1}(\Omega) \to H^1(\Omega)$ defines a continuous map, i.e.*

$$\|\widetilde{N}_0 f\|_{H^1(\Omega)} \le c\,\|f\|_{\widetilde{H}^{-1}(\Omega)}. \tag{6.3}$$

*Proof.* For $\varphi \in C_0^\infty(\Omega)$ we first have

$$\|\varphi\|^2_{H^{-1}(\mathbb{R}^d)} = \int_{\mathbb{R}^d} \frac{|\widehat{\varphi}(\xi)|^2}{1+|\xi|^2} d\xi$$

where the Fourier transform $\widehat{\varphi}$ is

$$\widehat{\varphi}(\xi) = (2\pi)^{-\frac{d}{2}} \int_{\mathbb{R}^d} e^{-i\langle x,\xi\rangle}\varphi(x)dx.$$

Due to $\operatorname{supp}\varphi \subset \Omega$ we have

$$\begin{aligned}
\|\varphi\|_{H^{-1}(\mathbb{R}^d)} &= \sup_{0\ne v\in H^1(\mathbb{R}^d)} \frac{\langle \varphi, v\rangle_{L_2(\mathbb{R}^d)}}{\|v\|_{H^1(\mathbb{R}^d)}} \\
&\le \sup_{0\ne v\in H^1(\Omega)} \frac{\langle \varphi, v\rangle_{L_2(\Omega)}}{\|v\|_{H^1(\Omega)}} = \|\varphi\|_{\widetilde{H}^{-1}(\Omega)}.
\end{aligned}$$

Moreover,

$$u(x) := (\widetilde{N}_0\varphi)(x) = \int_\Omega U^*(x,y)\varphi(y)dy \quad \text{for } x \in \mathbb{R}^d.$$

Let $\Omega \subset B_R(0)$, and let $\mu \in C_0^\infty([0,\infty))$ be a non–negative, monotone decreasing cut off function with compact support, and let $\mu(r) = 1$ for $r \in [0, 2R]$.

Define

$$u_\mu(x) := \int_\Omega \mu(|x-y|)U^*(x,y)\varphi(y)dy \quad \text{for } x \in \mathbb{R}^d.$$

Due to $\mu(|x-y|) = 1$ for $x, y \in \Omega$ we have

$$u_\mu(x) = u(x) \quad \text{for } x \in \Omega$$

and therefore

$$\|u\|_{H^1(\Omega)} = \|u_\mu\|_{H^1(\Omega)} \leq \|u_\mu\|_{H^1(\mathbb{R}^d)}$$

with

$$\|u_\mu\|^2_{H^1(\mathbb{R}^d)} = \int\limits_{\mathbb{R}^d} (1 + |\xi|^2)\, |\widehat{u}_\mu(\xi)|^2 d\xi.$$

For the computation of the Fourier transform $\widehat{u}_\mu$ we obtain

$$\begin{aligned}
\widehat{u}_\mu(\xi) &= (2\pi)^{-\frac{d}{2}} \int\limits_{\mathbb{R}^d} e^{-i\langle x,\xi\rangle} u_\mu(x)dx \\
&= (2\pi)^{-\frac{d}{2}} \int\limits_{\mathbb{R}^d} e^{-i\langle x,\xi\rangle} \int\limits_{\mathbb{R}^d} \mu(|x-y|)U^*(x,y)\varphi(y)dydx \\
&= (2\pi)^{-\frac{d}{2}} \int\limits_{\mathbb{R}^d}\int\limits_{\mathbb{R}^d} e^{-i\langle z+y,\xi\rangle} \mu(|z|)U^*(z+y,y)\varphi(y)dydz \\
&= (2\pi)^{-\frac{d}{2}} \int\limits_{\mathbb{R}^d} e^{-i\langle y,\xi\rangle}\varphi(y)dy \int\limits_{\mathbb{R}^d} e^{-i\langle z,\xi\rangle}\mu(|z|)U^*(z,0)dz \\
&= \widehat{\varphi}(\xi) \int\limits_{\mathbb{R}^d} e^{-i\langle z,\xi\rangle}\mu(|z|)U^*(z,0)dz\,.
\end{aligned}$$

Since the function $\mu(|z|)U^*(z,0)$ depends only on $|z|$, we can use Lemma 2.13, i.e. it is sufficient to evaluate the remaining integral in $\xi = (0,0,|\xi|)^\top$.

Let us now consider the case $d = 3$ only, for $d = 2$ the further steps are almost the same. Using spherical coordinates,

$$z_1 = r\cos\phi\sin\theta, \quad z_2 = r\sin\phi\sin\theta, \quad z_3 = r\cos\theta$$

for $r \in [0,\infty), \phi \in [0,2\pi), \theta \in [0,\pi)$, we obtain for the remaining integral

$$\begin{aligned}
I(|\xi|) &= \frac{1}{4\pi}\int\limits_{\mathbb{R}^d} e^{-i\langle z,\xi\rangle}\frac{\mu(|z|)}{|z|}dz \\
&= \frac{1}{4\pi}\int\limits_0^\infty\int\limits_0^{2\pi}\int\limits_0^\pi e^{-i|\xi|r\cos\theta}\frac{\mu(r)}{r}\,r^2\sin\theta d\theta\, d\phi\, dr \\
&= \frac{1}{2}\int\limits_0^\infty r\,\mu(r)\int\limits_0^\pi e^{-ir|\xi|\cos\theta}\sin\theta\, d\theta\, dr.
\end{aligned}$$

Using the transformation $u = \cos\theta$ we get for the inner integral

$$\int_0^\pi e^{-ir|\xi|\cos\theta}\sin\theta\,d\theta = \int_{-1}^1 e^{-ir|\xi|u}\,du = \left[-\frac{1}{ir|\xi|}e^{-ir|\xi|u}\right]_{-1}^1 = \frac{2\sin r|\xi|}{r|\xi|}$$

and therefore

$$I(|\xi|) = \frac{1}{|\xi|}\int_0^\infty \mu(r)\sin r|\xi|\,dr\,.$$

For $|\xi| > 1$ we use the transformation $s := r|\xi|$ to obtain

$$I(|\xi|) = \frac{1}{|\xi|^2}\int_0^\infty \mu\left(\frac{s}{|\xi|}\right)\sin s\,ds.$$

Due to $0 \leq \mu(r) \leq 1$ and since $\mu(r)$ has compact support, we further conclude

$$|I(|\xi|)| \leq c_1(R)\,\frac{1}{|\xi|^2} \quad \text{for } |\xi| \geq 1.$$

Note that

$$(1+|\xi|^2)^2 \leq 4\,|\xi|^4 \quad \text{for } |\xi| \geq 1.$$

Then,

$$\int_{|\xi|>1} (1+|\xi|^2)|\widehat{u}_\mu(\xi)|^2 d\xi = \int_{|\xi|>1} (1+|\xi|^2)|I(|\xi|)\widehat{\varphi}(\xi)|^2 d\xi$$
$$\leq [c_1(R)]^2 \int_{|\xi|>1} \frac{1+|\xi|^2}{|\xi|^4}\,|\widehat{\varphi}(\xi)|^2\,d\xi \leq 4[c_1(R)]^2 \int_{|\xi|>1} \frac{1}{1+|\xi|^2}\,|\widehat{\varphi}(\xi)|^2\,d\xi.$$

For $|\xi| \leq 1$ we have

$$I(|\xi|) = \int_0^\infty \mu(r)\frac{\sin r|\xi|}{|\xi|}dr$$

and therefore

$$|I(|\xi|)| \leq c_2(R) \quad \text{for } |\xi| \leq 1.$$

Hence we have

$$\int_{|\xi|\leq 1} (1+|\xi|^2)|\widehat{u}_\mu(\xi)|^2 d\xi = \int_{|\xi|\leq 1} (1+|\xi|^2)|I(|\xi|)\widehat{\varphi}(\xi)|^2 d\xi$$
$$\leq 2\,[c_2(R)]^2 \int_{|\xi|\leq 1} |\widehat{\varphi}(\xi)|^2 d\xi \leq 4[c_2(R)]^2 \int_{|\xi|\leq 1} \frac{1}{1+|\xi|^2}|\widehat{\varphi}(\xi)|^2 d\xi.$$

Taking the sum this gives

$$\|u_\mu\|^2_{H^1(\mathbb{R}^d)} = \int\limits_{\xi\in R^d} (1+|\xi|^2)|\widehat{u}_\mu(\xi)|^2 d\xi$$
$$\le c \int\limits_{\xi\in\mathbb{R}^d} \frac{1}{1+|\xi|^2}|\widehat{\varphi}(\xi)|^2 d\xi = c\,\|\varphi\|^2_{H^{-1}(\mathbb{R}^d)}$$

and therefore

$$\|\widetilde{N}_0\varphi\|_{H^1(\Omega)} \le c\,\|\varphi\|_{\widetilde{H}^{-1}(\Omega)}.$$

Hence we have

$$\frac{|\langle \widetilde{N}_0 f,\varphi\rangle_\Omega|}{\|\varphi\|_{\widetilde{H}^{-1}(\Omega)}} = \frac{|\langle f,\widetilde{N}_0\varphi\rangle_\Omega|}{\|\varphi\|_{\widetilde{H}^{-1}(\Omega)}} \le \frac{\|f\|_{\widetilde{H}^{-1}(\Omega)}\|\widetilde{N}_0\varphi\|_{H^1(\Omega)}}{\|\varphi\|_{\widetilde{H}^{-1}(\Omega)}} \le c\,\|f\|_{\widetilde{H}^{-1}(\Omega)}$$

for all $\varphi \in C_0^\infty(\Omega)$. When taking the closure with respect to the norm $\|\cdot\|_{\widetilde{H}^{-1}(\Omega)}$ and using a duality arguments gives (6.3). □

**Theorem 6.2.** *The volume potential $\widetilde{N}_0\widetilde{f}$ is a generalized solution of the partial differential equation*

$$-\Delta_x(\widetilde{N}_0\widetilde{f})(x) = \widetilde{f}(x) = \begin{cases} f(x) & \text{for } x \in \Omega, \\ 0 & \text{for } x \in \mathbb{R}^d\backslash\overline{\Omega}. \end{cases} \tag{6.4}$$

*Proof.* For $\varphi \in C_0^\infty(\mathbb{R}^d)$ we apply integration by parts, exchange the order of integration, and using the symmetry of the fundamental solution we obtain

$$\begin{aligned}
\int\limits_{\mathbb{R}^d} [-\Delta_x(\widetilde{N}_0\widetilde{f})(x)]\varphi(x)dx &= \int\limits_{\mathbb{R}^d} (\widetilde{N}_0\widetilde{f})(x)[-\Delta_x\varphi(x)]dx \\
&= \int\limits_{\mathbb{R}^d}\int\limits_{\mathbb{R}^d} U^*(x,y)\widetilde{f}(y)dy[-\Delta_x\varphi(x)]dx \\
&= \int\limits_{\mathbb{R}^d} \widetilde{f}(y)\int\limits_{\mathbb{R}^d} U^*(y,x)[-\Delta_x\varphi(x)]dxdy \\
&= \int\limits_{\mathbb{R}^d} \widetilde{f}(y)\int\limits_{\mathbb{R}^d} [-\Delta_x U^*(y,x)]\varphi(x)dxdy \\
&= \int\limits_{\mathbb{R}^d} \widetilde{f}(y)\int\limits_{\mathbb{R}^d} \delta_0(x-y)\varphi(x)dxdy \\
&= \int\limits_{\mathbb{R}^d} \widetilde{f}(y)\varphi(y)dy.
\end{aligned}$$

When taking the closure of $C_0^\infty(\mathbb{R}^d)$ with respect to the norm $\|\cdot\|_{H^1(\mathbb{R}^d)}$ this shows that the partial differential equation (6.4) is satisfied in the sense of $H^{-1}(\mathbb{R}^d)$. □

Considering the restriction to the bounded domain $\Omega \subset \mathbb{R}^d$ we further conclude:

**Corollary 6.3.** *The volume potential $\widetilde{N}_0 f$ is a generalized solution of the partial differential equation*

$$-\Delta_x \widetilde{N}_0 f(x) = f(x) \quad \textit{for } x \in \Omega.$$

The application of the interior trace operator

$$\gamma_0^{\text{int}}(\widetilde{N}_0 f)(x) = \lim_{\Omega \ni \widetilde{x} \to x \in \Gamma} (\widetilde{N}_0 f)(\widetilde{x}) \tag{6.5}$$

defines a linear bounded operator

$$N_0 = \gamma_0^{\text{int}} \widetilde{N}_0 : \widetilde{H}^{-1}(\Omega) \to H^{1/2}(\Gamma)$$

satisfying

$$\|N_0 f\|_{H^{1/2}(\Gamma)} \leq c_2^N \, \|f\|_{\widetilde{H}^{-1}(\Omega)} \quad \text{for all } f \in \widetilde{H}^{-1}(\Omega). \tag{6.6}$$

**Lemma 6.4.** *Let $f \in L_\infty(\Omega)$. Then there holds*

$$(N_0 f)(x) = \gamma_0^{int}(\widetilde{N}_0 f)(x) = \int_\Omega U^*(x,y) f(y) dy$$

*for $x \in \Gamma$ as a weakly singular surface integral.*

*Proof.* For an arbitrary given $\varepsilon > 0$ we consider $\widetilde{x} \in \Omega$ and $x \in \Gamma$ satisfying $|x - \widetilde{x}| < \varepsilon$. Then we have

$$\left| \int_\Omega U^*(\widetilde{x}, y) f(y) dy - \int_{y \in \Omega : |y-x| > \varepsilon} U^*(x,y) f(y) dy \right|$$

$$\leq \left| \int_{y \in \Omega : |y-x| > \varepsilon} [U^*(\widetilde{x}, y) - U^*(x,y)] f(y) dy \right| + \left| \int_{y \in \Omega : |y-x| \leq \varepsilon} U^*(\widetilde{x}, y) f(y) dy \right|,$$

and

$$\lim_{\Omega \ni \widetilde{x} \to x \in \Gamma} \left| \int_{y \in \Omega : |y-x| > \varepsilon} [U^*(\widetilde{x}, y) - U^*(x,y)] f(y) dy \right| = 0 .$$

For the remaining term we obtain

$$\left| \int_{y \in \Omega : |y-x| \leq \varepsilon} U^*(\widetilde{x}, y) f(y) dy \right| \leq \|f\|_{L_\infty(\Omega \cap B_\varepsilon(x))} \int_{\Omega \cap B_\varepsilon(x)} |U^*(\widetilde{x}, y)| dy$$

$$\leq \|f\|_{L_\infty(\Omega)} \int_{B_{2\varepsilon}(\widetilde{x})} |U^*(\widetilde{x}, y)| dy.$$

In the case $d = 2$ we get, by using polar coordinates,

$$\int\limits_{B_{2\varepsilon}(\widetilde{x})} |U^*(\widetilde{x}, y)| dy = \frac{1}{2\pi} \int\limits_{|y-\widetilde{x}|<2\varepsilon} |\log|y - \widetilde{x}|| \, dy$$

$$= \frac{1}{2\pi} \int\limits_0^{2\pi} \int\limits_0^{2\varepsilon} |\log r| \, r \, dr d\varphi = \varepsilon^2 \, [1 - 2\log(2\varepsilon)] \, .$$

In the same way we find for $d = 3$, by using spherical coordinates,

$$\int\limits_{B_{2\varepsilon}(\widetilde{x})} |U^*(\widetilde{x}, y)| dy = \frac{1}{4\pi} \int\limits_{|y-\widetilde{x}|<2\varepsilon} \frac{1}{|y - \widetilde{x}|} dy$$

$$= \frac{1}{4\pi} \int\limits_0^{2\pi} \int\limits_0^{\pi} \int\limits_0^{2\varepsilon} \frac{1}{r} r^2 \sin\psi \, dr d\psi d\varphi = 2\,\varepsilon^2 \, .$$

Taking the limits $\widetilde{x} \to x$ and $\varepsilon \to 0$ we finally get the assertion. □

**Lemma 6.5.** *The operator* $N_1 = \gamma_1^{int} \widetilde{N}_0 : \widetilde{H}^{-1}(\Omega) \to H^{-1/2}(\Gamma)$ *is bounded, i.e.*

$$\|N_1 f\|_{H^{-1/2}(\Gamma)} = \|\gamma_1^{int} \widetilde{N}_0 f\|_{H^{-1/2}(\Gamma)} \le c \|f\|_{\widetilde{H}^{-1}(\Omega)}$$

*is satisfied for all* $f \in \widetilde{H}^{-1}(\Omega)$.

*Proof.* First we note that $u = \widetilde{N}_0 f \in H^1(\Omega)$ is a generalized solution of the partial differential equation $-\Delta u(x) = f(x)$ for $x \in \Omega$. For an arbitrary given $w \in H^{1/2}(\Gamma)$ we apply the inverse trace theorem to obtain a bounded extension $\mathcal{E}w \in H^1(\Omega)$ satisfying

$$\|\mathcal{E}w\|_{H^1(\Omega)} \le c_{IT} \, \|w\|_{H^{1/2}(\Gamma)}.$$

Now, using Green's first formula,

$$\langle \gamma_1^{\text{int}} u, w \rangle_\Gamma = \int\limits_\Omega \nabla u(x) \nabla \mathcal{E}w(x) dx - \langle f, \mathcal{E}w \rangle_\Omega,$$

we get from Theorem 6.1

$$\left| \langle \gamma_1^{\text{int}} u, w \rangle_\Gamma \right| \le \left\{ \|u\|_{H^1(\Omega)} + \|f\|_{\widetilde{H}^{-1}(\Omega)} \right\} \|\mathcal{E}w\|_{H^1(\Omega)}$$

$$\le (c+1) c_{IT} \, \|f\|_{\widetilde{H}^{-1}(\Omega)} \|w\|_{H^{-1/2}(\Gamma)}. \qquad \square$$

## 6.2 Single Layer Potential

Let $w \in H^{-1/2}(\Gamma)$ be a given density function. Then we consider the single layer potential

$$u(x) := (\widetilde{V}w)(x) := \int_\Gamma U^*(x,y)w(y)ds_y \quad \text{for } x \in \Omega \cup \Omega^c. \tag{6.7}$$

**Lemma 6.6.** *The function $u(x) = (\widetilde{V}w)(x)$, $x \in \Omega \cup \Omega^c$, as defined in* (6.7) *is a solution of the homogeneous partial differential equation*

$$-\Delta u(x) = 0 \quad \textit{for}\, x \in \Omega \cup \Omega^c.$$

*For $w \in H^{-1/2}(\Gamma)$ we have $u \in H^1(\Omega)$ satisfying*

$$\|u\|_{H^1(\Omega)} = \|\widetilde{V}w\|_{H^1(\Omega)} \leq c\,\|w\|_{H^{-1/2}(\Gamma)}.$$

*Proof.* For $x \in \Omega \cup \Omega^c$ and $y \in \Gamma$ we notice that the fundamental solution $U^*(x,y)$ is $C^\infty$. Hence we can exchange differentiation and integration to obtain

$$-\Delta_x u(x) = -\Delta_x \int_\Omega U^*(x,y)f(y)dy = \int_\Omega [-\Delta_x U^*(x,y)]f(y)dy = 0.$$

Moreover, for $\varphi \in C^\infty(\Omega)$ we have

$$\begin{aligned}\int_\Omega u(x)\varphi(x)dx &= \int_\Omega \int_\Gamma U^*(x,y)w(y)ds_y\, \varphi(x)dx \\ &= \int_\Gamma w(y) \int_\Omega U^*(x,y)\varphi(x)dx\, ds_y = \int_\Gamma w(y)(N_0\varphi)(y)ds_y\end{aligned}$$

where

$$(N_0\varphi)(y) = \gamma_0^{\text{int}} \int_\Omega U^*(x,y)\varphi(x)dx \quad \text{for } y \in \Gamma.$$

By applying the estimate (6.6) we then obtain

$$\begin{aligned}\int_\Omega u(x)\varphi(x)dx &= \int_\Gamma w(y)(N_0\varphi)(y)ds_y \\ &\leq \|w\|_{H^{-1/2}(\Gamma)}\|N_0\varphi\|_{H^{1/2}(\Gamma)} \\ &\leq c_2^N\, \|w\|_{H^{-1/2}(\Gamma)}\|\varphi\|_{\widetilde{H}^{-1}(\Omega)}.\end{aligned}$$

Taking the closure of $C^\infty(\Omega)$ with respect to the norm $\|\cdot\|_{\widetilde{H}^{-1}(\Omega)}$ and using a duality argument finishes the proof. □

The single layer potential (6.7) defines a bounded linear map

$$\widetilde{V} : H^{-1/2}(\Gamma) \to H^1(\Omega).$$

Hence, the application of the interior trace operator to $\widetilde{V}w \in H^1(\Omega)$ is well defined. This defines a bounded linear operator

$$V = \gamma_0^{\text{int}} \widetilde{V} : H^{-1/2}(\Gamma) \to H^{1/2}(\Gamma)$$

satisfying

$$\|Vw\|_{H^{1/2}(\Gamma)} \leq c_2^V \, \|w\|_{H^{-1/2}(\Gamma)} \quad \text{for all } w \in H^{-1/2}(\Gamma). \tag{6.8}$$

**Lemma 6.7.** *Let $w \in L_\infty(\Gamma)$ be given. Then we have the representation*

$$(Vw)(x) = \gamma_0^{\text{int}}(\widetilde{V}w)(x) = \int_\Gamma U^*(x,y)w(y)ds_y$$

*for $x \in \Gamma$ as a weakly singular surface integral.*

*Proof.* For an arbitrary $\varepsilon > 0$ we consider $\widetilde{x} \in \Omega$ and $x \in \Gamma$ satisfying $|x - \widetilde{x}| < \varepsilon$. Then we have

$$\left| \int_\Gamma U^*(\widetilde{x},y)w(y)ds_y - \int_{y\in\Gamma:|y-x|>\varepsilon} U^*(x,y)w(y)ds_y \right|$$

$$\leq \left| \int_{y\in\Gamma:|y-x|>\varepsilon} [U^*(\widetilde{x},y) - U^*(x,y)]w(y)ds_y \right| + \left| \int_{y\in\Gamma:|y-x|\leq\varepsilon} U^*(\widetilde{x},y)w(y)ds_y \right|,$$

and for the first expression we obtain

$$\lim_{\Omega\ni\widetilde{x}\to x\in\Gamma} \left| \int_{y\in\Gamma:|y-x|>\varepsilon} [U^*(\widetilde{x},y) - U^*(x,y)]w(y)ds_y \right| = 0\,.$$

For the remaining term we have

$$\left| \int_{y\in\Gamma:|y-x|\leq\varepsilon} U^*(\widetilde{x},y)w(y)ds_y \right| \leq \|w\|_{L_\infty(\Gamma\cap B_\varepsilon(x))} \int_{\Gamma\cap B_\varepsilon(x)} |U^*(\widetilde{x},y)|ds_y$$

$$\leq \|w\|_{L_\infty(\Gamma)} \int_{\Gamma\cap B_\varepsilon(\widetilde{x})} |U^*(\widetilde{x},y)|ds_y.$$

The assertion now follows as in the proof of Lemma 6.4 for $\widetilde{x} \to x$ and $\varepsilon \to 0$, we skip the details. □

In the same way we obtain for the exterior trace

$$(Vw)(x) = \gamma_0^{\text{ext}}(\widetilde{V}w)(x) := \lim_{\Omega^c \ni \widetilde{x} \to x \in \Gamma} (\widetilde{V}w)(\widetilde{x}) \quad \text{for } x \in \Gamma.$$

Hence we get the jump relation of the single layer potential as

$$[\gamma_0 \widetilde{V}w] := \gamma_0^{\text{ext}}(\widetilde{V}w)(x) - \gamma_0^{\text{int}}(\widetilde{V}w)(x) = 0 \quad \text{for } x \in \Gamma. \tag{6.9}$$

## 6.3 Adjoint Double Layer Potential

For a given density $w \in H^{-1/2}(\Gamma)$ we can define $\widetilde{V}w \in H^1(\Omega)$ which is a solution of the homogeneous partial differential equation (cf. Lemma 6.6). Using Lemma 4.4 we can apply the interior conormal derivative to obtain a bounded linear operator

$$\gamma_1^{\text{int}} \widetilde{V} \, : \, H^{-1/2}(\Gamma) \to H^{-1/2}(\Gamma)$$

satisfying

$$\|\gamma_1^{\text{int}} \widetilde{V}w\|_{H^{-1/2}(\Gamma)} \leq c \, \|w\|_{H^{-1/2}(\Gamma)} \quad \text{for all } w \in H^{-1/2}(\Gamma).$$

**Lemma 6.8.** *For $w \in H^{-1/2}(\Gamma)$ we have the representation*

$$\gamma_1^{int}(\widetilde{V}w)(x) = \sigma(x)w(x) + (K'w)(x) \quad \textit{for } x \in \Gamma$$

*in the sense of $H^{-1/2}(\Gamma)$, i.e.*

$$\langle \gamma_1^{int} \widetilde{V}w, v \rangle_\Gamma = \langle \sigma w + K'w, v \rangle_\Gamma \quad \textit{for all } v \in H^{1/2}(\Gamma).$$

*Here we used the adjoint double layer potential*

$$(K'w)(x) := \lim_{\varepsilon \to 0} \int\limits_{y \in \Gamma : |y-x| \geq \varepsilon} \gamma_{1,x}^{int} U^*(x,y) w(y) ds_y \tag{6.10}$$

*and*

$$\sigma(x) := \lim_{\varepsilon \to 0} \frac{1}{2(d-1)\pi} \frac{1}{\varepsilon^{d-1}} \int\limits_{y \in \Omega : |y-x| = \varepsilon} ds_y \quad \textit{for } x \in \Gamma. \tag{6.11}$$

*Proof.* For $w \in H^{-1/2}(\Gamma)$ the single layer potential $\widetilde{V}w \in H^1(\Omega)$ is a solution of the homogeneous partial differential equation. Hence, from Green's first formula we find for $\varphi \in C^\infty(\Omega)$

$$\int_\Gamma \gamma_1^{\text{int}} u(x) \gamma_0^{\text{int}} \varphi(x) ds_x = \int_\Omega \nabla_x u(x) \nabla_x \varphi(x) dx$$
$$= \int_\Omega \nabla_x \int_\Gamma U^*(x,y) w(y) ds_y \nabla_x \varphi(x) dx.$$

Inserting the definition as weakly singular surface integrals and interchanging the order of integration this gives

$$\int_\Gamma \gamma_1^{\text{int}} u(x) \gamma_0^{\text{int}} \varphi(x) ds_x$$
$$= \int_\Omega \nabla_x \left( \lim_{\varepsilon \to 0} \int_{y \in \Gamma : |y-x| \geq \varepsilon} U^*(x,y) w(y) ds_y \right) \nabla_x \varphi(x) dx$$
$$= \int_\Gamma w(y) \lim_{\varepsilon \to 0} \int_{x \in \Omega : |x-y| \geq \varepsilon} \nabla_x U^*(x,y) \nabla_x \varphi(x) dx ds_y .$$

Using again Green's first formula we obtain for $y \in \Gamma$

$$\int_{x \in \Omega : |x-y| \geq \varepsilon} \nabla_x U^*(x,y) \nabla_x \varphi(x) dx = \int_{x \in \Gamma : |x-y| \geq \varepsilon} \gamma_{1,x}^{\text{int}} U^*(x,y) \gamma_0^{\text{int}} \varphi(x) ds_x$$
$$+ \int_{x \in \Omega : |x-y| = \varepsilon} \gamma_{1,x}^{\text{int}} U^*(x,y) \varphi(x) ds_x .$$

The first summand corresponds to the double layer potential operator $K'$ as defined in (6.10). The second term can be written as

$$\int_{x \in \Omega : |x-y| = \varepsilon} \gamma_{1,x}^{\text{int}} U^*(x,y) \varphi(x) ds_x = \int_{x \in \Omega : |x-y| = \varepsilon} \gamma_{1,x}^{\text{int}} U^*(x,y) [\varphi(x) - \varphi(y)] ds_x$$
$$+ \varphi(y) \int_{x \in \Omega : |x-y| = \varepsilon} \gamma_{1,x}^{\text{int}} U^*(x,y) ds_x$$

with

$$\left| \int_{x \in \Omega : |x-y| = \varepsilon} \gamma_{1,x}^{\text{int}} U^*(x,y) [\varphi(x) - \varphi(y)] ds_x \right|$$
$$\leq \max_{x \in \Omega : |x-y| = \varepsilon} |\varphi(x) - \varphi(y)| \int_{x \in \Omega : |x-y| = \varepsilon} |\gamma_{1,x}^{\text{int}} U^*(x,y)| ds_x .$$

For $d = 2$ we have

$$\int\limits_{x\in\Omega:|x-y|=\varepsilon} |\gamma_{1,x}^{\text{int}} U^*(x,y)| ds_x \leq \int\limits_{x\in\mathbb{R}^2:|x-y|=\varepsilon} |\gamma_{1,x}^{\text{int}} U^*(x,y)| ds_x = \frac{1}{2\pi} \int\limits_{x\in\mathbb{R}^2:|x-y|=\varepsilon} \frac{1}{|x-y|} ds_x = 1$$

while for $d = 3$

$$\int\limits_{x\in\Omega:|x-y|=\varepsilon} |\gamma_{1,x}^{\text{int}} U^*(x,y)| ds_x \leq \int\limits_{x\in\mathbb{R}^3:|x-y|=\varepsilon} |\gamma_{1,x}^{\text{int}} U^*(x,y)| ds_x = \frac{1}{4\pi} \int\limits_{x\in\mathbb{R}^3:|x-y|=\varepsilon} \frac{1}{|x-y|^2} ds_x = 1.$$

Taking the limit $\varepsilon \to 0$ this gives

$$\lim_{\varepsilon\to 0} \left| \int\limits_{x\in\Omega:|x-y|=\varepsilon} \gamma_{1,x}^{\text{int}} U^*(x,y)[\varphi(x) - \varphi(y)] ds_x \right| = 0\,.$$

For the remaining integral we find by using $n_x = \dfrac{y-x}{|y-x|}$ for $x \in \Omega$, $|y-x| = \varepsilon$,

$$\int\limits_{x\in\Omega:|x-y|=\varepsilon} \gamma_{1,x}^{\text{int}} U^*(x,y) ds_x = -\frac{1}{2(d-1)\pi} \int\limits_{x\in\Omega:|x-y|=\varepsilon} \frac{(n_x, x-y)}{|x-y|^d} ds_x$$
$$= \frac{1}{2(d-1)\pi} \int\limits_{x\in\Omega:|x-y|=\varepsilon} \frac{1}{|x-y|^{d-1}} ds_x = \frac{1}{2(d-1)\pi} \frac{1}{\varepsilon^{d-1}} \int\limits_{x\in\Omega:|x-y|=\varepsilon} ds_x.$$

Taking into account the definitions (6.10) and (6.11) we finally obtain

$$\int\limits_\Gamma \gamma_1^{\text{int}} u(x) \gamma_0^{\text{int}} \varphi(x) ds_x$$
$$= \int\limits_\Gamma w(y) \left[ \lim_{\varepsilon\to 0} \int\limits_{x\in\Gamma:|x-y|\geq\varepsilon} \gamma_{1,x}^{\text{int}} U^*(x,y) \gamma_0^{\text{int}} \varphi(x) ds_x + \gamma_0^{\text{int}} \varphi(y) \sigma(y) \right] ds_y$$
$$= \int\limits_\Gamma \gamma_0^{\text{int}} \varphi(x) \lim_{\varepsilon\to 0} \int\limits_{y\in\Gamma:|y-x|\geq\varepsilon} \gamma_{1,x}^{\text{int}} U^*(x,y) w(y) ds_y ds_x + \int\limits_\Gamma w(y)\sigma(y)\varphi(y) ds_y$$
$$= \int\limits_\Gamma [\sigma(x) w(x) + (K'w)(x)] \gamma_0^{\text{int}} \varphi(x) ds_x. \qquad \square$$

Let $\Gamma = \partial\Omega$ be at least differentiable within a vicinity of $x \in \Gamma$. From the definition (6.11) we then find

$$\sigma(x) \,=\, \frac{1}{2} \quad \text{for almost all } x \in \Gamma.$$

The boundary integral operator $K'$ which appears in the conormal derivative of the single layer potential is the adjoint double layer potential. The operator is linear and bounded, i.e.

$$\|K'w\|_{H^{-1/2}(\Gamma)} \,\leq\, c_2^{K'}\,\|w\|_{H^{-1/2}(\Gamma)} \quad \text{for } w \in H^{-1/2}(\Gamma).$$

As in the proof of Lemma 6.8 we obtain the following representation of the exterior conormal derivative of the single layer potential $\widetilde{V}$ in the sense of $H^{-1/2}(\Gamma)$,

$$\gamma_1^{\text{ext}}(\widetilde{V}w)(x) \,=\, [\sigma(x)-1]w(x) + (K'w)(x) \quad \text{for } x \in \Gamma.$$

**Lemma 6.9.** *For the conormal derivative of the single layer potential* $\widetilde{V}$ *there holds the jump relation*

$$[\gamma_1\widetilde{V}w] \,:=\, \gamma_1^{ext}(\widetilde{V}w)(x) - \gamma_1^{int}(\widetilde{V}w)(x) \,=\, -w(x) \quad \textit{for } x \in \Gamma \tag{6.12}$$

*in the sense of* $H^{-1/2}(\Gamma)$.

*Proof.* For $u = \widetilde{V}w$ and $\varphi \in C_0^\infty(\mathbb{R}^d)$ we first have

$$\begin{aligned}
\int\limits_{\mathbb{R}^d} [-\Delta u(x)]\varphi(x)dx &= \int\limits_{\mathbb{R}^d} -\Delta_x \int\limits_{\Gamma} U^*(x,y)w(y)ds_y\varphi(x)dx \\
&= \int\limits_{\Gamma} w(y) \int\limits_{\mathbb{R}^d} -\Delta_x U^*(x,y)\varphi(x)dxds_y \\
&= \int\limits_{\Gamma} w(y) \int\limits_{\mathbb{R}^d} \delta_0(x-y)\varphi(x)dxds_y \\
&= \int\limits_{\Gamma} w(y)\gamma_0^{\text{int}}\varphi(y)ds_y.
\end{aligned}$$

On the other hand,

$$\begin{aligned}
\int\limits_{\mathbb{R}^d} [-\Delta u(x)]\varphi(x)dx = a_{\mathbb{R}^d}(u,\varphi) \,&=\, a_\Omega(u,\varphi) + a_{\Omega^c}(u,\varphi) \\
&= \int\limits_{\Gamma} \gamma_1^{\text{int}}u(x)\gamma_0^{\text{int}}\varphi(x)ds_x - \int\limits_{\Gamma} \gamma_1^{\text{ext}}u(x)\gamma_0^{\text{ext}}\varphi(x)ds_x,
\end{aligned}$$

and therefore

$$\int_\Gamma w(x)\varphi(x)ds_x = \int_\Gamma [\gamma_1^{\text{int}} u(x) - \gamma_1^{\text{ext}} u(x)]\gamma_0^{\text{int}} \varphi(x) ds_x$$

holds for all $\varphi \in C_0^\infty(\mathbb{R}^d)$. The closure of $C_0^\infty(\mathbb{R}^d)$ with respect to $\|\cdot\|_{H^{1/2}(\Gamma)}$ and a duality argument then gives the assertion. □

## 6.4 Double Layer Potential

Let $v \in H^{1/2}(\Gamma)$ be a given density function. Then we consider the double layer potential

$$u(x) = (Wv)(x) := \int_\Gamma [\gamma_{1,y}^{\text{int}} U^*(x,y)]v(y)ds_y \quad \text{for } x \in \Omega \cup \Omega^c. \tag{6.13}$$

**Lemma 6.10.** *The function* $u(x) = (Wv)(x)$, $x \in \Omega \cup \Omega^c$, *as defined in* (6.13) *is a solution of the homogeneous partial differential equation*

$$-\Delta_x u(x) = 0 \quad \textit{for } x \in \Omega \cup \Omega^c.$$

*For* $v \in H^{1/2}(\Gamma)$ *we have* $u \in H^1(\Omega)$ *satisfying*

$$\|u\|_{H^1(\Omega)} = \|Wv\|_{H^1(\Omega)} \leq c\,\|v\|_{H^{1/2}(\Gamma)}.$$

*Proof.* For $x \in \Omega \cup \Omega^c$ and $y \in \Gamma$ we first notice that $x \neq y$. Hence we can interchange differentiation and integration and the first assertion follows from the properties of the fundamental solution $U^*(x,y)$.

For $\varphi \in C^\infty(\Omega)$ we then have

$$\begin{aligned}\langle Wv, \varphi\rangle_\Omega &= \int_\Omega \int_\Gamma [\gamma_{1,y}^{\text{int}} U^*(x,y)]v(y)ds_y \varphi(x)dx \\ &= \int_\Gamma v(y)\gamma_{1,y}^{\text{int}} \int_\Omega U^*(x,y)\varphi(x)dx ds_y \\ &= \int_\Gamma v(y)\gamma_{1,y}^{\text{int}}(\widetilde{N}_0\varphi)(y)ds_y = \langle v, \gamma_1^{\text{int}}\widetilde{N}_0\varphi\rangle_\Gamma.\end{aligned}$$

For $f \in \widetilde{H}^{-1}(\Omega)$ this gives

$$\langle Wv, f\rangle_\Omega = \langle v, \gamma_1^{\text{int}}\widetilde{N}_0 f\rangle_\Gamma$$

and, by applying Corollary 6.3, $\widetilde{N}_0 f \in H^1(\Omega)$ is a solution of the inhomogeneous partial differential equation

$$-\Delta_x(\widetilde{N}_0 f)(x) = f(x) \quad \text{for } x \in \Omega.$$

By Lemma 6.5 we further obtain $\gamma_1^{\text{int}}\widetilde{N}_0 f \in H^{-1/2}(\Gamma)$, and therefore

$$\begin{aligned} \|Wv\|_{H^1(\Omega)} &= \sup_{0\neq f\in \widetilde{H}^{-1}(\Omega)} \frac{\langle Wv, f\rangle_\Omega}{\|f\|_{\widetilde{H}^{-1}(\Omega)}} \\ &= \sup_{0\neq f\in \widetilde{H}^{-1}(\Omega)} \frac{\langle v, \gamma_1^{\text{int}}\widetilde{N}_0 f\rangle_\Gamma}{\|f\|_{\widetilde{H}^{-1}(\Omega)}} \le c\,\|v\|_{H^{1/2}(\Gamma)}. \quad \square \end{aligned}$$

The double layer potential (6.13) therefore defines a linear and bounded operator

$$W \,:\, H^{1/2}(\Gamma) \to H^1(\Omega).$$

When applying the interior trace operator $\gamma_0^{\text{int}} : H^1(\Omega) \to H^{1/2}(\Gamma)$ to the double layer potential $u = Wv \in H^1(\Omega)$ this declares, for $v \in H^{1/2}(\Gamma)$, a linear and bounded operator

$$\gamma_0^{\text{int}} W \,:\, H^{1/2}(\Gamma) \to H^{1/2}(\Gamma)$$

satisfying

$$\|\gamma_0^{\text{int}} Wv\|_{H^{1/2}(\Gamma)} \le c\,\|v\|_{H^{1/2}(\Gamma)} \quad \text{for } v \in H^{1/2}(\Gamma).$$

**Lemma 6.11.** *For $v \in H^{1/2}(\Gamma)$ we have the representation*

$$\gamma_0^{int}(Wv)(x) = [-1+\sigma(x)]v(x) + (Kv)(x) \quad \text{for } x \in \Gamma \tag{6.14}$$

*where $\sigma(x)$ is as defined in (6.11) and with the double layer potential*

$$(Kv)(x) := \lim_{\varepsilon\to 0} \int\limits_{y\in\Gamma:|y-x|\ge\varepsilon} [\gamma_{1,y}^{int} U^*(x,y)]v(y)ds_y \quad \text{for } x \in \Gamma.$$

*Proof.* Let $\varepsilon > 0$ be arbitrary but fixed. For the operator

$$(K_\varepsilon v)(x) = \int\limits_{y\in\Gamma:|y-x|\ge\varepsilon} [\gamma_{1,y}^{\text{int}} U^*(x,y)]v(y)ds_y$$

we first consider the limit $\Omega \ni \widetilde{x} \to x \in \Gamma$. Hence we assume $|\widetilde{x} - x| < \varepsilon$. Then,

$$\begin{aligned}
(Wv)(\widetilde{x}) - (K_\varepsilon v)(x) &= \int\limits_{y\in\Gamma:|y-x|\geq\varepsilon} \left[\gamma_{1,y}^{\text{int}} U^*(\widetilde{x},y) - \gamma_{1,y}^{\text{int}} U^*(x,y)\right] v(y) ds_y \\
&\quad + \int\limits_{y\in\Gamma:|y-x|<\varepsilon} [\gamma_{1,y}^{\text{int}} U^*(\widetilde{x},y)] v(y) ds_y \\
&= \int\limits_{y\in\Gamma:|y-x|\geq\varepsilon} \left[\gamma_{1,y}^{\text{int}} U^*(\widetilde{x},y) - \gamma_{1,y}^{\text{int}} U^*(x,y)\right] v(y) ds_y \\
&\quad + \int\limits_{y\in\Gamma:|y-x|<\varepsilon} [\gamma_{1,y}^{\text{int}} U^*(\widetilde{x},y)][v(y)-v(x)] ds_y \\
&\quad + v(x) \int\limits_{y\in\Gamma:|y-x|<\varepsilon} \gamma_{1,y}^{\text{int}} U^*(\widetilde{x},y) ds_y .
\end{aligned}$$

For all $\varepsilon > 0$ we have

$$\lim_{\Omega\ni\widetilde{x}\to x\in\Gamma} \int\limits_{y\in\Gamma:|y-x|\geq\varepsilon} \left[\gamma_{1,y}^{\text{int}} U^*(\widetilde{x},y) - \gamma_{1,y}^{\text{int}} U^*(x,y)\right] v(y) ds_y = 0$$

while the second term can be estimated by

$$\begin{aligned}
&\left| \int\limits_{y\in\Gamma:|y-x|<\varepsilon} [\gamma_{1,y}^{\text{int}} U^*(\widetilde{x},y)][v(y)-v(x)] ds_y \right| \\
&\qquad \leq \sup_{y\in\Gamma:|y-x|<\varepsilon} |v(x)-v(y)| \int\limits_{y\in\Gamma:|y-x|<\varepsilon} |\gamma_{1,y}^{\text{int}} U^*(\widetilde{x},y)| ds_y .
\end{aligned}$$

For $\widetilde{x} \in \Omega$ we further have

$$\int\limits_\Gamma |\gamma_{1,y}^{\text{int}} U^*(\widetilde{x},y)| ds_y \leq M .$$

Therefore, the second term vanishes when considering the limit $\varepsilon \to 0$. For the computation of the third term we consider

$$B_\varepsilon(x) = \{y \in \Omega : |y-x| < \varepsilon\} .$$

Then,

$$\int\limits_{y\in\Gamma:|y-x|<\varepsilon} \gamma_{1,y}^{\text{int}} U^*(\widetilde{x},y) ds_y = \int\limits_{\partial B_\varepsilon(x)} \gamma_{1,y}^{\text{int}} U^*(\widetilde{x},y) ds_y - \int\limits_{y\in\Omega:|y-x|=\varepsilon} \gamma_{1,y}^{\text{int}} U^*(\widetilde{x},y) ds_y .$$

Using the representation formula (6.1) for $u = 1$ and due to $\widetilde{x} \in B_\varepsilon(x)$ we obtain

$$\int_{\partial B_\varepsilon(x)} \gamma_1^{\text{int}} U^*(\widetilde{x}, y) ds_y = -1.$$

Moreover, inserting $\underline{n}(y) = \frac{1}{\varepsilon}(y-x)$ we get

$$\begin{aligned}\lim_{\varepsilon\to 0} \lim_{\Omega\ni\widetilde{x}\to x\in\Gamma} \int_{y\in\Omega:|y-x|=\varepsilon} \gamma_1^{\text{int}} U^*(\widetilde{x}, y) ds_y &= \lim_{\varepsilon\to 0} \int_{y\in\Omega:|y-x|=\varepsilon} \gamma_1^{\text{int}} U^*(x, y) ds_y \\ &= -\lim_{\varepsilon\to 0} \frac{1}{2(d-1)\pi} \int_{y\in\Omega:|y-x|=\varepsilon} \frac{(n_y, y-x)}{|x-y|^d} ds_y \\ &= -\lim_{\varepsilon\to 0} \frac{1}{2(d-1)\pi} \frac{1}{\varepsilon^{d-1}} \int_{y\in\Omega:|y-x|=\varepsilon} ds_y = -\sigma(x). \quad \square\end{aligned}$$

In the same way we obtain for the exterior trace

$$\gamma_0^{\text{ext}}(Wv)(x) = \sigma(x)v(x) + (Kv)(x) \quad \text{for } x \in \Gamma$$

and therefore the jump relation of the double layer potential,

$$[\gamma_0 Wv] := \gamma_0^{\text{ext}}(Wv)(x) - \gamma_0^{\text{int}}(Wv)(x) = v(x) \quad \text{for } x \in \Gamma.$$

**Lemma 6.12.** *For the jump of the conormal derivative of the double layer potential there holds*

$$[\gamma_1 Wv] = \gamma_1^{ext}(Wv)(x) - \gamma_1^{int}(Wv)(x) = 0 \quad \textit{for } x \in \Gamma.$$

*Proof.* For the double layer potential $u(x) = (Wv)(x)$, $x \in \mathbb{R}^n$ and for $\varphi \in C_0^\infty(\mathbb{R}^d)$ we first have

$$\begin{aligned}\int_{\mathbb{R}^d} [-\Delta u(x)]\varphi(x)dx &= \int_{\mathbb{R}^d} -\Delta_x \int_\Gamma \gamma_{1,y}^{\text{int}} U^*(x,y)w(y)ds_y \varphi(x)dx \\ &= \int_\Gamma w(y)\gamma_{1,y}^{\text{int}} \int_{\mathbb{R}^d} -\Delta_x U^*(x,y)\varphi(x)dxds_y \\ &= \int_\Gamma w(y)\gamma_{1,y}^{\text{int}} \int_{\mathbb{R}^d} \delta_0(x-y)\varphi(x)dxds_y = 0.\end{aligned}$$

On the other hand, using Green's first formula we have

$$\begin{aligned}0 = \int_{\mathbb{R}^d} [-\Delta u(x)]\varphi(x)dx &= a_{\mathbb{R}^d}(u,\varphi) = a_\Omega(u,\varphi) + a_{\Omega^c}(u,\varphi) \\ &= \int_\Gamma \gamma_1^{\text{int}} u(x)\gamma_0^{\text{int}}\varphi(x)ds_x - \int_\Gamma \gamma_1^{\text{ext}} u(x)\gamma_0^{\text{ext}}\varphi(x)ds_x\end{aligned}$$

and taking the closure of $C_0^\infty(\mathbb{R}^d)$ with respect to the $\|\cdot\|_{H^1(\mathbb{R}^d)}$ norm we obtain the assertion from $\gamma_0^{\text{int}}\varphi = \gamma_0^{\text{ext}}\varphi$. $\square$

## 6.5 Hypersingular Boundary Integral Operator

The conormal derivative of the double layer potential $Wv$ for $v \in H^{1/2}(\Gamma)$ defines a bounded operator

$$\gamma_1^{\text{int}} W : H^{1/2}(\Gamma) \to H^{-1/2}(\Gamma).$$

For

$$(Dv)(x) := -\gamma_1^{\text{int}}(Wv)(x) = -\lim_{\Omega \ni \widetilde{x} \to x \in \Gamma} n_x \cdot \nabla_{\widetilde{x}}(Wv)(\widetilde{x}) \quad \text{for } x \in \Gamma \quad (6.15)$$

we first have

$$\|Dv\|_{H^{-1/2}(\Gamma)} \leq c_2^D \, \|v\|_{H^{1/2}(\Gamma)} \quad \text{for } v \in H^{1/2}(\Gamma). \qquad (6.16)$$

In the two–dimensional case $d = 2$ the double layer potential reads

$$(Wv)(\widetilde{x}) = \frac{1}{2\pi} \lim_{\varepsilon \to 0} \int\limits_{y \in \Gamma : |y-x| \geq \varepsilon} \frac{(\widetilde{x} - y, n_y)}{|\widetilde{x} - y|^2} v(y) ds_y \quad \text{for } \widetilde{x} \in \Omega.$$

For a fixed $\varepsilon > 0$ we can interchange taking the limit $\widetilde{x} \to x \in \Gamma$ and computing the conormal derivative to obtain $(d = 2)$

$$(D_\varepsilon v)(x) = \frac{1}{2\pi} \int\limits_{y \in \Gamma : |y-x| \geq \varepsilon} \left[ -\frac{(n_x, n_y)}{|x - y|^2} + 2 \frac{(x - y, n_x)(x - y, n_y)}{|x - y|^4} \right] v(y) ds_y.$$

In the same way we find for $d = 3$

$$(D_\varepsilon v)(x) = \frac{1}{4\pi} \int\limits_{y \in \Gamma : |y-x| \geq \varepsilon} \left[ -\frac{(n_x, n_y)}{|x - y|^3} + 3 \frac{(y - x, n_y)(y - x, n_x)}{|x - y|^5} \right] v(y) ds_y.$$

However, when taking the limit $\varepsilon \to 0$ for $x \in \Gamma$ the integrals does not exist as Cauchy principal value. As a generalization of the Cauchy integral we therefore call $D$ a hypersingular boundary integral operator. To find an explicit representation of $D$ we therefore have to introduce a suitable regularisation. Inserting $u_0(x) \equiv 1$ into the representation formula (6.1) this gives

$$1 = -\int\limits_\Gamma \gamma_{1,y}^{\text{int}} U^*(\widetilde{x}, y) ds_y \quad \text{for } \widetilde{x} \in \Omega.$$

Hence we have

$$\nabla_{\widetilde{x}}(Wu_0)(\widetilde{x}) = \underline{0} \quad \text{for } \widetilde{x} \in \Omega,$$

and therefore

$$(Du_0)(x) = 0 \quad \text{for } x \in \Gamma. \qquad (6.17)$$

Moreover we can write

$$(Dv)(x) = -\lim_{\Omega\ni\widetilde{x}\to x\in\Gamma} n_x \cdot \nabla_{\widetilde{x}} \int_\Gamma \gamma_{1,y}^{\text{int}} U^*(\widetilde{x},y)[v(y)-v(x)]ds_y \quad \text{for } x\in\Gamma.$$

If the density $v$ is continuous, we can obtain for the hypersingular boundary integral operator $D$ the representation

$$(Dv)(x) = -\int_\Gamma \gamma_{1,x}^{\text{int}}\gamma_{1,y}^{\text{int}} U^*(x,y)[v(y)-v(x)]ds_y \quad \text{for } x\in\Gamma$$

as a Cauchy principal value integral.

In what follows we will describe alternative representations of the bilinear form which is induced by the hypersingular boundary integral operator $D$,

$$\langle Du,v\rangle_\Gamma = -\int_\Gamma v(x)\gamma_{1,x}^{\text{int}} \int_\Gamma \gamma_{1,y}^{\text{int}} U^*(x,y)u(y)ds_y ds_x.$$

In the two–dimensional case $d=2$ we assume that $\Gamma=\partial\Omega$ is piecewise smooth,

$$\Gamma = \bigcup_{k=1}^p \Gamma_k,$$

where each part $\Gamma_k$ is given by a local parametrization

$$\Gamma_k : y = y(t) = \begin{pmatrix} y_1(t)\\ y_2(t)\end{pmatrix} \quad \text{for } t\in(t_k,t_{k+1}). \tag{6.18}$$

Moreover,

$$ds_y = \sqrt{[y_1'(t)]^2+[y_2'(t)]^2}\,dt,$$

and the exterior normal vector is given by

$$n(y) = \frac{1}{\sqrt{[y_1'(t)]^2+[y_2'(t)]^2}}\begin{pmatrix} y_2'(t)\\ -y_1'(t)\end{pmatrix} \quad \text{for } y\in\Gamma_k.$$

For $x\in\mathbb{R}^2$, the rotation of a scalar function $\widetilde{v}$ is defined as

$$\underline{\text{curl}}\,\widetilde{v}(x) := \begin{pmatrix} \dfrac{\partial}{\partial x_2}\widetilde{v}(x)\\ -\dfrac{\partial}{\partial x_1}\widetilde{v}(x)\end{pmatrix}.$$

If $v(x)$, $x\in\Gamma_k$, is a given function, we may consider an appropriate extension $\widetilde{v}$ into a neighborhood of $\Gamma_k$, in particular we may define

$$\widetilde{v}(\widetilde{x}) = v(x) \quad \text{for } \widetilde{x} = x + (\widetilde{x}-x,\underline{n}(x))\underline{n}(x).$$

Then, for $x \in \Gamma_k$ we can introduce

$$\mathrm{curl}_{\Gamma_k} v(x) := \underline{n}(x) \cdot \underline{\mathrm{curl}}\, \widetilde{v}(x) = n_1(x) \frac{\partial}{\partial x_2} \widetilde{v}(x) - n_2(x) \frac{\partial}{\partial x_1} \widetilde{v}(x)$$

and we obtain

$$\begin{aligned}
\int\limits_{\Gamma_k} \mathrm{curl}_{\Gamma_k} v(y) ds_y &= \int\limits_{\Gamma_k} \left[ n_1(y) \frac{\partial}{\partial y_2} \widetilde{v}(y) - n_2(y) \frac{\partial}{\partial y_1} \widetilde{v}(y) \right] ds_y \\
&= \int\limits_{t_k}^{t_{k+1}} \left[ y_2'(t) \frac{\partial}{\partial y_2} v(y(t)) + y_1'(t) \frac{\partial}{\partial y_1} v(y(t)) \right] dt \\
&= \int\limits_{t_k}^{t_{k+1}} \frac{d}{dt} v(y(t))\, dt,
\end{aligned}$$

i.e. $\mathrm{curl}_{\Gamma_k} v$ does not depend on the chosen extension $\widetilde{v}$.

**Lemma 6.13.** *Let $\Gamma_k$ be an open boundary part which is given by a local parametrization* (6.18) *with continuously differentiable functions $y_i(t)$, $i = 1, 2$. If $v$ and $w$ are continuously differentiable, then we have the formula of integration by parts, i.e.*

$$\int\limits_{\Gamma_k} v(y)\, curl_{\Gamma_k} w(y)\, ds_y = - \int\limits_{\Gamma_k} curl_{\Gamma_k} v(y)\, w(y)\, ds_y + v(y(t)) w(y(t))|_{t_k}^{t_{k+1}}.$$

*Proof.* The assertion follows from

$$\int\limits_{\Gamma_k} \mathrm{curl}_{\Gamma_k} [v(y) w(y)] ds_y = \int\limits_{t_k}^{t_{k+1}} \frac{d}{dt} [v(y(t)) w(y(t))] dt = [v(y(t)) w(y(t))]_{t=t_k}^{t=t_{k+1}}$$

through differentiation by the product rule. □

For a function $v$ which is defined on a closed curve $\Gamma$ we define

$$\mathrm{curl}_\Gamma v(x) := \mathrm{curl}_{\Gamma_k} v(x) \quad \text{for } x \in \Gamma_k,\ k = 1, \ldots, p.$$

As a consequence of Lemma 6.13 we then have:

**Corollary 6.14.** *Let $\Gamma$ be a piecewise smooth closed curve. If $v$ and $w$ are piecewise continuously differentiable, then*

$$\int\limits_{\Gamma} v(y)\, curl_\Gamma w(y)\, ds_y = - \int\limits_{\Gamma} curl_\Gamma v(y)\, w(y)\, ds_y + \sum_{k=1}^{p} v(y(t)) w(y(t))|_{t_k}^{t_{k+1}}.$$

*If in addition $v$ and $w$ are globally continuous, then*

$$\int\limits_{\Gamma} v(y)\, curl_\Gamma w(y)\, ds_y = - \int\limits_{\Gamma} curl_\Gamma v(y)\, w(y)\, ds_y.$$

By applying integration by parts we can rewrite the bilinear form which is induced by the hypersingular boundary integral operator $D$ as a bilinear form which is induced by the single layer potential $V$. In case of the two–dimensional Laplace operator this relation already goes back to [101].

**Theorem 6.15.** *Let $\Gamma$ be a piecewise smooth closed curve and let $u$ and $v$ be globally continuous on $\Gamma$. Moreover, let $u$ and $v$ be continuously differentiable on the parts $\Gamma_k$. Then we can rewrite the bilinear form of the hypersingular boundary integral operator $D$ as*

$$\langle Du, v\rangle_\Gamma = -\frac{1}{2\pi}\int\limits_\Gamma \text{curl}_\Gamma v(x) \int\limits_\Gamma \log|x-y|\,\text{curl}_\Gamma u(y) ds_y ds_x\,. \tag{6.19}$$

*Proof.* The hypersingular boundary integral operator $D$ is defined as the negative conormal derivative of the double layer potential $W$, see (6.15). For $\widetilde{x} \in \Omega$ we have

$$w(\widetilde{x}) = (Wu)(\widetilde{x}) = -\frac{1}{2\pi}\int\limits_\Gamma u(y)\frac{\partial}{\partial n_y}\log|\widetilde{x}-y|ds_y.$$

Since $\widetilde{x} \in \Omega$ and $y \in \Gamma$ it follows that $\widetilde{x} \neq y$. With

$$\frac{\partial}{\partial y_i}\log|\widetilde{x}-y| = \frac{y_i-\widetilde{x}_i}{|\widetilde{x}-y|^2} = -\frac{\widetilde{x}_i-y_i}{|\widetilde{x}-y|^2} = -\frac{\partial}{\partial \widetilde{x}_i}\log|\widetilde{x}-y|$$

we obtain

$$\frac{\partial}{\partial \widetilde{x}_i}\left(\frac{\partial}{\partial n_y}\log|\widetilde{x}-y|\right) = -\underline{n}(y)\cdot\nabla_y\left(\frac{\partial}{\partial y_i}\log|\widetilde{x}-y|\right).$$

Due to $\Delta_y \log|\widetilde{x}-y| = 0$ for $y \neq \widetilde{x}$ we further get

$$\begin{aligned}
\text{curl}_{\Gamma,y}&\left(\frac{\partial}{\partial y_1}\log|\widetilde{x}-y|\right) = \\
&= n_1(y)\frac{\partial}{\partial y_2}\frac{\partial}{\partial y_1}\log|\widetilde{x}-y| - n_2(y)\frac{\partial}{\partial y_1}\frac{\partial}{\partial y_1}\log|\widetilde{x}-y| \\
&= n_1(y)\frac{\partial}{\partial y_2}\frac{\partial}{\partial y_1}\log|\widetilde{x}-y| + n_2(y)\frac{\partial}{\partial y_2}\frac{\partial}{\partial y_2}\log|\widetilde{x}-y| \\
&= \underline{n}(y)\cdot\nabla_y\left(\frac{\partial}{\partial y_2}\log|\widetilde{x}-y|\right),
\end{aligned}$$

and

$$\text{curl}_{\Gamma,y}\left(\frac{\partial}{\partial y_2}\log|\widetilde{x}-y|\right) = -n(y)\cdot\nabla_y\left(\frac{\partial}{\partial y_1}\log|\widetilde{x}-y|\right).$$

Hence we can write the partial derivatives of the double layer potential for a globally continuous function $u$ by applying integration by parts as

$$\begin{aligned}
\frac{\partial}{\partial \widetilde{x}_1} w(\widetilde{x}) &= -\frac{1}{2\pi} \int_\Gamma u(y) \frac{\partial}{\partial \widetilde{x}_1} \frac{\partial}{\partial n_y} \log|\widetilde{x} - y| \, ds_y \\
&= \frac{1}{2\pi} \int_\Gamma u(y) n(y) \cdot \nabla_y \left( \frac{\partial}{\partial y_1} \log|\widetilde{x} - y| \right) ds_y \\
&= -\frac{1}{2\pi} \int_\Gamma u(y) \, \mathrm{curl}_{\Gamma,y} \left( \frac{\partial}{\partial y_2} \log|\widetilde{x} - y| \right) ds_y \\
&= \frac{1}{2\pi} \int_\Gamma \mathrm{curl}_\Gamma u(y) \, \frac{\partial}{\partial y_2} \log|\widetilde{x} - y| \, ds_y,
\end{aligned}$$

and

$$\frac{\partial}{\partial \widetilde{x}_2} w(\widetilde{x}) = -\frac{1}{2\pi} \int_\Gamma \mathrm{curl}_\Gamma u(y) \, \frac{\partial}{\partial y_1} \log|\widetilde{x} - y| \, ds_y.$$

For the normal derivative of the double layer potential we then obtain

$$\begin{aligned}
&\underline{n}(x) \cdot \nabla_{\widetilde{x}} w(\widetilde{x}) = \\
&= \frac{1}{2\pi} \int_\Gamma \mathrm{curl}_\Gamma u(y) \left[ n_1(x) \frac{\partial}{\partial y_2} \log|\widetilde{x} - y| - n_2(x) \frac{\partial}{\partial y_1} \log|\widetilde{x} - y| \right] ds_y \\
&= \frac{1}{2\pi} \lim_{\varepsilon \to 0} \int_{y \in \Gamma, |y-x| \geq \varepsilon} \mathrm{curl}_\Gamma u(y) \left( n_1(x) \frac{\partial}{\partial y_2} \log|\widetilde{x} - y| - n_2(x) \frac{\partial}{\partial y_1} \log|\widetilde{x} - y| \right) ds_y
\end{aligned}$$

and taking the limit $\Omega \ni \widetilde{x} \to x \in \Gamma$ this gives

$$\begin{aligned}
&\frac{\partial}{\partial n_x} w(x) = \\
&= \frac{1}{2\pi} \lim_{\varepsilon \to 0} \int_{y \in \Gamma : |y-x| \geq \varepsilon} \mathrm{curl}_\Gamma u(y) \left( n_1(x) \frac{\partial}{\partial y_2} \log|x - y| - n_2(x) \frac{\partial}{\partial y_1} \log|x - y| \right) ds_y \\
&= -\frac{1}{2\pi} \lim_{\varepsilon \to 0} \int_{y \in \Gamma : |y-x| \geq \varepsilon} \mathrm{curl}_\Gamma u(y) \left( n_1(x) \frac{\partial}{\partial x_2} \log|x - y| - n_2(x) \frac{\partial}{\partial x_1} \log|x - y| \right) ds_y \\
&= -\frac{1}{2\pi} \lim_{\varepsilon \to 0} \int_{y \in \Gamma : |y-x| \geq \varepsilon} \mathrm{curl}_\Gamma u(y) \, \mathrm{curl}_{\Gamma,x} \log|x - y| \, ds_y.
\end{aligned}$$

Therefore,

$$\begin{aligned}\int_\Gamma v(x)\frac{\partial}{\partial n_x}w(x)ds_x &= -\frac{1}{2\pi}\int_{x\in\Gamma} v(x) \lim_{\varepsilon\to 0}\int_{y\in\Gamma:|y-x|\ge\varepsilon} \text{curl}_\Gamma u(y)\,\text{curl}_{\Gamma,x}\log|x-y|\,ds_y ds_x \\ &= -\frac{1}{2\pi}\int_{y\in\Gamma}\text{curl}_\Gamma u(y)\lim_{\varepsilon\to 0}\int_{x\in\Gamma:|x-y|\ge\varepsilon} v(x)\,\text{curl}_{\Gamma,x}\log|x-y|\,ds_x ds_y \\ &= \frac{1}{2\pi}\int_{y\in\Gamma}\text{curl}_\Gamma u(y)\lim_{\varepsilon\to 0}\int_{x\in\Gamma:|x-y|\ge\varepsilon}\text{curl}_\Gamma v(x)\,\log|x-y|\,ds_x ds_y,\end{aligned}$$

from which we finally obtain (6.19). □

The representation of the hypersingular boundary integral operator $D$ via integration by parts can be applied correspondingly to the three–dimensional case [50]. Let

$$\Gamma = \bigcup_{k=1}^{p}\Gamma_k$$

be a piecewise smooth surface where each piece $\Gamma_k$ can be described via a parametrization

$$y\in\Gamma_k\,:\,y(s,t) = \begin{pmatrix} y_1(s,t)\\ y_2(s,t)\\ y_3(s,t)\end{pmatrix} \quad \text{for } (s,t)\in\tau$$

where $\tau$ is some reference element. The rotation or curl of a vector–valued function $\underline{v}$ is defined as

$$\underline{\text{curl}}\,\underline{v} := \nabla\times\underline{v}(x) \quad \text{for } x\in\mathbb{R}^3.$$

If $u$ is a scalar function given on $\Gamma_k$, then

$$\underline{\text{curl}}_{\Gamma_k} u(x) := \underline{n}(x)\times\nabla\widetilde{u}(x) \quad \text{for } x\in\Gamma_k$$

defines the surface curl, where $\widetilde{u}$ is a suitable extension of the given $u$ on $\Gamma_k$ into a three–dimensional neighborhood of $\Gamma_k$. Finally we introduce

$$\text{curl}_{\Gamma_k}\underline{v}(x) := \underline{n}(x)\cdot\underline{\text{curl}}\,\widetilde{\underline{v}}(x) \quad \text{for } x\in\Gamma_k.$$

**Lemma 6.16.** *Let $\Gamma$ be a piecewise smooth closed Lipschitz surface in $\mathbb{R}^3$. Assume that each surface part $\Gamma_k$ is smooth having a piecewise smooth boundary curve $\partial\Gamma_k$. Let $u$ and $v$ be globally continuous, but locally bounded and smooth on each $\Gamma_k$. Then, applying integration by parts,*

$$\int_\Gamma \underline{\text{curl}}_\Gamma u(x)\cdot\underline{v}(x)\,ds_x = -\int_\Gamma u(x)\text{curl}_\Gamma\underline{v}(x)ds_x\,.$$

*Proof.* Using the product rule

$$\nabla \times [\widetilde{u}(x)\underline{v}(x)] \;=\; \nabla\widetilde{u}(x) \times \underline{v}(x) + \widetilde{u}(x)\,[\nabla \times \underline{v}(x)]$$

we obtain

$$\begin{aligned}\int\limits_{\Gamma_k} \underline{\mathrm{curl}}_{\Gamma_k} u(x) \cdot \underline{v}(x) ds_x &= \int\limits_{\Gamma_k} [\underline{n}(x) \times \nabla\widetilde{u}(x)] \cdot \underline{v}(x) ds_x \\ &= \int\limits_{\Gamma_k} [\nabla\widetilde{u}(x) \times \underline{v}(x)] \cdot n(x)\, ds_x \\ &= \int\limits_{\Gamma_k} [\nabla \times [\widetilde{u}(x)\underline{v}(x)] - \widetilde{u}(x)\,[\nabla \times \underline{v}(x)]] \cdot n(x)\, ds_x \\ &= \int\limits_{\partial\Gamma_k} u(x)\underline{v}(x)\underline{t}(x) d\sigma - \int\limits_{\Gamma_k} u(x)\, \mathrm{curl}_{\Gamma_k} \underline{v}(x) ds_x\end{aligned}$$

by applying the integral theorem of Stokes. □

**Theorem 6.17.** *Let $\Gamma$ be a piecewise smooth closed surface, and let $u$ and $v$ be globally continuous functions defined on $\Gamma$ which are differentiable on $\Gamma_k$. Then the bilinear form of the hypersingular boundary integral operator $D$ can be written as*

$$\langle Du, v\rangle_\Gamma \;=\; \frac{1}{4\pi} \int\limits_\Gamma \int\limits_\Gamma \frac{\underline{\mathrm{curl}}_\Gamma u(y) \cdot \underline{\mathrm{curl}}_\Gamma v(x)}{|x-y|} ds_x ds_y.$$

*Proof.* The proof follows essentially as in the two–dimensional case. For $\widetilde{x} \in \Omega$ and using the definition (6.15) of the hypersingular boundary integral operator $D$ we have to consider the double layer potential

$$w(\widetilde{x}) \;:=\; -\frac{1}{4\pi} \int\limits_\Gamma u(y) \frac{\partial}{\partial n_y} \frac{1}{|\widetilde{x}-y|} ds_y.$$

Using

$$\frac{\partial}{\partial y_i} \frac{1}{|\widetilde{x}-y|} \;=\; \frac{\widetilde{x}_i - y_i}{|\widetilde{x}-y|^3} \;=\; -\frac{y_i - \widetilde{x}_i}{|\widetilde{x}-y|} \;=\; -\frac{\partial}{\partial \widetilde{x}_i} \frac{1}{|\widetilde{x}-y|}$$

we obtain for the partial derivatives of the kernel function

$$\frac{\partial}{\partial \widetilde{x}_i} \left( \frac{\partial}{\partial n_y} \frac{1}{|\widetilde{x}-y|} \right) \;=\; -\underline{n}(y) \cdot \nabla_y \left( \frac{\partial}{\partial y_i} \frac{1}{|\widetilde{x}-y|} \right).$$

Let $\underline{e}_i$ be the $i$–th unit vector of $\mathbb{R}^3$. Due to $\widetilde{x} \neq y$ we can expand the vector product as

$$\begin{aligned}
\underline{\operatorname{curl}}_y \left( \underline{e}_i \times \nabla_y \frac{1}{|\widetilde{x}-y|} \right) &= \nabla_y \times \left( \underline{e}_i \times \nabla_y \frac{1}{|\widetilde{x}-y|} \right) \\
&= \left( \nabla_y \cdot \nabla_y \frac{1}{|\widetilde{x}-y|} \right) \underline{e}_i - (\nabla_y \cdot \underline{e}_i)\, \nabla_y \frac{1}{|\widetilde{x}-y|} \\
&= \Delta_y \frac{1}{|\widetilde{x}-y|}\, \underline{e}_i - \frac{\partial}{\partial y_i} \left( \nabla_y \frac{1}{|\widetilde{x}-y|} \right) = -\frac{\partial}{\partial y_i} \left( \nabla_y \frac{1}{|\widetilde{x}-y|} \right).
\end{aligned}$$

When exchanging the order of differentiation and integration we then obtain the partial derivatives of the double layer potential as

$$\begin{aligned}
\frac{\partial}{\partial \widetilde{x}_i} w(\widetilde{x}) &= -\frac{1}{4\pi} \int_\Gamma u(y) \frac{\partial}{\partial \widetilde{x}_i} \left( \frac{\partial}{\partial n_y} \frac{1}{|\widetilde{x}-y|} \right) ds_y \\
&= \frac{1}{4\pi} \int_\Gamma u(y)\, n_y \cdot \nabla_y \left( \frac{\partial}{\partial y_i} \frac{1}{|\widetilde{x}-y|} \right) ds_y \\
&= -\frac{1}{4\pi} \int_\Gamma u(y)\, n_y \cdot \underline{\operatorname{curl}}_y \left( \underline{e}_i \times \nabla_y \frac{1}{|\widetilde{x}-y|} \right) ds_y \\
&= -\frac{1}{4\pi} \int_\Gamma u(y) \operatorname{curl}_{\Gamma,y} \left( \underline{e}_i \times \nabla_y \frac{1}{|\widetilde{x}-y|} \right) ds_y.
\end{aligned}$$

By using Lemma 6.16 we have

$$\begin{aligned}
\frac{\partial}{\partial \widetilde{x}_i} w(\widetilde{x}) &= \frac{1}{4\pi} \int_\Gamma \underline{\operatorname{curl}}_{\Gamma,y} u(y) \cdot \left( \underline{e}_i \times \nabla_y \frac{1}{|\widetilde{x}-y|} \right) ds_y \\
&= -\frac{1}{4\pi} \int_\Gamma \underline{e}_i \cdot \left( \underline{\operatorname{curl}}_{\Gamma,y} u(y) \times \nabla_y \frac{1}{|\widetilde{x}-y|} \right) ds_y,
\end{aligned}$$

and hence we can write the gradient of the double layer potential as

$$\begin{aligned}
\nabla_{\widetilde{x}} w(\widetilde{x}) &= -\frac{1}{4\pi} \int_\Gamma \left( \underline{\operatorname{curl}}_{\Gamma,y} u(y) \times \nabla_y \frac{1}{|\widetilde{x}-y|} \right) ds_y \\
&= \frac{1}{4\pi} \int_\Gamma \left( \underline{\operatorname{curl}}_{\Gamma,y} u(y) \times \nabla_{\widetilde{x}} \frac{1}{|\widetilde{x}-y|} \right) ds_y.
\end{aligned}$$

Multiplying this with the normal vector $\underline{n}(x)$ this gives

$$\begin{aligned}
\underline{n}(x) \cdot \nabla_{\widetilde{x}} w(\widetilde{x}) &= \frac{1}{4\pi} \int_\Gamma \left( \underline{\operatorname{curl}}_{\Gamma,y} u(y) \times \nabla_{\widetilde{x}} \frac{1}{|\widetilde{x}-y|} \right) \cdot \underline{n}(x)\, ds_y \\
&= -\frac{1}{4\pi} \int_\Gamma \underline{\operatorname{curl}}_{\Gamma,y} u(y) \cdot \left( \underline{n}(x) \times \nabla_{\widetilde{x}} \frac{1}{|\widetilde{x}-y|} \right) ds_y \\
&= -\frac{1}{4\pi} \lim_{\varepsilon \to 0} \int_{y \in \Gamma : |y-x| \geq \varepsilon} \underline{\operatorname{curl}}_{\Gamma,y} u(y) \cdot \left( \underline{n}(x) \times \nabla_{\widetilde{x}} \frac{1}{|\widetilde{x}-y|} \right) ds_y.
\end{aligned}$$

Taking the limit $\Omega \ni \widetilde{x} \to x \in \Gamma$ we find

$$(Du)(x) = -\frac{1}{4\pi} \lim_{\varepsilon \to 0} \int\limits_{y \in \Gamma : |y-x| \geq \varepsilon} \underline{\operatorname{curl}}_{\Gamma,y} u(y) \cdot \left( \underline{n}(x) \times \nabla_x \frac{1}{|x-y|} \right) ds_y$$
$$= -\frac{1}{4\pi} \lim_{\varepsilon \to 0} \int\limits_{y \in \Gamma : |y-x| \geq \varepsilon} \underline{\operatorname{curl}}_{\Gamma,y} u(y) \cdot \underline{\operatorname{curl}}_{\Gamma,x} \frac{1}{|x-y|} ds_y .$$

For the bilinear form of the hypersingular boundary integral operator we therefore obtain

$$\langle Du, v \rangle_\Gamma = -\frac{1}{4\pi} \int\limits_\Gamma v(x) \lim_{\varepsilon \to 0} \int\limits_{y \in \Gamma : |y-x| \geq \varepsilon} \underline{\operatorname{curl}}_{\Gamma,y} u(y) \cdot \underline{\operatorname{curl}}_{\Gamma,x} \frac{1}{|x-y|} ds_y ds_x$$
$$= -\frac{1}{4\pi} \int\limits_\Gamma \lim_{\varepsilon \to 0} \int\limits_{x \in \Gamma : |x-y| \geq \varepsilon} \left( v(x) \underline{\operatorname{curl}}_{\Gamma,y} u(y) \right) \cdot \underline{\operatorname{curl}}_{\Gamma,x} \frac{1}{|x-y|} ds_x ds_y$$
$$= \frac{1}{4\pi} \int\limits_\Gamma \lim_{\varepsilon \to 0} \int\limits_{x \in \Gamma : |x-y| \geq \varepsilon} \operatorname{curl}_{\Gamma,x} \left( v(x) \underline{\operatorname{curl}}_{\Gamma,y} u(y) \right) \frac{1}{|x-y|} ds_x ds_y .$$

By using

$$\begin{aligned} \operatorname{curl}_{\Gamma,x} [v(x) \underline{\operatorname{curl}}_{\Gamma,y} u(y)] &= \underline{n}(x) \cdot \left[ \nabla_x \times [v(x) \underline{\operatorname{curl}}_{\Gamma,y} u(y)] \right] \\ &= \underline{n}(x) \cdot \left[ \nabla_x v(x) \times \underline{\operatorname{curl}}_{\Gamma,y} u(y) \right] \\ &= [\underline{n}(x) \times \nabla_x v(x)] \cdot \underline{\operatorname{curl}}_{\Gamma,y} u(y) \\ &= \underline{\operatorname{curl}}_{\Gamma,x} v(x) \cdot \underline{\operatorname{curl}}_{\Gamma,y} u(y) \end{aligned}$$

we finally conclude the assertion. □

## 6.6 Properties of Boundary Integral Operators

Before proving the ellipticity of the single layer potential $V$ and of the hypersingular boundary integral operator $D$ we will derive some basic relations of boundary integral operators. For this we first consider the representation formula (6.1) for $\widetilde{x} \in \Omega$,

$$u(\widetilde{x}) = \int\limits_\Gamma U^*(\widetilde{x}, y) \gamma_1^{\text{int}} u(y) ds_y - \int\limits_\Gamma \gamma_{1,y}^{\text{int}} U^*(\widetilde{x}, y) \gamma_0^{\text{int}} u(y) ds_y$$
$$+ \int\limits_\Omega U^*(\widetilde{x}, y) f(y) dy .$$

Taking the limit $\Omega \ni \widetilde{x} \to x \in \Gamma$ we find from all properties of boundary and volume potentials as already considered in this chapter a boundary integral equation for $x \in \Gamma$,

$$\gamma_0^{\text{int}} u(x) = (V\gamma_1^{\text{int}} u)(x) + [1-\sigma(x)]\gamma_0^{\text{int}} u(x) - (K\gamma_0^{\text{int}} u)(x) + N_0 f(x). \quad (6.20)$$

The application of the conormal derivative to the function $u$ defined by the representation formula yields a second boundary integral equation for $x \in \Gamma$,

$$\gamma_1^{\text{int}} u(x) = \sigma(x)\gamma_1^{\text{int}} u(x) + (K'\gamma_1^{\text{int}} u)(x) + (D\gamma_0^{\text{int}} u)(x) + N_1 f(x). \quad (6.21)$$

With (6.20) and (6.21) we have obtained a system of two boundary integral equations which can be written for $x \in \Gamma$ as

$$\begin{pmatrix} \gamma_0^{\text{int}} u \\ \gamma_1^{\text{int}} u \end{pmatrix} = \begin{pmatrix} (1-\sigma)I - K & V \\ D & \sigma I + K' \end{pmatrix} \begin{pmatrix} \gamma_0^{\text{int}} u \\ \gamma_1^{\text{int}} u \end{pmatrix} + \begin{pmatrix} N_0 f \\ N_1 f \end{pmatrix} \quad (6.22)$$

where

$$\mathcal{C} = \begin{pmatrix} (1-\sigma)I - K & V \\ D & \sigma I + K' \end{pmatrix} \quad (6.23)$$

is the Calderón projection.

**Lemma 6.18.** *The operator* $\mathcal{C}$ *as defined in* (6.23) *is a projection, i.e.,* $\mathcal{C} = \mathcal{C}^2$.

*Proof.* Let $(\psi, \varphi) \in H^{-1/2}(\Gamma) \times H^{1/2}(\Gamma)$ be arbitrary but fixed. The function

$$u(\widetilde{x}) := (\widetilde{V}\psi)(\widetilde{x}) - (W\varphi)(\widetilde{x}) \quad \text{for } \widetilde{x} \in \Omega$$

is then a solution of the homogeneous partial differential equation. For the trace and for the conormal derivative of $u$ we find from the properties of the boundary potentials for $x \in \Gamma$

$$\begin{aligned} \gamma_0^{\text{int}} u(x) &= (V\psi)(x) + (1-\sigma(x))\varphi(x) - (K\varphi)(x), \\ \gamma_1^{\text{int}} u(x) &= \sigma\psi(x) + (K'\psi)(x) + (D\varphi)(x). \end{aligned}$$

The function $u$ is therefore a solution of the homogeneous partial differential equation whereas the associated Cauchy data are determined for $x \in \Gamma$ by $[\gamma_0^{\text{int}} u(x), \gamma_1^{\text{int}} u(x)]$. These Cauchy data are therefore solutions of the boundary integral equations (6.20) and (6.21), i.e. for $x \in \Gamma$

$$\begin{aligned} (V\gamma_1^{\text{int}} u)(x) &= (\sigma I + K)\gamma_0^{\text{int}} u(x), \\ (D\gamma_0^{\text{int}} u)(x) &= ((1-\sigma)I - K')\gamma_1^{\text{int}} u(x). \end{aligned}$$

This is equivalent to

$$\begin{pmatrix} \gamma_0^{\text{int}} u(x) \\ \gamma_1^{\text{int}} u(x) \end{pmatrix} = \begin{pmatrix} (1-\sigma)I - K & V \\ D & \sigma I + K' \end{pmatrix} \begin{pmatrix} \gamma_0^{\text{int}} u(x) \\ \gamma_1^{\text{int}} u(x) \end{pmatrix}.$$

Inserting

$$\begin{pmatrix} \gamma_0^{\text{int}} u(x) \\ \gamma_1^{\text{int}} u(x) \end{pmatrix} = \begin{pmatrix} (1-\sigma)I - K & V \\ D & \sigma I + K' \end{pmatrix} \begin{pmatrix} \varphi(x) \\ \psi(x) \end{pmatrix}$$

this gives the assertion. □

From the projection property $\mathcal{C} = \mathcal{C}^2$ we can immediately conclude the following relations of boundary integral operators.

**Corollary 6.19.** *For all boundary integral operators there hold the relations*

$$VD = (\sigma I + K)((1-\sigma)I - K), \tag{6.24}$$

$$DV = (\sigma I + K')((1-\sigma)I - K'), \tag{6.25}$$

$$VK' = KV, \tag{6.26}$$

$$K'D = DK. \tag{6.27}$$

Note that (6.26) describes the symmetrization of the double layer potential $K$, which is in general not self–adjoint, by the single layer potential $V$. This property was already described by J. Plemelj in 1911 in the case of the two–dimensional Laplace operator [112].

From the system (6.22) of boundary integral equations we may also find a suitable representation of the Newton potential $N_1 f$ when assuming the invertibility of the single layer potential $V$, see also Subsection 6.6.1.

**Lemma 6.20.** *For the volume potential* $(N_1 f)(x)$, $x \in \Gamma$, *there holds the representation*

$$(N_1 f)(x) = ([\sigma - 1]I + K')V^{-1}(N_0 f)(x)\,.$$

*Proof.* Using the first boundary integral equation in (6.22) and assuming the invertibility of the single layer potential $V$ we first obtain

$$\gamma_1^{\text{int}} u(x) = V^{-1}(\sigma I + K)\gamma_0^{\text{int}} u(x) - V^{-1}(N_0 f)(x) \quad \text{for } x \in \Gamma.$$

Inserting this into the second boundary integral equation of (6.22) we get

$$\begin{aligned}
\gamma_1^{\text{int}} u(x) &= (D\gamma_0^{\text{int}} u)(x) + (\sigma I + K')\gamma_1^{\text{int}} u(x) + (N_1 f)(x) \\
&= (D\gamma_0^{\text{int}} u)(x) + (\sigma I + K')[V^{-1}(\sigma(x) + K)\gamma_0^{\text{int}} u(x) - V^{-1}(N_0 f)(x)] \\
&\qquad +(N_1 f)(x) \\
&= \left[D + (\sigma I + K')V^{-1}(\sigma I + K)\right] \gamma_0^{\text{int}} u(x) \\
&\qquad -(\sigma I + K')V^{-1}(N_0 f)(x) + (N_1 f)(x)
\end{aligned}$$

and therefore the equality

$$-V^{-1}(N_0 f)(x) = -(\sigma I + K')V^{-1}(N_0 f)(x) + (N_1 f)(x) \quad \text{for } x \in \Gamma.$$

From this we immediately find the assertion. □

### 6.6.1 Ellipticity of the Single Layer Potential

By using Theorem 3.4 (Lax–Milgram theorem) we can ensure the invertibility of the single layer potential $V : H^{-1/2}(\Gamma) \to H^{1/2}(\Gamma)$. Hence we need to prove the $H^{-1/2}(\Gamma)$–ellipticity of $V$.

The function

$$u(x) = (\widetilde{V}w)(x) \quad \text{for } x \in \Omega$$

is a solution of the interior Dirichlet boundary value problem

$$-\Delta u(x) = 0 \quad \text{for } x \in \Omega, \quad u(x) = \gamma_0^{\text{int}}(\widetilde{V}w)(x) = (Vw)(x) \quad \text{for } x \in \Gamma.$$

Assuming $w \in H^{-1/2}(\Gamma)$ we have $u = \widetilde{V}w \in H^1(\Omega)$ and by choosing $v \in H^1(\Omega)$ we obtain, by applying Green's first formula (1.5),

$$a_\Omega(u,v) := \int_\Omega \nabla u(x) \nabla v(x)\, dx = \langle \gamma_1^{\text{int}} u, \gamma_0^{\text{int}} v\rangle_\Gamma . \tag{6.28}$$

Moreover, inequality (4.17) implies

$$c_1^{\text{int}}\, \|\gamma_1^{\text{int}} u\|^2_{H^{-1/2}(\Gamma)} \le a_\Omega(u,u) . \tag{6.29}$$

To obtain a corresponding result for the exterior conormal derivative $\gamma_1^{\text{ext}} u \in H^{-1/2}(\Gamma)$ we need to investigate the far field behavior of the single layer potential $(\widetilde{V}w)(x)$ as $|x| \to \infty$. For this we first introduce the subspace

$$H_*^{-1/2}(\Gamma) := \Big\{ w \in H^{-1/2}(\Gamma) \, : \, \langle w, 1\rangle_\Gamma = 0 \Big\} \tag{6.30}$$

of functions which are orthogonal to the constants.

**Lemma 6.21.** *For $y_0 \in \Omega$ and $x \in \mathbb{R}^d$ we assume*

$$|x - y_0| > \max\{1, 2\,\text{diam}(\Omega)\}$$

*to be satisfied. Let $w \in H^{-1/2}(\Gamma)$ for $d = 3$ and $w \in H_*^{-1/2}(\Gamma)$ for $d = 2$, respectively. For $u = \widetilde{V}w$ we then have the bounds*

$$|u(x)| = |(\widetilde{V}w)(x)| \le c_1(w) \frac{1}{|x - y_0|} \tag{6.31}$$

*and*

$$|\nabla u(x)| = |\nabla(\widetilde{V}w)(x)| \le c_2(w) \frac{1}{|x - y_0|^2} . \tag{6.32}$$

*Proof.* By using the triangle inequality we have for $y \in \Omega$

$$|x - y_0| \le |x - y| + |y - y_0| \le |x - y| + \text{diam}(\Omega) \le |x - y| + \frac{1}{2}|x - y_0|$$

and therefore

$$|x-y| \geq \frac{1}{2}|x-y_0|.$$

In the case $d=3$ we find the estimate (6.31) from

$$\begin{aligned}|u(x)| &= \left|\langle \gamma_0^{\text{int}} U^*(x,\cdot), w\rangle_\Gamma\right| \\ &\leq \|\gamma_0^{\text{int}} U^*(x,\cdot)\|_{H^{1/2}(\Gamma)} \|w\|_{H^{-1/2}(\Gamma)} \leq c_T \,||U^*(x,\cdot)||_{H^1(\Omega)} ||w||_{H^{-1/2}(\Gamma)}\end{aligned}$$

and by using

$$\begin{aligned}||U^*(x,\cdot)||^2_{H^1(\Omega)} &= \frac{1}{16\pi^2}\int_\Omega \frac{1}{|x-y|^2}dy + \frac{1}{16\pi^2}\int_\Omega \frac{1}{|x-y|^4}dy \\ &\leq \frac{1}{4\pi^2}\int_\Omega \frac{1}{|x-y_0|^2}dy + \frac{1}{\pi^2}\int_\Omega \frac{1}{|x-y_0|^4}dy \leq \frac{5}{4}\frac{|\Omega|}{\pi^2}\frac{1}{|x-y_0|^2}.\end{aligned}$$

The estimate (6.32) follows correspondingly from

$$\frac{\partial}{\partial x_i}u(x) = \frac{1}{4\pi}\int_\Gamma \frac{y_i - x_i}{|x-y|^3} w(y) ds_y.$$

In the case $d=2$ we consider the Taylor expansion

$$\log|y-x| = \log|y_0 - x| + \frac{(y-y_0, \bar{y}-x)}{|\bar{y}-x|^2}$$

with an appropriate $\bar{y} \in \Omega$. Due to $w \in H_*^{-1/2}(\Gamma)$ we then obtain

$$u(x) = -\frac{1}{2\pi}\int_\Gamma \frac{(y-y_0, \bar{y}-x)}{|\bar{y}-x|^2} w(y) ds_y,$$

and therefore the estimate (6.31) follows as in the three–dimensional case. The estimate (6.32) follows in the same way. □

For $y_0 \in \Omega$ and $R > 2\,\mathrm{diam}(\Omega)$ we define

$$B_R(y_0) := \left\{x \in \mathbb{R}^d \,:\, |x-y_0| < R\right\}.$$

Then, $u(x) = (\widetilde{V}w)(x)$ for $x \in \Omega^c$ is the unique solution of the Dirichlet boundary value problem

$$\begin{aligned} -\Delta u(x) &= 0 && \text{for } x \in B_R(y_0)\backslash\overline{\Omega}, \\ u(x) &= \gamma_0^{\text{ext}}(\widetilde{V}w)(x) = (Vw)(x) && \text{for } x \in \Gamma, \\ u(x) &= (\widetilde{V}w)(x) && \text{for } x \in \partial B_R(y_0). \end{aligned}$$

Using Green's first formula with respect to the bounded domain $B_R(y_0)\backslash\overline{\Omega}$ this gives

$$a_{B_R(y_0)\backslash\overline{\Omega}}(u,v) \;=\; -\langle\gamma_1^{\text{ext}}u,\gamma_0^{\text{ext}}v\rangle_\Gamma + \langle\gamma_1^{\text{int}}u,\gamma_0^{\text{int}}v\rangle_{\partial B_R(y_0)}$$

where we have used the opposite direction of the exterior normal vector along $\Gamma$. Choosing $v = u$ and using Lemma 6.21 we have

$$\Big|\langle\gamma_1^{\text{int}}u,\gamma_0^{\text{int}}u\rangle_{\partial B_R(y_0)}\Big| \le c_1(w)c_2(w) \int\limits_{|x-y_0|=R} \frac{1}{|x-y_0|^3}ds_x \;\le\; c\,R^{d-4}.$$

Hence we can consider the limit $R \to \infty$ to obtain Green's first formula for $u = \widetilde{V}w$ with respect to the exterior domain as

$$a_{\Omega^c}(u,u) := \int\limits_{\Omega^c} \nabla u(x)\nabla u(x)dx \;=\; -\langle\gamma_1^{\text{ext}}u,\gamma_0^{\text{ext}}u\rangle_\Gamma\,. \qquad (6.33)$$

Note that in the two–dimensional case the assumption $w \in H_*^{-1/2}(\Gamma)$ is essential to ensure the above result. In analogy to the estimate (4.17) for the solution of the interior Dirichlet boundary value problem we find

$$c_1^{\text{ext}}\,\|\gamma_1^{\text{ext}}u\|^2_{H^{-1/2}(\Gamma)} \;\le\; a_{\Omega^c}(u,u). \qquad (6.34)$$

**Theorem 6.22.** *Let $w \in H^{-1/2}(\Gamma)$ for $d = 3$ and $w \in H_*^{-1/2}(\Gamma)$ for $d = 2$, respectively. Then there holds*

$$\langle Vw,w\rangle_\Gamma \;\ge\; c_1^V\,\|w\|^2_{H^{-1/2}(\Gamma)}$$

*with a positive constant $c_1^V > 0$.*

*Proof.* For $u = \widetilde{V}w$ we can apply both the interior and exterior Green's formulae, i.e. (6.28) and (6.33) to obtain

$$a_\Omega(u,u) = \langle\gamma_1^{\text{int}}u,\gamma_0^{\text{int}}u\rangle_\Gamma,$$
$$a_{\Omega^c}(u,u) = -\langle\gamma_1^{\text{ext}}u,\gamma_0^{\text{ext}}u\rangle_\Gamma.$$

Taking the sum of the above equations we obtain from the jump relation (6.9) of the single layer potential

$$a_\Omega(u,u) + a_{\Omega^c}(u,u) \;=\; \langle[\gamma_1^{\text{int}}u - \gamma_1^{\text{ext}}u],\gamma_0 u\rangle_\Gamma.$$

The jump relation (6.12) of the conormal derivate of the single layer potential reads

$$\gamma_1^{\text{int}}u(x) - \gamma_1^{\text{ext}}u(x) \;=\; w(x) \quad \text{for } x \in \Gamma$$

and therefore we have

$$a_\Omega(u,u) + a_{\Omega^c}(u,u) \;=\; \langle Vw,w\rangle_\Gamma.$$

Using the inequalities (6.29) and (6.34) this gives

$$\begin{aligned}\langle Vw, w\rangle_\Gamma &= a_\Omega(u,u) + a_{\Omega^c}(u,u)\\ &\geq c_1^{\text{int}}\, \|\gamma_1^{\text{int}} u\|^2_{H^{-1/2}(\Gamma)} + c_1^{\text{ext}}\, \|\gamma_1^{\text{ext}} u\|^2_{H^{-1/2}(\Gamma)}\\ &\geq \min\{c_1^{\text{int}}, c_1^{\text{ext}}\} \left[\|\gamma_1^{\text{int}} u\|^2_{H^{-1/2}(\Gamma)} + \|\gamma_1^{\text{ext}} u\|^2_{H^{-1/2}(\Gamma)}\right].\end{aligned}$$

On the other hand, the $H^{-1/2}(\Gamma)$ norm of $w$ can be estimated as

$$\begin{aligned}\|w\|^2_{H^{-1/2}(\Gamma)} &= \|\gamma_1^{\text{int}} u - \gamma_1^{\text{ext}} u\|^2_{H^{-1/2}(\Gamma)}\\ &\leq \left[\|\gamma_1^{\text{int}} u\|_{H^{-1/2}(\Gamma)} + \|\gamma_1^{\text{ext}} u\|_{H^{-1/2}(\Gamma)}\right]^2\\ &\leq 2\left[\|\gamma_1^{\text{int}} u\|^2_{H^{-1/2}(\Gamma)} + \|\gamma_1^{\text{ext}} u\|^2_{H^{-1/2}(\Gamma)}\right]. \qquad \square\end{aligned}$$

In the two–dimensional case we only have the $H_*^{-1/2}(\Gamma)$ ellipticity of the single layer potential when using the previous theorem. To obtain a more general result we first consider the following saddle point problem, $d = 2, 3$, to find $(t, \lambda) \in H^{-1/2}(\Gamma) \times \mathbb{R}$ such that

$$\begin{aligned}\langle Vt, \tau\rangle_\Gamma - \lambda\langle 1, \tau\rangle_\Gamma &= 0 \quad \text{for all } \tau \in H^{-1/2}(\Gamma),\\ \langle t, 1\rangle_\Gamma &= 1.\end{aligned} \tag{6.35}$$

If we consider the ansatz $t := \widetilde{t} + 1/|\Gamma|$ for an arbitrary $\widetilde{t} \in H_*^{-1/2}(\Gamma)$ the second equation is always satisfied. Hence, to find $\widetilde{t} \in H_*^{-1/2}(\Gamma)$ the first equation reads

$$\langle V\widetilde{t}, \tau\rangle_\Gamma = -\frac{1}{|\Gamma|}\langle V1, \tau\rangle_\Gamma \quad \text{for all } \tau \in H_*^{-1/2}(\Gamma).$$

The unique solvability of the variational problem follows from the $H_*^{-1/2}(\Gamma)$–ellipticity of the single layer potential $V$, see Theorem 6.22. The resulting solution $w_{\text{eq}} := \widetilde{t} + 1/|\Gamma|$ is denoted as the natural density. By choosing $\tau = w_{\text{eq}}$ we can finally compute the Lagrange parameter

$$\lambda = \langle Vw_{\text{eq}}, w_{\text{eq}}\rangle_\Gamma.$$

In the three–dimensional case $d = 3$ it follows from Theorem 6.22 that $\lambda > 0$ is strictly positive. In this case the Lagrange parameter $\lambda$ is called the capacity of $\Gamma$. In the two–dimensional case $d = 2$ we define by

$$\text{cap}_\Gamma := e^{-2\pi\lambda}$$

the logarithmic capacity. For a positive number $r \in \mathbb{R}_+$ we may define the parameter dependent fundamental solution

$$U_r^*(x, y) := \frac{1}{2\pi}\log r - \frac{1}{2\pi}\log|x - y|$$

which induces an associated boundary integral operator

$$(V_r w)(x) := \int_\Gamma U_r^*(x,y)w(y)ds_y \quad \text{for } x \in \Gamma$$

satisfying

$$(V_r w_{\text{eq}})(x) = \frac{1}{2\pi}\log r + \lambda = \frac{1}{2\pi}\log\frac{r}{\text{cap}_\Gamma}.$$

In particular for $r = 1$ we obtain

$$\lambda := \frac{1}{2\pi}\log\frac{1}{\text{cap}_\Gamma}.$$

If the logarithmic capacity $\text{cap}_\Gamma < 1$ is strictly less than one, we conclude $\lambda > 0$. To ensure $\text{cap}_\Gamma < 1$ a sufficient criteria is to assume $\text{diam}\,\Omega < 1$ [81, 157]. This assumption can be always guaranteed when considering a suitable scaling of the domain $\Omega \subset \mathbb{R}^2$.

**Theorem 6.23.** *For $d = 2$ let $diam(\Omega) < 1$ and therefore $\lambda > 0$ be satisfied. The single layer potential $V$ is then $H^{-1/2}(\Gamma)$–elliptic, i.e.,*

$$\langle Vw, w\rangle_\Gamma \geq \widetilde{c}_1^V \, \|w\|^2_{H^{-1/2}(\Gamma)} \quad \textit{for all } w \in H^{-1/2}(\Gamma).$$

*Proof.* For an arbitrary $w \in H^{-1/2}(\Gamma)$ we consider the unique decomposition

$$w = \widetilde{w} + \alpha\, w_{\text{eq}}, \quad \widetilde{w} \in H_*^{-1/2}(\Gamma), \quad \alpha = \langle w, 1\rangle_\Gamma$$

satisfying

$$\begin{aligned}
\|w\|^2_{H^{-1/2}(\Gamma)} &= \|\widetilde{w} + \alpha w_{\text{eq}}\|^2_{H^{-1/2}(\Gamma)} \\
&\leq \left[\|\widetilde{w}\|_{H^{-1/2}(\Gamma)} + \alpha\,\|w_{\text{eq}}\|_{H^{-1/2}(\Gamma)}\right]^2 \\
&\leq 2\left[\|\widetilde{w}\|^2_{H^{-1/2}(\Gamma)} + \alpha^2\,\|w_{\text{eq}}\|^2_{H^{-1/2}(\Gamma)}\right] \\
&\leq 2\max\{1, \|w_{\text{eq}}\|^2_{H^{-1/2}(\Gamma)}\}\left[\|\widetilde{w}\|^2_{H^{-1/2}(\Gamma)} + \alpha^2\right].
\end{aligned}$$

On the other hand we have by using $\langle Vw_{\text{eq}}, \widetilde{w}\rangle_\Gamma = 0$

$$\begin{aligned}
\langle Vw, w\rangle_\Gamma &= \langle V(\widetilde{w} + \alpha w_{\text{eq}}), \widetilde{w} + \alpha w_{\text{eq}}\rangle_\Gamma \\
&= \langle V\widetilde{w}, \widetilde{w}\rangle_\Gamma + 2\alpha\,\langle Vw_{\text{eq}}, \widetilde{w}\rangle_\Gamma + \alpha^2\,\langle Vw_{\text{eq}}, w_{\text{eq}}\rangle_\Gamma \\
&\geq c_1^V\,\|\widetilde{w}\|^2_{H^{-1/2}(\Gamma)} + \alpha^2\,\lambda \\
&\geq \min\{c_1^V, \lambda\}\left[\|\widetilde{w}\|^2_{H^{-1/2}(\Gamma)} + \alpha^2\right],
\end{aligned}$$

and therefore the ellipticity estimate follows. □

The natural density $w_{\text{eq}} \in H^{-1/2}(\Gamma)$ is a solution of an operator equation with a constraint,

$$(Vw_{\rm eq})(x) = \lambda \quad \text{for } x \in \Gamma, \quad \langle w_{\rm eq}, 1\rangle_\Gamma = 1.$$

By introducing the scaling

$$w_{\rm eq} := \lambda\, \widetilde{w}_{\rm eq}$$

we obtain

$$(V\widetilde{w}_{\rm eq})(x) = 1 \quad \text{for } x \in \Gamma, \quad \frac{1}{\lambda} = \langle \widetilde{w}_{\rm eq}, 1\rangle_\Gamma \,. \tag{6.36}$$

Instead of the saddle point problem (6.35) we may solve the boundary integral equation (6.36) to find the natural density $w_{\rm eq}$ and afterwards we can compute the capacity $\lambda$ by integrating the natural density $\widetilde{w}_{\rm eq}$.

The boundary integral operator $V : H^{-1/2}(\Gamma) \to H^{1/2}(\Gamma)$ is due to (6.8) bounded and $H^{-1/2}(\Gamma)$–elliptic, see Theorem 6.22 for $d = 3$ and Theorem 6.23 for $d = 2$ where we assume $\operatorname{diam}(\Omega) < 1$. By the Lax–Milgram theorem (Theorem 3.4) we therefore conclude the invertibility of the single layer potential $V$, i.e. $V^{-1} : H^{1/2}(\Gamma) \to H^{-1/2}(\Gamma)$ is bounded satisfying (see (3.13))

$$\|V^{-1}v\|_{H^{-1/2}(\Gamma)} \leq \frac{1}{c_1^V}\, \|v\|_{H^{1/2}(\Gamma)} \quad \text{for all } v \in H^{1/2}(\Gamma).$$

For $w \in H_*^{-1/2}(\Gamma)$ we have

$$\langle Vw, w_{\rm eq}\rangle_\Gamma = \langle w, Vw_{\rm eq}\rangle_\Gamma = \lambda\langle w, 1\rangle_\Gamma = 0$$

and therefore $Vw \in H_*^{1/2}(\Gamma)$ where

$$H_*^{1/2}(\Gamma) := \Big\{ v \in H^{1/2}(\Gamma) \,:\, \langle v, w_{\rm eq}\rangle_\Gamma = 0 \Big\}.$$

Thus, $V : H_*^{-1/2}(\Gamma) \to H_*^{1/2}(\Gamma)$ is an isomorphism.

### 6.6.2 Ellipticity of the Hypersingular Boundary Integral Operator

Due to (6.17) we have $(Du_0)(x) = 0$ with the eigensolution $u_0(x) \equiv 1$ for $x \in \Gamma$. Hence we can not ensure the ellipticity of the hypersingular boundary integral operator $D$ on $H^{1/2}(\Gamma)$. Instead we have to consider a suitable subspace.

**Theorem 6.24.** *The hypersingular boundary integral operator $D$ is $H_*^{1/2}(\Gamma)$–elliptic, i.e.,*

$$\langle Dv, v\rangle_\Gamma \geq c_1^D\, \|v\|^2_{H^{1/2}(\Gamma)} \quad \textit{for all } v \in H_*^{1/2}(\Gamma).$$

*Proof.* For $v \in H_*^{1/2}(\Gamma)$ we consider the double layer potential

$$u(x) := -(Wv)(x) \quad \text{for } x \in \Omega \cup \Omega^c$$

which is a solution of the homogeneous partial differential equation. The application of the trace operators gives

$$\gamma_0^{\text{int}} u(x) = (1-\sigma(x))v(x) - (Kv)(x), \quad \gamma_1^{\text{int}} u(x) = (Dv)(x) \quad \text{for } x \in \Gamma$$

and

$$\gamma_0^{\text{ext}} u(x) = -\sigma(x)v(x) - (Kv)(x), \quad \gamma_1^{\text{ext}} u(x) = (Dv)(x) \quad \text{for } x \in \Gamma.$$

The function $u = -Wv$ is therefore the unique solution of the interior Dirichlet boundary value problem

$$-\Delta u(x) = 0 \quad \text{for } x \in \Omega, \quad \gamma_0^{\text{int}} u(x) = (1-\sigma(x))v(x) - (Kv)(x) \quad \text{for } x \in \Gamma$$

and we have, by applying Green's first formula (1.5),

$$\int_\Omega \nabla u(x) \nabla w(x) dx = \langle \gamma_1^{\text{int}} u, \gamma_0^{\text{int}} w \rangle_\Gamma$$

for all $w \in H^1(\Omega)$.

For $y_0 \in \Omega$ let $B_R(y_0)$ be a ball of radius $R > 2\,\text{diam}(\Omega)$ which circumscribes $\Omega$, $\Omega \subset B_R(y_0)$. Then, $u = -Wv$ is also the unique solution of the Dirichlet boundary value problem

$$\begin{aligned} -\Delta u(x) &= 0 && \text{for } x \in B_R(y_0)\backslash\overline{\Omega}, \\ \gamma_0^{\text{ext}} u(x) &= -\sigma(x)v(x) - (Kv)(x) && \text{for } x \in \Gamma = \partial\Omega, \\ \gamma_0 u(x) &= -(Wv)(x) && \text{for } x \in \partial B_R(y_0) \end{aligned}$$

and the corresponding Green's first formula reads

$$\int_{B_R(y_0)\backslash\overline{\Omega}} \nabla u(x) \nabla w(x) dx = -\langle \gamma_1^{\text{ext}} u, \gamma_0^{\text{ext}} w \rangle_\Gamma + \langle \gamma_1 u, \gamma_0 w \rangle_{\partial B_R(y_0)}$$

for all $w \in H^1(B_R(y_0)\backslash\overline{\Omega})$. For $x \notin \Gamma$ we have by definition

$$u(x) = \frac{1}{2(d-1)\pi} \int_\Gamma \frac{(y-x, n_y)}{|x-y|^d} v(y) ds_y .$$

In particular for $x \in \partial B_R(y_0)$ we then obtain the estimates

$$|u(x)| \leq c_1(v)\, R^{1-d}, \quad |\nabla u(x)| \leq c_2(v)\, R^{-d} .$$

By choosing $w = u = -Wv$ and taking the limit $R \to \infty$ we finally obtain Green's first formula with respect to the exterior domain,

$$\int_{\Omega^c} |\nabla u(x)|^2 dx = -\langle \gamma_1^{\text{ext}} u, \gamma_0^{\text{ext}} u \rangle_\Gamma .$$

By taking the sum of both Green's formulae with respect to the interior and to the exterior domain, and considering the jump relations of the boundary integral operators involved, we obtain for the bilinear form of the hypersingular boundary integral operator

$$\begin{aligned}\langle Dv, v\rangle_\Gamma &= \langle \gamma_1^{\text{int}} u, [\gamma_0^{\text{int}} u - \gamma_0^{\text{ext}} u]\rangle_\Gamma = \langle \gamma_1^{\text{int}} u, \gamma_0^{\text{int}} u\rangle_\Gamma - \langle \gamma_1^{\text{ext}} u, \gamma_0^{\text{ext}} u\rangle_\Gamma \\ &= \int_\Omega |\nabla u(x)|^2 dx + \int_{\Omega^c} |\nabla u(x)|^2 dx = |u|^2_{H^1(\Omega)} + |u|^2_{H^1(\Omega^c)} \,.\end{aligned}$$

For the exterior domain $\Omega^c$ we find from the far field behavior of the double layer potential $u(x) = -(Wv)(x)$ as $|x| \to \infty$ the norm equivalence

$$c_1 \,\|u\|^2_{H^1(\Omega^c)} \le |u|^2_{H^1(\Omega^c)} \le c_2 \,\|u\|^2_{H^1(\Omega^c)}.$$

For $v \in H^{1/2}_*(\Gamma)$, for the natural density $w_{\text{eq}} \in H^{-1/2}(\Gamma)$, $Vw_{\text{eq}} = 1$, and by using the symmetry relation (6.26) we further obtain

$$\begin{aligned}\langle \gamma_0^{\text{int}} u, w_{\text{eq}}\rangle_\Gamma = \langle (\tfrac{1}{2} I - K)v, w_{\text{eq}}\rangle_\Gamma &= \langle v, w_{\text{eq}}\rangle_\Gamma - \langle (\tfrac{1}{2} I + K)v, w_{\text{eq}}\rangle_\Gamma \\ &= -\langle (\tfrac{1}{2} I + K)v, V^{-1}1\rangle_\Gamma = -\langle V^{-1}(\tfrac{1}{2} I + K)v, 1\rangle_\Gamma \\ &= -\langle (\tfrac{1}{2} I + K')V^{-1}v, 1\rangle_\Gamma = -\langle V^{-1}v, (\tfrac{1}{2} I + K)1\rangle_\Gamma = 0\end{aligned}$$

and therefore $\gamma_0^{\text{int}} u \in H^{1/2}_*(\Gamma)$. By using the norm equivalence theorem of Sobolev (Theorem 2.6) we find

$$\|u\|_{H^1_*(\Omega)} := \left\{ [\langle \gamma_0^{\text{int}} u, w_{\text{eq}}\rangle_\Gamma]^2 + \|\nabla u\|^2_{L_2(\Omega)} \right\}^{1/2}$$

to be an equivalent norm in $H^1(\Omega)$. For $v \in H^{1/2}_*(\Gamma)$ we have $\gamma_0^{\text{int}} u \in H^{1/2}_*(\Gamma)$ and therefore

$$|u|^2_{H^1(\Omega)} = [\langle \gamma_0^{\text{int}} u, w_{\text{eq}}\rangle_\Gamma]^2 + \|\nabla u\|^2_{L_2(\Omega)} = \|u\|^2_{H^1_*(\Omega)} \ge c\,\|u\|^2_{H^1(\Omega)} \,.$$

By using the trace theorem and the jump relation of the double layer potential we obtain

$$\begin{aligned}\langle Dv, v\rangle_\Gamma &\ge c \left\{ \|u\|^2_{H^1(\Omega)} + \|u\|^2_{H^1(\Omega^c)} \right\} \\ &\ge \tilde{c} \left\{ \|\gamma_0^{\text{int}} u\|^2_{H^{1/2}(\Gamma)} + \|\gamma_0^{\text{ext}} u\|^2_{H^{1/2}(\Gamma)} \right\} \\ &\ge \frac{1}{2} \tilde{c} \,\|\gamma_0^{\text{int}} u - \gamma_0^{\text{ext}} u\|^2_{H^{1/2}(\Gamma)} = c_1^D \,\|v\|^2_{H^{1/2}(\Gamma)}\end{aligned}$$

for all $v \in H^{1/2}_*(\Gamma)$ and therefore the $H^{1/2}_*(\Gamma)$–ellipticity of the hypersingular boundary integral operator $D$. □

To prove the ellipticity of the hypersingular boundary integral operator $D$ we have to restrict the functions to a suitable subspace, i.e. orthogonal to the constants. When considering the orthogonality with respect to different inner products this gives the ellipticity of the hypersingular boundary integral operator $D$ with respect to different subspaces.

As in the norm equivalence theorem of Sobolev (cf. Theorem 2.6) we define

$$\|v\|_{H_*^{1/2}(\Gamma)} := \left\{ \left[ \langle v, w_{\text{eq}} \rangle_\Gamma \right]^2 + |v|^2_{H^{1/2}(\Gamma)} \right\}^{1/2}$$

to be an equivalent norm in $H^{1/2}(\Gamma)$. Here, $w_{\text{eq}} \in H^{-1/2}(\Gamma)$ is the natural density as defined in (6.36).

**Corollary 6.25.** *The hypersingular boundary integral operator $D$ is $H^{1/2}(\Gamma)$–semi–elliptic, i.e.*

$$\langle Dv, v \rangle_\Gamma \geq \bar{c}_1^D \, |v|^2_{H^{1/2}(\Gamma)} \quad \textit{for all } v \in H^{1/2}(\Gamma). \tag{6.37}$$

The definition of $H_*^{1/2}(\Gamma)$ involves the natural density $w_{\text{eq}} \in H^{-1/2}(\Gamma)$ as the unique solution of the boundary integral equation (6.36). From a practical point of view, this seems not to be very convenient for a computational realization. Hence we may use a subspace which is induced by a much simpler inner product. For this we define

$$H_{**}^{1/2}(\Gamma) := \left\{ v \in H^{1/2}(\Gamma) \,:\, \langle v, 1 \rangle_\Gamma = 0 \right\}.$$

From (6.37) we then have for $v \in H_{**}^{1/2}(\Gamma)$

$$\begin{aligned}\langle Dv, v \rangle_\Gamma &\geq \bar{c}_1^D \, |v|^2_{H^{1/2}(\Gamma)} \\ &= \bar{c}_1^D \left\{ |v|^2_{H^{1/2}(\Gamma)} + [\langle v, 1 \rangle_\Gamma]^2 \right\} \geq \tilde{c}_1^D \, \|v\|^2_{H^{1/2}(\Gamma)} \end{aligned} \tag{6.38}$$

the $H_{**}^{1/2}(\Gamma)$–ellipticity of $D$ where we again used the norm equivalence theorem of Sobolev (cf. Theorem 2.6).

We finally consider an open surface $\Gamma_0 \subset \Gamma$. For a given $v \in \widetilde{H}^{1/2}(\Gamma_0)$ let $\widetilde{v} \in H^{1/2}(\Gamma)$ denote the extension defined by

$$\widetilde{v}(x) = \begin{cases} v(x) & \text{for } x \in \Gamma_0, \\ 0 & \text{elsewhere.} \end{cases}$$

As in the norm equivalence theorem of Sobolev (cf. Theorem 2.6) we define

$$\|w\|_{H^{1/2}(\Gamma),\Gamma_0} := \left\{ \|w\|^2_{L_2(\Gamma\setminus\Gamma_0)} + |w|^2_{H^{1/2}(\Gamma)} \right\}^{1/2}.$$

to be an equivalent norm in $H^{1/2}(\Gamma)$. Hence we have for $v \in \widetilde{H}^{1/2}(\Gamma_0)$

$$\begin{aligned}\langle Dv, v\rangle_{\Gamma_0} = \langle D\widetilde{v}, \widetilde{v}\rangle_\Gamma \geq \bar{c}_1^D\, |\widetilde{v}|^2_{H^{1/2}(\Gamma)} &= \bar{c}_1^D \left[ \|\widetilde{v}\|^2_{L_2(\Gamma\setminus\Gamma_0)} + |\widetilde{v}|^2_{H^{1/2}(\Gamma)} \right] \\ &= \bar{c}_1^D\, \|\widetilde{v}\|^2_{H^{1/2}(\Gamma),\Gamma_0} \geq \hat{c}_1^D\, \|\widetilde{v}\|^2_{H^{1/2}(\Gamma)} = \hat{c}_1^D\, \|v\|^2_{\widetilde{H}^{1/2}(\Gamma_0)} \qquad (6.39)\end{aligned}$$

and therefore the $\widetilde{H}^{1/2}(\Gamma_0)$–ellipticity of the hypersingular boundary integral operator $D$.

### 6.6.3 Steklov–Poincaré Operator

When considering the solution of boundary value problems the interaction of the Cauchy data $\gamma_0^{\text{int}}u$ and $\gamma_1^{\text{int}}u$ plays an important role. Let us consider the system (6.22) of boundary integral equations for a homogeneous partial differential equation, i.e., $f \equiv 0$:

$$\begin{pmatrix} \gamma_0^{\text{int}}u \\ \gamma_1^{\text{int}}u \end{pmatrix} = \begin{pmatrix} (1-\sigma)I - K & V \\ D & \sigma I + K' \end{pmatrix} \begin{pmatrix} \gamma_0^{\text{int}}u \\ \gamma_1^{\text{int}}u \end{pmatrix}.$$

Since the single layer potential $V$ is invertible, we get from the first boundary integral equation a representation for the Dirichlet to Neumann map,

$$\gamma_1^{\text{int}}u(x) \;=\; V^{-1}(\sigma I + K)\gamma_0^{\text{int}}u(x) \quad \text{for } x \in \Gamma. \qquad (6.40)$$

The operator

$$S \;:=\; V^{-1}(\sigma I + K) \,:\, H^{1/2}(\Gamma) \to H^{-1/2}(\Gamma) \qquad (6.41)$$

is bounded, and $S$ is called Steklov–Poincaré operator. Inserting (6.40) into the second equation of the Calderón projection this gives

$$\begin{aligned}\gamma_1^{\text{int}}u(x) &= (D\gamma_0^{\text{int}}u)(x) + (\sigma I + K')\gamma_1^{\text{int}}u(x) \\ &= \left[D + (\sigma I + K')V^{-1}(\sigma I + K)\right] \gamma_0^{\text{int}}u(x) \quad \text{for } x \in \Gamma. \qquad (6.42)\end{aligned}$$

Hence we have obtained a symmetric representation of the Steklov–Poincaré operator which is equivalent to (6.41),

$$S \;:=\; D + (\sigma I + K')V^{-1}(\sigma I + K) \,:\, H^{1/2}(\Gamma) \to H^{-1/2}(\Gamma). \qquad (6.43)$$

With (6.40) and (6.42) we have described the Dirichlet to Neumann map

$$\gamma_1^{\text{int}}u(x) \;=\; (S\gamma_0^{\text{int}})u(x) \quad \text{for } x \in \Gamma \qquad (6.44)$$

which maps some given Dirichlet datum $\gamma_0^{\text{int}}u \in H^{1/2}(\Gamma)$ to the corresponding Neumann datum $\gamma_1^{\text{int}}u \in H^{-1/2}(\Gamma)$ of the harmonic function $u \in H^1(\Omega)$ satisfying $Lu = 0$.

By using the $H^{1/2}(\Gamma)$–ellipticity of the inverse single layer potential $V^{-1}$ we obtain

$$\langle Sv, v\rangle_\Gamma = \langle Dv, v\rangle_\Gamma + \langle V^{-1}(\sigma I + K)v, (\sigma I + K)v\rangle_\Gamma \geq \langle Dv, v\rangle_\Gamma \quad (6.45)$$

for all $v \in H^{1/2}(\Gamma)$. Therefore, the Steklov–Poincaré operator $S$ admits the same ellipticity estimates as the hypersingular boundary integral operator $D$. In particular we have

$$\langle Sv, v\rangle_\Gamma \geq c_1^D \, \|v\|^2_{H^{1/2}(\Gamma)} \quad \text{for all } v \in H^{1/2}_*(\Gamma) \quad (6.46)$$

as well as

$$\langle Sv, v\rangle_\Gamma \geq \widetilde{c}_1^D \, \|v\|^2_{H^{1/2}(\Gamma)} \quad \text{for all } v \in H^{1/2}_{**}(\Gamma) \quad (6.47)$$

while for $\Gamma_0 \subset \Gamma$ we get

$$\langle Sv, v\rangle_{\Gamma_0} \geq \hat{c}_1 \, \|v\|^2_{\widetilde{H}^{1/2}(\Gamma_0)} \quad \text{for all } v \in \widetilde{H}^{1/2}(\Gamma_0). \quad (6.48)$$

### 6.6.4 Contraction Estimates of the Double Layer Potential

It is possible to transfer the ellipticity estimates of the single layer potential $V$ and of the hypersingular boundary integral operator $D$ to the double layer potential $\sigma I + K : H^{1/2}(\Gamma) \to H^{1/2}(\Gamma)$, see [145]. Since the single layer potential $V : H^{-1/2}(\Gamma) \to H^{1/2}(\Gamma)$ is bounded and $H^{-1/2}(\Gamma)$–elliptic, we may define

$$\|u\|_{V^{-1}} := \sqrt{\langle V^{-1}u, u\rangle_\Gamma} \quad \text{for all } u \in H^{1/2}(\Gamma)$$

to be an equivalent norm in $H^{1/2}(\Gamma)$.

**Theorem 6.26.** *For $u \in H^{1/2}_*(\Gamma)$ we have*

$$(1 - c_K)\, \|u\|_{V^{-1}} \leq \|(\sigma I + K)u\|_{V^{-1}} \leq c_K \, \|u\|_{V^{-1}} \quad (6.49)$$

*with*

$$c_K = \frac{1}{2} + \sqrt{\frac{1}{4} - c_1^V c_1^D} < 1 \quad (6.50)$$

*where $c_1^V$ and $c_1^D$ are the ellipticity constants of the single layer potential $V$ and of the hypersingular boundary integral operator $D$, respectively.*

*Proof.* Using the symmetric representation (6.43) of the Steklov–Poincaré operator $S$ we have

$$\begin{aligned}\|(\sigma I + K)u\|^2_{V^{-1}} &= \langle V^{-1}(\sigma I + K)u, (\sigma I + K)u\rangle_\Gamma \\ &= \langle Su, u\rangle_\Gamma - \langle Du, u\rangle_\Gamma.\end{aligned}$$

Let $J : H^{-1/2}(\Gamma) \to H^{1/2}(\Gamma)$ be the Riesz map which is defined via

$$\langle Jw, v\rangle_{H^{1/2}(\Gamma)} = \langle w, v\rangle_\Gamma \quad \text{for all } v \in H^{1/2}(\Gamma).$$

Then, $A := JV^{-1} : H^{1/2}(\Gamma) \to H^{1/2}(\Gamma)$ is self–adjoint and $H^{1/2}(\Gamma)$–elliptic. Using the first representation (6.41) of the Steklov–Poincaré operator $S$ and considering the splitting $A = A^{1/2}A^{1/2}$ we conclude the inequality

$$\begin{aligned}\langle Su, u\rangle_\Gamma &= \langle V^{-1}(\sigma I + K)u, u\rangle_\Gamma \\ &= \langle JV^{-1}(\sigma I + K)u, u\rangle_{H^{1/2}(\Gamma)} \\ &= \langle A^{1/2}(\sigma I + K)u, A^{1/2}u\rangle_{H^{1/2}(\Gamma)} \\ &\leq \|A^{1/2}(\sigma I + K)u\|_{H^{1/2}(\Gamma)}\|A^{1/2}u\|_{H^{1/2}(\Gamma)}\end{aligned}$$

and with

$$\begin{aligned}\|A^{1/2}v\|^2_{H^{1/2}(\Gamma)} &= \langle A^{1/2}v, A^{1/2}v\rangle_{H^{1/2}(\Gamma)} \\ &= \langle JV^{-1}v, v\rangle_{H^{1/2}(\Gamma)} \\ &= \langle V^{-1}v, v\rangle_\Gamma \\ &= \|v\|^2_{V^{-1}}\end{aligned}$$

it follows that

$$\langle Su, u\rangle_\Gamma \leq \|(\sigma I + K)u\|_{V^{-1}}\|u\|_{V^{-1}}.$$

Since the hypersingular integral operator $D$ is elliptic for $u \in H^{1/2}_*(\Gamma)$ we find from the mapping properties of the inverse single layer potential $V^{-1}$ the lower estimate

$$\langle Du, u\rangle_\Gamma \geq c_1^D\,\|u\|^2_{H^{1/2}(\Gamma)} \geq c_1^D c_1^V\,\langle V^{-1}u, u\rangle_\Gamma = c_1^D c_1^V\,\|u\|^2_{V^{-1}}.$$

Hence we have obtained

$$\begin{aligned}\|(\sigma I + K)u\|^2_{V^{-1}} &= \langle Su, u\rangle_\Gamma - \langle Du, u\rangle_\Gamma \\ &\leq \|(\sigma I + K)u\|_{V^{-1}}\|u\|_{V^{-1}} - c_1^V c_1^D\,\|u\|^2_{V^{-1}}.\end{aligned}$$

Denoting

$$a := \|(\sigma I + K)u\|_{V^{-1}} \geq 0, \quad b := \|u\|_{V^{-1}} > 0$$

we conclude

$$\left(\frac{a}{b}\right)^2 - \frac{a}{b} + c_1^V c_1^D \leq 0$$

which is equivalent to

$$\frac{1}{2} - \sqrt{\frac{1}{4} - c_1^V c_1^D} \leq \frac{a}{b} \leq \frac{1}{2} + \sqrt{\frac{1}{4} - c_1^V c_1^D}$$

and therefore to the assertion. □

The contraction property of $\sigma I + K$, in particular the upper estimate in (6.49), can be extended to hold in $H^{1/2}(\Gamma)$.

**Corollary 6.27.** *For $u \in H^{1/2}(\Gamma)$ there holds*

$$\|(\sigma I + K)u\|_{V^{-1}} \leq c_K \, \|u\|_{V^{-1}} \tag{6.51}$$

*where the contraction rate $c_K < 1$ is given as in* (6.50).

*Proof.* For an arbitrary $u \in H^{1/2}(\Gamma)$ we can write

$$u = \widetilde{u} + \frac{\langle u, w_{\text{eq}} \rangle_\Gamma}{\langle 1, w_{\text{eq}} \rangle_\Gamma} u_0$$

where $\widetilde{u} \in H_*^{1/2}(\Gamma)$ and $u_0 \equiv 1$. Due to $(\sigma I + K)u_0 = 0$ we have by using Theorem 6.26

$$\|(\sigma I + K)u\|_{V^{-1}} = \|(\sigma I + K)\widetilde{u}\|_{V^{-1}} \leq c_K \, \|\widetilde{u}\|_{V^{-1}} .$$

On the other hand,

$$\|u\|^2_{V^{-1}} = \|\widetilde{u}\|^2_{V^{-1}} + \frac{[\langle u, w_{\text{eq}} \rangle_\Gamma]^2}{\langle 1, w_{\text{eq}} \rangle_\Gamma} \geq \|\widetilde{u}\|^2_{V^{-1}}$$

which implies the contraction estimate (6.51). □

Note that a similar result as in Theorem 6.26 can be shown for the shifted operator$(1-\sigma)I - K$.

**Corollary 6.28.** *For $v \in H_*^{1/2}(\Gamma)$ there holds*

$$(1 - c_K) \, \|v\|_{V^{-1}} \leq \|([1-\sigma]I - K)v\|_{V^{-1}} \leq c_K \, \|v\|_{V^{-1}} \tag{6.52}$$

*where the contraction rate $c_K < 1$ is given as in* (6.50).

*Proof.* By using both the triangle inequality and the contraction estimate of $\sigma I + K$ we obtain with

$$\begin{aligned}
\|v\|_{V^{-1}} &= \|([1-\sigma]I - K)v + (\sigma I + K)v\|_{V^{-1}} \\
&\leq \|([1-\sigma]I - K)v\|_{V^{-1}} + \|(\sigma I + K)v\|_{V^{-1}} \\
&\leq \|([1-\sigma]I - K)v\|_{V^{-1}} + c_K \, \|v\|_{V^{-1}}
\end{aligned}$$

the lower estimate in (6.52). Moreover, using both representations (6.41) and (6.43) of the Steklov–Poincaré operator $S$, we conclude

$$\begin{aligned}
\|((1-\sigma)I - K)v\|^2_{V^{-1}} &= \|[I - (\sigma I + K)]v\|^2_{V^{-1}} \\
&= \|v\|^2_{V^{-1}} + \|(\sigma I + K)v\|^2_{V^{-1}} - 2\langle V^{-1}(\sigma I + K)v, v\rangle_\Gamma \\
&= \|v\|^2_{V^{-1}} + \|(\sigma I + K)v\|^2_{V^{-1}} - 2\langle Sv, v\rangle_\Gamma \\
&= \|v\|^2_{V^{-1}} - \|(\sigma I + K)v\|^2_{V^{-1}} - 2\langle Dv, v\rangle_\Gamma \\
&\leq \left[1 - (1-c_K)^2 - 2c_1^V c_1^D\right] \|v\|^2_{V^{-1}} = c_K^2 \, \|v\|^2_{V^{-1}}.
\end{aligned}$$

This gives the upper estimate in (6.52). □

Let $H_*^{-1/2}(\Gamma)$ be the subspace as defined in (6.30). Due to $V : H_*^{-1/2}(\Gamma) \to H_*^{1/2}(\Gamma)$ we can transfer the estimates (6.49) of the double layer potential $\sigma I + K : H^{1/2}(\Gamma) \to H^{1/2}(\Gamma)$ immediately to the adjoint double layer potential $\sigma I + K' : H^{-1/2}(\Gamma) \to H^{-1/2}(\Gamma)$.

**Corollary 6.29.** *For the adjoint double layer potential and for $w \in H_*^{-1/2}(\Gamma)$ there holds*

$$(1 - c_K)\, \|w\|_V \;\leq\; \|(\sigma I + K')w\|_V \;\leq\; c_K\, \|w\|_V \tag{6.53}$$

*where the contraction rate $c_K < 1$ is given as in* (6.50) *and $\|\cdot\|_V$ is the norm which is induced by the single layer potential $V$.*

*Proof.* For $w \in H_*^{-1/2}(\Gamma)$ there exists a uniquely determined $v \in H_*^{1/2}(\Gamma)$ satisfying $v = Vw$ or $w = V^{-1}v$. Using the symmetry property (6.26) we first have

$$\begin{aligned}\|(\sigma I + K')w\|_V^2 &= \langle V(\sigma I + K')V^{-1}v, (\sigma I + K')V^{-1}v\rangle_\Gamma \\ &= \langle V^{-1}(\sigma I + K)v, (\sigma I + K)v\rangle_\Gamma \;=\; \|(\sigma I + K)v\|_{V^{-1}}^2,\end{aligned}$$

as well as

$$\|w\|_V^2 \;=\; \langle Vw, w\rangle_\Gamma \;=\; \langle V^{-1}v, v\rangle_\Gamma \;=\; \|v\|_{V^{-1}}^2.$$

Therefore, (6.53) is equivalent to (6.49). □

As in Corollary 6.27 we can extend the contraction property of $\sigma I + K'$ to $H^{-1/2}(\Gamma)$.

**Corollary 6.30.** *For $w \in H^{-1/2}(\Gamma)$ there holds the contraction estimate*

$$\|(\sigma I + K')w\|_V \;\leq\; c_K\, \|w\|_V. \tag{6.54}$$

To prove related properties of the shifted adjoint double layer potential $(1-\sigma)I - K'$ again we need to bear in mind the correct subspaces.

**Corollary 6.31.** *For $w \in H_*^{-1/2}(\Gamma)$ we have*

$$(1 - c_K)\, \|w\|_V \;\leq\; \|((1-\sigma)I - K')w\|_V \;\leq\; c_K\, \|w\|_V \tag{6.55}$$

*where the contraction rate $c_K < 1$ is given as in* (6.50).

### 6.6.5 Mapping Properties

All mapping properties of boundary integral operators considered up to now are based on the mapping properties of the Newton potential $\widetilde{N}_0 : \widetilde{H}^{-1}(\Omega) \to H^1(\Omega)$, and on the application of trace theorems and on duality arguments. But even for Lipschitz domains more general results hold.

**Theorem 6.32.** *The Newton potential $\widetilde{N}_0 : \widetilde{H}^s(\Omega) \to H^{s+2}(\Omega)$ is a continuous map for all $s \in [-2, 0]$, i.e.*

$$\|\widetilde{N}_0 f\|_{H^{s+2}(\Omega)} \leq c \|f\|_{\widetilde{H}^s(\Omega)} \quad \textit{for all } f \in \widetilde{H}^s(\Omega).$$

*Proof.* Let $s \in (-1, 0]$ and consider $\widetilde{f}$ to be the extension of $f \in H^s(\Omega)$ as defined in (6.2). Then,

$$\|\widetilde{f}\|_{H^s(\mathbb{R}^d)} = \sup_{0 \neq v \in H^{-s}(\mathbb{R}^d)} \frac{\langle \widetilde{f}, v\rangle_{\mathbb{R}^d}}{\|v\|_{H^{-s}(\mathbb{R}^d)}} \leq \sup_{0 \neq v \in H^{-s}(\Omega)} \frac{\langle f, v\rangle_\Omega}{\|v\|_{H^{-s}(\Omega)}} = \|f\|_{H^s(\Omega)},$$

and the assertion follows as in the proof of Theorem 6.1, i.e.

$$\|\widetilde{N}_0 f\|_{H^{s+2}(\Omega)} \leq c \|f\|_{\widetilde{H}^s(\Omega)} \quad \text{for all } f \in \widetilde{H}^s(\Omega).$$

Since the Newton potential $\widetilde{N}_0$ is self–adjoint, for $s \in [-2, -1)$ we obtain by duality

$$\|\widetilde{N}_0 f\|_{H^{s+2}(\Omega)} = \sup_{0 \neq g \in \widetilde{H}^{-2-s}(\Omega)} \frac{\langle \widetilde{N}_0 f, g\rangle_\Omega}{\|g\|_{\widetilde{H}^{-2-s}(\Omega)}} = \sup_{0 \neq g \in \widetilde{H}^{-2-s}(\Omega)} \frac{\langle f, \widetilde{N}_0 g\rangle_\Omega}{\|g\|_{\widetilde{H}^{-2-s}(\Omega)}}$$

$$\leq \|f\|_{\widetilde{H}^s(\Omega)} \sup_{0 \neq g \in \widetilde{H}^{-2-s}(\Omega)} \frac{\|\widetilde{N}_0 g\|_{H^{-s}(\Omega)}}{\|g\|_{\widetilde{H}^{-2-s}(\Omega)}} \leq c \|f\|_{\widetilde{H}^s(\Omega)}. \quad \square$$

By the application of Theorem 6.32 we can deduce corresponding mapping properties for the single layer potential $\widetilde{V}$ as defined in (6.7) and for the boundary integral operator $V := \gamma_0^{\text{int}} \widetilde{V}$ by considering the trace of $\widetilde{V}$.

**Theorem 6.33.** *The single layer potential $V : H^{-\frac{1}{2}+s}(\Gamma) \to H^{\frac{1}{2}+s}(\Gamma)$ is bounded for $|s| < \frac{1}{2}$, i.e.,*

$$\|Vw\|_{H^{1/2+s}(\Gamma)} \leq c \|w\|_{H^{-1/2+s}(\Gamma)}$$

*for all $w \in H^{-1/2+s}(\Gamma)$.*

*Proof.* For $\varphi \in C^\infty(\Omega)$ we first consider

$$\langle \widetilde{V} w, \varphi\rangle_\Omega = \int_\Omega \varphi(x) \int_\Gamma U^*(x,y) w(y) ds_y dx = \int_\Gamma w(y) \int_\Omega U^*(x,y)\varphi(x) dx ds_y$$

$$= \langle w, \gamma_0^{\text{int}} \widetilde{N}_0 \varphi\rangle_\Gamma \leq \|w\|_{H^{-1/2+s}(\Gamma)} \|\gamma_0^{\text{int}} \widetilde{N}_0 \varphi\|_{H^{1/2-s}(\Gamma)}$$

$$\leq c_T \|w\|_{H^{-1/2+s}(\Gamma)} \|\widetilde{N}_0 \varphi\|_{H^{1-s}(\Omega)}$$

where the application of the trace theorem requires $\frac{1}{2} - s > 0$, see Theorem 2.21. With Theorem 6.32 we then obtain

$$\langle \widetilde{V}w, \varphi\rangle_\Omega \le c\,\|w\|_{H^{-1/2+s}(\Gamma)}\|\varphi\|_{\widetilde{H}^{-1-s}(\Omega)} \quad \text{for all } \varphi \in C^\infty(\Omega).$$

By using a density argument we conclude $\widetilde{V}w \in H^{1+s}(\Omega)$. Taking the trace this gives $Vw := \gamma_0^{\text{int}}\widetilde{V}w \in H^{1/2+s}(\Gamma)$ when assuming $\frac{1}{2} + s > 0$. □

In the case of a Lipschitz domain $\Omega$ we can prove as in Theorem 6.33 corresponding mapping properties for all boundary integral operators.

**Theorem 6.34.** [44] *Let $\Gamma := \partial\Omega$ be the boundary of a Lipschitz domain $\Omega$. Then, the boundary integral operators*

$$\begin{aligned} V &: H^{-1/2+s}(\Gamma) \to H^{1/2+s}(\Gamma), \\ K &: H^{1/2+s}(\Gamma) \to H^{1/2+s}(\Gamma), \\ K' &: H^{-1/2+s}(\Gamma) \to H^{-1/2+s}(\Gamma), \\ D &: H^{1/2+s}(\Gamma) \to H^{-1/2+s}(\Gamma) \end{aligned}$$

*are bounded for all $s \in [-\frac{1}{2}, \frac{1}{2}]$.*

*Proof.* For the single layer potential $V$ and for $|s| < \frac{1}{2}$ the assertion was already shown in Theorem 6.33. This remains true for $|s| = \frac{1}{2}$ [152], see also the discussion in [103].

For all other boundary integral operators the assertion follows from the mapping properties of the conormal derivative operator.
First we consider the adjoint double layer potential $K'$. Recall that the single layer potential $u(x) = (\widetilde{V}w)(x)$, $x \in \Omega$, is a solution of the homogeneous partial differential equation with Dirichlet data $\gamma_0^{\text{int}}u(x) = (Vw)(x)$ for $x \in \Gamma$. By using Theorem 4.6 and the continuity of the single layer potential $V : L_2(\Gamma) \to H^1(\Gamma)$ we obtain

$$\|\gamma_1^{\text{int}}u\|_{L_2(\Gamma)} \le c\,\|Vw\|_{H^1(\Gamma)} \le \widetilde{c}\,\|w\|_{L_2(\Gamma)}$$

and therefore the continuity of $\gamma_1^{\text{int}}\widetilde{V} = \sigma I + K' : L_2(\Gamma) \to L_2(\Gamma)$. On the other hand we have by duality

$$\|\gamma_1^{\text{int}}u\|_{H^{-1}(\Gamma)} = \sup_{0 \ne \varphi \in H^1(\Gamma)} \frac{\langle \gamma_1^{\text{int}}u, \varphi\rangle_\Gamma}{\|\varphi\|_{H^1(\Gamma)}}.$$

For an arbitrary $\varphi \in H^1(\Gamma)$ let $v \in H^{3/2}(\Omega)$ be the unique solution of the Dirichlet boundary value problem $Lv(x) = 0$ for $x \in \Omega$ and $\gamma_0^{\text{int}}v(x) = \varphi(x)$ for $x \in \Gamma$. By using Theorem 4.6 this gives

$$\|\gamma_1^{\text{int}}v\|_{L_2(\Gamma)} \le c\,\|\gamma_0^{\text{int}}v\|_{H^1(\Gamma)} = c\,\|\varphi\|_{H^1(\Gamma)}.$$

Since both $u = \widetilde{V}w$ and $v$ are solutions of a homogeneous partial differential equation, we obtain by applying Green's first formula (1.5) twice and taking into account the symmetry of the bilinear form

$$\begin{aligned}\langle \gamma_1^{\text{int}} u, \varphi\rangle_\Gamma = a(u,v) &= a(v,u) = \langle \gamma_1^{\text{int}} v, \gamma_0^{\text{int}} u\rangle_\Gamma \\ &\leq \|\gamma_1^{\text{int}} v\|_{L_2(\Gamma)} \|\gamma_0^{\text{int}} u\|_{L_2(\Gamma)} \leq c\,\|\varphi\|_{H^1(\Gamma)} \|Vw\|_{L_2(\Gamma)}.\end{aligned}$$

From the continuity of the single layer potential $V : H^{-1}(\Gamma) \to L_2(\Gamma)$ we now conclude

$$\|\gamma_1^{\text{int}} u\|_{H^{-1}(\Gamma)} \leq c\,\|Vw\|_{L_2(\Gamma)} \leq \widetilde{c}\,\|w\|_{H^{-1}(\Gamma)}$$

and therefore the continuity of $\gamma_1^{\text{int}}\widetilde{V} = \sigma I + K' : H^{-1}(\Gamma) \to H^{-1}(\Gamma)$. Using an interpolation argument we obtain $K' : H^{-1/2+s}(\Gamma) \to H^{-1/2+s}(\Gamma)$ for all $|s| \leq \frac{1}{2}$. Due to

$$\begin{aligned}\|Kv\|_{H^{1/2+s}(\Gamma)} &= \sup_{0 \neq w \in H^{-1/2-s}(\Gamma)} \frac{\langle Kv, w\rangle_\Gamma}{\|w\|_{H^{-1/2-s}(\Gamma)}} \\ &= \sup_{0 \neq w \in H^{-1/2-s}(\Gamma)} \frac{\langle v, K'w\rangle_\Gamma}{\|w\|_{H^{-1/2-s}(\Gamma)}} \\ &= \|v\|_{H^{1/2+s}(\Gamma)} \sup_{0 \neq w \in H^{-1/2-s}(\Gamma)} \frac{\|K'w\|_{H^{-1/2-s}(\Gamma)}}{\|w\|_{H^{-1/2-s}(\Gamma)}} \\ &\leq c\,\|v\|_{H^{1/2+s}(\Gamma)}\end{aligned}$$

we immediately conclude $K : H^{1/2+s}(\Gamma) \to H^{1/2+s}(\Gamma)$ for $|s| \leq \frac{1}{2}$.

It remains to prove the assertion for the hypersingular boundary integral operator $D$. The double layer potential $u(x) = (Wv)(x)$, $x \in \Omega$, yields, by the application of Theorem 4.6,

$$\|Dv\|_{L_2(\Gamma)} = \|\gamma_1^{\text{int}} u\|_{L_2(\Gamma)} \leq c\,\|\gamma_0^{\text{int}} u\|_{H^1(\Gamma)} = c\,\|([\sigma - 1]I + K)v\|_{H^1(\Gamma)}$$

and therefore $D : H^1(\Gamma) \to L_2(\Gamma)$. Again, using duality and interpolation arguments completes the proof. □

If the boundary $\Gamma = \partial\Omega$ of the bounded domain $\Omega \subset \mathbb{R}^d$ is piecewise smooth, Theorem 6.34 remains true for larger values of $|s|$. For example, if $\Omega \subset \mathbb{R}^2$ is polygonal bounded with $J$ corner points and associated interior angles $\alpha_j$ we may define

$$\sigma_0 := \min_{j=1,\ldots,J} \left\{ \min\left[\frac{\pi}{\alpha_j}, \frac{\pi}{2\pi - \alpha_j}\right]\right\}.$$

Then, Theorem 6.34 holds for all $|s| < \sigma_0$ [45]. If the boundary $\Gamma$ is $C^\infty$, then Theorem 6.34 remains true for all $s \in \mathbb{R}$.

## 6.7 Linear Elasticity

All mapping properties of boundary integral operators as shown above for the model problem of the Laplace equation can be transfered to general second order partial differential equations, when a fundamental solution is known. In what follows we will consider the system of linear elastostatics which reads for $d = 2, 3$ and $x \in \Omega \subset \mathbb{R}^d$ as

$$-\frac{E}{2(1+\nu)}\Delta \underline{u}(x) - \frac{E}{2(1+\nu)(1-2\nu)}\operatorname{grad}\operatorname{div}\underline{u}(x) \;=\; \underline{f}(x). \qquad (6.56)$$

The associated fundamental solution is the Kelvin tensor (5.9)

$$U^*_{ij}(x,y) \;=\; \frac{1}{4(d-1)\pi}\,\frac{1}{E}\,\frac{1+\nu}{1-\nu}\left[(3-4\nu)E(x,y)\delta_{ij} + \frac{(x_i-y_i)(x_j-y_j)}{|x-y|^d}\right]$$

for $i, j = 1, \ldots, d$ where

$$E(x,y) \;=\; \begin{cases} -\log|x-y| & \text{for } d = 2,\\[4pt] \dfrac{1}{|x-y|} & \text{for } d = 3. \end{cases}$$

For the components $u_i$ of the solution there holds the representation formula (5.10) (Somigliana identity), $\widetilde{x} \in \Omega$, $i = 1, \ldots, d$,

$$u_i(\widetilde{x}) = \int_\Gamma \sum_{j=1}^d U^*_{ij}(\widetilde{x},y)t_j(y)ds_y - \int_\Gamma \sum_{j=1}^d T^*_{ij}(\widetilde{x},y)u_j(y)ds_y$$
$$+ \int_\Omega \sum_{j=1}^d U^*_{ij}(\widetilde{x},y)f_j(y)dy. \qquad (6.57)$$

As in (6.4) we define the Newton potential

$$(\widetilde{N}_0 f)_i(\widetilde{x}) \;=\; \int_\Omega \sum_{j=1}^d U^*_{ij}(\widetilde{x},y)f_j(y)dy \quad \text{for } \widetilde{x} \in \Omega, i = 1, \ldots, d$$

which is a generalized solution of (6.56). Moreover, as in Theorem 6.1,

$$\|\widetilde{N}_0 \underline{f}\|_{[H^1(\Omega)]^d} \;\leq\; c\,\|\underline{f}\|_{[\widetilde{H}^{-1}(\Omega)]^d}.$$

By taking the interior traces of $u_i$,

$$(N_0\underline{f})_i(x) \;:=\; \gamma_0^{\text{int}}(\widetilde{N}_0\underline{f})_i(x) \;=\; \lim_{\Omega\ni\widetilde{x}\to x\in\Gamma}(\widetilde{N}_0\underline{f})_i(\widetilde{x}) \quad \text{for } i = 1, \ldots, d,$$

this defines a linear and bounded operator

$$N_0 \;:=\; \gamma_0^{\text{int}}\widetilde{N}_0 \,:\, [\widetilde{H}^{-1}(\Omega)]^d \to [H^{1/2}(\Gamma)]^d.$$

In addition, by applying the boundary stress operator (1.23),

$$(\gamma_1^{\rm int}\widetilde{N}_0\underline{f})_i(x) = \lim_{\Omega\ni\widetilde{x}\to x\in\Gamma}\sum_{j=1}^d \sigma_{ij}(\widetilde{N}_0\underline{f},\widetilde{x})n_j(x) \quad \text{for } i=1,\ldots,d,$$

we introduce a second linear and bounded operator

$$N_1 := \gamma_1^{\rm int}\widetilde{N}_0 : [\widetilde{H}^{-1}(\Omega)]^d \to [H^{-1/2}(\Gamma)]^d.$$

For $\widetilde{x} \in \Omega \cup \Omega^c$ the single layer potential

$$(\widetilde{V}\underline{w})_i(\widetilde{x}) = \int_\Gamma \sum_{j=1}^d U_{ij}^*(\widetilde{x},y)w_j(y)ds_y \quad \text{for } i=1,\ldots,d$$

is a solution of the homogeneous system of linear elasticity (6.56) with $\underline{f}=\underline{0}$. Hence we have

$$\widetilde{V} : [H^{1/2}(\Gamma)]^d \to [H^1(\Omega)]^d.$$

When considering the interior and exterior traces of $\widetilde{V}$ this defines a bounded linear operator

$$V := \gamma_0^{\rm int}\widetilde{V} = \gamma_0^{\rm ext}\widetilde{V} : [H^{-1/2}(\Gamma)]^d \to [H^{1/2}(\Gamma)]^d$$

with a representation as a weakly singular surface integral,

$$(V\underline{w})_i(x) = \int_\Gamma \sum_{j=1}^d U_{ij}^*(x,y)w_j(y)ds_y \quad \text{for } x\in\Gamma, i=1,\ldots,d. \tag{6.58}$$

If $\underline{w} \in [H^{-1/2}(\Gamma)]^d$ is given, the single layer potential $\widetilde{V}\underline{w} \in [H^1(\Omega)]^d$ is a solution of the homogeneous system (6.56) of linear elastostatics. Then the application of the interior boundary stress operator (1.23),

$$(\gamma_1^{\rm int}\widetilde{V}\underline{w})_i(x) = \lim_{\Omega\ni\widetilde{x}\to x\in\Gamma}\sum_{j=1}^d \sigma_{ij}(\widetilde{V}\underline{w},\widetilde{x})n_j(x) \quad \text{for } i=1,\ldots,d,$$

defines a bounded linear operator

$$\gamma_1^{\rm int}\widetilde{V} : [H^{-1/2}(\Gamma)]^d \to [H^{-1/2}(\Gamma)]^d$$

with the representation

$$(\gamma_1^{\rm int}\widetilde{V}\underline{w})_i(x) = \frac{1}{2}w_i(x) + (K'\underline{w})_i(x) \quad \text{for almost all } x\in\Gamma, i=1,\ldots,d,$$

where

$$(K'\underline{w})_k(x) = \lim_{\varepsilon\to 0}\int_{y\in\Gamma:|y-x|\geq\varepsilon}\sum_{j=1}^d\sum_{\ell=1}^d \sigma_{k\ell}(\underline{U}_j^*(x,y),x)n_\ell(x)w_j(y)ds_y$$

is the adjoint double layer potential, $k = 1, \ldots, d$. For simplicity we may assume that $x \in \Gamma$ is on a smooth part of the surface, in particular we do not consider the case when $x \in \Gamma$ is either a corner point or on an edge. Correspondingly, the application of the exterior boundary stress operator gives

$$(\gamma_1^{\mathrm{ext}} \widetilde{V} \underline{w})_i(x) = -\frac{1}{2} w_i(x) + (K' \underline{w})_i(x) \quad \text{for almost all } x \in \Gamma, i = 1, \ldots, d.$$

Hence we obtain the jump relation for the boundary stress of the single layer potential as

$$[\gamma_1 \widetilde{V} \underline{w}] := (\gamma_1^{\mathrm{ext}} \widetilde{V} \underline{w})(x) - (\gamma_1^{\mathrm{int}} \widetilde{V} \underline{w})(x) = -\underline{w}(x) \quad \text{for } x \in \Gamma$$

in the sense of $[H^{-1/2}(\Gamma)]^d$.

As for the Laplace operator the far field behavior of the single layer potential $\widetilde{V}$ is essential when investigating the ellipticity of the single layer potential $V$. The approach as considered for the Laplace equation can be applied as well for the system of linear elastostatics. Note that the related subspace is now induced by the rigid body motions (translations). Hence, for $d = 2$, we define

$$[H_+^{-1/2}(\Gamma)]^2 := \left\{ \underline{w} \in [H^{-1/2}(\Gamma)]^2 : \langle w_i, 1 \rangle_\Gamma = 0 \quad \text{for } i = 1, 2 \right\}.$$

**Lemma 6.35.** *For $y_0 \in \Omega$ and $x \in \mathbb{R}^3$ let $|x - y_0| > 2\, diam(\Omega)$ be satisfied. Assume $\underline{w} \in [H^{-1/2}(\Gamma)]^3$ for $d = 3$ and $\underline{w} \in [H_+^{-1/2}(\Gamma)]^2$ for $d = 2$. For $\underline{u}(x) = (\widetilde{V} \underline{w})(x)$ we then have*

$$|u_i(x)| \leq c \frac{1}{|x - y_0|}, \; i = 1, \ldots, d.$$

*Proof.* Let $d = 3$. Due to

$$|u_i(x)| = \frac{1}{4\pi} \frac{1}{E} \frac{1+\nu}{1-\nu} \left| \int_\Gamma \left[ (3 - 4\nu) \frac{\delta_{ij}}{|x - y|} + \frac{(x_i - y_i)(x_j - y_j)}{|x - y|^3} \right] w_j(y) ds_y \right|$$

$$\leq \frac{1}{4\pi} \frac{1}{E} \frac{1+\nu}{1-\nu} \int_\Gamma \left[ (3 - 4\nu) \frac{\delta_{ij}}{|x - y|} + \frac{1}{|x - y|} \right] |w_j(y)| ds_y$$

we obtain the assertion as in the proof of Lemma 6.21. For $d = 2$ we consider the Taylor expansion of the fundamental solution to conclude the result as in the proof of Lemma 6.21. □

As for the single layer potential of the Laplace operator (Theorem 6.22) we now can prove the following ellipticity result.

**Theorem 6.36.** *Assume $\underline{w} \in [H^{-1/2}(\Gamma)]^d$ for $d = 3$ and $\underline{w} \in [H_+^{-1/2}(\Gamma)]^d$ for $d = 2$. Then we have*

$$\langle V \underline{w}, \underline{w} \rangle_\Gamma \geq c_1^V \, \|\underline{w}\|^2_{[H^{-1/2}(\Gamma)]^d}$$

*with a positive constant $c_1^V$.*

*Proof.* The ellipticity estimate follows as is the proof of Theorem 6.22 by using Lemma 4.19. □

To prove the $[H^{-1/2}(\Gamma)]^2$–ellipticity of the two–dimensional single layer potential $V$ we first introduce the generalized fundamental solution

$$U^\alpha_{ij}(x,y) = \frac{1}{4\pi}\frac{1}{E}\frac{1+\nu}{1-\nu}\left[(4\nu-3)\log(\alpha|x-y|)\delta_{ij} + \frac{(x_i-y_i)(x_j-y_j)}{|x-y|^2}\right]$$

for $i,j=1,2$ which depends on a real parameter $\alpha \in \mathbb{R}_+$, and we consider the corresponding single layer potential $V_\alpha : [H^{-1/2}(\Gamma)]^2 \to [H^{-1/2}(\Gamma)]^2$. Note that this approach corresponds to some scaling of the computational domain $\Omega \subset \mathbb{R}^2$ and its boundary $\Gamma$, respectively.

For $\underline{w} \in [H_+^{-1/2}(\Gamma)]^2$ we have by using Theorem 6.36 the ellipticity estimate

$$\langle V_\alpha \underline{w}, \underline{w}\rangle_\Gamma = \langle V\underline{w}, \underline{w}\rangle_\Gamma \geq c_1^V \, \|\underline{w}\|^2_{[H^{-1/2}(\Gamma)]^2}.$$

The further approach now corresponds to the case of the scalar single layer potential of the Laplace operator [142] to find $(\underline{w}^1, \underline{\lambda}^1) \in [H^{-1/2}(\Gamma)]^2 \times \mathbb{R}^2$ as the solution of the saddle point problem

$$\begin{aligned}
\langle V_\alpha \underline{w}^1, \underline{\tau}\rangle_\Gamma - \lambda_1^1\langle 1, \tau_1\rangle_\Gamma - \lambda_2^1\langle 1, \tau_2\rangle_\Gamma &= 0\\
\langle w_1^1, 1\rangle_\Gamma &= 1\\
\langle w_2^1, 1\rangle_\Gamma &= 0
\end{aligned}$$

to be satisfied for all $\underline{\tau} \in [H^{-1/2}(\Gamma)]^2$. By introducing $w_1^1 := \widetilde{w}_1^1 + 1/|\Gamma|$ and $w_2^1 := \widetilde{w}_2^1$ it remains to find $\underline{\widetilde{w}}^1 \in [H_+^{-1/2}(\Gamma)]^2$ as the unique solution of the variational problem

$$\langle V_\alpha \underline{\widetilde{w}}^1, \underline{\tau}\rangle_\Gamma = -\frac{1}{|\Gamma|}\langle V_\alpha(1,0)^\top, \underline{\tau}\rangle_\Gamma \quad \text{for all } \underline{\tau} \in [H_+^{-1/2}(\Gamma)]^2.$$

When $\underline{\widetilde{w}}^1 \in [H_+^{-1/2}(\Gamma)]^2$ and therefore $\underline{w}^1 \in [H^{-1/2}(\Gamma)]^2$ is known we can compute

$$\lambda_1^1 = \langle V_\alpha \underline{w}^1, \underline{w}^1\rangle_\Gamma.$$

In the same way we find $(\underline{w}^2, \underline{\lambda}^2) \in [H^{-1/2}(\Gamma)]^2 \times \mathbb{R}^2$ as the solution of the saddle point problem

$$\begin{aligned}
\langle V_\alpha \underline{w}^2, \underline{\tau}\rangle_\Gamma - \lambda_1^2\langle 1, \tau_1\rangle_\Gamma - \lambda_2^2\langle 1, \tau_2\rangle_\Gamma &= 0\\
\langle w_1^2, 1\rangle_\Gamma &= 0\\
\langle w_2^2, 1\rangle_\Gamma &= 1
\end{aligned}$$

to be satisfied for all $\underline{\tau} \in [H^{-1/2}(\Gamma)]^2$. Moreover, we obtain

$$\lambda_2^2 = \langle V_\alpha \underline{w}^2, \underline{w}^2\rangle_\Gamma,$$

as well as

$$\lambda_2^1 = \lambda_1^2 = \langle V_\alpha \underline{w}^1, \underline{w}^2\rangle_\Gamma.$$

**Lemma 6.37.** *For the Lagrange multiplier* $\lambda_i^1$ $(i = 1, 2)$ *we have the representation*

$$\lambda_i^i = \langle V\underline{w}^i, \underline{w}^i\rangle_\Gamma + \frac{1}{4\pi}\frac{1}{E}\frac{1+\nu}{1-\nu}(4\nu-3)\log\alpha,$$

*while the Lagrange multiplier* $\lambda_2^1 = \lambda_1^2$ *is independent of* $\alpha \in \mathbb{R}_+$,

$$\lambda_2^1 = \lambda_1^2 = \langle V\underline{w}^1, \underline{w}^2\rangle_\Gamma.$$

*Proof.* For $i = 1$, a direct computation gives, by splitting the fundamental solution $\log(\alpha|x-y|)$,

$$\begin{aligned}
\lambda_1^1 &= \langle V_\alpha \underline{w}^1, \underline{w}^1\rangle_\Gamma \\
&= \frac{1}{4\pi}\frac{1}{E}\frac{1+\nu}{1-\nu}\int_\Gamma\int_\Gamma \sum_{i=1}^2 (4\nu-3)\log(\alpha|x-y|)w_i^1(y)w_i^1(x)ds_xds_y \\
&\quad +\frac{1}{4\pi}\frac{1}{E}\frac{1+\nu}{1-\nu}\int_\Gamma\int_\Gamma \sum_{i,j=1}^2 \frac{(x_i-y_i)(x_j-y_j)}{|x-y|^2}w_i^1(y)w_j^1(x)ds_xds_y \\
&= \frac{1}{4\pi}\frac{1}{E}\frac{1+\nu}{1-\nu}\int_\Gamma\int_\Gamma \sum_{i=1}^2 (4\nu-3)\log|x-y|w_i^1(y)w_i^1(x)ds_xds_y \\
&\quad +\frac{1}{4\pi}\frac{1}{E}\frac{1+\nu}{1-\nu}\int_\Gamma\int_\Gamma \sum_{i,j=1}^2 \frac{(x_i-y_i)(x_j-y_j)}{|x-y|^2}w_i^1(y)w_j^1(x)ds_xds_y \\
&\quad +\frac{1}{4\pi}\frac{1}{E}\frac{1+\nu}{1-\nu}(4\nu-3)\log\alpha\sum_{i=1}^2\left[\langle w_i^1, 1\rangle_\Gamma\right]^2 \\
&= \langle V_1\underline{w}^1, \underline{w}^1\rangle_\Gamma + \frac{1}{4\pi}\frac{1}{E}\frac{1+\nu}{1-\nu}(4\nu-3)\log\alpha
\end{aligned}$$

due to $\langle w_1^1, 1\rangle_\Gamma = 1$ and $\langle w_2^1, 1\rangle_\Gamma = 0$. For $\lambda_2^2$ the assertion follows in the same way. Finally, for $\lambda_1^2 = \lambda_2^1$ we have

$$\begin{aligned}
\lambda_1^2 &= \langle V_\alpha \underline{w}^1, \underline{w}^2\rangle_\Gamma \\
&= \langle V_1\underline{w}^1, \underline{w}^2\rangle_\Gamma + \frac{1}{4\pi}\frac{1}{E}\frac{1+\nu}{1-\nu}(4\nu-3)\log\alpha\sum_{i=1}^2\langle w_i^1, 1\rangle_\Gamma\langle w_i^2, 1\rangle_\Gamma \\
&= \langle V_1\underline{w}^1, \underline{w}^2\rangle_\Gamma
\end{aligned}$$

due to $\langle w_1^2, 1\rangle_\Gamma = \langle w_2^1, 1\rangle_\Gamma = 0$. □

Hence we can choose the scaling parameter $\alpha \in \mathbb{R}_+$ such that

$$\min\{\lambda_1^1, \lambda_2^2\} \geq 2\,|\lambda_2^1| \tag{6.59}$$

is satisfied. An arbitrary given $\underline{w} \in [H^{-1/2}(\Gamma)]^2$ can be written as

$$\underline{w} = \widetilde{\underline{w}} + \alpha_1\underline{w}^1 + \alpha_2\underline{w}^2, \quad \alpha_i = \langle w_i, 1\rangle_\Gamma \quad (i = 1, 2) \tag{6.60}$$

where $\widetilde{\underline{w}} \in [H_+^{-1/2}(\Gamma)]^2$.

**Theorem 6.38.** *Let the scaling parameter $\alpha \in \mathbb{R}_+$ be chosen such that (6.59) is satisfied. Then the single layer potential $V_\alpha$ is $[H^{-1/2}(\Gamma)]^2$–elliptic, i.e.*

$$\langle V_\alpha \underline{w}, \underline{w}\rangle_\Gamma \geq \widetilde{c}_1^V \, \|\underline{w}\|^2_{[H^{-1/2}(\Gamma)]^2} \quad \textit{for all } \underline{w} \in [H^{-1/2}(\Gamma)]^2 .$$

*Proof.* For an arbitrary $\underline{w} \in [H^{-1/2}(\Gamma)]^2$ we consider the splitting (6.60). By using the triangle inequality as well as the Cauchy–Schwarz inequality we obtain

$$\begin{aligned}
\|\underline{w}\|^2_{[H^{-1/2}(\Gamma)]^2} &= \|\underline{\widetilde{w}} + \alpha_1 \underline{w}^1 + \alpha_2 \underline{w}^2\|^2_{[H^{-1/2}(\Gamma)]^2} \\
&\leq \Big[ \|\underline{\widetilde{w}}\|_{[H^{-1/2}(\Gamma)]^2} + |\alpha_1| \, \|\underline{w}^1\|^2_{[H^{-1/2}(\Gamma)]^2} + |\alpha_2| \, \|\underline{w}^2\|^2_{[H^{-1/2}(\Gamma)]^2} \Big]^2 \\
&\leq 3 \Big[ \|\underline{\widetilde{w}}\|^2_{[H^{-1/2}(\Gamma)]^2} + \alpha_1^2 \, \|\underline{w}^1\|^2_{[H^{-1/2}(\Gamma)]^2} + \alpha_2^2 \, \|\underline{w}^2\|^2_{[H^{-1/2}(\Gamma)]^2} \Big] \\
&\leq 3 \max\Big\{1, \|\underline{w}^1\|^2_{[H^{-1/2}(\Gamma)]^2}, \|\underline{w}^2\|^2_{[H-1/2(\Gamma)]^2}\Big\} \Big[ \|\underline{\widetilde{w}}\|^2_{[H^{-1/2}(\Gamma)]^2} + \alpha_1^2 + \alpha_2^2 \Big] .
\end{aligned}$$

Moreover, by the construction of $\underline{w}^1$ and $\underline{w}^2$ we have

$$\begin{aligned}
\langle V_\alpha \underline{w}, \underline{w}\rangle_\Gamma &= \langle V_\alpha [\underline{\widetilde{w}} + \alpha_1 \underline{w}^1 + \alpha_2 \underline{w}^2], \underline{\widetilde{w}} + \alpha_1 \underline{w}^1 + \alpha_2 \underline{w}^2\rangle_\Gamma \\
&= \langle V_\alpha \underline{\widetilde{w}}, \underline{\widetilde{w}}\rangle_\Gamma + \alpha_1^2 \, \langle V_\alpha \underline{w}^1, \underline{w}^1\rangle_\Gamma + \alpha_2^2 \, \langle V_\alpha \underline{w}^2, \underline{w}^2\rangle_\Gamma \\
&\quad + 2\alpha_1 \langle V_\alpha \underline{w}^1, \underline{\widetilde{w}}\rangle_\Gamma + 2\alpha_2 \langle V_\alpha \underline{w}^2, \underline{\widetilde{w}}\rangle_\Gamma + 2\alpha_1\alpha_2 \langle V_\alpha \underline{w}^1, \underline{w}^2\rangle_\Gamma \\
&= \langle V_\alpha \underline{\widetilde{w}}, \underline{\widetilde{w}}\rangle_\Gamma + \alpha_1^2 \, \lambda_1^1 + \alpha_2^2 \, \lambda_2^2 + 2\alpha_1\alpha_2 \, \lambda_1^2 .
\end{aligned}$$

From the $[H_+^{-1/2}(\Gamma)]^2$–ellipticity of $V_\alpha$ and by using the scaling condition (6.59) we finally get

$$\begin{aligned}
\langle V_\alpha \underline{w}, \underline{w}\rangle_\Gamma &\geq c_1^V \, \|\underline{\widetilde{w}}\|^2_{[H^{-1/2}(\Gamma)]^2} + \alpha_1^2 \lambda_1^1 + \alpha_2^2 \, \lambda_2^2 - 2|\alpha_1| \, |\alpha_2| \, |\lambda_1^2| \\
&\geq c_1^V \, \|\underline{\widetilde{w}}\|^2_{[H^{-1/2}(\Gamma)]^2} + \min\{\lambda_1^1, \lambda_2^2\} \, [\alpha_1^2 + \alpha_2^2 - |\alpha_1| \, |\alpha_2|] \\
&\geq c_1^V \, \|\underline{\widetilde{w}}\|^2_{[H^{-1/2}(\Gamma)]^2} + \frac{1}{2} \min\{\lambda_1^1, \lambda_2^2\} \, [\alpha_1^2 + \alpha_2^2] \\
&\geq \min\Big\{c_1^V, \frac{1}{2}\lambda_1^1, \frac{1}{2}\lambda_2^2\Big\} \Big[ \|\underline{\widetilde{w}}\|^2_{[H^{-1/2}(\Gamma)]^2} + \alpha_1^2 + \alpha_2^2 \Big] . \qquad \square
\end{aligned}$$

Therefore, the single layer potential $V : [H^{-1/2}(\Gamma)]^d \to [H^{1/2}(\Gamma)]^d$ is bounded and $[H^{-1/2}(\Gamma)]^d$–elliptic, where for $d = 2$ we have to assume a suitable scaling of the domain $\Omega$, see (6.59). By using the Lax–Milgram lemma (Theorem 3.4) we therefore conclude the existence of the inverse operator $V^{-1} : [H^{1/2}(\Gamma)]^d \to [H^{-1/2}(\Gamma)]^d$.

By $\mathcal{R}$ we denote the set of rigid body motions, i.e. (1.36) for $d = 2$ and (1.29) for $d = 3$, respectively. Define

$$[H_*^{-1/2}(\Gamma)]^d := \Big\{ \underline{w} \in [H^{-1/2}(\Gamma)]^d : \langle \underline{w}, \underline{v}_k\rangle_\Gamma = 0 \quad \text{for } \underline{v}_k \in \mathcal{R} \Big\} ,$$

and

$$[H_*^{1/2}(\Gamma)]^d := \left\{ \underline{v} \in [H^{1/2}(\Gamma)]^d : \langle V^{-1}\underline{v}, \underline{v}_k \rangle_\Gamma = 0 \quad \text{for } \underline{v}_k \in \mathcal{R} \right\}.$$

Obviously, $V : [H_*^{-1/2}(\Gamma)]^d \to [H_*^{1/2}(\Gamma)]^d$ is an isomorphism.

For $\widetilde{x} \in \Omega \cup \Omega^c$ we define by

$$(W\underline{v})_i(\widetilde{x}) := \int_\Gamma \sum_{j=1}^d T_{ij}^*(\widetilde{x}, y) u_j(y) ds_y, \quad i = 1, \ldots, d,$$

the double layer potential of linear elastostatics satisfying

$$W : [H^{1/2}(\Gamma)]^d \to [H^1(\Gamma)]^d.$$

The application of the interior trace operator defines a bounded linear operator

$$\gamma_0^{\text{int}} W : [H^{1/2}(\Gamma)]^d \to [H^{1/2}(\Gamma)]^d$$

with the representation

$$(\gamma_0^{\text{int}} W\underline{v})_i(x) = -\frac{1}{2} v_i(x) + (K\underline{v})_i(x) \quad \text{for almost all } x \in \Gamma, i = 1, \ldots, d, \tag{6.61}$$

where

$$(K\underline{v})_i(x) := \lim_{\varepsilon \to 0} \int_{y \in \Gamma : |y-x| \geq \varepsilon} \sum_{j=1}^d T_{ij}^*(\widetilde{x}, y) u_j(y) ds_y, \quad i = 1, \ldots, d$$

is the double layer potential. Correspondingly, the application of the exterior trace operator gives

$$(\gamma_0^{\text{ext}} W\underline{v})_i(x) = \frac{1}{2} v_i(x) + (K\underline{v})_i(x) \quad \text{for almost all } x \in \Gamma, i = 1, \ldots, d.$$

Hence, we obtain the jump relation of the double layer potential as

$$[\gamma_0 W\underline{v}] = \gamma_0^{\text{ext}}(W\underline{v})(x) - \gamma_0^{\text{int}}(W\underline{v})(x) = \underline{v}(x) \quad \text{for } x \in \Gamma.$$

From the Somigliana identity (6.57) we get for $\Omega \ni \widetilde{x} \to x \in \Gamma$ by using (6.58) and (6.61) the boundary integral equation

$$(V\underline{t})(x) = \left( \frac{1}{2} I + K \right) \underline{u}(x) - (N_0 \underline{f})(x) \quad \text{for almost all } x \in \Gamma. \tag{6.62}$$

Inserting the rigid body motions (1.36) for $d = 2$ and (1.29) for $d = 3$ this gives

$$\left( \frac{1}{2} I + K \right) \underline{v}_k(x) = \underline{0} \quad \text{for } x \in \Gamma \text{ and } \underline{v}_k \in \mathcal{R}.$$

The application of the interior boundary stress operator $\gamma_1^{\text{int}}$ on the double layer potential $W\underline{v}$ defines a bounded linear operator

$$\gamma_1^{\text{int}} W = \gamma_1^{\text{ext}} W : [H^{1/2}(\Gamma)]^d \to [H^{-1/2}(\Gamma)]^d.$$

As in the case of the Laplace operator we denote by $D := -\gamma_1^{\text{int}} W$ the hypersingular boundary integral operator.

When applying the boundary stress operator $\gamma_1^{\text{int}}$ on the Somigliana identity (6.57) this gives the hypersingular boundary integral equation

$$(D\underline{u})(x) = \left(\frac{1}{2}I - K'\right)\underline{t}(x) - (N_1\underline{f})(x) \quad \text{for } x \in \Gamma \tag{6.63}$$

in the sense of $[H^{-1/2}(\Gamma)]^d$.

As in (6.22) we can write both boundary integral equations (6.62) and (6.63) as a system with the Calderón projection (6.23). Note that the projection property of the Calderón projection (Lemma 6.18) as well as all relations of Corollary 6.19 remain valid as for the scalar Laplace equation.

Analogous to the Laplace operator we can rewrite the bilinear form of the hypersingular boundary integral operator by using integration by parts as a sum of weakly singular bilinear forms. In particular for $d = 2$ we have the representation [107]

$$\langle D\underline{u}, \underline{v}\rangle_\Gamma = \sum_{i,j=1}^{2} \int_\Gamma \text{curl}_\Gamma v_j(x) \int_\Gamma G_{ij}(x,y) \text{curl}_\Gamma u_i(y) ds_y ds_x$$

for all $\underline{u}, \underline{v} \in [H^{1/2}(\Gamma) \cap C(\Gamma)]^2$ where

$$G_{ij}(x,y) = \frac{1}{4\pi}\frac{E}{1-\nu^2}\left[-\log|x-y|\,\delta_{ij} + \frac{(x_i-y_i)(x_j-y_j)}{|x-y|^2}\right], \quad i,j = 1,2.$$

Here, $\text{curl}_\Gamma$ denotes the derivative with respect to the arc length. Note that the kernel functions $G_{ij}(x,y)$ correspond, up to constants, to the kernel functions of the Kelvin fundamental solution (5.9).

In the three–dimensional case $d = 3$ and $i, j = 1, \ldots, 3$ we define

$$M_{ij}(\partial_x, n(x)) := n_j(x)\frac{\partial}{\partial x_i} - n_i(x)\frac{\partial}{\partial x_j}$$

and

$$\frac{\partial}{\partial S_1(x)} := M_{32}(\partial_x, n(x)),$$
$$\frac{\partial}{\partial S_2(x)} := M_{13}(\partial_x, n(x)),$$
$$\frac{\partial}{\partial S_3(x)} := M_{21}(\partial_x, n(x)).$$

The bilinear form of the hypersingular boundary integral operator $D$ can then be written as [75]

$$\begin{aligned}\langle D\underline{u},\underline{v}\rangle_\Gamma &= \frac{\mu}{4\pi}\int_\Gamma\int_\Gamma \frac{1}{|x-y|}\left(\sum_{k=1}^{3}\frac{\partial}{\partial S_k(y)}\underline{u}(y)\cdot\frac{\partial}{\partial S_k(x)}\underline{v}(x)\right)ds_y ds_x \\ &+ \iint_{\Gamma\Gamma}(M(\partial_x,n(x))\underline{v}(x))^\top\left(\frac{\mu}{2\pi}\frac{I}{|x-y|}-4\mu^2U^*(x,y)\right)M(\partial_y,n(y))\underline{u}(y)ds_y ds_x \\ &+\frac{\mu}{4\pi}\int_\Gamma\int_\Gamma\sum_{i,j,k=1}^{3}M_{kj}(\partial_x,n(x))v_i(x)\frac{1}{|x-y|}M_{ki}(\partial_y,n(y))v_j(y)ds_y ds_x. \quad (6.64)\end{aligned}$$

Hence we can express the bilinear form of the hypersingular boundary integral operator $D$ by components of the single layer potential $V$ only.

Moreover, for $d=3$ there holds a related representation of the double layer potential $K$, see [89],

$$\begin{aligned}(K\underline{u})(x) &= \frac{1}{4\pi}\int_\Gamma \frac{\partial}{\partial n(y)}\frac{1}{|x-y|}\underline{u}(y)ds_y - \frac{1}{4\pi}\int_\Gamma\frac{1}{|x-y|}M(\partial_y,n(y))\underline{u}(y)ds_y \\ &+\frac{E}{1+\nu}(V(M(\partial.,n(\cdot))\underline{u}(\cdot)))(x)\end{aligned}$$

where the evaluation of the Laplace single and double layer potentials has to be taken componentwise.

Hence we can reduce all boundary integral operators of linear elastostatics to the single and double layer potentials of the Laplace equation. These relations can be used when considering the Galerkin discretization of boundary integral equations, where only weakly singular surface integrals have to be computed.

Inserting the rigid body motions (1.36) for $d=2$ and (1.29) for $d=3$ into the representation formula (6.57) this gives

$$\underline{v}_k(\widetilde{x}) \;=\; -(W\underline{v}_k)(\widetilde{x}) \quad \text{for } \widetilde{x}\in\Omega \text{ and } \underline{v}_k\in\mathcal{R}.$$

The application of the boundary stress operator $\gamma_1^{\text{int}}$ yields

$$(D\underline{v}_k)(x) \;=\; \underline{0} \quad \text{for } x\in\Gamma \text{ and } \underline{v}_k\in\mathcal{R}.$$

As in Theorem 6.24 we can prove the $[H_*^{1/2}(\Gamma)]^d$–ellipticity of the hypersingular boundary integral operator $D$,

$$\langle D\underline{v},\underline{v}\rangle_\Gamma \;\geq\; c_1^D\,\|\underline{v}\|^2_{[H^{1/2}(\Gamma)]^d} \quad \text{for all } \underline{v}\in[H_*^{1/2}(\Gamma)]^d.$$

In addition, the ellipticity of the hypersingular boundary integral operator $D$ can be formulated also in the subspace

$$[H^{1/2}_{**}(\Gamma)]^d := \left\{ \underline{v} \in [H^{1/2}(\Gamma)]^d \, : \, \langle \underline{v}, \underline{v}_k \rangle_\Gamma = 0 \quad \text{for } \underline{v}_k \in \mathcal{R} \right\}$$

of functions which are orthogonal to the rigid body motions. Then we have, as in (6.38),

$$\langle D\underline{v}, \underline{v} \rangle_\Gamma \geq \widetilde{c}_1^D \, \|\underline{v}\|^2_{[H^{1/2}(\Gamma)]^d} \quad \text{for all } \underline{v} \in [H^{1/2}_{**}(\Gamma)]^d.$$

As in Subsection 6.6.3 we can define the Dirichlet to Neumann map, which relates given boundary displacements to the associated boundary stresses via the Steklov–Poincaré operator. Moreover, all results on the contraction property of the double layer potential (see Subsection 6.6.4) as well as all mapping properties of boundary integral operators (see Subsection 6.6.5) remain valid for the system of linear elastostatics.

## 6.8 Stokes System

Now we consider the homogeneous Stokes system (1.38) where we assume $\mu = 1$ for simplicity, i.e.,

$$-\Delta \underline{u}(x) + \nabla p(x) = \underline{0}, \quad \operatorname{div} \underline{u}(x) = 0 \quad \text{for } x \in \Omega.$$

Since the fundamental solution of the Stokes system coincides with the Kelvin fundamental solution of linear elastostatics when considering $\nu = \frac{1}{2}$ and $E = 3$ as material parameters, we can write the representation formula (6.57) and all related boundary integral operators of linear elastostatics for $\nu = \frac{1}{2}$ and $E = 3$ to obtain the Stokes case. However, since the analysis of the mapping properties of all boundary integral operators of linear elastostatics assumes $\nu \in (0, \frac{1}{2})$ we can not transfer the boundedness and ellipticity estimates from linear elastostatics to the Stokes system. These results will be shown by considering the Stokes single layer potential which is also of interest for the case of almost incompressible linear elasticity ($\nu = \frac{1}{2}$) [141].

Let $\Omega \subset \mathbb{R}^d$ be a simple connected domain with boundary $\Gamma = \partial\Omega$. The single layer potential $\widetilde{V} : [H^{-1/2}(\Gamma)]^d \to [H^1(\Omega)]^d$ induces a function

$$u_i(\widetilde{x}) := (\widetilde{V}\underline{w})_i(\widetilde{x}) = \int_\Gamma \sum_{j=1}^d U^*_{ij}(\widetilde{x}, y) w_j(y) ds_y \quad \text{for } \widetilde{x} \in \Omega, i = 1, \ldots, d$$

which is divergence–free in $\Omega$, and satisfies Green's first formula

$$2\mu \int_\Omega \sum_{i,j=1}^d e_{ij}(\underline{u}, \underline{v}) e_{ij}(\underline{v}, x) dx = \int_\Gamma \underline{v}(x)^\top (T\underline{u})(x) ds_x \tag{6.65}$$

for all $\underline{v} \in [H^1(\Omega)]^d$ with $\operatorname{div} \underline{v} = 0$. The application of the interior trace operator defines the single layer potential

$$V := \gamma_0^{\text{int/ext}} \widetilde{V} : [H^{-1/2}(\Gamma)]^d \to [H^{1/2}(\Gamma)]^d$$

which allows a representation as given in (6.58). To investigate the ellipticity of the single layer potential $V$ we first note that $\underline{u}^* = \underline{0}$ and $p = -1$ defines a solution of the homogeneous Stokes system. From the boundary integral equation (6.62) we then obtain

$$(V\underline{t}^*)(x) = (\frac{1}{2}I + K)\underline{u}^*(x) = \underline{0} \quad \text{for } x \in \Gamma$$

with the associated boundary stress

$$\underline{t}^*(x) = -p^*(x)\,\underline{n}(x) + 2\left(\sum_{j=1}^d e_{ij}(\underline{u}^*, x)n_j(x)\right)_{i=1}^d = \underline{n}(x) \quad \text{for } x \in \Gamma.$$

Hence we can expect the ellipticity of the Stokes single layer potential $V$ only in a subspace which is orthogonal to the exterior normal vector $\underline{n}$.

Let $V^L : H^{-1/2}(\Gamma) \to H^{1/2}(\Gamma)$ be the Laplace single layer potential which is $H^{-1/2}(\Gamma)$–elliptic. Hence we can define

$$\langle \underline{w}, \underline{\tau} \rangle_{V^L} := \sum_{i=1}^d \langle V^L w_i, \tau_i \rangle_\Gamma$$

as an inner product in $[H^{-1/2}(\Gamma)]^d$. When considering the subspace

$$[H^{-1/2}_{V^L}(\Gamma)]^d := \left\{ \underline{w} \in [H^{-1/2}(\Gamma)]^d : \langle \underline{w}, \underline{n} \rangle_{V^L} = 0 \right\}$$

we can prove the following result [50, 159]:

**Theorem 6.39.** *The Stokes single layer potential $V$ is $[H^{-1/2}_{V^L}(\Gamma)]^d$–elliptic, i.e.*

$$\langle V\underline{w}, \underline{w} \rangle_\Gamma \ge c_1^V \, \|\underline{w}\|^2_{[H^{-1/2}(\Gamma)]^d} \quad \textit{for all } \underline{w} \in [H^{-1/2}_{V^L}(\Gamma)]^d.$$

As for the homogeneous Neumann boundary value problem for the Laplace equation we can introduce an extended bilinear form

$$\langle \widetilde{V}\underline{w}, \underline{\tau} \rangle_\Gamma := \langle V\underline{w}, \underline{\tau} \rangle_\Gamma + \langle \underline{w}, \underline{n} \rangle_{V^L} \langle \underline{\tau}, \underline{n} \rangle_{V^L}$$

which defines an $[H^{-1/2}(\Gamma)]^d$–elliptic boundary integral operator $\widetilde{V}$.

When the computational domain $\Omega$ is multiple connected, the dimension of the kernel of the Stokes single layer potential is equal to the number of closed sub–boundaries. Then we have to modify the stabilization in a corresponding manner, see [116].

## 6.9 Helmholtz Equation

Finally we consider the interior Helmholtz equation

$$-\Delta u(x) - k^2 u(x) = 0 \quad \text{for } x \in \Omega \subset \mathbb{R}^d$$

where the fundamental solution is, see (5.20) for $d = 2$ and (5.18) for $d = 3$,

$$U_k^*(x,y) = \begin{cases} \dfrac{1}{2\pi} Y_0(k|x-y|) & \text{for } d = 2, \\ \dfrac{1}{4\pi} \dfrac{e^{ik|x-y|}}{|x-y|} & \text{for } d = 3. \end{cases}$$

Then we can define the standard boundary integral operators for $x \in \Gamma$, i.e. the single layer potential

$$(V_k w)(x) = \int_\Gamma U_k^*(x,y) w(y) ds_y,$$

the double layer potential

$$(K_k v)(x) = \int_\Gamma \frac{\partial}{\partial n_y} U_k^*(x,y) v(y) ds_y,$$

the adjoint double layer potential

$$(K_k' v)(x) = \int_\Gamma \frac{\partial}{\partial n_x} U_k^*(x,y) v(y) ds_y,$$

and the hypersingular boundary integral operator

$$(D_k v)(x) = -\frac{\partial}{\partial n_x} \int_\Gamma \frac{\partial}{\partial n_y} U_k^*(x,y) v(y) ds_y.$$

As for the Laplace operator there hold all the mapping properties as given in Theorem 6.34. In particular, $V_k : H^{-1/2}(\Gamma) \to H^{1/2}(\Gamma)$ is bounded, but not $H^{-1/2}(\Gamma)$–elliptic. However, the single layer potential is coercive satisfying a Gårdings inequality.

For $x \in \Omega$ we consider the function

$$u(x) = (\widetilde{V}_k w)(x) - (\widetilde{V} w)(x) = \int_\Omega [U_k^*(x,y) - U^*(x,y)] w(y) ds_y \qquad (6.66)$$

where $U^*(x,y)$ is the fundamental solution of the Laplace operator. In particular we have for $y \in \Gamma$ and $x \in \Omega$

$$-\Delta_x U_k^*(x,y) - k^2 U_k^*(x,y) = 0, \quad -\Delta_x U^*(x,y) = 0.$$

Then, by interchanging differentiation and integration, we obtain

$$\begin{aligned}[-\Delta_x - k^2]u(x) &= \int_\Gamma [-\Delta_x - k^2][U_k^*(x,y) - U^*(x,y)]w(y)ds_y \\ &= k^2 \int_\Gamma U^*(x,y)w(y)ds_y.\end{aligned}$$

Moreover,

$$-\Delta_x[-\Delta_x - k^2]u(x) \,=\, -k^2\Delta_x \int_\Gamma U^*(x,y)w(y)ds_y \,=\, 0,$$

i.e. the function $u$ as defined in (6.66) solves the partial differential equation

$$-\Delta_x[-\Delta_x - k^2]u(x) \,=\, 0 \quad \text{for } x \in \Omega$$

which is of fourth order. Hence, we obtain as in the case of the Laplace operator, by considering the corresponding Newton potentials, that

$$\widetilde{V}_k - \widetilde{V} \,:\, H^{-1/2}(\Gamma) \to H^3(\Omega).$$

Thus,

$$V_k - V \,=\, \gamma_0^{\text{int}}[\widetilde{V}_k - \widetilde{V}] \,:\, H^{-1/2}(\Gamma) \to H^{5/2}(\Gamma),$$

and by the compact imbedding of $H^{5/2}(\Gamma)$ in $H^{1/2}(\Gamma)$ we conclude that

$$V_k - V \,:\, H^{-1/2}(\Gamma) \to H^{1/2}(\Gamma) \tag{6.67}$$

is compact.

**Theorem 6.40.** *The single layer potential $V_k \,:\, H^{-1/2}(\Gamma) \to H^{1/2}(\Gamma)$ is coercive, i.e. there exists a compact operator $C : H^{-1/2}(\Gamma) \to H^{1/2}(\Gamma)$ such that the Gårdings inequality*

$$\langle V_k w, w\rangle_\Gamma + \langle Cw, w,\rangle_\Gamma \,\geq\, c_1^V \, \|w\|^2_{H^{-1/2}(\Gamma)} \quad \textit{for } w \in H^{-1/2}(\Gamma)$$

*is satisfied. For $d = 2$ we have to assume $\operatorname{diam}\Omega < 1$.*

*Proof.* By considering the compact operator

$$C = V - V_k : H^{-1/2}(\Gamma) \to H^{1/2}(\Gamma)$$

we have

$$\langle V_k w, w\rangle_\Gamma + \langle Cw, w\rangle_\Gamma \,=\, \langle Vw, w\rangle_\Gamma \,\geq\, c_1^V \, \|w\|^2_{H^{-1/2}(\Gamma)}$$

by using Theorem 6.22 for $d = 3$ and Theorem 6.23 for $d = 2$. □

Note that also the operators

$$\begin{aligned} D_k - D &: H^{1/2}(\Gamma) \to H^{-1/2}(\Gamma), \\ K_k - K &: H^{1/2}(\Gamma) \to H^{1/2}(\Gamma), \\ K_K' - K' &: H^{-1/2}(\Gamma) \to H^{-1/2}(\Gamma) \end{aligned}$$

are compact where $D, K, K'$ are the hypersingular integral operator, the double layer potential and its adjoint of the Laplace operator, respectively.

As for the Laplace operator the bilinear form of the hypersingular boundary integral operator $D_k$ can be written as, by using integration by parts [107],

$$\begin{aligned} \langle D_k u, v \rangle_\Gamma = \frac{1}{4\pi} \int_\Gamma \int_\Gamma \frac{e^{\imath k|x-y|}}{|x-y|} \big(\underline{\mathrm{curl}}_\Gamma u(y), \underline{\mathrm{curl}}_\Gamma v(x)\big) ds_y ds_x \\ - \frac{k^2}{4\pi} \int_\Gamma \int_\Gamma \frac{e^{\imath k|x-y|}}{|x-y|} u(y) v(x) \big(\underline{n}(x), \underline{n}(y)\big) ds_y ds_x \,. \end{aligned} \quad (6.68)$$

## 6.10 Exercises

Let $\Gamma = \partial\Omega$ be the boundary of the circle $\Omega = B_r(0) \subset \mathbb{R}^2$ which can be described by using polar coordinates as

$$x(t) = r \begin{pmatrix} \cos 2\pi t \\ \sin 2\pi t \end{pmatrix} \in \Gamma \quad \text{for } t \in [0,1).$$

**6.1** Find a representation of the two–dimensional single layer potential

$$(Vw)(x) = -\frac{1}{2\pi} \int_\Gamma \log|x-y| w(y) ds_y \quad \text{for } x \in \Gamma = \partial B_r(0)$$

when using polar coordinates $x = x(\tau)$ and $y = y(t)$, respectively.

**6.2** Find a representation of the two–dimensional double layer potential

$$(Kv)(x) = -\frac{1}{2\pi} \int_\Gamma \frac{(y-x, \underline{n}(y))}{|x-y|^2} v(y) ds_y \quad \text{for } x \in \Gamma = \partial B_r(0)$$

when using polar coordinates $x = x(\tau)$ and $y = y(t)$, respectively.

**6.3** The eigenfunctions of the double layer potential as considered in Exercise 6.2 are given by

$$v_k(t) = e^{i2\pi kt} \quad \text{for } k \in \mathbb{N}_0.$$

Compute the associated eigenvalues.

**6.4** By using the eigenfunctions as given in Exercise 6.3 compute all eigenvalues of the single layer potential as given in Exercise 6.1. Give sufficient conditions such that the single layer potential is invertible, and positive definite.

**6.5** Prove Corollary 6.19.

**6.6** Determine the eigenfunctions of the hypersingulur boundary integral operator $D$ for $x \in \partial B_r(0)$ and compute the corresponding eigenvalues.

**6.7** Let now $\Gamma$ be the boundary of an ellipse given by the parametrization

$$x(t) \; = \; \begin{pmatrix} a \cos 2\pi t \\ b \sin 2\pi t \end{pmatrix} \in \Gamma \quad \text{for } t \in [0, 2\pi).$$

Find a representation of the corresponding double layer potential

$$(Kv)(x) \; = \; -\frac{1}{2\pi} \int_\Gamma \frac{(y - x, \underline{n}(y))}{|x - y|^2} v(y) ds_y \quad \text{for } x \in \Gamma.$$

**6.8** Prove that the eigenfunctions of the double layer potential as considered in Exercise 6.7 are given by

$$v_k(t) \; = \; \begin{cases} \cos 2\pi k t & \text{for } k > 0, \\ 1 & \text{for } k = 0, \\ \sin 2\pi k t & \text{for } k < 0. \end{cases}$$

Compute the corresponding eigenvalues. Describe the behavior of the maximal eigenvalue as $\frac{a}{b} \to \infty$.

**6.9** Prove for the double layer potential of the Helmholtz equation that

$$K_k \; = \; K'_{-k}$$

is satisfied when considering the complex inner product

$$\langle w, v \rangle_\Gamma \; = \; \int_\Gamma w(x) \overline{v(x)} ds_x.$$

# 7

# Boundary Integral Equations

In this chapter we consider boundary value problems for scalar homogeneous partial differential equations

$$(Lu)(x) = 0 \quad \text{for } x \in \Omega \tag{7.1}$$

where $L$ is an elliptic and self–adjoint partial differential operator of second order, and $\Omega$ is a bounded and simple connected domain with Lipschitz boundary $\Gamma = \partial\Omega$. In particular we focus on the Laplace and on the Helmholtz equations. Note that boundary integral equations for boundary value problems in linear elasticity can be formulated and analyzed as for the Laplace equation. To handle inhomogeneous partial differential equations, Newton potentials have to be considered in addition. By computing particular solutions of the inhomogeneous partial differential equations all Newton potentials can be reduced to surface potentials only, see, for example, [86, 136].

Any solution $u$ of the homogeneous partial differential equation (7.1) is given for $\widetilde{x} \in \Omega$ by the representation formula (5.2),

$$u(\widetilde{x}) = \int_\Gamma U^*(\widetilde{x}, y)\gamma_1^{\text{int}} u(y) ds_y - \int_\Gamma \gamma_{1,y}^{\text{int}} U^*(\widetilde{x}, y)\gamma_0^{\text{int}} u(y) ds_y. \tag{7.2}$$

Hence we have to find the complete Cauchy data $\gamma_0^{\text{int}} u(x)$ and $\gamma_1^{\text{int}} u(x)$ for $x \in \Gamma$, which are given by boundary conditions only partially. For this we will describe appropriate boundary integral equations. The starting point is the representation formula (7.2) and the related system (6.22) of boundary integral equations,

$$\begin{pmatrix} \gamma_0^{\text{int}} u \\ \gamma_1^{\text{int}} u \end{pmatrix} = \begin{pmatrix} [1-\sigma]I - K & V \\ D & \sigma I + K' \end{pmatrix} \begin{pmatrix} \gamma_0^{\text{int}} u \\ \gamma_1^{\text{int}} u \end{pmatrix}. \tag{7.3}$$

This approach is called direct where the density functions of all boundary integral operators are just the Cauchy data $[\gamma_0^{\text{int}} u(x), \gamma_1^{\text{int}} u(x)]$, $x \in \Gamma$. When describing the solution of boundary value problems by using suitable potentials, we end up with the so called indirect approach. For example, solutions of the homogeneous partial differential equation (7.1) are given either by the single layer potential

$$u(\widetilde{x}) = \int_\Gamma U^*(\widetilde{x},y)w(y)ds_y \quad \text{for } \widetilde{x} \in \Omega, \tag{7.4}$$

or by the double layer potential

$$u(\widetilde{x}) = -\int_\Gamma \gamma_{1,y}^{\text{int}} U^*(\widetilde{x},y)v(y)ds_y \quad \text{for } \widetilde{x} \in \Omega. \tag{7.5}$$

It is worth to mention that in general the density functions $w$ and $v$ of the indirect approach have no physical meaning.

In this chapter we will consider different boundary integral equations to find the unknown Cauchy data to describe the solution of several boundary value problems with different boundary conditions.

## 7.1 Dirichlet Boundary Value Problem

First we consider the Dirichlet boundary value problem

$$(Lu)(x) = 0 \quad \text{for } x \in \Omega, \quad \gamma_0^{\text{int}} u(x) = g(x) \quad \text{for } x \in \Gamma. \tag{7.6}$$

When using the direct approach (7.2) we obtain the representation formula

$$u(\widetilde{x}) = \int_\Gamma U^*(\widetilde{x},y)\gamma_1^{\text{int}} u(y)ds_y - \int_\Gamma \gamma_{1,y}^{\text{int}} U^*(\widetilde{x},y)g(y)ds_y \quad \text{for } \widetilde{x} \in \Omega \tag{7.7}$$

where we have to find the yet unknown Neumann datum $\gamma_1^{\text{int}} u \in H^{-1/2}(\Gamma)$. By using the first boundary integral equation in (7.3) we obtain with

$$(V\gamma_1^{\text{int}} u)(x) = \sigma(x)g(x) + (Kg)(x) \quad \text{for } x \in \Gamma \tag{7.8}$$

a first kind Fredholm boundary integral equation. Since the single layer potential $V : H^{-1/2}(\Gamma) \to H^{1/2}(\Gamma)$ is bounded (see (6.8)) and $H^{-1/2}(\Gamma)$–elliptic (see Theorem 6.22 for $d = 3$ and Theorem 6.23 for $d = 2$ when assuming $\text{diam}(\Omega) < 1$), we conclude the unique solvability of the boundary integral equation (7.8) when applying the Lax–Milgram lemma (Theorem 3.4). Moreover, the unique solution $\gamma_1^{\text{int}} u \in H^{-1/2}(\Gamma)$ satisfies

$$\|\gamma_1^{\text{int}} u\|_{H^{-1/2}(\Gamma)} \le \frac{1}{c_1^V} \|(\sigma I + K)g\|_{H^{1/2}(\Gamma)} \le \frac{c_2^W}{c_1^V} \|g\|_{H^{1/2}(\Gamma)}.$$

Since the boundary integral equation (7.8) is formulated in $H^{1/2}(\Gamma)$, this gives

$$\begin{aligned} 0 &= \|V\gamma_1^{\mathrm{int}}u - (\sigma I + K)g\|_{H^{1/2}(\Gamma)} \\ &= \sup_{0\neq\tau\in H^{-1/2}(\Gamma)} \frac{\langle V\gamma_1^{\mathrm{int}}u - (\sigma I + K)g, \tau\rangle_\Gamma}{\|\tau\|_{H^{-1/2}(\Gamma)}}. \end{aligned}$$

and therefore, instead of (7.8) we may consider the equivalent variational problem to find $\gamma_1^{\mathrm{int}}u \in H^{-1/2}(\Gamma)$ such that

$$\langle V\gamma_1^{\mathrm{int}}u, \tau\rangle_\Gamma = \langle(\frac{1}{2}I + K)g, \tau\rangle_\Gamma \tag{7.9}$$

is satisfied for all $\tau \in H^{-1/2}(\Gamma)$. Note that the definition of $\sigma(x)$ gives $\sigma(x) = \frac{1}{2}$ for almost all $x \in \Gamma$.

Instead of (7.8) we may also use the second boundary integral equation in (7.3) to find the unknown Neumann datum $\gamma_1^{\mathrm{int}}u \in H^{-1/2}(\Gamma)$, i.e.

$$([1-\sigma]I - K')\gamma_1^{\mathrm{int}}u(x) = (Dg)(x) \quad \text{for } x \in \Gamma \tag{7.10}$$

which is a second kind Fredholm boundary integral equation. The solution of this boundary integral equation is given by the Neumann series

$$\gamma_1^{\mathrm{int}}u(x) = \sum_{\ell=0}^{\infty}(\sigma I + K')^\ell(Dg)(x) \quad \text{for } x \in \Gamma. \tag{7.11}$$

The convergence of the series (7.11) in $H^{-1/2}(\Gamma)$ follows from the contraction property (6.54) of $\sigma I + K'$ when considering the equivalent Sobolev norm $\|\cdot\|_V$ which is induced by the single layer potential $V$.

When using the indirect single layer potential ansatz (7.4) to find the unknown density $w \in H^{-1/2}(\Gamma)$ we have to solve the boundary integral equation

$$(Vw)(x) = g(x) \quad \text{for } x \in \Gamma. \tag{7.12}$$

Note that the boundary integral equation (7.12) differs from the boundary integral equation (7.8) of the direct approach only in the definition of the right hand side. Hence we can conclude the unique solvability of the boundary integral equation (7.12) as for (7.8).

By using the double layer potential (7.5) to describe the solution of the homogeneous partial differential equation we obtain from the jump relation (6.14) of the double layer potential the boundary integral equation

$$[1-\sigma(x)]v(x) - (Kv)(x) = g(x) \quad \text{for } x \in \Gamma \tag{7.13}$$

to compute the density $v \in H^{1/2}(\Gamma)$ via the Neumann series

$$v(x) = \sum_{\ell=0}^{\infty}(\sigma I + K)^\ell g(x) \quad \text{for } x \in \Gamma. \tag{7.14}$$

The convergence of the series (7.14) in $H^{1/2}(\Gamma)$ follows from the contraction property (6.51) of $\sigma I + K$ when considering the equivalent Sobolev norm $\|\cdot\|_{V^{-1}}$ which is induced by the inverse single layer potential $V^{-1}$.

To obtain a variational formulation of the boundary integral equation (7.13) in $H^{1/2}(\Gamma)$ we first consider

$$0 = \|[1-\sigma]v - Kv - g\|_{H^{1/2}(\Gamma)} = \sup_{0\neq\tau\in H^{-1/2}(\Gamma)} \frac{\langle \frac{1}{2}v - Kv - g, \tau\rangle_\Gamma}{\|\tau\|_{H^{-1/2}(\Gamma)}}$$

where we have used $\sigma(x) = \frac{1}{2}$ for almost all $x \in \Gamma$.

This gives a variational formulation to find $v \in H^{1/2}(\Gamma)$ such that

$$\langle (\frac{1}{2}I - K)v, \tau\rangle_\Gamma = \langle g, \tau\rangle_\Gamma \tag{7.15}$$

is satisfied for all $\tau \in H^{-1/2}(\Gamma)$.

**Lemma 7.1.** *There holds the stability condition*

$$c_S\,\|v\|_{H^{1/2}(\Gamma)} \leq \sup_{0\neq\tau\in H^{-1/2}(\Gamma)} \frac{\langle(\frac{1}{2}I - K)v, \tau\rangle_\Gamma}{\|\tau\|_{H^{-1/2}(\Gamma)}} \quad \textit{for all } v \in H^{1/2}(\Gamma)$$

*with a positive constant* $c_S > 0$.

*Proof.* Let $v \in H^{1/2}(\Gamma)$ be arbitrary but fixed. For $\tau_v := V^{-1}v \in H^{-1/2}(\Gamma)$ we then have

$$\|\tau_v\|_{H^{-1/2}(\Gamma)} = \|V^{-1}v\|_{H^{-1/2}(\Gamma)} \leq \frac{1}{c_1^V}\,\|v\|_{H^{1/2}(\Gamma)}.$$

By using the contraction estimate (6.51) and the mapping properties of the single layer potential $V$ we obtain

$$\begin{aligned}
\langle(\frac{1}{2}I - K)v, \tau_v\rangle_\Gamma &= \langle(\frac{1}{2}I - K)v, V^{-1}v\rangle_\Gamma \\
&= \langle V^{-1}v, v\rangle_\Gamma - \langle V^{-1}(\frac{1}{2}I + K)v, v,\rangle_\Gamma \\
&\geq \|v\|^2_{V^{-1}} - \|(\frac{1}{2}I + K)v\|_{V^{-1}}\|v\|_{V^{-1}} \\
&\geq (1 - c_K)\,\|v\|^2_{V^{-1}} \\
&= (1 - c_K)\,\langle V^{-1}v, v\rangle_\Gamma \\
&\geq (1 - c_K)\frac{1}{c_2^V}\,\|v\|^2_{H^{1/2}(\Gamma)} \\
&\geq (1 - c_K)\frac{c_1^V}{c_2^V}\,\|v\|_{H^{1/2}(\Gamma)}\|\tau_v\|_{H^{-1/2}(\Gamma)}
\end{aligned}$$

from which the stability condition follows immediately. □

Hence we conclude the unique solvability of the variational problem (7.15) by applying Theorem 3.7.

*Remark 7.2.* To describe the solution of the Dirichlet boundary value problem (7.6) we have described four different boundary integral equations, and we have shown their unique solvability. Depending on the application and on the discretization scheme to be used, each of the above formulations may have their advantages or disadvantages. In this book, we will mainly consider the approximate solution of the variational formulation (7.9).

## 7.2 Neumann Boundary Value Problem

When considering the scalar Neumann boundary value problem

$$(Lu)(x) = 0 \quad \text{for } x \in \Omega, \quad \gamma_1^{\text{int}} u(x) = g(x) \quad \text{for } x \in \Gamma \tag{7.16}$$

we have to assume the solvability condition (1.17),

$$\int_\Gamma g(x) ds_x = 0 \,. \tag{7.17}$$

The representation formula (7.2) then yields

$$u(\widetilde{x}) = \int_\Gamma U^*(\widetilde{x}, y) g(y) ds_y - \int_\Gamma \gamma_1^{\text{int}} U^*(\widetilde{x}, y) \gamma_0^{\text{int}} u(y) ds_y \quad \text{for } \widetilde{x} \in \Omega \tag{7.18}$$

where we have to find the yet unknown Dirichlet datum $\gamma_0^{\text{int}} u \in H^{1/2}(\Gamma)$. From the second boundary integral equation of the Calderon system (7.3) we obtain

$$(D\gamma_0^{\text{int}} u)(x) = (1 - \sigma(x)) g(x) - (K' g)(x) \quad \text{for } x \in \Gamma \tag{7.19}$$

which is a first kind Fredholm boundary integral equation. Due to (6.17) we have that $u_0 \equiv 1$ is an eigensolution of the hypersingular boundary integral operator, i.e. $(Du_0)(x) = 0$. Hence we have $\ker D = \text{span}\,\{u_0\}$, and to ensure the solvability of the boundary integral equation (7.19) we need to assume, by applying Theorem 3.6, the solvability condition

$$(1 - \sigma) g - K' g \in \text{Im}(D) = (\ker D)^0 .$$

Note that $(\ker D)^0$ is the orthogonal space which is induced by $\ker D$, see (3.15). From

$$\begin{aligned} \langle [1 - \sigma] g - K' g, u_0 \rangle_\Gamma &= \langle g, 1 \rangle_\Gamma - \langle (\sigma I + K') g, u_0 \rangle_\Gamma \\ &= \langle g, 1 \rangle_\Gamma - \langle g, (\sigma I + K) u_0 \rangle_\Gamma = 0 \end{aligned} \tag{7.20}$$

we then conclude the solvability of the boundary integral equation (7.19). The hypersingular boundary integral operator $D : H^{1/2}(\Gamma) \to H^{-1/2}(\Gamma)$ is bounded (see (6.16)) and $H^{1/2}_*(\Gamma)$–elliptic (see Theorem 6.24). Then, applying the Lax–Milgram lemma (Theorem 3.4), there exists a unique solution $\gamma_0^{\text{int}} u \in H^{1/2}_*(\Gamma)$ of the hypersingular boundary integral equation (7.19). The equivalent varitional problem is to find $\gamma_0^{\text{int}} u \in H^{1/2}_*(\Gamma)$ such that

$$\langle D\gamma_0^{\text{int}} u, v\rangle_\Gamma \;=\; \langle (\frac{1}{2}I - K')g, v\rangle_\Gamma \tag{7.21}$$

is satisfied for all $v \in H^{1/2}_*(\Gamma)$.

Instead of the variational problem (7.21) with a constraint we may also consider a saddle point problem to find $(\gamma_0^{\text{int}} u, \lambda) \in H^{1/2}(\Gamma) \times \mathbb{R}$ such that

$$\begin{aligned} \langle D\gamma_0^{\text{int}} u, v\rangle_\Gamma + \lambda\,\langle v, w_{\text{eq}}\rangle_\Gamma &= \langle (\tfrac{1}{2}I - K')g, v\rangle_\Gamma \\ \langle \gamma_0^{\text{int}} u, w_{\text{eq}}\rangle_\Gamma &= 0 \end{aligned} \tag{7.22}$$

is satisfied for all $v \in H^{1/2}(\Gamma)$.

When inserting $v = u_0 \in H^{1/2}(\Gamma)$ as a test function of the first equation in the saddle point problem (7.22) this gives $Du_0 = 0$ and from the orthogonality (7.20) we get

$$0 \;=\; \lambda\,\langle 1, w_{\text{eq}}\rangle_\Gamma \;=\; \lambda\,\langle 1, V^{-1}1\rangle_\Gamma$$

and therefore $\lambda = 0$, since the inverse single layer potential $V^{-1}$ is elliptic. The saddle point problem (7.22) is therefore equivalent to finding $(\gamma_0^{\text{int}} u, \lambda) \in H^{1/2}(\Gamma) \times \mathbb{R}$ such that

$$\begin{aligned} \langle D\gamma_0^{\text{int}} u, v\rangle_\Gamma \;+\; \lambda\,\langle v, w_{\text{eq}}\rangle_\Gamma &= \langle (\tfrac{1}{2}I - K')g, v\rangle_\Gamma \\ \langle \gamma_0^{\text{int}} u, w_{\text{eq}}\rangle_\Gamma \;-\; \lambda/\alpha \quad &= 0 \end{aligned} \tag{7.23}$$

is satisfied for all $v \in H^{1/2}(\Gamma)$. Here, $\alpha \in \mathbb{R}_+$ is some parameter to be chosen. Hence we can eliminate the Lagrange multiplier $\lambda \in \mathbb{R}$ to obtain a modified variational problem to find $\gamma_0^{\text{int}} u \in H^{1/2}(\Gamma)$ such that

$$\langle D\gamma_0^{\text{int}} u, v\rangle_\Gamma + \alpha\,\langle \gamma_0^{\text{int}} u, w_{\text{eq}}\rangle_\Gamma \langle v, w_{\text{eq}}\rangle_\Gamma \;=\; \langle (\frac{1}{2}I - K')g, v\rangle_\Gamma \tag{7.24}$$

is satisfied for all $v \in H^{1/2}(\Gamma)$. The modified hypersingular boundary integral operator $\widetilde{D} : H^{1/2}(\Gamma) \to H^{-1/2}(\Gamma)$ which is defined via the bilinear form

$$\langle \widetilde{D}w, v\rangle_\Gamma \;:=\; \langle Dw, v\rangle_\Gamma + \alpha\,\langle w, w_{\text{eq}}\rangle_\Gamma \langle v, w_{\text{eq}}\rangle_\Gamma$$

for all $v, w \in H^{1/2}(\Gamma)$ is bounded, and $H^{1/2}(\Gamma)$–elliptic, due to

$$\begin{aligned} \langle \widetilde{D}v, v\rangle_\Gamma &= \langle Dv, v\rangle_\Gamma + \alpha\,[\langle v, w_{\text{eq}}\rangle_\Gamma]^2 \\ &\geq \bar{c}_1^D\,|v|^2_{H^{1/2}(\Gamma)} + \alpha\,[\langle v, w_{\text{eq}}\rangle_\Gamma]^2 \\ &\geq \min\{\bar{c}_1^D, \alpha\}\,\Big\{|v|^2_{H^{1/2}(\Gamma)} + [\langle v, w_{\text{eq}}\rangle_\Gamma]^2\Big\} \\ &= \min\{\bar{c}_1^D, \alpha\}\,\|v\|^2_{H^{1/2}_*(\Gamma)} \;\geq\; \hat{c}_1^D\,\|v\|^2_{H^{1/2}(\Gamma)} \end{aligned}$$

for all $v \in H^{1/2}(\Gamma)$. This estimate also indicates an appropriate choice of the parameter $\alpha \in \mathbb{R}_+$. Note that the modified variational problem (7.24) admits a unique solution for any right hand side, and therefore for any given Neumann datum $g \in H^{-1/2}(\Gamma)$. If the given Neumann datum $g$ satisfies the solvability condition (7.17), then we conclude, by inserting $v = u_0 \equiv 1$ as a test function, from the variational problem (7.24)

$$\alpha \, \langle \gamma_0^{\text{int}} u, w_{\text{eq}} \rangle_\Gamma \langle 1, w_{\text{eq}} \rangle_\Gamma = 0, \quad \langle 1, w_{\text{eq}} \rangle_\Gamma = \langle 1, V^{-1} 1 \rangle_\Gamma > 0$$

and therefore $\gamma_0^{\text{int}} u \in H_*^{1/2}(\Gamma)$. The modified variational problem (7.24) is thus equivalent to the original variational problem (7.21).

Since the hypersingular boundary integral operator $D$ is also $H_{**}^{1/2}(\Gamma)$–elliptic (see (6.38)), there also exists a unique solution $\gamma_0^{\text{int}} u \in H_{**}^{1/2}(\Gamma)$ of the boundary integral equation (7.19). In analogy to the above treatment we obtain a modified variational problem to find $\gamma_0^{\text{int}} u \in H^{1/2}(\Gamma)$ such that

$$\langle D \gamma_0^{\text{int}} u, v \rangle_\Gamma + \bar{\alpha} \langle \gamma_0^{\text{int}} u, 1 \rangle_\Gamma \langle v, 1 \rangle_\Gamma = \langle (\frac{1}{2} I - K') g, v \rangle_\Gamma \tag{7.25}$$

is satisfied for all $v \in H^{1/2}(\Gamma)$. Again, $\bar{\alpha} \in \mathbb{R}_+$ is some parameter to be chosen. Moreover, if we assume the solvability condition (7.17) this gives $\gamma_0^{\text{int}} u \in H_{**}^{1/2}(\Gamma)$. By the bilinear form

$$\langle \hat{D} w, v \rangle_\Gamma := \langle D w, v \rangle_\Gamma + \bar{\alpha} \, \langle w, 1 \rangle_\Gamma \langle v, 1 \rangle_\Gamma \tag{7.26}$$

for all $w, v \in H^{1/2}(\Gamma)$ we define a modified hypersingular boundary integral operator $\hat{D} : H^{1/2}(\Gamma) \to H^{-1/2}(\Gamma)$ which is bounded and $H^{1/2}(\Gamma)$–elliptic.

When using the indirect double layer potential (7.5) to find the unknown density $v \in H_*^{1/2}(\Gamma)$ we obtain the hypersingular boundary integral equation

$$(Dv)(x) = g(x) \quad \text{for } x \in \Gamma, \tag{7.27}$$

which can be analyzed as the boundary integral equation (7.19).

If we consider the representation formula (7.18) of the direct approach, and use the first boundary integral equation of the resulting Calderon projection (7.3), we find the yet unknown Dirichlet datum as the solution of the boundary integral equation

$$(\sigma I + K) \gamma_0^{\text{int}} u(x) = (Vg)(x) \quad \text{for } x \in \Gamma. \tag{7.28}$$

The solution of the boundary integral equation (7.28) is given by the Neumann series

$$\gamma_0^{\text{int}} u(x) = \sum_{\ell=0}^{\infty} ([1-\sigma] I - K)^\ell (Vg)(x) \quad \text{for } x \in \Gamma. \tag{7.29}$$

The convergence of the Neumann series (7.29) follows from the contraction property (6.52) of $([1-\sigma]I - K)$ in $H_*^{1/2}(\Gamma)$ when considering the equivalent

Sobolev norm $\|\cdot\|_{V^{-1}}$ which is induced by the inverse single layer potential $V^{-1}$. The variational formulation of the boundary integral equation (7.28) needs therefore to be considered in $H^{1/2}(\Gamma)$. Since the single layer potential $V : H^{-1/2}(\Gamma) \to H^{1/2}(\Gamma)$ is bounded and $H^{-1/2}(\Gamma)$–elliptic, we can define

$$\langle w, v\rangle_{V^{-1}} := \langle V^{-1}w, v\rangle_\Gamma \quad \text{for } w, v \in H^{1/2}(\Gamma)$$

to be an inner product in $H^{1/2}(\Gamma)$. The variational formulation of the boundary integral equation (7.28) with respect to the inner product $\langle\cdot,\cdot\rangle_{V^{-1}}$ then reads to find $\gamma_0^{\text{int}}u \in H_*^{1/2}(\Gamma)$ such that

$$\langle(\sigma I + K)\gamma_0^{\text{int}}u, v\rangle_{V^{-1}} = \langle Vg, v\rangle_{V^{-1}} \tag{7.30}$$

is satisfied for all $v \in H_*^{1/2}(\Gamma)$. The variational problem (7.30) is equivalent to finding $\gamma_0^{\text{int}}u \in H_*^{1/2}(\Gamma)$ such that

$$\langle S\gamma_0^{\text{int}}u, v\rangle_\Gamma = \langle V^{-1}(\sigma I + K)\gamma_0^{\text{int}}u, v\rangle_\Gamma = \langle g, v\rangle_\Gamma \tag{7.31}$$

is satisfied for all $v \in H_*^{1/2}(\Gamma)$. Since the Steklov–Poincaré operator $S : H^{1/2}(\Gamma) \to H^{-1/2}(\Gamma)$ admits the same mapping properties as the hypersingular boundary integral operator $D : H^{1/2}(\Gamma) \to H^{-1/2}(\Gamma)$, the unique solvability of the variational problem (7.31) follows as for the variational formulation (7.21). When using the symmetric representation (6.42) of the Steklov–Poincaré operator $S$, the variational problem (7.31) is equivalent to

$$\langle S\gamma_0^{\text{int}}u, v\rangle_\Gamma = \left\langle\left[D + (\sigma I + K')V^{-1}(\sigma I + K)\right]\gamma_0^{\text{int}}u, v\right\rangle_\Gamma = \langle g, v\rangle_\Gamma \tag{7.32}$$

When using the indirect single layer potential (7.4) we finally obtain the boundary integral equation

$$(\sigma I + K')w(x) = g(x) \quad \text{for } x \in \Gamma \tag{7.33}$$

to find the unknown density $w \in H^{-1/2}(\Gamma)$ which is given via the Neumann series

$$w(x) = \sum_{\ell=0}^{\infty}((1-\sigma)I - K')^\ell g(x) \quad \text{for } x \in \Gamma. \tag{7.34}$$

The convergence of the Neumann series (7.34) follows from the contraction property (6.55) of $((1-\sigma)I - K')$ in $H_*^{-1/2}(\Gamma)$ when considering the equivalent Sobolev norm $\|\cdot\|_V$.

*Remark 7.3.* For the solution of the Neumann boundary value problem (7.16) again we have given and analyzed four different formulations of boundary integral equations. As for the Dirichlet boundary value problem each of them may have their advantages and disadvantages. Here we will mainly consider the approximate solution of the modified variational formulation (7.25).

## 7.3 Mixed Boundary Conditions

In addition to the standard boundary value problems (7.6) and (7.16) with either Dirichlet or Neumann boundary conditions, boundary value problems with mixed boundary conditions are of special interest,

$$\begin{aligned} (Lu)(x) &= 0 && \text{for } x \in \Omega, \\ \gamma_0^{\text{int}} u(x) &= g_D(x) && \text{for } x \in \Gamma_D, \\ \gamma_1^{\text{int}} u(x) &= g_N(x) && \text{for } x \in \Gamma_N. \end{aligned} \tag{7.35}$$

From the representation formula (7.2) we get for $\widetilde{x} \in \Omega$

$$\begin{aligned} u(\widetilde{x}) = & \int_{\Gamma_N} U^*(\widetilde{x}, y) g_N(y) ds_y + \int_{\Gamma_D} U^*(\widetilde{x}, y) \gamma_{1,y}^{\text{int}} u(y) ds_y \\ & - \int_{\Gamma_D} \gamma_{1,y}^{\text{int}} U^*(\widetilde{x}, y) g_D(y) ds_y - \int_{\Gamma_N} \gamma_1^{\text{int}} U^*(\widetilde{x}, y) \gamma_0^{\text{int}} u(y) ds_y. \end{aligned} \tag{7.36}$$

Hence we have to find the yet unknown Dirichlet datum $\gamma_0^{\text{int}} u(x)$ for $x \in \Gamma_N$ and the Neumann datum $\gamma_1^{\text{int}} u(x)$ for $x \in \Gamma_D$. Keeping in mind the different boundary integral formulations for both the Dirichlet and the Neumann problems, there seems to be a wide variety of different boundary integral formulations to solve the mixed boundary value problem (7.35). Here we will only consider two formulations which are based on the representation formula (7.36) of the direct approach.

The symmetric formulation [134] is based on the use of the first boundary integral equation in (7.3) for $x \in \Gamma_D$ while the second boundary integral equation in (7.3) is considered for $x \in \Gamma_N$,

$$\begin{aligned} (V\gamma_1^{\text{int}} u)(x) &= (\sigma I + K)\gamma_0^{\text{int}} u(x) && \text{for } x \in \Gamma_D, \\ (D\gamma_0^{\text{int}} u)(x) &= ((1-\sigma)I - K')\gamma_1^{\text{int}} u(x) && \text{for } x \in \Gamma_N. \end{aligned} \tag{7.37}$$

Let $\widetilde{g}_D \in H^{1/2}(\Gamma)$ and $\widetilde{g}_N \in H^{-1/2}(\Gamma)$ be suitable extensions of the given boundary data $g_D \in H^{1/2}(\Gamma_D)$ and $g_N \in H^{-1/2}(\Gamma_N)$ satisfying

$$\widetilde{g}_D(x) = g_D(x) \quad \text{for } x \in \Gamma_D, \quad \widetilde{g}_N(x) = g_N(x) \quad \text{for } x \in \Gamma_N.$$

Inserting these extensions into the system (7.37) this gives the symmetric formulation to find

$$\widetilde{u} := \gamma_0^{\text{int}} u - \widetilde{g}_D \in \widetilde{H}^{1/2}(\Gamma_N), \quad \widetilde{t} := \gamma_1^{\text{int}} u - \widetilde{g}_N \in \widetilde{H}^{-1/2}(\Gamma_D)$$

such that

$$\begin{aligned} (V\widetilde{t})(x) - (K\widetilde{u})(x) &= (\sigma I + K)\widetilde{g}_D(x) - (V\widetilde{g}_N)(x) && \text{for } x \in \Gamma_D, \\ (D\widetilde{u})(x) + (K'\widetilde{t})(x) &= ((1-\sigma)I - K')\widetilde{g}_N(x) - (D\widetilde{g}_D)(x) && \text{for } x \in \Gamma_N. \end{aligned} \tag{7.38}$$

The related variational formulation is to find $(\widetilde{t},\widetilde{u}) \in \widetilde{H}^{-1/2}(\Gamma_D) \times \widetilde{H}^{1/2}(\Gamma_N)$ such that

$$a(\widetilde{t},\widetilde{u};\tau,v) \;=\; F(\tau,v) \tag{7.39}$$

is satisfied for all $(\tau,v) \in \widetilde{H}^{-1/2}(\Gamma_D) \times \widetilde{H}^{1/2}(\Gamma_N)$ where

$$a(\widetilde{t},\widetilde{u};\tau,v) = \langle V\widetilde{t},\tau\rangle_{\Gamma_D} - \langle K\widetilde{u},\tau\rangle_{\Gamma_D} + \langle K'\widetilde{t},v\rangle_{\Gamma_N} + \langle D\widetilde{u},v\rangle_{\Gamma_N},$$

$$F(\tau,v) = \langle(\frac{1}{2}I + K)\widetilde{g}_D - V\widetilde{g}_N,\tau\rangle_{\Gamma_D} + \langle(\frac{1}{2}I - K')\widetilde{g}_N - D\widetilde{g}_D,v\rangle_{\Gamma_N}.$$

**Lemma 7.4.** *The bilinear form $a(\cdot;\cdot)$ of the symmetric boundary integral formulation is bounded and $\widetilde{H}^{-1/2}(\Gamma_D) \times \widetilde{H}^{1/2}(\Gamma_N)$–elliptic, i.e.*

$$a(t,u;\tau,v) \;\leq\; c_2^A \,\|(t,u)\|_{\widetilde{H}^{-1/2}(\Gamma_D)\times\widetilde{H}^{1/2}(\Gamma_N)}\|(\tau,v)\|_{\widetilde{H}^{-1/2}(\Gamma_D)\times\widetilde{H}^{1/2}(\Gamma_N)}$$

*and*

$$a(\tau,v;\tau,v) \geq \min\{c_1^V,\hat{c}_1^D\}\,\|(\tau,v)\|^2_{\widetilde{H}^{-1/2}(\Gamma_D)\times\widetilde{H}^{1/2}(\Gamma_N)}$$

*for all $(t,u),(\tau,v) \in \widetilde{H}^{-1/2}(\Gamma_D) \times \widetilde{H}^{1/2}(\Gamma_N)$ where the norm is defined by*

$$\|(\tau,v)\|^2_{\widetilde{H}^{-1/2}(\Gamma_D)\times\widetilde{H}^{1/2}(\Gamma_N)} \;:=\; \|\tau\|^2_{\widetilde{H}^{-1/2}(\Gamma_D)} + \|v\|^2_{\widetilde{H}^{1/2}(\Gamma_N)}.$$

*Proof.* By using

$$\begin{aligned} a(\tau,v;\tau,v) &= \langle V\tau,\tau\rangle_{\Gamma_D} - \langle Kv,\tau\rangle_{\Gamma_D} + \langle K'\tau,v\rangle_{\Gamma_N} + \langle Dv,v\rangle_{\Gamma_N} \\ &= \langle V\tau,\tau\rangle_{\Gamma_D} + \langle Dv,v\rangle_{\Gamma_N} \\ &\geq c_1^V\,\|\tau\|^2_{\widetilde{H}^{-1/2}(\Gamma_D)} + \hat{c}_1^D\,\|v\|^2_{\widetilde{H}^{1/2}(\Gamma_N)} \end{aligned}$$

we conclude the ellipticity of the bilinear form $a(\cdot,\cdot;\cdot,\cdot)$ from the ellipticity estimates of the boundary integral operators $V$ and $D$, see Theorem 6.22 for $d=3$ and Theorem 6.23 for $d=2$, as well as (6.39). The boundedness of the bilinear form $a(\cdot,\cdot;\cdot,\cdot)$ is a direct consequence of the boundedness of all boundary integral operators. □

Since the linear form $F(\tau,v)$ is bounded for $(\tau,v) \in \widetilde{H}^{-1/2}(\Gamma_D) \times \widetilde{H}^{1/2}(\Gamma_N)$, the unique solvability of the variational formulation (7.39) follows from the Lax–Milgram lemma (Theorem 3.4).

To obtain a second boundary integral equation to solve the mixed boundary value problem (7.35) we consider the Dirichlet to Neumann map (6.44) to find $\gamma_0^{\text{int}}u \in H^{1/2}(\Gamma)$ such that

$$\begin{aligned} \gamma_0^{\text{int}}u(x) &= g_D(x) &&\text{for } x \in \Gamma_D, \\ \gamma_1^{\text{int}}u(x) \;=\; (S\gamma_0^{\text{int}}u)(x) &= g_N(x) &&\text{for } x \in \Gamma_N. \end{aligned}$$

Let $\widetilde{g}_D \in H^{1/2}(\Gamma)$ be some arbitrary but fixed extension of the given Dirichlet datum $g_D \in H^{1/2}(\Gamma_D)$. Then we have to find $\widetilde{u} := \gamma_0^{\text{int}}u - \widetilde{g}_D \in \widetilde{H}^{1/2}(\Gamma_N)$ such that

$$\langle S\widetilde{u}, v\rangle_{\Gamma_N} = \langle g_N - S\widetilde{g}_D, v\rangle_{\Gamma_N} \tag{7.40}$$

is satisfied for all $v \in \widetilde{H}^{1/2}(\Gamma_N)$. Since the Steklov–Poincaré operator $S : H^{1/2}(\Gamma) \to H^{-1/2}(\Gamma)$ is bounded and $\widetilde{H}^{1/2}(\Gamma_N)$–elliptic (see (6.48)) we conclude the unique solvability of the variational problem (7.40) from the Lax–Milgram lemma (Theorem 3.4). If the Dirichlet datum $\gamma_0^{\text{int}} u \in H^{1/2}(\Gamma)$ is known, we can compute the complete Neumann datum $\gamma_1^{\text{int}} u \in H^{-1/2}(\Gamma)$ by solving a Dirichlet boundary value problem.

## 7.4 Robin Boundary Conditions

Next we consider the Robin boundary value problem

$$(Lu)(x) = 0 \quad \text{for } x \in \Omega, \quad \gamma_1^{\text{int}} u(x) + \kappa(x)\gamma_0^{\text{int}} u(x) = g(x) \quad \text{for } x \in \Gamma.$$

To formulate a boundary integral equation to find the yet unknown Dirichlet datum $\gamma_0^{\text{int}} u \in H^{1/2}(\Gamma)$ again we can use the Dirichlet to Neumann map (6.44), i.e.

$$\gamma_1^{\text{int}} u(x) \;=\; (S\gamma_0^{\text{int}} u)(x) \;=\; g(x) - \kappa(x)\gamma_0^{\text{int}} u(x) \quad \text{for } x \in \Gamma.$$

The related variational problem is to find $\gamma_0^{\text{int}} u \in H^{1/2}(\Gamma)$ such that

$$\langle S\gamma_0^{\text{int}} u, v\rangle_\Gamma + \langle \kappa\gamma_0^{\text{int}} u, v\rangle_\Gamma \;=\; \langle g, v\rangle_\Gamma \tag{7.41}$$

is satisfied for all $v \in H^{1/2}(\Gamma)$. By using (6.45) and the $H^{1/2}(\Gamma)$–semi–ellipticity of the hypersingular boundary integral operator $D$ and assuming $\kappa(x) \geq \kappa_0$ for $x \in \Gamma$ we conclude

$$\begin{aligned} a(v, v) &:= \langle Sv, v\rangle_\Gamma + \langle \kappa v, v\rangle_\Gamma \\ &\geq \bar{c}_1^D \, |v|^2_{H^{1/2}(\Gamma)} + \kappa_0 \, \|v\|^2_\Gamma \;=\; \min\{\bar{c}_1^D, \kappa_0\} \, \|v\|^2_{H^{1/2}(\Gamma)} \end{aligned}$$

and therefore the $H^{1/2}(\Gamma)$–ellipticity of the bilinear form $a(\cdot,\cdot)$. Again we obtain the unique solvability of the variational problem (7.41) from the Lax–Milgram lemma (Theorem 3.4).

## 7.5 Exterior Boundary Value Problems

An advantage of boundary integral equation methods is the explicit consideration of far field boundary conditions when solving boundary value problems in the exterior domain $\Omega^c := \mathbb{R}^d \backslash \overline{\Omega}$. As a model problem we consider the exterior Dirichlet boundary value problem for the Laplace equation,

$$-\Delta u(x) = 0 \quad \text{for } x \in \Omega^c, \quad \gamma_0^{\text{ext}} u(x) = g(x) \quad \text{for } x \in \Gamma \tag{7.42}$$

together with the far field boundary condition

$$|u(x) - u_0| = \mathcal{O}\left(\frac{1}{|x|}\right) \quad \text{as } |x| \to \infty. \tag{7.43}$$

where $u_0 \in \mathbb{R}$ is some given number.

First we consider Green's first formula for the exterior domain. For $y_0 \in \Omega$ and $R \geq 2\,\mathrm{diam}(\Omega)$ let $B_R(y_0)$ be a ball with center $y_0$, which circumscribes $\Omega$. Using the representation formula (6.1) for $x \in B_R(y_0)\backslash\overline{\Omega}$ this gives

$$u(x) = -\int_\Gamma U^*(x,y)\gamma_1^{\text{ext}}u(y)ds_y + \int_\Gamma \gamma_{1,y}^{\text{ext}}U^*(x,y)\gamma_0^{\text{ext}}u(y)ds_y$$
$$+ \int_{\partial B_R(y_0)} U^*(x,y)\gamma_1^{\text{int}}u(y)ds_y - \int_{\partial B_R(y_0)} \gamma_{1,y}^{\text{int}}U^*(x,y)\gamma_0^{\text{int}}u(y)ds_y.$$

Inserting the far field boundary condition (7.43) and taking the limit $R \to \infty$ this results in the representation formula for $x \in \Omega^c$,

$$u(x) = u_0 - \int_\Gamma U^*(x,y)\gamma_1^{\text{ext}}u(y)ds_y + \int_\Gamma \gamma_{1,y}^{\text{ext}}U^*(x,y)\gamma_0^{\text{ext}}u(y)ds_y.$$

To find the unknown Cauchy data again we can formulate different boundary integral equations. The application of the exterior trace operator gives

$$\gamma_0^{\text{ext}}u(x) = u_0 - (V\gamma_1^{\text{ext}}u)(x) + \sigma(x)\gamma_0^{\text{ext}}u(x) + (K\gamma_0^{\text{ext}}u)(x) \quad \text{for } x \in \Gamma,$$

while the application of the exterior conormal derivative yields

$$\gamma_1^{\text{ext}}u(x) = [1-\sigma(x)]\gamma_1^{\text{ext}}u(x) - (K'\gamma_1^{\text{ext}}u)(x) - (D\gamma_0^{\text{ext}}u)(x) \quad \text{for } x \in \Gamma.$$

As in (6.22) we obtain a system of boundary integral equations,

$$\begin{pmatrix} \gamma_0^{\text{ext}}u \\ \gamma_1^{\text{ext}}u \end{pmatrix} = \begin{pmatrix} \sigma I + K & -V \\ -D & [1-\sigma]I - K' \end{pmatrix} \begin{pmatrix} \gamma_0^{\text{ext}}u \\ \gamma_1^{\text{ext}}u \end{pmatrix} + \begin{pmatrix} u_0 \\ 0 \end{pmatrix}.$$

Using the boundary integral equations of this system, the exterior Calderon projection, we can formulate different boundary integral equations to handle exterior boundary value problems with different boundary conditions. In particular for the exterior Dirichlet boundary value problem (7.42) and (7.43) we can find the yet unknown Neumann datum $\gamma_1^{\text{ext}}u \in H^{-1/2}(\Gamma)$ as the unique solution of the boundary integral equation

$$(V\gamma_1^{\text{ext}}u)(x) = -[1-\sigma(x)]g_D(x) + (Kg_D)(x) + u_0 \quad \text{for } x \in \Gamma. \tag{7.44}$$

Note that the unique solvability of the boundary integral equation (7.44) follows as for interior boundary value problems from the mapping properties of the single layer potential $V : H^{-1/2}(\Gamma) \to H^{1/2}(\Gamma)$.

## 7.6 Helmholtz Equation

Finally we consider boundary value problems for the Helmholtz equation, i.e. the interior Dirichlet boundary value problem

$$-\Delta u(x) - k^2 u(x) = 0 \quad \text{for } x \in \Omega, \quad \gamma_0^{\text{int}} u(x) = g(x) \quad \text{for } x \in \Gamma \tag{7.45}$$

where the solution is given by the representation formula

$$u(x) = \int_\Gamma U_k^*(x,y)\gamma_1^{\text{int}} u(y) ds_y - \int_\Gamma \gamma_{1,y}^{\text{int}} U_k^*(x,y) g(y) ds_y \quad \text{for } x \in \Omega.$$

The unknown Neumann datum $t = \gamma_1^{\text{int}} u \in H^{-1/2}(\Gamma)$ solves the boundary integral equation

$$(V_k t)(x) = (\frac{1}{2} I + K_k) g(x) \quad \text{for } x \in \Gamma. \tag{7.46}$$

Since the single layer potential $V_k : H^{-1/2}(\Gamma) \to H^{1/2}(\Gamma)$ is coercive, see Theorem 6.40, we can apply Theorem 3.15 to investigate the solvability of the boundary integral equation (7.46).

**Lemma 7.5.** *If $k^2 = \lambda$ is an eigenvalue of the Dirichlet eigenvalue problem of the Laplace equation,*

$$-\Delta u_\lambda(x) = \lambda\, u_\lambda(x) \quad \textit{for } x \in \Omega, \quad \gamma_0^{int} u_\lambda(x) = 0 \quad \textit{for } x \in \Gamma, \tag{7.47}$$

*then the single layer potential $V_k : H^{-1/2}(\Gamma) \to H^{1/2}(\Gamma)$ is not injective, i.e.*

$$(V_k \gamma_1^{int} u_\lambda)(x) = 0 \quad \textit{for } x \in \Gamma.$$

*Moreover,*

$$(\frac{1}{2} I - K_k') \gamma_1^{int} u_\lambda(x) = 0 \quad \textit{for } x \in \Gamma.$$

*Proof.* The assertion immediately follows from the direct boundary integral equations

$$(V_k \gamma_i^{\text{int}} u_\lambda)(x) = (\frac{1}{2} I + K_k) \gamma_0^{\text{int}} u_\lambda(x) = 0 \quad \text{for } x \in \Gamma$$

and

$$(\frac{1}{2} I - K_k') \gamma_i^{\text{int}} u_\lambda(x) = (D_k \gamma_0^{\text{int}} u_\lambda)(x) = 0\,. \qquad \square$$

Hence we conclude, that if $k^2$ is not an eigenvalue of the Dirichlet eigenvalue problem of the Laplace equation, the Helmholtz single layer potential $V_k : H^{-1/2}(\Gamma) \to H^{1/2}(\Gamma)$ is coercive and injective, i.e. the boundary integral equation (7.46) admits a unique solution, see Theorem 3.15.

Next we consider the exterior Dirichlet boundary value problem

$$-\Delta u(x) - k^2 u(x) = 0 \quad \text{for } x \in \Omega^c, \quad \gamma_0^{\text{ext}} u(x) = g(x) \quad \text{for } x \in \Gamma \tag{7.48}$$

where, in addition, we have to require the Sommerfeld radiation condition

$$\left| \left( \frac{x}{|x|}, \nabla u(x) \right) - iku(x) \right| = \mathcal{O}\left( \frac{1}{|x|^2} \right) \quad \text{as } |x| \to \infty$$

Note that the exterior Dirichlet boundary value problem is uniquely solvable due to the Sommerfeld radiation condition. The solution is given by the representation formula

$$u(x) = -\int_\Gamma U_k^*(x,y)\gamma_1^{\text{ext}} u(y) ds_y + \int_\Gamma \gamma_{1,y}^{\text{ext}} U_k^*(x,y) g(y) ds_y \quad \text{for } x \in \Omega^c.$$

To find the unknown Neumann datum $t = \gamma_1^{\text{ext}} u \in H^{-1/2}(\Gamma)$ we consider the direct boundary integral equation

$$(V_k t)(x) = (-\frac{1}{2} I + K_k) g(x) \quad \text{for } x \in \Gamma. \tag{7.49}$$

Since the single layer potential $V_k$ of the exterior Dirichlet boundary value problem coincides with the single layer potential of the interior problem, $V_k$ is not invertible when $k^2 = \lambda$ is an eigenvalue of the Dirichlet eigenvalue problem (7.47). However, due to

$$\langle (-\frac{1}{2} g, \gamma_1^{\text{int}}) u_\lambda \rangle_\Gamma = -\langle g, (\frac{1}{2} I - K'_{-k}) \gamma_1^{\text{int}} u_\lambda \rangle_\Gamma = 0$$

we conclude

$$(-\frac{1}{2} I + K_k) g \in \operatorname{Im} V_k.$$

In fact, the boundary integral equation (7.49) of the direct approach is solvable but not unique.

Instead of a direct approach, we may also consider an indirect single layer potential approach

$$u(x) = \int_\Gamma U_k^*(x,y) w(y) ds_y \quad \text{for } x \in \Omega^c$$

which leads to the boundary integral equation to find $w \in H^{-1/2}(\Gamma)$ such that

$$(V_k w)(x) = g(x) \quad \text{for } x \in \Gamma. \tag{7.50}$$

As before, we have unique solvability of the boundary integral equation (7.50) only for those wave numbers $k^2$ which are not eigenvalues of the interior Dirichlet eigenvalue problem (7.47).

When using an indirect double layer potential ansatz

$$u(x) = \int_\Gamma \gamma_{1,y}^{\rm ext} U_k^*(x,y)v(y)ds_y \quad \text{for } x \in \Omega^c$$

the unknown density function $v \in H^{1/2}(\Gamma)$ solves the boundary integral equation

$$(\frac{1}{2}I + K_k)v(x) = g(x) \quad \text{for } x \in \Gamma. \tag{7.51}$$

**Lemma 7.6.** *If $k^2 = \mu$ is an eigenvalue of the interior Neumann eigenvalue problem of the Laplace equation,*

$$-\Delta u_\mu(x) = \mu u_\mu(x) \quad \text{for } x \in \Omega, \quad \gamma_1^{int} u_\mu(x) = 0 \quad \text{for } x \in \Gamma, \tag{7.52}$$

*then*

$$(\frac{1}{2}I + K_k)\gamma_0^{int} u_\mu(x) = 0 \quad \text{for } x \in \Gamma.$$

*Proof.* The assertion immediately follows from the direct boundary integral equation

$$(\frac{1}{2}I + K_k)\gamma_0^{\rm int} u_\mu(x) = (V_k \gamma_1^{\rm int} u_\mu)(x) = 0 \quad \text{for } x \in \Gamma. \quad \Box$$

The boundary integral equation (7.51) is therefore uniquely solvable if $k^2$ is not an eigenvalue of the Neumann eigenvalue problem (7.52).

Although the exterior Dirichlet boundary value problem for the Helmholtz equation is uniquely solvable, the related boundary integral equations may not be solvable, in particular, when $k^2$ is either an eigenvalue of the interior Dirichlet eigenvalue problem (7.47), or of the interior Neumann eigenvalue problem (7.52). Since $k^2$ can not be an eigenvalue of both the interior Dirichlet and the interior Neumann boundary value problem, one may combine both the indirect single and double layer potential formulations to derive a boundary integral equation which is uniquely solvable for all wave numbers. This leads to the well known Brakhage–Werner formulation [23]

$$u(x) = \int_\Gamma \gamma_{1,y}^{\rm ext} U_k^*(x,y)w(y)ds_y - i\eta \int_\Gamma U_k^*(x,y)w(y)ds_y \quad \text{for } x \in \Omega^c$$

where $\eta \in \mathbb{R}_+$ is some real parameter. This leads to a boundary integral equation to find $w \in L_2(\Gamma)$ such that

$$(\frac{1}{2}I + K_k)w(x) - i\eta(V_k w)(x) = g(x) \quad \text{for } x \in \Gamma. \tag{7.53}$$

Instead of considering the boundary integral equation (7.53) in $L_2(\Gamma)$, one may formulate some modified boundary integral equations to be considered in the energy space $H^{-1/2}(\Gamma)$, i.e. find $w \in H^{-1/2}(\Gamma)$ such that

$$\left[V_k + i\eta\left(\frac{1}{2}I + K_k\right)\widetilde{D}^{-1}\left(\frac{1}{2}I + K'_{-k}\right)\right] w(x) = g(x) \quad \text{for } x \in \Gamma \qquad (7.54)$$

where $\widetilde{D}$ is the modified hypersingular integral operator of the Laplace equation as defined in (7.24). The modified boundary integral equation (7.54) admits a unique solution for all wave number $k$ for general Lipschitz domains [54], for other regularizations, see [33, 34].

## 7.7 Exercises

**7.1** Consider the mixed boundary value problem

$$\begin{aligned} -\Delta u(x) &= 0 && \text{for } x \in \Omega\backslash\overline{\Omega_0}, \Omega_0 \subset \Omega, \\ \gamma_0^{\text{int}} u(x) &= g(x) && \text{for } x \in \Gamma = \partial\Omega, \\ \gamma_1^{\text{int}} u(x) &= 0 && \text{for } x \in \Gamma_0 = \partial\Omega_0. \end{aligned}$$

Derive the symmetric formulation of boundary integral equations to find the complete Cauchy data. Discuss the solvability of the resulting variational problem.

**7.2** Discuss boundary integral formulations to solve the exterior Neumann boundary value problem

$$-\Delta u(x) - k^2 u(x) = 0 \quad \text{for } x \in \Omega^c, \quad \gamma_1^{\text{ext}} u(x) = g(x) \quad \text{for } x \in \Gamma$$

and

$$\left|\left(\frac{x}{|x|}, \nabla u(x)\right) - iku(x)\right| = \mathcal{O}\left(\frac{1}{|x|^2}\right) \quad \text{as } |x| \to \infty.$$

# 8

# Approximation Methods

In this chapter we describe and analyze approximation methods to solve the variational problems for operator equations as formulated in Chapter 3. This is done by introducing conforming finite dimensional trial spaces leading to linear systems of algebraic equations.

## 8.1 Galerkin–Bubnov Methods

Let $A : X \to X'$ be a bounded and $X$–elliptic linear operator satisfying

$$\langle Av, v\rangle \geq c_1^A \, \|v\|_X^2, \quad \|Av\|_{X'} \leq c_2^A \, \|v\|_X \quad \text{for all } v \in X.$$

For a given $f \in X'$ we want to find the solution $u \in X$ of the variational problem (3.4),

$$\langle Au, v\rangle \;=\; \langle f, v\rangle \quad \text{for all } v \in X. \tag{8.1}$$

Due to the Lax–Milgram theorem 3.4 there exists a unique solution of the variational problem (8.1) satisfying

$$\|u\|_X \;\leq\; \frac{1}{c_1^A} \, \|f\|_{X'}.$$

For $M \in \mathbb{N}$ we consider a sequence

$$X_M \;:=\; \operatorname{span}\{\varphi_k\}_{k=1}^M \subset X$$

of conforming trial spaces. The approximate solution

$$u_M \;:=\; \sum_{k=1}^M u_k \varphi_k \in X_M \tag{8.2}$$

is defined as the solution of the Galerkin–Bubnov variational problem

$$\langle Au_M, v_M \rangle = \langle f, v_M \rangle \quad \text{for all } v_M \in X_M. \tag{8.3}$$

Note that we have used the same trial and test functions for the Galerkin–Bubnov method.

It remains to investigate the unique solvability of the variational problem (8.3), the stability of the approximate solutions $u_M \in X_M$ as well as their convergence for $M \to \infty$ to the unique solution $u \in X$ of the variational problem (8.1). Due to $X_M \subset X$ we can choose $v = v_M \in X_M$ in the variational formulation (8.1). Subtracting the Galerkin–Bubnov problem (8.3) from the continuous variational formulation (8.1) this gives the Galerkin orthogonality

$$\langle A(u - u_M), v_M \rangle = 0 \quad \text{for all } v_M \in X_M. \tag{8.4}$$

Inserting the approximate solution (8.2) into the Galerkin–Bubnov formulation (8.3) we obtain, by using the linearity of the operator $A$, the finite dimensional variational problem

$$\sum_{k=1}^{M} u_k \langle A\varphi_k, \varphi_\ell \rangle = \langle f, \varphi_\ell \rangle \quad \text{for } \ell = 1, \ldots, M.$$

With

$$A_M[\ell, k] := \langle A\varphi_k, \varphi_\ell \rangle, \quad f_\ell := \langle f, \varphi_\ell \rangle$$

for $k, \ell = 1, \ldots, M$ this is equivalent to the linear system of algebraic equations

$$A_M \underline{u} = \underline{f} \tag{8.5}$$

to find the coefficient vector $\underline{u} \in \mathbb{R}^M$. For any arbitrary vector $\underline{v} \in \mathbb{R}^M$ we can define a function

$$v_M = \sum_{k=1}^{M} v_k \varphi_k \in X_M$$

and vice versa. For arbitrary given vectors $\underline{u}, \underline{v} \in \mathbb{R}^M$ we then have

$$\begin{aligned}
(A_M \underline{u}, \underline{v}) &= \sum_{k=1}^{M} \sum_{\ell=1}^{M} A_M[\ell, k] u_k v_\ell = \sum_{k=1}^{M} \sum_{\ell=1}^{M} \langle A\varphi_k, \varphi_\ell \rangle u_k v_\ell \\
&= \langle A \sum_{k=1}^{M} u_k \varphi_k, \sum_{\ell=1}^{M} v_\ell \varphi_\ell \rangle = \langle Au_M, v_M \rangle.
\end{aligned}$$

Hence, all properties of the operator $A : X \to X'$ are inherited by the stiffness matrix $A_M \in \mathbb{R}^{M \times M}$. In particular, the matrix $A_M$ is symmetric and positive definite, since the operator $A$ is self–adjoint and $X$–elliptic, respectively. Indeed,

$$(A_M \underline{v}, \underline{v}) = \langle Av_M, v_M \rangle \geq c_1^A \, \|v_M\|_X^2$$

for all $\underline{v} \in \mathbb{R}^M \leftrightarrow v_M \in X_M$ implies that $A_M$ is positive definite. Therefore, the $X$–ellipticity of the operator $A$ implies the unique solvability of the variational problem (8.1) as well as the unique solvability of the Galerkin–Bubnov formulation (8.3) and hence, of the equivalent linear system (8.5).

**Theorem 8.1 (Cea's Lemma).** *Let $A : X \to X'$ be a bounded and $X$–elliptic linear operator. For the unique solution $u_M \in X_M$ of the variational problem* (8.3) *there holds the stability estimate*

$$\|u_M\|_X \le \frac{1}{c_1^A} \|f\|_{X'} \tag{8.6}$$

*as well as the error estimate*

$$\|u - u_M\|_X \le \frac{c_2^A}{c_1^A} \inf_{v_M \in X_M} \|u - v_M\|_X. \tag{8.7}$$

*Proof.* The unique solvability of the variational problem (8.3) was already discussed before. For the approximate solution $u_M \in X_M$ of (8.3) we conclude from the $X$–ellipticity of $A$

$$c_1^A \|u_M\|_X^2 \le \langle Au_M, u_M\rangle = \langle f, u_M\rangle \le \|f\|_{X'}\|u_M\|_X$$

and therefore we obtain the stability estimate (8.6). By using the $X$–ellipticity and the boundedness of the linear operator $A$, and by using the Galerkin orthogonality (8.4) we get for any arbitrary $v_M \in X_M$

$$\begin{aligned} c_1^A \|u - u_M\|_X^2 &\le \langle A(u - u_M), u - u_M\rangle \\ &= \langle A(u - u_M), u - v_M\rangle + \langle A(u - u_M), v_M - u_M\rangle \\ &= \langle A(u - u_M), u - v_M\rangle \\ &\le c_2^A \|u - u_M\|_X \|u - v_M\|_X \end{aligned}$$

and therefore the error estimate (8.7). □

The convergence of the approximate solution $u_M \to u \in X$ as $M \to \infty$ then follows from the approximation property of the trial space $X_M$,

$$\lim_{M\to\infty} \inf_{v_M \in X_M} \|v - v_M\|_X = 0 \quad \text{for all } v \in X. \tag{8.8}$$

The sequence of conforming trial spaces $\{X_M\}_{M\in\mathbb{N}} \subset X$ has to be constructed in such a way that the approximation property (8.8) can be ensured. In Chapter 9 we will consider the construction of local polynomial basis functions for finite elements, while in Chapter 10 we will do the same for boundary elements. Assuming additional regularity of the solution, we then prove also corresponding approximation properties.

## 8.2 Approximation of the Linear Form

In different applications the right hand side $f \in X'$ is given as $f = Bg$ where $g \in Y$ is prescribed, and $B : Y \to X'$ is a bounded linear operator satisfying

$$\|Bg\|_{X'} \leq c_2^B \, \|g\|_Y \quad \text{for all } g \in Y.$$

Hence we have to find $u \in X$ as the solution of the variational problem

$$\langle Au, v \rangle = \langle Bg, v \rangle \quad \text{for all } v \in X. \tag{8.9}$$

The approximate solution $u_M \in X_M$ is then given as in (8.3) as the unique solution of the variational problem

$$\langle Au_M, v_M \rangle = \langle Bg, v_M \rangle \quad \text{for all } v_M \in X_M. \tag{8.10}$$

The generation of the linear system (8.5) then requires the computation of

$$f_\ell = \langle Bg, \varphi_\ell \rangle = \langle g, B'\varphi_\ell \rangle \quad \text{for } \ell = 1, \ldots, M,$$

i.e. we have to evaluate the application of the operator $B : Y \to X'$ or of the adjoint operator $B' : X \to Y'$. In what follows we will replace the given function $g$ by an approximation

$$g_N = \sum_{i=1}^N g_i \psi_i \in Y_N = \text{span}\{\psi_i\}_{i=1}^N \subset Y.$$

Then we have to find an approximate solution $\widetilde{u}_M \in X_M$ of the perturbed variational problem

$$\langle A\widetilde{u}_M, v_M \rangle = \langle Bg_N, v_M \rangle \quad \text{for all } v_M \in X_M. \tag{8.11}$$

This is equivalent to the linear system

$$A_M \underline{\widetilde{u}} = B_N \underline{g} \tag{8.12}$$

with matrices defined by

$$A_M[\ell, k] = \langle A\varphi_k, \varphi_\ell \rangle, \quad B_N[\ell, i] = \langle B\psi_i, \varphi_\ell \rangle$$

for $i = 1, \ldots, N$ and $k, \ell = 1, \ldots, M$, as well as with the vector $\underline{g}$ describing the approximation $g_N$. The matrix $B_N$ can hereby be computed independently of the given approximate function $g_N$. From the $X$–ellipticity of the operator $A$ we find the positive definiteness of the matrix $A_M$, and therefore the unique solvability of the linear system (8.12) and therefore of the equivalent variational problem (8.11). Obviously, we have to recognize the error which is introduced by the approximation of the given data in the linear form of the right hand side.

**Theorem 8.2 (Strang Lemma).** *Let $A : X \to X'$ be a bounded linear and $X$–elliptic operator. Let $u \in X$ be the unique solution of the continuous variational problem (8.9), and let $u_M \in X_M$ be the unique solution of the Galerkin variational problem (8.10). For the unique solution $\widetilde{u}_M \in X_M$ of the perturbed variational problem (8.11) there holds the error estimate*

$$\|u - \widetilde{u}_M\|_X \le \frac{1}{c_1^A} \left\{ c_2^A \inf_{v_M \in X_M} \|u - v_M\|_X + c_2^B \, \|g - g_N\|_Y \right\}.$$

*Proof.* When subtracting the perturbed variational problem (8.11) from the Galerkin variational problem (8.10) this gives

$$\langle A(u_M - \widetilde{u}_M), v_M \rangle = \langle B(g - g_N), v_M \rangle \quad \text{for all } v_M \in X_M.$$

In particular for the test function $v_M := u_M - \widetilde{u}_M \in X_M$ we obtain from the $X$–ellipticity of $A$ and using the boundedness of $B$

$$\begin{aligned} c_1^A \, \|u_M - \widetilde{u}_M\|_X^2 &\le \langle A(u_M - \widetilde{u}_M), u_M - \widetilde{u}_M \rangle \\ &= \langle B(g - g_N), u_M - \widetilde{u}_M \rangle \\ &\le \|B(g - g_N)\|_{X'} \|u_M - \widetilde{u}_M\|_X \\ &\le c_2^B \, \|g - g_N\|_Y \|u_M - \widetilde{u}_M\|_X. \end{aligned}$$

Hence we get the estimate

$$\|u_M - \widetilde{u}_M\|_X \le \frac{c_2^B}{c_1^A} \, \|g - g_N\|_Y.$$

Applying the triangle inequality

$$\|u - \widetilde{u}_M\|_X \le \|u - u_M\|_X + \|u_M - \widetilde{u}_M\|_X$$

we finally obtain the assertion from Theorem 8.1 (Cea's Lemma). □

## 8.3 Approximation of the Operator

Besides an approximation of the given right hand side we also have to consider an approximation of the given operator, e.g. when applying numerical integration schemes. Instead of the Galerkin variational problem (8.3) we then have to find the solution $\widetilde{u}_M \in X_M$ of the perturbed variational problem

$$\langle \widetilde{A}\widetilde{u}_M, v_M \rangle = \langle f, v_M \rangle \quad \text{for all } v_M \in X_M. \tag{8.13}$$

In (8.13) $\widetilde{A} : X \to X'$ is a bounded linear operator satisfying

$$\|\widetilde{A}v\|_{X'} \le \widetilde{c}_2^A \, \|v\|_X \quad \text{for all } v \in X. \tag{8.14}$$

Subtracting the perturbed variational problem (8.13) from the Galerkin variational problem (8.3) we find

$$\langle Au_M - \widetilde{A}\widetilde{u}_M, v_M \rangle = 0 \quad \text{for all } v_M \in X_M. \tag{8.15}$$

To ensure the unique solvability of the perturbed variational problem (8.13) we have to assume the discrete stability of the approximate operator $\widetilde{A}$. From this we then also obtain an error estimate for the approximate solution $\widetilde{u}_M \in X_M$ of (8.13).

**Theorem 8.3 (Strang Lemma).** *Assume that the approximate operator $\widetilde{A} : X \to X'$ is $X_M$–elliptic, i.e.*

$$\langle \widetilde{A}v_M, v_M \rangle \geq \widetilde{c}_1^A \, \|v_M\|_X^2 \quad \textit{for all } v_M \in X_M. \tag{8.16}$$

*Then there exists a unique solution $\widetilde{u}_M \in X_M$ of the perturbed variational problem* (8.13) *satisfying the error estimate*

$$\|u - \widetilde{u}_M\|_M \leq \left[1 + \frac{1}{\widetilde{c}_1^A}(c_2^A + \widetilde{c}_2^A)\right] \frac{c_2^A}{c_1^A} \inf_{v_M \in X_M} \|u - v_M\|_X + \frac{1}{\widetilde{c}_1^A} \|(A - \widetilde{A})u\|_{X'}. \tag{8.17}$$

*Proof.* The unique solvability of the variational problem (8.13) is a direct consequence of the $X_M$–ellipticity of the approximate operator $\widetilde{A}$, since the associated stiffness matrix $\widetilde{A}_M$ is positive definite.

Let $u_M \in X_M$ be the unique solution of the variational problem (8.3). Using again the assumption that $\widetilde{A}$ is $X_M$–elliptic, and using the orthogonality relation (8.15) we obtain

$$\begin{aligned} \widetilde{c}_1^A \, \|u_M - \widetilde{u}_M\|_X^2 &\leq \langle \widetilde{A}(u_M - \widetilde{u}_M), u_M - \widetilde{u}_M \rangle \\ &= \langle (\widetilde{A} - A)u_M, u_M - \widetilde{u}_M \rangle \\ &\leq \|(\widetilde{A} - A)u_M\|_{X'} \|u_M - \widetilde{u}_M\|_X \end{aligned}$$

and therefore

$$\|u_M - \widetilde{u}_M\|_X \leq \frac{1}{\widetilde{c}_1^A} \|(A - \widetilde{A})u_M\|_{X^*}.$$

Since both operators $A, \widetilde{A} : X \to X'$ are bounded, this gives

$$\begin{aligned} \|(A - \widetilde{A})u_M\|_{X'} &\leq \|(A - \widetilde{A})u\|_{X'} + \|(A - \widetilde{A})(u - u_M)\|_{X'} \\ &\leq \|(A - \widetilde{A})u\|_{X'} + [c_2^A + \widetilde{c}_2^A] \, \|u - u_M\|_X. \end{aligned}$$

Applying the triangle inequality we obtain

$$\begin{aligned}\|u-\widetilde{u}_M\|_X &\leq \|u-u_M\|_X + \|u_M-\widetilde{u}_M\|_X \\ &\leq \|u-u_M\|_X + \frac{1}{\widetilde{c}_1^A}\,\|(A-\widetilde{A})u_M\|_{X'} \\ &\leq \|u-u_M\|_X + \frac{1}{\widetilde{c}_1^A}\,\|(A-\widetilde{A})u\|_{X'} + \frac{1}{\widetilde{c}_1^A}[c_2^A+\widetilde{c}_2^A]\,\|u-u_M\|_X\end{aligned}$$

and the assertion finally follows from Theorem 8.1 (Cea's Lemma). □

## 8.4 Galerkin–Petrov Methods

Let $B : X \to \Pi'$ be a bounded linear operator, and let us assume that the stability condition

$$c_S\,\|v\|_X \leq \sup_{0\neq q\in\Pi}\frac{\langle Bv,q\rangle}{\|q\|_\Pi} \quad \text{for all } v\in(\ker B)^\perp\subset X \tag{8.18}$$

is satisfied. Then, for a given $g\in \operatorname{Im}_X(B)$ there exists a unique solution $u\in(\ker B)^\perp$ of the operator equation $Bu=g$ (cf. Theorem 3.7) satisfying

$$\langle Bu,q\rangle = \langle g,q\rangle \quad \text{for all } q\in\Pi.$$

For $M\in\mathbb{N}$ we introduce two sequences of conforming trial spaces

$$X_M = \operatorname{span}\{\varphi_k\}_{k=1}^M \subset (\ker B)^\perp, \quad \Pi_M = \operatorname{span}\{\psi_k\}_{k=1}^M\subset\Pi.$$

Using (8.2) we can define an approximate solution $u_M\in X_M$ as the solution of the Galerkin–Petrov variational problem

$$\langle Bu_M,q_M\rangle = \langle g,q_M\rangle \quad \text{for all } q_M\in\Pi_M. \tag{8.19}$$

In contrast to the Galerkin–Bubnov variational problem (8.3) we now have two different test and trial spaces.

Due to $\Pi_M\subset\Pi$ we have the Galerkin orthogonality

$$\langle B(u-u_M),q_M\rangle = 0 \quad \text{for all } q_M\in\Pi_M.$$

The variational problem (8.19) is equivalent to the linear system $B_M\underline{u}_M=\underline{g}$ with the matrix $B_M$ defined by

$$B_M[\ell,k] = \langle B\varphi_k,\psi_\ell\rangle$$

for $k,\ell=1,\ldots,M$ and with the right hand side $\underline{g}$ given by

$$g_\ell = \langle g,\psi_\ell\rangle$$

for $\ell=1,\ldots,M$. As in the continuous case we obtain the unique solvability of the linear system when assuming a discrete stability condition,

$$\widetilde{c}_S\,\|v_M\|_X \leq \sup_{0\neq q_M\in\Pi_M}\frac{\langle Bv_M,q_M\rangle}{\|q_M\|_\Pi} \quad \text{for all } v_M\in X_M. \tag{8.20}$$

**Theorem 8.4.** *Let $u \in (\ker B)^\perp$ be the unique solution of the operator equation $Bu = g$, and let $u_M \in X_M$ be the unique solution of the variational problem (8.19). We further assume the discrete stability condition (8.20). Then there holds the error estimate*

$$\|u - u_M\|_X \leq \left(1 + \frac{c_2^B}{\widetilde{c}_B}\right) \inf_{v_M \in X_M} \|u - v_M\|_X .$$

*Proof.* For an arbitrary $v \in (\ker B)^\perp \subset X$ there exists a uniquely determined $v_M = P_M v \in X_M$ as the unique solution of the variational problem

$$\langle Bv_M, q_M \rangle = \langle Bv, q_M \rangle \quad \text{for all } q_M \in \Pi_M .$$

For the solution $v_M \in X_M$ we obtain from the discrete stability condition

$$\widetilde{c}_S \, \|v_M\|_X \leq \sup_{0 \neq q_M \in \Pi_M} \frac{\langle Bv_M, q_M \rangle}{\|q_M\|_\Pi} = \sup_{0 \neq q_M \in \Pi_M} \frac{\langle Bv, q_M \rangle}{\|q_M\|_\Pi} \leq c_2^B \, \|v\|_X .$$

For any $v \in (\ker B)^\perp$ we therefore obtain a unique $v_M = P_M v \in \Pi_M$ satisfying

$$\|P_M v\|_X \leq \frac{c_2^B}{\widetilde{c}_S} \, \|v\|_X .$$

In particular, for the unique solution $u_M \in X_M$ of the variational problem (8.19) we obtain $u_M = P_M u$. On the other hand we have $v_M = P_M v_M$ for all $v_M \in X_M$. Hence we have for an arbitrary $v_M \in X_M$

$$\begin{aligned} \|u - u_M\|_X &= \|u - v_M + v_M - u_M\|_X = \|u - v_M - P_M(u - v_M)\|_X \\ &\leq \|u - v_M\|_X + \|P_M(u - v_M)\|_X \leq \left(1 + \frac{c_2^B}{\widetilde{c}_S}\right) \|u - v_M\|_X \end{aligned}$$

and therefore the assertion. □

The convergence of the approximate solution $u_M \to u \in X$ as $M \to \infty$ then follows as for a Galerkin–Bubnov method from an approximation property of the trial space $X_M$.

It remains to establish the discrete stability condition (8.20). A possible criterion is the following result due to Fortin [58].

**Lemma 8.5 (Criterium of Fortin).** *Let $B : X \to \Pi'$ be a bounded linear operator, and let the continuous stability condition (8.18) be satisfied. If there exist a bounded projection operator $R_M : \Pi \to \Pi_M$ satisfying*

$$\langle Bv_M, q - R_M q \rangle = 0 \quad \textit{for all } v_M \in X_M$$

*and*

$$\|R_M q\|_\Pi \leq c_R \, \|q\|_\Pi \quad \textit{for all } q \in \Pi,$$

*then there holds the discrete stability condition (8.20) with $\widetilde{c}_S = c_S / c_R$.*

*Proof.* Using the stability condition (8.18) we have for $q_N \in \Pi_N \subset \Pi$

$$\begin{aligned} c_S \, \|v_M\|_X \leq \sup_{0 \neq q \in \Pi} \frac{\langle Bv_M, q \rangle}{\|q\|_\Pi} &= \sup_{0 \neq q \in \Pi} \frac{\langle Bv_M, R_M q \rangle}{\|q\|_\Pi} \\ \leq c_R \sup_{0 \neq q \in \Pi} \frac{\langle Bv_M, R_M q \rangle}{\|R_M q\|_\Pi} &\leq c_R \sup_{0 \neq q_M \in \Pi_M} \frac{\langle Bv_M, q_M \rangle}{\|q_M\|_\Pi}, \end{aligned}$$

and therefore the discrete stability condition (8.20). □

## 8.5 Mixed Formulations

Now we consider the approximate solution of the saddle point problem (3.22) to find $(u, p) \in X \times \Pi$ such that

$$\begin{aligned} \langle Au, v \rangle + \langle Bv, p \rangle &= \langle f, v \rangle \\ \langle Bu, q \rangle &= \langle g, q \rangle \end{aligned} \tag{8.21}$$

is satisfied for all $(v, q) \in X \times \Pi$.

We assume that $A : X \to X'$ and $B : X \to \Pi'$ are bounded linear operators, and that $A$ is $X$–elliptic. For example, the last assumption is satisfied when considering the Stokes system and the modified variational formulation (4.22) and (4.23) for a Dirichlet boundary value problem with Lagrange multipliers. We further assume the stability condition (8.18). Hence, all assumptions of Theorem 3.11 and of Theorem 3.13 are satisfied, and there exists a unique solution $(u, p) \in X \times \Pi$ of the saddle point problem (8.21).

For $N, M \in \mathbb{N}$ we define two sequences of conforming trial spaces

$$X_M = \operatorname{span}\{\varphi_k\}_{k=1}^M \subset X, \quad \Pi_N = \operatorname{span}\{\psi_i\}_{i=1}^N \subset \Pi.$$

Then the Galerkin variational formulation of the saddle point problem (8.21) is to find $(u_M, p_N) \in X_M \times \Pi_N$ such that

$$\begin{aligned} \langle Au_M, v_M \rangle + \langle Bv_M, p_N \rangle &= \langle f, v_M \rangle \\ \langle Bu_M, q_N \rangle &= \langle g, q_N \rangle \end{aligned} \tag{8.22}$$

is satisfied for all $(v_M, q_N) \in X_M \times \Pi_N$. With the matrices $A_M$ and $B_N$ defined by

$$A_M[\ell, k] = \langle A\varphi_k, \varphi_\ell \rangle, \quad B_N[j, k] = \langle B\varphi_k, \psi_j \rangle$$

for $k, \ell = 1, \ldots, M$, $j = 1, \ldots, N$, and with the vectors $\underline{f}$ and $\underline{g}$ given by

$$f_\ell = \langle f, \varphi_\ell \rangle, \quad g_j = \langle g, \psi_j \rangle$$

for $\ell = 1, \ldots, M$, $j = 1, \ldots, N$, the variational formulation (8.22) is equivalent to the linear system

$$\begin{pmatrix} A_M & B_N^\top \\ B_N & 0 \end{pmatrix} \begin{pmatrix} \underline{u} \\ \underline{p} \end{pmatrix} = \begin{pmatrix} \underline{f} \\ \underline{g} \end{pmatrix} . \tag{8.23}$$

We first consider the unique solvability of the linear system (8.23) from an algebraic point of view. The dimension of the system matrix $K$ in (8.23) is $N + M$. With

$$\operatorname{rang} A_M \leq M, \quad \operatorname{rang} B_N \leq \min\{M, N\}$$

we find

$$\operatorname{rang} K \leq M + \min\{M, N\} .$$

In particular for $M < N$ we obtain

$$\operatorname{rang} K \leq 2M < M + N = \dim K .$$

i.e. the linear system (8.23) is in general not solvable. Hence we have to define the trial spaces $X_M$ and $\Pi_N$ with care. The necessary condition $M \geq N$ shows, that the trial space $X_M$ has to be rich enough compared with the trial space $\Pi_N$.

To investigate the unique solvability of the Galerkin variational problem (8.22) we will make use of Theorem 3.13. When considering the conforming trial space $X_M \subset X$ we obtain from the $X$–ellipticity of $A$ the positive definiteness of the matrix $A_M$, i.e.

$$(A_M \underline{v}, \underline{v}) = \langle A v_M, v_M \rangle \geq c_1^A \, \|v_M\|_X^2 > 0$$

for all $\underline{0} \neq \underline{v} \in \mathbb{R}^M \leftrightarrow v_M \in X_M$. The matrix $A_M$ is therefore invertible and the linear system (8.23) can be transformed into the Schur complement system

$$B_N A_M^{-1} B_N^\top \underline{p} = B_N A_M^{-1} \underline{f} - \underline{g} . \tag{8.24}$$

It remains to investigate the unique solvability of the linear system (8.24). For this we assume the discrete stability or Babuška–Brezzi–Ladyshenskaya (BBL) condition

$$\widetilde{c}_S \, \|q_N\|_\Pi \leq \sup_{0 \neq v_M \in X_M} \frac{\langle B v_M, q_N \rangle}{\|v_M\|_X} \quad \text{for all } q_N \in \Pi_N . \tag{8.25}$$

It is worth to remark, that the discrete stability condition (8.25) is in general not an immediate consequence of the continuous stability condition (3.25).

**Lemma 8.6.** *Let $A : X \to X'$ and $B : X \to \Pi'$ be bounded linear operators, and let $A$ be $X$–elliptic. For the conforming trial spaces $X_M \subset X$ and $\Pi_N \subset \Pi$ we assume the discrete stability condition* (8.25). *The symmetric Schur complement matrix $S_N := B_N A_M^{-1} B_N^\top$ of the Schur complement system* (8.24) *is then positive definite, i.e.*

$$(S_N \underline{q}, \underline{q}) \geq c_1^{S_N} \, \|q_N\|_\Pi^2$$

*for all $\underline{0} \neq \underline{q} \in \Pi_N \leftrightarrow q_N \in \Pi_N$.*

*Proof.* For an arbitrary but fixed $\underline{q} \in \mathbb{R}^N$ we define $\underline{\bar{u}} := A_M^{-1} B_N^\top \underline{q}$, i.e. for the associated functions $q_N \in \Pi_N$ and $\bar{u}_M \in X_M$ we have

$$\langle A\bar{u}_M, v_M \rangle = \langle Bv_M, q_N \rangle \quad \text{for all } v_M \in X_M.$$

Using the $X$–ellipticity of $A$ we then obtain

$$c_1^A \, \|\bar{u}_M\|_X^2 \le \langle A\bar{u}_M, \bar{u}_M \rangle = \langle B\bar{u}_M, q_N \rangle = (B_N \underline{\bar{u}}, \underline{q}) = (B_N A_M^{-1} B_N^\top \underline{q}, \underline{q}).$$

On the other hand, the discrete stability condition (8.25) gives

$$c_S \, \|q_N\|_\Pi \le \sup_{\underline{0} \ne \underline{v}_M \in \mathbb{R}^M} \frac{\langle Bv_M, q_N \rangle}{\|v_M\|_X} = \sup_{\underline{0} \ne \underline{v}_M \in \mathbb{R}^M} \frac{\langle A\bar{u}_M, v_M \rangle}{\|v_M\|_X} \le c_2^A \, \|\bar{u}_M\|_X$$

and with

$$\|q_N\|_\Pi^2 \le \left( \frac{c_2^A}{c_S} \right)^2 \, \|\bar{u}_M\|_X^2 \le \frac{1}{c_1^A} \left( \frac{c_2^A}{c_S} \right)^2 \, (B_N A_M^{-1} B_N^\top \underline{q}, \underline{q})$$

we finally get the assertion. □

Hence we have the unique solvability of the Schur complement system (8.24) and therefore of the linear system (8.23). Moreover, we also have the following stability estimate.

**Theorem 8.7.** *Let $A : X \to X'$ and $B : X \to \Pi'$ be bounded and linear operators, and let $A$ be $X$–elliptic. For the conforming trial spaces $X_M \subset X$ and $\Pi_N \subset \Pi$ we assume the discrete stability condition (8.25). For the unique solution $(u_M, p_N) \in X_M \times \Pi_N$ of the saddle point problem (8.22) we then have the stability estimates*

$$\|p_N\|_\Pi \le \frac{1}{c_1^{S_N}} \frac{c_2^B}{c_1^A} \, \|f\|_{X'} + \frac{1}{c_1^{S_N}} \, \|g\|_{\Pi'} \tag{8.26}$$

*and*

$$\|u_M\|_X \le \left( 1 + \frac{c_2^B}{c_1^{S_N}} \frac{c_2^B}{c_1^A} \right) \|f\|_{X'} + \frac{1}{c_1^{S_N}} \frac{c_2^B}{c_1^A} \|g\|_{\Pi'}. \tag{8.27}$$

*Proof.* Let $(\underline{u}, \underline{p}) \in \mathbb{R}^M \times \mathbb{R}^N \leftrightarrow (u_M, p_N) \in X_M \times \Pi_N$ be the unique solution of the linear system (8.23) and of the saddle point problem (8.22), respectively. Using Lemma 8.6 we have

$$\begin{aligned} c_1^{S_N} \, \|p_N\|_\Pi^2 \le (S_N \underline{p}, \underline{p}) &= (B_N A_M^{-1} B_N^\top \underline{p}, \underline{p}) = (B_N A_M^{-1} \underline{f} - \underline{g}, \underline{p}) \\ &= \langle B\bar{u}_M - g, p_N \rangle \le \left[ c_2^B \, \|\bar{u}_M\|_X + \|g\|_{\Pi'} \right] \|p_N\|_\Pi \end{aligned}$$

and therefore

$$\|p_N\|_\Pi \le \frac{1}{c_1^{S_N}} \left[ c_2^B \, \|\bar{u}_M\|_X + \|g\|_{\Pi'} \right].$$

Hereby, $\underline{\bar{u}} = A_M^{-1} \underline{f} \in \mathbb{R}^M \leftrightarrow \bar{u}_M \in X_M$ is the unique solution of the variational problem

$$\langle A\bar{u}_M, v_M \rangle = \langle f, v_M \rangle \quad \text{for all } v_M \in X_M.$$

From the $X$–ellipticity of $A$ we then find

$$\|\bar{u}_M\|_X \leq \frac{1}{c_1^A} \|f\|_{X'}.$$

Moreover,

$$\begin{aligned} c_1^A \|u_M\|_X^2 &\leq \langle Au_M, u_M \rangle \\ &= \langle f, u_M \rangle - \langle Bu_M, p_N \rangle \leq \left[ \|f\|_{X'} + c_2^B \|p_N\|_\Pi \right] \|u_M\|_X \end{aligned}$$

and therefore

$$\|u_M\|_X \leq \frac{1}{c_1^A} \left[ \|f\|_{X'} + c_2^B \|p_N\|_\Pi \right].$$

With (8.26) we then obtain (8.27) □.

Using the stability estimates (8.26) and (8.27) we also obtain an error estimate for the approximate solution $(u_M, p_N) \in X_M \times \Pi_N$.

**Theorem 8.8.** *Let all assumptions of Theorem* 8.7 *be valid. For the unique approximate solution* $(u_M, p_N) \in X_M \times \Pi_N$ *of the saddle point problem* (8.22) *there holds the error estimate*

$$\|u - u_M\|_X + \|p - p_N\|_\Pi \leq c \left\{ \inf_{v_M \in X_M} \|u - v_M\|_X + \inf_{q_N \in \Pi_N} \|p - q_N\|_\Pi \right\}.$$

*Proof.* When taking the difference of the continuous saddle point formulation (8.21) with the Galerkin variational problem (8.22) for the conforming trial spaces $X_M \times \Pi_N \subset X \times \Pi$ we obtain the Galerkin orthogonalities

$$\begin{aligned} \langle A(u - u_M), v_M \rangle + \langle Bv_M, p - p_N \rangle &= 0 \\ \langle B(u - u_M), q_N \rangle &= 0 \end{aligned}$$

for all $(v_M, q_N) \in X_M \times \Pi_N$. For arbitrary $(\bar{u}_M, \bar{p}_N) \in X_M \times \Pi_N$ we then obtain

$$\begin{aligned} \langle A(\bar{u}_M - u_M), v_M \rangle + \langle Bv_M, \bar{p}_N - p_N \rangle &= \langle A(\bar{u}_M - u) + \langle B'(\bar{p}_N - p), v_M \rangle \\ \langle B(\bar{u}_M - u_M), q_N \rangle &= \langle B(\bar{u}_M - u), q_N \rangle \end{aligned}$$

for all $(v_M, q_N) \in X_M \times \Pi_N$. Using Theorem 8.7 we find the unique solution $(\bar{u}_M - u_M, \bar{p}_N - p_M) \in X_M \times \Pi_N$, and we obtain the stability estimates

$$\|\bar{p}_N - p_N\|_\Pi \leq c_1 \|A(\bar{u}_M - u) + B'(\bar{p}_N - p)\|_{X'} + c_2 \|B(\bar{u}_M - u)\|_{\Pi'},$$

$$\|\bar{u}_M - u_M\|_X \leq c_3 \|A(\bar{u}_M - u) + B'(\bar{p}_N - p)\|_{X'} + c_4 \|B(\bar{u}_M - u)\|_{\Pi'}$$

for arbitrary $(\bar{u}_M, \bar{p}_N) \in X_M \times \Pi_N$. Due to the mapping properties of the bounded operators $A$, $B$ and $B'$ we get with the triangle inequality

$$\begin{aligned}\|p - p_N\|_\Pi &\leq \|p - \bar{p}_N\|_\Pi + \|\bar{p}_N + p_N\|_\Pi \\ &\leq (1 + c_1 c_2^B)\, \|p - \bar{p}_N\|_\Pi + (c_1 c_2^A + c_2 c_2^B)\, \|u - \bar{u}_M\|_X\end{aligned}$$

for arbitrary $(\bar{u}_M, \bar{p}_N) \in X_M \times \Pi_N$. The estimate for $\|u - u_M\|_X$ follows in the same way. □

It remains to validate the discrete stability condition (8.25). As in Lemma 8.5 we can use the criterion of Fortin to establish (8.25).

**Lemma 8.9 (Criteria of Fortin).** *Let $B : X \to \Pi'$ be a bounded linear operator, and let the continuous stability condition (3.25) be satisfied. If there exists a bounded projection operator $P_M : X \to X_M$ satisfying*

$$\langle B(v - P_M v), q_N \rangle = 0 \quad \textit{for all } q_N \in \Pi_N$$

*and*

$$\|P_M v\|_X \leq c_P \, \|v\|_X \quad \textit{for all } v \in X,$$

*then there holds the discrete stability condition (8.25) with $\tilde{c}_S = c_S / c_P$.*

## 8.6 Coercive Operators

We finally consider an approximate solution of the operator equation $Au = f$ when $A : X \to X'$ is assumed to be coercive, i.e. there exists a compact operator $C : X \to X'$ such that Gårding's inequality (3.32) is satisfied,

$$\langle (A + C)v, v \rangle \geq c_1^A \, \|v\|_X^2 \quad \text{for all } v \in X.$$

For a sequence $X_M \subset X$ of finite dimensional trial spaces we consider the Galerkin variational problem to find $u_M \in X_M$ such that

$$\langle A u_M, v_M \rangle = \langle f, v_M \rangle \tag{8.28}$$

is satisfied for all $v_M \in X_M$. Note that the variational problem (8.28) formally coincides with the Galerkin–Bubnov formulation (8.3). However, since we now consider the more general case of a coercive operator instead of an elliptic operator, the numerical analysis to establish suitable stability and error estimates is different.

**Theorem 8.10 (Cea's Lemma).** *Let $A : X \to X_M$ be a bounded linear coercive operator and let the stability condition*

$$c_S \, \|w_M\|_X \leq \sup_{v_M \in X_M, \|v_M\|_X > 0} \frac{\langle A w_M, v_M \rangle}{\|v_M\|_X} \tag{8.29}$$

*be satisfied for all $w_M \in X_M$. Then there exists a unique solution $u_M \in X_M$ of the Galerkin variational problem (8.28) satisfying the stability estimate*

$$\|u_M\|_X \leq \frac{1}{c_S} \|f\|_{X'} \tag{8.30}$$

*and the error estimate*

$$\|u - u_M\|_X \leq \left(1 + \frac{c_2^A}{c_S}\right) \inf_{v_M \in X_M} \|u - v_M\|_X . \tag{8.31}$$

*Proof.* We consider the homogeneous linear system $A_M \underline{\bar{w}} = \underline{0}$ to find an approximate solution $\bar{w}_M \in X_M$ of the homogeneous variational problem

$$\langle A\bar{w}_M, v_M \rangle = 0 \quad \text{for all } v_M \in X_M .$$

Using the stability condition (8.29) we then obtain

$$c_S \|\bar{w}_M\|_X \leq \sup_{v_M \in X_M, \|v_M\|_X > 0} \frac{\langle A\bar{w}_M, v_M \rangle}{\|v_M\|_X} = 0$$

and therefore $\bar{w}_M = 0 \leftrightarrow \underline{\bar{w}} = \underline{0}$. This ensures the unique solvability of the linear system $A_M \underline{u} = \underline{f}$ and therefore of the variational problem (8.28).

Let $\underline{u} \in \mathbb{R}^M \leftrightarrow u_M \in X_M$ be the unique solution of the Galerkin variational problem (8.28). Again, applying the stability condition (8.29) this gives

$$\begin{aligned} c_S \|u_M\|_X &\leq \sup_{v_M \in X_M, \|v_M\|_X > 0} \frac{\langle Au_M, v_M \rangle}{\|v_M\|_X} \\ &= \sup_{v_M \in X_M, \|v_M\|_X > 0} \frac{\langle f, v_M \rangle}{\|v_M\|_X} \leq \|f\|_{X'} \end{aligned}$$

and therefore the stability estimate (8.30).

For an arbitrary $w \in X$ we define an approximate solution $w_M \in X_M$ of the Galerkin variational formulation

$$\langle Aw_M, v_M \rangle = \langle Aw, v_M \rangle \quad \text{for all } v_M \in X_M .$$

This defines the projection operator $G_M : X \to X_M$ by $w_M = G_M w$ satisfying

$$\|G_M w\|_X = \|w_M\|_X \leq \frac{1}{c_S} \|Aw\|_{X'} \leq \frac{c_2^A}{c_S} \|w\|_X .$$

In particular, we have $u_M = G_M u$ for the solution of the Galerkin variational formulation (8.28). Since $G_M$ is a projection, $G_M v_M = v_M$ for all $v_M \in X_M$, we then find

$$\begin{aligned} \|u - u_M\|_X &= \|u - v_M + G_M v_M - u_M\|_X \\ &\leq \|u - v_M\|_X + \|G_M(u - v_M)\|_X \leq \left(1 + \frac{c_2^A}{c_S}\right) \|u - v_M\|_X \end{aligned}$$

for all $v_M \in X_M$. From this, the error estimate (8.31) follows. □

It remains to validate the discrete stability condition (8.29). Note, that for an $X$–elliptic operator $A$ we then obtain

$$c_1^A \|w_M\|_X \leq \frac{\langle Aw_M, w_M\rangle}{\|w_M\|_X} \leq \sup_{v_M \in X_M, \|v_M\|_X > 0} \frac{\langle Aw_M, v_M\rangle}{\|v_M\|_X}$$

for all $w_M \in X_M$, i.e. (8.29). In what follows we consider the case of an coercive operator $A$.

**Theorem 8.11.** *Let $A : X \to X'$ be a bounded linear operator which is assumed to be coercive and injective. Let $X_M \subset X$ be a dense sequence of conforming trial spaces. Then there exists an index $M_0 \in \mathbb{N}$ such that the discrete stability condition (8.29) is satisfied for $M \geq M_0$.*

*Proof.* Let $w_M \in X_M$ be arbitrary but fixed. Since $A : X \to X'$ is assumed to be coercive, there is a compact operator $C : X \to X'$ such that the bounded operator $D = A + C : X \to X'$ is $X$–elliptic. Hence we can set $\bar{v} = D^{-1}Cw_M$ as the unique solution $\bar{v} \in X$ of the variational problem

$$\langle D\bar{v}, v\rangle = \langle Cw_M, v\rangle \quad \text{for all } v \in X.$$

Moreover we can define an approximate solution $\bar{v}_M \in X_M$ as the unique solution of the Galerkin variational problem

$$\langle D\bar{v}_M, v_M\rangle = \langle Cw_M, v_M\rangle \quad \text{for all } v_M \in X_M.$$

Hence we have the Galerkin orthogonality

$$\langle D(\bar{v} - \bar{v}_M), v_M\rangle = 0 \quad \text{for all } v_M \in X_M.$$

Applying Cea's lemma (cf. Theorem 8.1) for the $X$–elliptic operator $D$ we also find the stability estimate

$$\|\bar{v}_M\|_X \leq \frac{1}{c_1^D} \|Cw_M\|_{X'} \leq \frac{c_2^C}{c_1^D} \|w_M\|_X$$

and therefore

$$\|w_M - \bar{v}_M\|_X \leq \|w_M\|_X + \|v_M\|_X \leq \left(1 + \frac{c_2^C}{c_1^D}\right) \|w_M\|_X$$

as well as the error estimate

$$\|\bar{v} - \bar{v}_M\|_X \leq \frac{c_2^D}{c_1^D} \inf_{v_M \in X_M} \|\bar{v} - v_M\|_X.$$

Hence, the approximation property (8.8) of the trial space gives the convergence $\bar{v}_M \in \bar{v}$ in $X$ for $M \to \infty$.

Considering as test function $v_M = w_M - \bar{v}_M$ we obtain

$$\begin{aligned}\langle Aw_M, w_M - \bar{v}_M\rangle &= \langle Aw_M, w_M - \bar{v}\rangle + \langle Aw_M, \bar{v} - \bar{v}_M\rangle \\ &= \langle Aw_M, w_M - D^{-1}Cw_M\rangle + \langle Aw_M, \bar{v} - \bar{v}_M\rangle.\end{aligned}$$

For the first summand we further get

$$\begin{aligned}\langle Aw_M, w_M - D^{-1}Cw_M\rangle &= \langle Aw_M, D^{-1}(D - C)w_M\rangle \\ &= \langle Aw_M, D^{-1}Aw_M\rangle \\ &\geq c_1^D \, \|Aw_M\|_{X'}^2 \geq c_1^D c_A \, \|w_M\|_X^2\end{aligned}$$

since $A : X \to X'$ has a bounded inverse. On the other hand, using the Galerkin orthogonality we get

$$\begin{aligned}|\langle Aw_M, \bar{v} - \bar{v}_M\rangle| &= |\langle Dw_M, \bar{v} - \bar{v}_M\rangle - \langle Cw_M, \bar{v} - \bar{v}_M\rangle| \\ &= |\langle w_M, D(\bar{v} - \bar{v}_M)\rangle - \langle w_M, C(\bar{v} - \bar{v}_M)\rangle| \\ &= |\langle w_M, C(\bar{v} - \bar{v}_M)\rangle| \leq \|w_M\|_X \|C(\bar{v} - \bar{v}_M)\|_{X'}\end{aligned}$$

Since $C : X \to X'$ is a compact operator, there exists a subsequence $\{\bar{v}_M\}_{M\in\mathbb{N}}$ satisfying

$$\lim_{M\to\infty} \frac{\|C(\bar{v} - \bar{v}_M)\|_{X'}}{\|\bar{v}\|_X} = 0.$$

Hence there exists an index $M_0 \in \mathbb{N}$ such that

$$\begin{aligned}\langle Aw_M, w_M - \bar{v}_M\rangle &\geq \frac{1}{2} c_1^D c_A \, \|w_M\|_X^2 \\ &\geq \frac{1}{2} c_1^D c_A \left(1 + \frac{c_2^D}{c_1^D}\right)^{-1} \|w_M\|_X \|w_M - \bar{v}_M\|_X\end{aligned}$$

is satisfied for all $M \geq M_0$ which implies the stability condition (8.29). □

## 8.7 Exercises

**8.1** Let $X$ be a Hilbert space and let $a(\cdot,\cdot) : X \times X \to \mathbb{R}$ be a symmetric and positive definite bilinear form. For the approximation of the minimization problem

$$F(u) = \min_{v\in X} F(v), \quad F(v) = \frac{1}{2}a(v,v) - \langle f, v\rangle$$

we introduce a finite–dimensional trial space

$$X_M = \text{span}\{\varphi_k\}_{k=1}^M \subset X.$$

Derive the variational problem to find the approximate solution $u_M \in X_M$.

# 9

# Finite Elements

For the approximate solution of variational formulations as described in Chapter 4 we introduce appropriate finite–dimensional trial spaces, and prove certain approximation properties in Sobolev spaces. For simplicity we just consider lowest order polynomial basis functions. For an introduction of more general finite elements we refer, for example, to [31, 41, 85].

## 9.1 Reference Elements

Let $\Omega \subset \mathbb{R}^d$ $(d = 1, 2, 3)$ be a bounded domain with a polygonal $(d = 2)$ or with a polyhedral $(d = 3)$ boundary. We consider a sequence $\{\mathcal{T}_N\}_{N \in \mathbb{N}}$ of decompositions (meshes)

$$\overline{\Omega} = \overline{\mathcal{T}}_N = \bigcup_{\ell=1}^{N} \tau_\ell \tag{9.1}$$

with finite elements $\tau_\ell$. In the simplest case we have an interval $(d = 1)$, a triangle $(d = 2)$, or a tetrahedron $(d = 3)$. Further we denote by $\{x_k\}_{k=1}^{M}$ the set of all nodes of the decomposition $\mathcal{T}_N$, see Fig. 9.1 for a finite element $\tau_\ell$ and the corresponding nodes $x_k$. In addition, for $d = 2, 3$ we have by $\{k_j\}_{j=1}^{K}$ the set of all edges.

By $I(k)$ we denote the index set of all elements $\tau_\ell$ where $x_k \in \overline{\tau}_\ell$ is a node,

$$I(k) := \{\ell \in \mathbb{N} : x_k \in \overline{\tau}_\ell\} \quad \text{for } k = 1, \ldots, M.$$

Moreover,

$$J(\ell) := \{k \in \mathbb{N} : x_k \in \overline{\tau}_\ell\} \quad \text{for } \ell = 1, \ldots, N$$

is the index set of all nodes $x_k$ with $x_k \in \overline{\tau}_\ell$. Note that $\dim J(\ell) = d + 1$ in the case of the finite elements $\tau_\ell$ considered here. Finally,

$$K(j) := \{\ell \in \mathbb{N} : k_j \in \overline{\tau}_\ell\} \quad \text{for } j = 1, \ldots, K$$

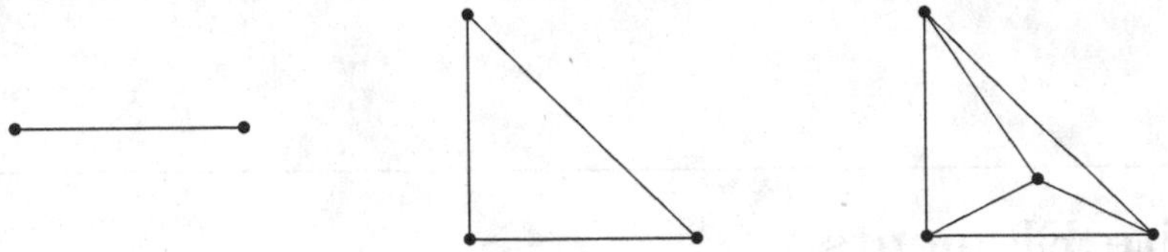


**Fig. 9.1.** Finite element $\tau_\ell$ and related nodes $x_k$.

is the index set of all elements $\tau_\ell$ with the edge $k_j$.

The decomposition (9.1) is called admissible, if two neighboring elements join either a node ($d = 1, 2, 3$), an edge ($d = 2, 3$), or a triangle ($d = 3$), see Fig. 9.2. In particular, we avoid hanging nodes as in the inadmissible decomposition of Fig. 9.2.

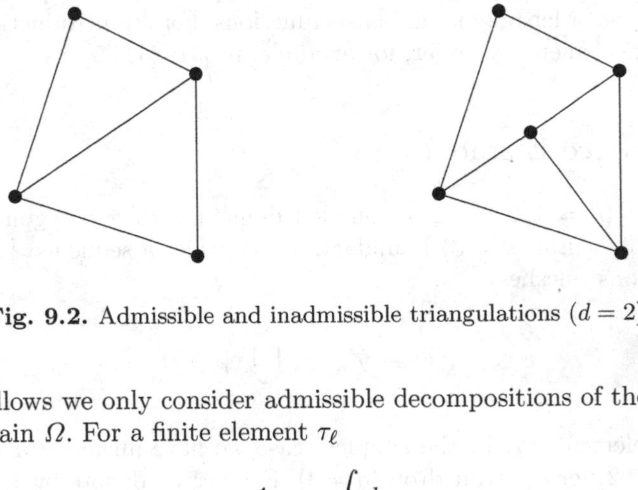

**Fig. 9.2.** Admissible and inadmissible triangulations ($d = 2$).

In what follows we only consider admissible decompositions of the computational domain $\Omega$. For a finite element $\tau_\ell$

$$\Delta_\ell := \int_{\tau_\ell} dx \tag{9.2}$$

is the volume, while

$$h_\ell := \Delta_\ell^{1/d}$$

is the local mesh size. Moreover,

$$d_\ell := \sup_{x,y \in \tau_\ell} |x - y|$$

is the diameter of the finite element $\tau_\ell$, which coincides with the longest edge of the element $\tau_\ell$. Obviously, for $d = 1$ we have

$$\Delta_\ell = h_\ell = d_\ell.$$

Finally, $r_\ell$ is the radius of the largest circle ($d = 2$) or sphere ($d = 3$) which can be inscribed in the finite element $\tau_\ell$. A finite element $\tau_\ell$ of the decomposition (9.1) is called shape regular, if the diameter $d_\ell$ of the finite element $\tau_\ell$ is bounded uniformly by the radius $r_\ell$, i.e.

$$d_\ell \leq c_F \, r_\ell \quad \text{for } \ell = 1, \ldots, N$$

where the constant $c_F$ does not depend on $\mathcal{T}_N$. For the two–dimensional case $d = 2$ we then have

$$\pi r_\ell^2 \leq \Delta_\ell = h_\ell^2 \leq d_\ell^2 \leq c_F^2 \, r_\ell^2,$$

and therefore the equivalence relations

$$\sqrt{\pi} \, r_\ell \leq h_\ell \leq d_\ell \leq c_F \, r_\ell.$$

Correspondingly, for $d = 3$ we have

$$\frac{4}{3}\pi r_\ell^3 \leq \Delta_\ell = h_\ell^3 \leq d_\ell^3 \leq c_F^3 \, r_\ell^3$$

and therefore

$$\sqrt[3]{\frac{4}{3}\pi} \, r_\ell \leq h_\ell \leq d_\ell \leq c_F \, r_\ell.$$

The global mesh size $h$ is defined by

$$h = h_{\max} := \max_{\ell=1,\ldots,N} h_\ell$$

while

$$h_{\min} := \min_{\ell=1,\ldots,N} h_\ell$$

is the minimal local mesh size. The family of decompositions $\mathcal{T}_N$ is called globally quasi–uniform, if

$$\frac{h_{\max}}{h_{\min}} \leq c_G$$

is bounded by a global constant $c_G \geq 1$ which is independent of $N \in \mathbb{N}$. The family $\mathcal{T}_N$ is called locally quasi–uniform, if

$$\frac{h_\ell}{h_j} \leq c_L \quad \text{for } \ell = 1, \ldots, N$$

holds for all neighboring finite elements $\tau_j$ and $\tau_\ell$. Here, two finite elements $\tau_\ell$ and $\tau_j$ are called neighboring, if the average $\overline{\tau}_\ell \cap \overline{\tau}_j$ consists either of a node, an edge, or a triangle.

In the one–dimensional case $d = 1$ each finite element $\tau_\ell$ can be described via a local parametrization, in particular for $x \in \tau_\ell$ and $\ell_1, \ell_2 \in J(\ell)$ we have

$$x = x_{\ell_1} + \xi \, (x_{\ell_2} - x_{\ell_1}) = x_{\ell_1} + \xi \, h_\ell \quad \text{for } \xi \in (0,1).$$

Here, the element

$$\tau := (0,1) \tag{9.3}$$

is called reference element. If we consider a function $v(x)$ for $x \in \tau_\ell$ we can write

$$v(x) \,=\, v(x_{\ell_1} + \xi\, h_\ell) \,=:\, \widetilde{v}_\ell(\xi) \quad \text{for } \xi \in \tau,$$

in particular we can identify a function $v(x)$ for $x \in \tau_\ell$ with a function in the reference element, $\widetilde{v}_\ell(\xi)$ for $\xi \in \tau$. It follows that

$$\|v\|^2_{L_2(\tau_\ell)} \,=\, \int\limits_{\tau_\ell} |v(x)|^2 dx \,=\, \int\limits_{\tau} |\widetilde{v}_\ell(\xi)|^2\, h_\ell\, d\xi \,=\, h_\ell\, \|\widetilde{v}_\ell\|^2_{L_2(\tau)}.$$

For the first derivative we have, by applying the chain rule,

$$\frac{d}{d\xi}\widetilde{v}_\ell(\xi) \,=\, h_\ell \frac{d}{dx} v(x) \quad \text{for } x \in \tau_\ell, \xi \in \tau$$

and therefore

$$\frac{d}{dx} v(x) \,=\, \frac{1}{h_\ell} \frac{d}{d\xi}\widetilde{v}_\ell(\xi) \quad \text{for } x \in \tau_\ell, \xi \in \tau.$$

For $m \in \mathbb{N}$ the recursive application of this result gives

$$\frac{d^m}{dx^m} v(x) \,=\, h_\ell^{-m} \frac{d^m}{d\xi^m}\widetilde{v}_\ell(\xi) \quad \text{for } x \in \tau_\ell, \xi \in \tau.$$

Hence we obtain for the local norms of $v$ and $\widetilde{v}_\ell$

$$\left\| \frac{d^m}{dx^m} v \right\|^2_{L_2(\tau_\ell)} \,=\, h_\ell^{1-2m} \left\| \frac{d^m}{d\xi^m} \widetilde{v}_\ell \right\|^2_{L_2(\tau)} \qquad \text{for } m \in \mathbb{N}_0\,. \tag{9.4}$$

In the two–dimensional case $d = 2$ the reference element $\tau$ is given by the triangle

$$\tau \,=\, \left\{ \xi \in \mathbb{R}^2 \,:\, 0 \leq \xi_1 \leq 1,\; 0 \leq \xi_2 \leq 1 - \xi_1 \right\}. \tag{9.5}$$

Then we find the local parametrization for $x \in \tau_\ell$ as

$$x = x_{\ell_1} + \sum_{i=1}^{2} \xi_i (x_{\ell_{i+1}} - x_{\ell_1}) \,=\, x_{\ell_1} + J_\ell \xi \quad \text{for } \xi \in \tau$$

with the Jacobian

$$J_\ell \,=\, \begin{pmatrix} x_{\ell_2,1} - x_{\ell_1,1} & x_{\ell_3,1} - x_{\ell_1,1} \\ x_{\ell_2,2} - x_{\ell_1,2} & x_{\ell_3,2} - x_{\ell_1,2} \end{pmatrix}.$$

To compute the area (volume) of the finite element $\tau_\ell$ we obtain

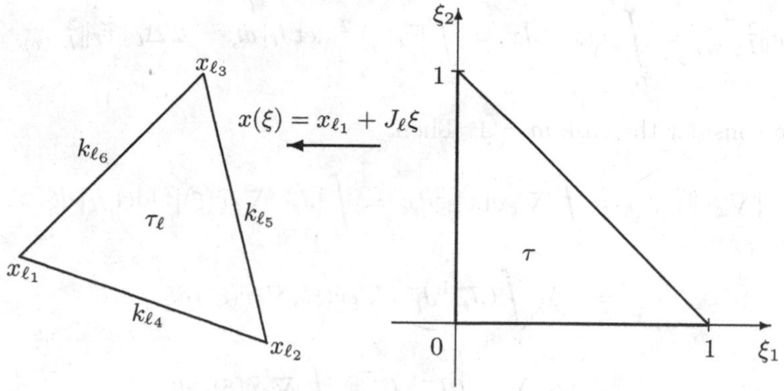


**Fig. 9.3.** Finite element $\tau_\ell$ and reference element $\tau$ $(d = 2)$.

$$\Delta_\ell = \int_{\tau_\ell} ds_x = \int_\tau |\det J_\ell|\, d\xi \;=\; |\det J_\ell| \int_0^1 \int_0^{1-\xi_1} d\xi_2 d\xi_1 \;=\; \frac{1}{2}\,|\det J_\ell|$$

and therefore

$$|\det J_\ell| \;=\; 2\,\Delta_\ell. \tag{9.6}$$

If we consider a function $v(x)$ for $x \in \tau_\ell$ we can write

$$v(x) \;=\; v(x_{\ell_1} + J_\ell \xi) \;=\; \widetilde{v}_\ell(\xi) \quad \text{for } \xi \in \tau.$$

Then, by applying the chain rule we get

$$\nabla_\xi \widetilde{v}_\ell(\xi) \;=\; J_\ell^\top \nabla_x v(x)$$

and therefore

$$\nabla_x v(x) \;=\; J_\ell^{-\top} \nabla_\xi \widetilde{v}_\ell(\xi).$$

As for the one–dimensional case $d = 1$ we can show the following norm equivalence estimates:

**Lemma 9.1.** *For $d = 2$ and $m \in \mathbb{N}_0$ there hold the norm equivalence inequalities*

$$\frac{1}{c_m}(2\Delta_\ell)^{1-m}\,\|\nabla_\xi^m \widetilde{v}_\ell\|^2_{L_2(\tau)} \;\le \|\nabla_x^m v\|^2_{L_2(\tau_\ell)} \le\; c_m(2\Delta_\ell)^{1-m}\,\|\nabla_\xi^m \widetilde{v}_\ell\|^2_{L_2(\tau)} \tag{9.7}$$

*where*

$$c_m \;=\; \left(\frac{c_F^2}{\pi}\right)^m .$$

*Proof.* For $m = 0$ the assertion follows directly from

$$\|v\|^2_{L_2(\mathcal{T}_\ell)} = \int\limits_{\mathcal{T}_\ell} |v(x)|^2 dx = \int\limits_{\mathcal{T}} |\widetilde{v}_\ell(\xi)|^2 |\det J_\ell|\, d\xi = 2\,\Delta_\ell\, \|\widetilde{v}_\ell\|^2_{L_2(\mathcal{T})}.$$

Now we consider the case $m = 1$. Then,

$$\begin{aligned}
\|\nabla_x v\|^2_{L_2(\mathcal{T}_\ell)} &= \int\limits_{\mathcal{T}_\ell} |\nabla_x v(x)|^2 dx = \int\limits_{\mathcal{T}} |J_\ell^{-\top} \nabla_\xi \widetilde{v}(\xi)|^2\, |\det J_\ell|\, d\xi \\
&= 2\Delta_\ell \int\limits_{\mathcal{T}} (J_\ell^{-1} J_\ell^{-\top} \nabla_\xi \widetilde{v}(\xi), \nabla_\xi \widetilde{v}(\xi))\, d\xi \\
&\le 2\Delta_\ell\, \lambda_{\max}(J_\ell^{-1} J_\ell^{-\top}) \int\limits_{\mathcal{T}} |\nabla_\xi \widetilde{v}(\xi)|^2 d\xi \\
&= 2\Delta_\ell\, \lambda_{\max}(J_\ell^{-1} J_\ell^{-\top})\, \|\nabla_\xi \widetilde{v}\|^2_{L_2(\mathcal{T})}
\end{aligned}$$

as well as

$$\|\nabla_x v\|^2_{L_2(\mathcal{T}_\ell)} \ge 2\Delta_\ell\, \lambda_{\min}(J_\ell^{-1} J_\ell^{-\top})\, \|\nabla_\xi \widetilde{v}\|^2_{L_2(\mathcal{T})}.$$

It is therefore sufficient to estimate the eigenvalues of the matrix $J_\ell^\top J_\ell$. With

$$a := |x_{\ell_2} - x_{\ell_1}|, \quad b := |x_{\ell_3} - x_{\ell_1}|, \quad \alpha = \sphericalangle(x_{\ell_3} - x_{\ell_1}, x_{\ell_2} - x_{\ell_1})$$

we have

$$J_\ell^\top J_\ell = \begin{pmatrix} a^2 & a\,b\cos\alpha \\ a\,b\cos\alpha & b^2 \end{pmatrix},$$

and the eigenvalues of $J_\ell^\top J_\ell$ are

$$\lambda_{1/2} = \frac{1}{2}\left[a^2 + b^2 \pm \sqrt{(a^2 - b^2)^2 + 4a^2 b^2 \cos^2\alpha}\right].$$

Obviously, for the maximal eigenvalue $\lambda_1$ we have the inclusion

$$\frac{1}{2}(a^2 + b^2) \le \lambda_1 \le a^2 + b^2$$

while for the product of the eigenvalues $\lambda_{1/2}$ we have with (9.6)

$$\lambda_1 \lambda_2 = \det(J_\ell^\top J_\ell) = |\det J_\ell|^2 = 4\Delta_\ell^2.$$

The minimal eigenvalue $\lambda_2$ admits the lower estimate

$$\lambda_2 = \frac{4\Delta_\ell^2}{\lambda_1} \ge \frac{4\Delta_\ell^2}{a^2 + b^2}$$

and therefore we conclude

$$\frac{4\Delta_\ell^2}{a^2 + b^2} \le \lambda_{\min}(J_\ell^\top J_\ell) \le \lambda_{\max}(J_\ell^\top J_\ell) \le a^2 + b^2.$$

Moreover,

$$a^2 + b^2 \le 2\, d_\ell^2 \le 2\, c_F^2\, r_\ell^2 \le \frac{2c_F^2}{\pi}\, \Delta_\ell.$$

Hence we have

$$\frac{2\pi}{c_F^2}\, \Delta_\ell \le \lambda_{\min}(J_\ell^\top J_\ell) \le \lambda_{\max}(J_\ell^\top J_\ell) \le \frac{2c_F^2}{\pi}\, \Delta_\ell$$

and the eigenvalues of the inverse matrix $J_\ell^{-1} J_\ell^{-\top}$ can be estimated as

$$\frac{\pi}{c_F^2}(2\Delta_\ell)^{-1} \le \lambda_{\min}(J_\ell^{-1} J_\ell^{-\top}) \le \lambda_{\max}(J_\ell^{-1} J_\ell^{-\top}) \le \frac{c_F^2}{\pi}(2\Delta_\ell)^{-1}.$$

Hence we conclude the norm equivalence inequalities for $m = 1$. For $m > 1$, the assertion follows by recursive applications of the above estimates. □

In the three–dimensional case $d = 3$ the reference element $\tau$ is given by the tetrahedron

$$\tau = \left\{\xi \in \mathbb{R}^3 : 0 \le \xi_1 \le 1, 0 \le \xi_2 \le 1 - \xi_1, 0 \le \xi_3 \le 1 - \xi_1 - \xi_2\right\}. \tag{9.8}$$

For $x \in \tau_\ell$ we then have the local parametrization

$$x = x_{\ell_1} + \sum_{i=1}^{3} \xi_i (x_{\ell_{i+1}} - x_{\ell_1}) = x_{\ell_1} + J_\ell \xi \quad \text{for } \xi \in \tau$$

with the Jacobian

$$J_\ell = \begin{pmatrix} x_{\ell_2,1} - x_{\ell_1,1} & x_{\ell_3,1} - x_{\ell_1,1} & x_{\ell_4,1} - x_{\ell_1,1} \\ x_{\ell_2,2} - x_{\ell_1,2} & x_{\ell_3,2} - x_{\ell_1,2} & x_{\ell_4,2} - x_{\ell_1,2} \\ x_{\ell_2,3} - x_{\ell_1,3} & x_{\ell_3,3} - x_{\ell_1,3} & x_{\ell_4,3} - x_{\ell_1,3} \end{pmatrix}.$$

For the volume of the finite element $\tau_\ell$ we find

$$\begin{aligned} \Delta_\ell &= \int_{\tau_\ell} ds_x = \int_\tau |\det J_\ell|\, d\xi \\ &= |\det J_\ell| \int_0^1 \int_0^{1-\xi_1} \int_0^{1-\xi_1-\xi_2} d\xi_3 d\xi_2 d\xi_1 = \frac{1}{6}\, |\det J_\ell| \end{aligned} \tag{9.9}$$

and therefore

$$|\det J_\ell| = 6\, \Delta_\ell. \tag{9.10}$$

As for the two–dimensional case we can write a function $v(x)$ for $x \in \tau_\ell$ as

$$v(x) = v(x_{\ell_1} + J_\ell \xi) = \widetilde{v}_\ell(\xi) \quad \text{for } \xi \in \tau.$$

Again, the application of the chain rule gives

$$\nabla_\xi \widetilde{v}_\ell(\xi) = J_\ell^\top \nabla_x v(x), \quad \nabla_x v(x) = J_\ell^{-\top} \nabla_\xi \widetilde{v}_\ell(\xi)$$

and in analogy to Lemma 9.1 we have:

**Lemma 9.2.** *For $d = 3$ and $m \in \mathbb{N}_0$ there hold the norm equivalence inequalities*

$$c_1 \,\Delta_\ell \, h_\ell^{-2m} \, \|\nabla_\xi^m \widetilde{v}_\ell\|^2_{L_2(\tau)} \,\leq\, \|\nabla_x^m v\|^2_{L_2(\tau_\ell)} \,\leq\, c_2 \,\Delta_\ell \, h_\ell^{-2m} \, \|\nabla_\xi^m \widetilde{v}_\ell\|^2_{L_2(\tau)}$$

*with positive constants $c_1$ and $c_2$ which may depend on $m$ and on $c_F$.*

*Proof.* For $m = 0$ a direct computation gives

$$\|v(x)\|^2_{L_2(\tau_\ell)} \;=\; \int_{\tau_\ell} |v(x)|^2 dx \;=\; \int_{\tau} |\widetilde{v}_\ell(\xi)|^2 |\det J_\ell| \, d\xi \;=\; 6\Delta_\ell \, \|\widetilde{v}_\ell\|^2_{L_2(\tau)}.$$

For $m = 1$ we first obtain, as in the proof of Lemma 9.1, the equivalence inequalities

$$\begin{aligned} 6\Delta_\ell \, \lambda_{\min}\,(J_\ell^{-1} J_\ell^{-\top}) \, \|\nabla_\xi \widetilde{v}_\ell\|^2_{L_2(\tau)} &\leq \|\nabla_x v\|^2_{L_2(\tau_\ell)} \\ &\leq 6\Delta_\ell \, \lambda_{\max}\,(J_\ell^{-1} J_\ell^{-\top}) \, \|\nabla_\xi \widetilde{v}_\ell\|^2_{L_2(\tau)}. \end{aligned}$$

Hence we have to estimate the eigenvalues of the symmetric and positive definite matrix

$$J_\ell^\top J_\ell \;=\; \begin{pmatrix} a^2 & ab\cos\alpha & ac\cos\beta \\ ab\cos\alpha & b^2 & bc\cos\gamma \\ ac\cos\beta & bc\cos\gamma & c^2 \end{pmatrix}$$

where

$$a := |x_{\ell_2} - x_{\ell_1}|, \quad b := |x_{\ell_3} - x_{\ell_1}|, \quad c := |x_{\ell_4} - x_{\ell_1}|$$

and

$$\begin{aligned} \alpha &:= \sphericalangle(x_{\ell_2} - x_{\ell_1}, x_{\ell_3} - x_{\ell_1}), \\ \beta &:= \sphericalangle(x_{\ell_2} - x_{\ell_1}, x_{\ell_4} - x_{\ell_1}), \\ \gamma &:= \sphericalangle(x_{\ell_3} - x_{\ell_1}, x_{\ell_4} - x_{\ell_1}). \end{aligned}$$

From $0 < \lambda_i$ for $i = 1, 2, 3$ and

$$\lambda_1 + \lambda_2 + \lambda_3 \;=\; a^2 + b^2 + c^2$$

we can estimate the maximal eigenvalue by

$$\lambda_{\max}\,(J_\ell^\top J_\ell) \;\leq\; a^2 + b^2 + c^2 \,.$$

The product of all eigenvalues can be written by using (9.10) as

$$\lambda_1 \lambda_2 \lambda_3 \;=\; \det(J_\ell^\top J_\ell) \;=\; |\det J_\ell|^2 \;=\; 36\,\Delta_\ell^2$$

and hence we obtain an estimate for the minimal eigenvalue

$$\lambda_{\min}\,(J_\ell^\top J_\ell) \;\geq\; \frac{36\Delta_\ell^2}{[\lambda_{\max}\,(J_\ell^\top J_\ell)]^2} \;\geq\; \frac{36\Delta_\ell^2}{[a^2 + b^2 + c^2]^2}.$$

Altogether we therefore have

$$\frac{36\Delta_\ell^2}{[a^2+b^2+c^2]^2} \leq \lambda_{\min}(J_\ell^\top J_\ell) \leq \lambda_{\max}(J_\ell^\top J_\ell) \leq a^2+b^2+c^2.$$

Since the finite element $\tau_\ell$ is assumed to be shape regular, we can estimate the length of all edges by

$$a^2+b^2+c^2 \leq 3d_\ell^2 \leq 3c_F^2 r_\ell^2 \leq 3\sqrt[3]{\frac{9}{16\pi^2}}\, c_F^2\, h_\ell^2$$

and hence we obtain

$$\frac{4}{c_F^4}\sqrt[3]{\frac{256\pi^4}{81}}\, h_\ell^2 \leq \lambda_{\min}(J_\ell^\top J_\ell) \leq \lambda_{\max}(J_\ell^\top J_\ell) \leq 3\sqrt[3]{\frac{9}{16\pi^2}}\, c_F^2\, h_\ell^2.$$

The assertion now follows as in the proof of Lemma 9.1. □

By using the norm equivalence estimates (9.4) for $d=1$ as well as Lemma 9.1 for $d=2$ and Lemma 9.2 for $d=3$ we can formulate the following result:

**Theorem 9.3.** *Let $\tau_\ell \subset \mathbb{R}^d$ be a finite element of a shape regular and admissible decomposition $\mathcal{T}_N$. If $v$ is sufficiently smooth we then have for $m \in \mathbb{N}_0$*

$$c_1\, \Delta_\ell\, h_\ell^{-2m}\, \|\nabla_\xi^m \widetilde{v}_\ell\|_{L_2(\tau)}^2 \leq \|\nabla_x^m v\|_{L_2(\tau_\ell)}^2 \leq c_2\, \Delta_\ell\, h_\ell^{-2m}\, \|\nabla_\xi^m \widetilde{v}_\ell\|_{L_2(\tau)}^2$$

*with positive constants $c_1$ and $c_2$ which may depend on $m$ and on $c_F$.*

## 9.2 Form Functions

With respect to the decomposition $\mathcal{T}_N$ as defined in (9.1) we now introduce trial spaces of piecewise polynomial functions. The related basis functions, which are associated to global degrees of freedom, are defined locally by using suitable form functions which are formulated with respect to an element $\tau_\ell$.

We consider a reference element $\tau$ which is either an interval (9.3) for $d=1$, a triangle (9.5) for $d=2$, or a tetrahedron (9.8) for $d=3$.

The simplest form functions are the constant functions

$$\psi_1^0(\xi) = 1 \quad \text{for } \xi \in \tau.$$

If we consider a function $v_h(x)$ which is constant for $x \in \tau_\ell$ we then have the representation

$$v_h(x) = v_h(x_{\ell_1} + J_\ell \xi) = v_\ell\, \psi_1^0(\xi) \quad \text{for } x \in \tau_\ell, \xi \in \tau, \tag{9.11}$$

where $v_\ell$ is the associated coefficient describing the value of $v_h$ on $\tau_\ell$. Moreover, we have

$$\|v_h\|_{L_2(\tau_\ell)}^2 = \Delta_\ell\, v_\ell^2\,. \tag{9.12}$$

If we consider a function $v_h(x)$ which is linear for $x \in \tau_\ell$, then this function is uniquely determined by the values $\widetilde{v}_k$ at the nodes of the reference element $\tau$,

$$\widetilde{v}_h(\xi) = \sum_{k=1}^{d+1} \widetilde{v}_k \psi_k^1(\xi) \quad \text{for } \xi \in \tau. \tag{9.13}$$

Here, the linear form functions are given for $d = 1$

$$\psi_1^1(\xi) := 1 - \xi, \quad \psi_2^1(\xi) := \xi,$$

for $d = 2$

$$\psi_1^1(\xi) := 1 - \xi_1 - \xi_2, \quad \psi_2^1(\xi) := \xi_1, \quad \psi_3^1(\xi) := \xi_2,$$

and for $d = 3$

$$\psi_1^1(\xi) := 1 - \xi_1 - \xi_2 - \xi_3, \quad \psi_2^1(\xi) := \xi_1, \quad \psi_3^1(\xi) := \xi_2, \quad \psi_4^1(\xi) := \xi_3.$$

Let $\tau_\ell$ be an arbitrary finite element with nodes $x_{\ell_k}$, $\ell_k \in J(\ell)$. If $v_h$ is a linear function on $\tau_\ell$, then we can write

$$v_h(x) = v_h(x_{\ell_1} + J_\ell \xi) = \sum_{k=1}^{d+1} v_{\ell_k} \psi_k^1(\xi) \quad \text{for } x \in \tau_\ell, \xi \in \tau. \tag{9.14}$$

As in (9.12) we can estimate the $L_2$ norm $\|v_h\|_{L_2(\tau_\ell)}$ by the Euclidean norm of the nodal values.

**Lemma 9.4.** *Let $v_h$ be a linear function as given in* (9.14). *Then,*

$$\frac{\Delta_\ell}{(d+1)(d+2)} \sum_{k=1}^{d+1} v_{\ell_k}^2 \leq \|v_h\|_{L_2(\tau_\ell)}^2 \leq \frac{\Delta_\ell}{d+1} \sum_{k=1}^{d+1} v_{\ell_k}^2.$$

*Proof.* We can compute the local $L_2$ norm of the linear function $v_h$ as

$$\|v_h\|_{L_2(\tau_\ell)}^2 = \langle v_h, v_h \rangle_{L_2(\tau_\ell)} = \sum_{i=1}^{d+1} \sum_{j=1}^{d+1} v_i v_j \int_\tau \psi_i(\xi) \psi_j(\xi) |\det J_\ell| d\xi = (G_\ell \underline{v}^\ell, \underline{v}^\ell)$$

where

$$G_\ell = \frac{\Delta_\ell}{(d+1)(d+2)} (I_{d+1} + \underline{e}_{d+1} \underline{e}_{d+1}^\top)$$

is the local mass matrix and $\underline{e}_{d+1} = \underline{1} \in \mathbb{R}^{d+1}$. From the eigenvalues of the matrix $I_{d+1} + \underline{e}_{d+1} \underline{e}_{d+1}^\top$,

$$\lambda_1 = d + 2, \quad \lambda_2 = \dots = \lambda_{d+1} = 1,$$

the assertion follows. $\square$

**Corollary 9.5.** *Let $v_h$ be a linear function as given in* (9.14). *Then,*

$$\frac{\Delta_\ell}{(d+1)(d+2)}\,\|v_h\|^2_{L_\infty(\tau_\ell)} \le \|v_h\|^2_{L_2(\tau_\ell)} \le \Delta_\ell\,\|v_h\|^2_{L_\infty(\tau_\ell)}.$$

*Proof.* Obviously, the maximal value of $v_h$ and therefore the maximum norm $\|v_h\|_{L_\infty(\tau_\ell)}$ is equal some nodal value $v_{k^*} = v_h(x_{k^*})$. The assertion then follows from Lemma 9.4. □

In many applications it is essential to bound the norm of the gradient of a piecewise polynomial function by the norm of this function itself.

**Lemma 9.6.** *Let $v_h$ be a linear function as given in* (9.14). *Then there holds the local inverse inequality*

$$\|\nabla_x v_h\|_{L_2(\tau_\ell)} \le c_I\, h_\ell^{-1}\,\|v_h\|_{L_2(\tau_\ell)} \tag{9.15}$$

*where $c_I$ is some positive constant.*

*Proof.* The application of Theorem 9.3 gives first

$$\|\nabla_x v_h\|^2_{L_2(\tau_\ell)} \le c_2\,\Delta_\ell\,h_\ell^{-2}\,\|\nabla_\xi \widetilde{v}_\ell\|^2_{L_2(\tau)}.$$

To compute the gradient of the linear function

$$\widetilde{v}_\ell(\xi) = \sum_{k=1}^{d+1} v_{\ell_k}\psi^1_k(\xi)$$

we obtain for $d = 1$

$$\nabla_\xi \widetilde{v}_\ell = v_{\ell_2} - v_{\ell_1}$$

and therefore

$$\|\nabla_\xi \widetilde{v}_\ell\|^2_{L_2(\tau)} = (v_{\ell_2} - v_{\ell_1})^2 \le 2\,[v^2_{\ell_1} + v^2_{\ell_2}] \le 4\,\|v_h\|^2_{L_\infty(\tau_\ell)}.$$

In the two–dimensional case $d = 2$ the gradient is

$$\nabla_\xi \widetilde{v}_\ell = \begin{pmatrix} v_{\ell_2} - v_{\ell_1} \\ v_{\ell_3} - v_{\ell_1} \end{pmatrix}$$

and therefore we obtain

$$\begin{aligned}\|\nabla_\xi \widetilde{v}_\ell\|^2_{L_2(\tau)} &= \frac{1}{2}\left[(v_{\ell_2} - v_{\ell_1})^2 + (v_{\ell_3} - v_{\ell_1})^2\right] \\ &\le \frac{1}{2}\left[2v^2_{\ell_2} + 2v^2_{\ell_3} + 4v^2_{\ell_1}\right] \le 4\,\|v_h\|^2_{L_\infty(\tau_\ell)}.\end{aligned}$$

Finally, for $d = 3$ we have

$$\nabla_\xi \widetilde{v}_\ell = \begin{pmatrix} v_{\ell_2} - v_{\ell_1} \\ v_{\ell_3} - v_{\ell_1} \\ v_{\ell_4} - v_{\ell_1} \end{pmatrix}$$

and thus

$$\begin{aligned}\|\nabla_\xi \widetilde{v}_\ell\|^2_{L_2(\tau)} &= \frac{1}{6}\left[(v_{\ell_2} - v_{\ell_1})^2 + (v_{\ell_3} - v_{\ell_1})^2 + (v_{\ell_4} - v_{\ell_1})^2\right] \\ &\le \frac{1}{6}\left[2v^2_{\ell_2} + 2v^2_{\ell_3} + 2v^2_{\ell_4} + 6v^2_{\ell_1}\right] \le 2\,\|v_h\|^2_{L_\infty(\tau_\ell)}.\end{aligned}$$

Altogether we therefore have

$$\|\nabla_x v_h\|^2_{L_2(\tau_\ell)} \le 4\,c_2\,\Delta_\ell\,h_\ell^{-2}\,\|v_h\|^2_{L_\infty(\tau_\ell)}$$

and the inverse inequality now follows from Corollary 9.5. □

Form functions of locally higher polynomial degree can be defined hierarchically based on piecewise linear form functions. We define quadratic form functions for $d = 1$ by

$$\psi^2_1(\xi) = 1 - \xi, \quad \psi^2_2(\xi) = \xi, \quad \psi^2_3(\xi) = 4\xi(1 - \xi),$$

for $d = 2$ by

$$\begin{aligned}&\psi^2_1(\xi) = 1 - \xi_1 - \xi_2, \qquad \psi^2_2(\xi) = \xi_1, \qquad \psi^2_3(\xi) = \xi_2, \\ &\psi^2_4(\xi) = 4\xi_1(1 - \xi_1 - \xi_2),\, \psi^2_5(\xi) = 4\xi_1\xi_2,\, \psi^2_6(\xi) = 4\xi_2(1 - \xi_1 - \xi_2),\end{aligned}$$

and for $d = 3$ by

$$\begin{aligned}&\psi^2_1(\xi) = 1 - \xi_1 - \xi_2 - \xi_3, && \psi^2_5(\xi) = 4\xi_1(1 - \xi_1 - \xi_2 - \xi_3), \\ &\psi^2_2(\xi) = \xi_1, && \psi^2_6(\xi) = 4\xi_1\xi_2, \\ &\psi^2_3(\xi) = \xi_2, && \psi^2_7(\xi) = 4\xi_2(1 - \xi_1 - \xi_2 - \xi_3), \\ &\psi^2_4(\xi) = \xi_3, && \psi^2_8(\xi) = 4\xi_3(1 - \xi_1 - \xi_2 - \xi_3), \\ & && \psi^2_9(\xi) = 4\xi_3\xi_1, \\ & && \psi^2_{10}(\xi) = 4\xi_2\xi_3.\end{aligned}$$

Note that linear form functions are associated to degrees of freedom at the nodes $x_k \in \overline{\tau}_\ell$, while the quadratic form functions are associated to the edge mid points $x^*_{k_j}$. If the function $v_h$ is quadratic on $\tau_\ell$ then we can write

$$v_h(x) = v_h(x_{\ell_1} + J_\ell\xi) = \sum_{k=1}^{\frac{1}{2}(d+1)(d+2)} v_{\ell_k}\psi^2_k(\xi) \quad \text{for } x \in \tau_\ell, \xi \in \tau. \tag{9.16}$$

As in the proof of Lemma 9.4 we have

$$\|v_h\|^2_{L_2(\tau_\ell)} = \sum_{i,j=1}^{\frac{1}{2}(d+1)(d+2)} v_i v_j \int_\tau \psi^2_i(\xi)\psi^2_j(\xi)|\det J_\ell|d\xi = (G_\ell \underline{v}^\ell, \underline{v}^\ell).$$

In particular for $d = 1$ the local mass matrix is

$$G_\ell = \Delta_\ell \begin{pmatrix} 1/3 & 1/6 & 1/3 \\ 1/6 & 1/3 & 1/3 \\ 1/3 & 1/3 & 8/15 \end{pmatrix}$$

where the eigenvalues of $G_\ell$ are

$$\lambda_1 = \frac{\Delta_\ell}{6}, \quad \lambda_{2/3} = \Delta_\ell \left[ \frac{31}{60} \pm \frac{\sqrt{89}}{20} \right].$$

By using a similar approach in the two–dimensional case $d = 2$ as well as in the three–dimensional case $d = 3$ we can prove equivalence estimates which correspond to the results of Lemma 9.4,

$$c_1 \, \Delta_\ell \sum_{k=1}^{\frac{1}{2}(d+1)(d+2)} v_{\ell_k}^2 \le \|v_h\|^2_{L_2(\tau_\ell)} \le c_2 \, \Delta_\ell \sum_{k=1}^{\frac{1}{2}(d+1)(d+2)} v_{\ell_k}^2 \tag{9.17}$$

where $v_h$ is a quadratic function as defined in (9.16). Moreover, as for linear functions also the inverse inequality (9.15) remains valid for quadratic functions.

Finally we will discuss bubble functions $\varphi_\ell^B$ and their associated form functions $\psi_B$ which are needed, for example, for a stable discretization of the Stokes problem. The basis functions $\varphi_\ell^B$ are polynomial in the finite element $\tau_\ell$, and zero on the element boundary $\partial\tau_\ell$. Hence we can extend them by zero outside of the finite element $\tau_\ell$. Later we will make use of an inverse inequality for the induced trial space $S_h^B(\mathcal{T}_N)$ which is spanned by the bubble functions. For $d = 1$ we have the form function

$$\psi_B(\xi) = \xi(1-\xi) \quad \text{for } \xi \in \tau$$

and for the associated basis function $\varphi_\ell^B$ it follows that

$$\|\varphi_\ell^B\|^2_{L_2(\tau_\ell)} = \Delta_\ell \int_0^1 [\xi(1-\xi)]^2 d\xi = \frac{1}{30} \, h_\ell .$$

Moreover,

$$\|\nabla_x \varphi_\ell^B\|^2_{L_2(\tau_\ell)} = h_\ell^{-1} \int_0^1 \left[ \frac{d}{d\xi} [\xi(1-\xi)] \right]^2 d\xi = \frac{1}{3} h_\ell^{-1}.$$

Hence we conclude for the one–dimensional bubble function the local inverse inequality

$$\|\nabla_x \varphi_\ell^B\|_{L_2(\tau_\ell)} = \sqrt{10} \, h_\ell^{-1} \, \|\varphi_\ell^B\|_{L_2(\tau_\ell)}.$$

In the two–dimensional case the form function $\psi_B$ reads

$$\psi_B = \xi_1\xi_2(1-\xi_1-\xi_2) \quad \text{for } \xi \in \tau$$

and for the associated basis function $\varphi_\ell^B$ it follows that

$$\|\varphi_\ell^B\|^2_{L_2(\tau_\ell)} = 2\Delta_\ell \int_\tau [\xi_1\xi_2(1-\xi_1-\xi_2)]^2 d\xi = \frac{1}{2520}\Delta_\ell.$$

Then, by using Lemma 9.1 we conclude

$$\|\nabla_x \varphi_\ell^B\|^2_{L_2(\tau_\ell)} \le c\,\|\nabla_\xi \psi_B\|^2_{L_2(\tau)} = c \int_\tau \left| \begin{pmatrix} \xi_2(1-\xi_2-2\xi_1) \\ \xi_1(1-\xi_1-2\xi_2) \end{pmatrix} \right|^2 d\xi = \frac{c}{90}$$

and therefore we obtain the local inverse inequality

$$\|\nabla_x \varphi_\ell^B\|_{L_2(\tau_\ell)} \le \tilde{c}\, h_\ell^{-1}\, \|\varphi_\ell^B\|_{L_2(\tau_\ell)}. \tag{9.18}$$

In the three–dimensional case $d = 3$ we finally have

$$\psi(\xi) = \xi_1\xi_2\xi_3(1-\xi_1-\xi_2-\xi_3)$$

and therefore

$$\|v_\ell^B\|_{L_2(\tau_\ell)} = 6\Delta_\ell \int_\tau [\psi_B(\xi)]^2 d\xi = \frac{1}{415800}\Delta_\ell$$

as well as

$$\|\nabla_x v_\ell^B\|^2_{L_2(\tau_\ell)} \le c\,\Delta_\ell\, h_\ell^{-2}\, \|\nabla_\xi \psi_B\|^2_{L_2(\tau)} = \frac{c}{15120}\,\Delta_\ell\, h_\ell^{-2},$$

in particular we conclude the local inverse inequality (9.18) also for $d = 3$.

## 9.3 Trial Spaces

The standard trial space to construct an approximate solution of boundary value problems with second order partial differential equations is the space $S_h^1(\mathcal{T}_N)$ of piecewise linear and globally continuous functions. When considering an admissible decomposition (9.1) those functions are uniquely determined by the nodal function values $v_k = v_h(x_k)$ which are given at the nodes $x_k$ of the decomposition. Therefore, in the finite element $\tau_\ell$ we then have a local representation by using local form functions. The dimension $\dim S_h^1(\mathcal{T}_N) = M$ of the global trial space $S_h^1(\mathcal{T}_N)$ is obviously equal to the number of nodes in the decomposition. A basis of the trial space $S_h^1(\mathcal{T}_N)$ is given by, see Fig. 9.4,

$$\varphi_k^1(x) := \begin{cases} 1 & \text{for } x = x_k, \\ 0 & \text{for } x = x_\ell \ne x_k, \\ \text{linear} & \text{elsewhere.} \end{cases}$$

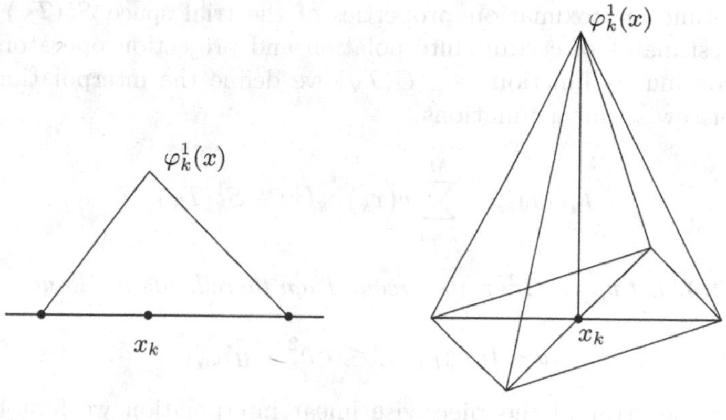


**Fig. 9.4.** Linear basis functions for $d = 1, 2$.

If $v_h \in S_h^1(\mathcal{T}_N)$ is piecewise linear, then we can write

$$v_h(x) = \sum_{k=1}^{M} v_k \varphi_k^1(x).$$

**Lemma 9.7.** *For $v_h \in S_h^1(\mathcal{T}_N)$ there hold the spectral equivalence inequalities*

$$\frac{1}{(d+1)(d+2)} \sum_{k=1}^{M} \left( \sum_{\ell \in I(k)} \Delta_\ell \right) v_k^2 \leq \|v_h\|^2_{L_2(\mathcal{T}_N)} \leq \frac{1}{d+1} \sum_{k=1}^{M} \left( \sum_{\ell \in I(k)} \Delta_\ell \right) v_k^2 .$$

*Proof.* By using Lemma 9.4 we have with

$$\|v_h\|^2_{L_2(\mathcal{T}_N)} = \sum_{\ell=1}^{N} \|v_h\|^2_{L_2(\tau_\ell)} \leq \sum_{\ell=1}^{N} \frac{\Delta_\ell}{d+1} \sum_{k=1}^{d+1} v_{\ell_k}^2 = \frac{1}{d+1} \sum_{k=1}^{M} \left( \sum_{\ell \in I(k)} \Delta_\ell \right) v_k^2$$

the upper estimate. The lower estimate follows in the same way. □

**Lemma 9.8.** *For a piecewise linear function $v_h \in S_h^1(\mathcal{T}_N)$ there holds the inverse inequality*

$$\|\nabla_x v_h\|^2_{L_2(\mathcal{T}_N)} \leq c_I \sum_{\ell=1}^{N} h_\ell^{-2} \|v_h\|^2_{L_2(\tau_\ell)}.$$

*If the decomposition $\mathcal{T}_N$ is globally quasi–uniform, then we have*

$$\|\nabla_x v_h\|_{L_2(\mathcal{T}_N)} \leq c\, h^{-1} \|v_h\|_{L_2(\mathcal{T}_N)}. \tag{9.19}$$

*Proof.* Both estimates follow immediately from Lemma 9.6. □

To prove some approximation properties of the trial space $\mathcal{S}_h^1(\mathcal{T}_N)$ we will use error estimates of certain interpolation and projection operators. For a globally continuous function $v \in C(\mathcal{T}_N)$ we define the interpolation in the space of piecewise linear functions,

$$I_h v(x) := \sum_{k=1}^{M} v(x_k)\varphi_k(x) \in \mathcal{S}_h^1(\mathcal{T}_N). \tag{9.20}$$

**Lemma 9.9.** *Let $v_{|\tau_\ell} \in H^2(\tau_\ell)$be given. Then there holds the local error estimate*

$$\|v - I_h v\|_{L_2(\tau_\ell)} \leq c\, h_\ell^2\, |v|_{H^2(\tau_\ell)}.$$

*Proof.* For the error of the piecewise linear interpolation we first have, by using the norm equivalence inequalities of Theorem 9.3,

$$\|v - I_h v\|_{L_2(\tau_\ell)} \leq c\,\Delta_\ell\, \|\widetilde{v}_\ell - I_\tau \widetilde{v}_\ell\|_{L_2(\tau)},$$

where $I_\tau$ is the linear interpolation operator with respect to the reference element $\tau$. Then,

$$\|I_\tau \widetilde{v}_\ell\|_{L_2(\tau)} \leq \text{meas}\,(\tau)\, \|\widetilde{v}_\ell\|_{L_\infty(\tau)}$$

and the use of the Sobolev imbedding theorem (Theorem 2.5) gives

$$\|\widetilde{v}_\ell\|_{L_\infty(\tau)} \leq c\, \|\widetilde{v}_\ell\|_{H^2(\tau)}\,.$$

Therefore we conclude that the linear operator

$$I_\tau \,:\, H^2(\tau) \to L_2(\tau)$$

is bounded. For an arbitrary but fixed $w \in L_2(\tau)$ we define the linear functional

$$f(u) \,:=\, \int_\tau [(I - I_\tau)u(\xi)]w(\xi)d\xi\,.$$

If $u \in H^2(\tau)$ is given, then we have

$$|f(u)| = \left|\int_\tau [(I - I_\tau)u(\xi)]w(\xi)d\xi\right|$$
$$\leq \|(I - I_\tau)u\|_{L_2(\tau)}\|w\|_{L_2(\tau)} \leq c\,\|u\|_{H^2(\tau)}\|w\|_{L_2(\tau)}\,.$$

For any linear function $q \in P_1(\tau)$ we have $I_\tau q = q$ and therefore $f(q) = 0$ for all $q \in P_1(\tau)$. Thus, all assumptions of the Bramble–Hilbert lemma (Theorem 2.8) are satisfied implying

$$|f(u)| \leq \widetilde{c}\,\|w\|_{L_2(\tau)}|u|_{H^2(\tau)}\,.$$

When choosing $u := \widetilde{v}_\ell$ and $w := (I - I_\tau)\widetilde{v}_\ell$ we obtain

$$\begin{aligned}\|(I-I_\tau)\widetilde{v}_\ell\|^2_{L_2(\tau)} &= \int_\tau [(I-I_\ell)\widetilde{v}_\ell(\xi)]^2 d\xi = \int_\tau [(I-I_\ell)\widetilde{v}_\ell]w(\xi)d\xi = |f(\widetilde{v}_\ell)| \\ &\le \widetilde{c}\,\|w\|_{L_2(\tau)}|\widetilde{v}_\ell|_{H^2(\tau)} \le \widetilde{c}\,\|(I-I_\tau)\widetilde{v}_\ell\|_{L_2(\tau)}|\widetilde{v}_\ell|_{H^2(\tau)}\end{aligned}$$

and hence the estimate

$$\|(I-I_\tau)\widetilde{v}_\ell\|_{L_2(\tau)} \le \widetilde{c}\,|\widetilde{v}_\ell|_{H^2(\tau)}$$

follows. Altogether we therefore have

$$\|v - I_h v\|_{L_2(\tau)} \le c\,\Delta_\ell\,|\widetilde{v}_\ell|_{H^2(\tau)} \le \hat{c}\,h_\ell^2\,|v|_{H^2(\tau_\ell)}$$

by applying the norm equivalence theorem (Theorem 9.3). □

As a direct consequence of the above we conclude the global error estimate

$$\|v - I_h v\|^2_{L_2(\mathcal{T}_N)} \le c \sum_{\ell=1}^N h_\ell^4\,|v|^2_{H^2(\tau_\ell)}\,. \tag{9.21}$$

In the same way we obtain also the error estimate

$$\|v - I_h v\|^2_{H^1(\mathcal{T}_N)} \le c \sum_{\ell=1}^N h_\ell^2\,|v|^2_{H^2(\tau_\ell)}\,. \tag{9.22}$$

The application of the interpolation operator requires the global continuity of the given function to be interpolated. To weaken this strong assumption we now consider projection operators which are defined via variational problems. For a given $u \in L_2(\mathcal{T}_N)$ we define the $L_2$ projection $Q_h u \in S_h^1(\mathcal{T}_N)$ as the unique solution of the variational problem

$$\langle Q_h u, v_h\rangle_{L_2(\mathcal{T}_N)} = \langle u, v_h\rangle_{L_2(\mathcal{T}_N)} \quad \text{for all } v_h \in S_h^1(\mathcal{T}_N). \tag{9.23}$$

When choosing $v_h = Q_h u$ as a test function we obtain the stability estimate

$$\|Q_h u\|_{L_2(\mathcal{T}_N)} \le \|u\|_{L_2(\mathcal{T}_N)} \quad \text{for all } u \in L_2(\mathcal{T}_N), \tag{9.24}$$

and by using the Galerkin orthogonality

$$\langle u - Q_h u, v_h\rangle_{L_2(\mathcal{T}_N)} = 0 \quad \text{for all } v_h \in S_h^1(\mathcal{T}_N) \tag{9.25}$$

we conclude

$$\begin{aligned}\|u - Q_h u\|^2_{L_2(\mathcal{T}_N)} &= \langle u - Q_h u, u - Q_h u\rangle_{L_2(\mathcal{T}_N)} \\ &= \langle u - Q_h u, u\rangle_{L_2(\mathcal{T}_N)} \\ &\le \|u - Q_h u\|_{L_2(\mathcal{T}_N)}\|u\|_{L_2(\mathcal{T}_N)}\end{aligned}$$

and therefore

$$\|u - Q_h u\|_{L_2(\mathcal{T}_N)} \leq \|u\|_{L_2(\mathcal{T}_N)} \quad \text{for all } u \in L_2(\mathcal{T}_N). \tag{9.26}$$

On the other hand, again by using the Galerkin orthogonality (9.25) we have

$$\begin{aligned} \|u - Q_h u\|^2_{L_2(\mathcal{T}_N)} &= \langle u - Q_h u, u - Q_h u \rangle_{L_2(\mathcal{T}_N)} \\ &= \langle u - Q_h u, u - I_h u \rangle_{L_2(\mathcal{T}_N)} \\ &\leq \|u - Q_h u\|_{L_2(\mathcal{T}_N)} \|u - I_h u\|_{L_2(\mathcal{T}_N)} \end{aligned}$$

and therefore the error estimate

$$\|u - Q_h u\|^2_{L_2(\mathcal{T}_N)} \leq \|u - I_h u\|^2_{L_2(\mathcal{T}_N)} \leq c \sum_{\ell=1}^{N} h_\ell^4 \, |v|^2_{H^2(\tau_\ell)} \tag{9.27}$$

as well as

$$\|u - Q_h u\|_{L_2(\mathcal{T}_N)} \leq c\, h^2 \, \|v\|_{H^2(\mathcal{T}_N)}.$$

By interpolating this estimate with the error estimate (9.26) this yields the error estimate

$$\|u - Q_h u\|_{L_2(\mathcal{T}_N)} \leq c\, h \, \|v\|_{H^1(\mathcal{T}_N)}. \tag{9.28}$$

By $Q_h^1 : H^1(\mathcal{T}_N) \to S_h^1(\mathcal{T}_N)$ we denote the $H^1$ projection which is defined as the unique solution of the variational problem

$$\langle Q_h^1 u, v_h \rangle_{H^1(\mathcal{T}_N)} = \langle u, v_h \rangle_{H^1(\mathcal{T}_N)} \quad \text{for all } v_h \in S_h^1(\mathcal{T}_N). \tag{9.29}$$

As above we find the stability estimate

$$\|Q_h^1 u\|_{H^1(\mathcal{T}_N)} \leq \|u\|_{H^1(\mathcal{T}_N)} \quad \text{for all } u \in H^1(\mathcal{T}_N) \tag{9.30}$$

and the error estimate

$$\|u - Q_h^1 u\|_{H^1(\mathcal{T}_N)} \leq \|u\|_{H^1(\mathcal{T}_N)} \tag{9.31}$$

as well as

$$\|u - Q_h^1 u\|^2_{H^1(\mathcal{T}_N)} \leq \|u - I_h u\|^2_{H^1(\mathcal{T}_N)} \leq c \sum_{\ell=1}^{N} h_\ell^2 \, |u|^2_{H^2(\tau_\ell)} \,. \tag{9.32}$$

Hence we obtain the approximate property of the trial space $S_h^1(\mathcal{T}_N)$ of piecewise linear and continuous functions.

**Theorem 9.10.** *Let $u \in H^s(\mathcal{T}_N)$ with $s \in [\sigma, 2]$ and $\sigma = 0, 1$. Then there holds the approximation property*

$$\inf_{v_h \in S_h^1(\mathcal{T}_N)} \|u - v_h\|_{H^\sigma(\mathcal{T}_N)} \leq c\, h^{s-\sigma} \, |u|_{H^s(\mathcal{T}_N)}. \tag{9.33}$$

*Proof.* For $\sigma = 0$ and $s = 2$ the assertion is a direct consequence of the error estimate (9.27). For $s = 0$ the approximation property is just the error estimate (9.26). For $s \in (0, 2)$ we then apply the interpolation theorem (Theorem 2.18). For $\sigma = 1$ we use the error estimates (9.32) and (9.31) to obtain the result in the same way. □

In what follows we will investigate further properties of the $H^1$ projection $Q_h^1$ which are needed later on.

**Lemma 9.11.** *For $s \in (0, 1]$ let $w \in H_0^1(\mathcal{T}_N)$ be the uniquely determined solution of the variational problem*

$$\langle w, v\rangle_{H^1(\mathcal{T}_N)} = \langle u - Q_h^1 u, v\rangle_{H^{1-s}(\mathcal{T}_N)} \quad \textit{for all } v \in H^1(\mathcal{T}_N). \tag{9.34}$$

*If we assume $w \in H^{1+s}(\mathcal{T}_N)$ satisfying*

$$\|w\|_{H^{1+s}(\mathcal{T}_N)} \le c\,\|u - Q_h^1 u\|_{H^{1-s}(\mathcal{T}_N)},$$

*then there holds the error estimate*

$$\|u - Q_h^1 u\|_{H^{1-s}(\mathcal{T}_N)} \le c\,h^s\,\|u - Q_h^1 u\|_{H^1(\mathcal{T}_N)}\,. \tag{9.35}$$

*Proof.* Using the assumptions we conclude

$$\begin{aligned}
\|u - Q_h^1 u\|^2_{H^{1-s}(\mathcal{T}_N)} &= \langle u - Q_h^1 u, u - Q_h^1 u\rangle_{H^{1-s}(\mathcal{T}_N)} \\
&= \langle w, u - Q_h^1 u\rangle_{H^1(\mathcal{T}_N)} \\
&= \langle w - Q_h^1 w, u - Q_h^1 u\rangle_{H^1(\mathcal{T}_N)} \\
&\le \|w - Q_h^1 w\|_{H^1(\mathcal{T}_N)} \|u - Q_h^1 u\|_{H^1(\mathcal{T}_N)} \\
&\le c\,h^s\,|w|_{H^{1+s}(\mathcal{T}_N)}\,\|u - Q_h^1 u\|_{H^1(\mathcal{T}_N)} \\
&\le \tilde{c}\,h^s\,\|u - Q_h^1 u\|_{H^{1-s}(\mathcal{T}_N)} \|u - Q_h^1 u\|_{H^1(\mathcal{T}_N)}
\end{aligned}$$

from which the error estimate follows. □

*Remark 9.12.* In Lemma 9.11, the best possible value of $s \in (0, 1]$ depends on the regularity of the decomposition $\mathcal{T}_N$. If, for example, $\mathcal{T}_N$ is convex, then we obtain $s = 1$ [66]. In the case of a corner domain, see, for example, [49].

Note that due to $S_h^1(\mathcal{T}_N) \subset H^{1+s}(\mathcal{T}_N)$ for $s \in (0, \frac{1}{2})$ the $H^1$ projection $Q_h^1 u$ is well defined also for functions $u \in H^{1-s}(\mathcal{T}_N)$ and $s \in (0, \frac{1}{2})$. As in the proof of Lemma 9.11 we then can conclude the stability estimate

$$\|Q_h^1 u\|_{H^{1-s}(\mathcal{T}_N)} \le c\,\|u\|_{H^{1-s}(\mathcal{T}_N)} \quad \text{for all } u \in H^{1-s}(\mathcal{T}_N). \tag{9.36}$$

Using the error estimates of Lemma 9.11 for $s = 1$ we can show the stability of the $L_2$ projection in $H^1(\mathcal{T}_N)$.

**Lemma 9.13.** *Let the assumptions of Lemma* 9.11 *be satisfied for* $s = 1$. *Then, the* $L_2$ *projection* $Q_h : H^1(\mathcal{T}_N) \to \mathcal{S}_h^1(\mathcal{T}_N) \subset H^1(\mathcal{T}_N)$ *is bounded, i.e.*

$$\|Q_h v\|_{H^1(\mathcal{T}_N)} \le c\,\|v\|_{H^1(\mathcal{T}_N)} \quad \textit{for all } v \in H^1(\mathcal{T}_N).$$

*Proof.* Let $Q_h^1 : H^1(\mathcal{T}_N) \to \mathcal{S}_h^1(\mathcal{T}_N) \subset H^1(\mathcal{T}_N)$ be the $H^1$ projection as defined in (9.29). By using the triangle inequality, the stability estimate (9.30), the global inverse inequality (9.19), and the projection property $Q_h v_h = v_h$ for all $v_h \in \mathcal{S}_h^1(\mathcal{T}_N)$ we obtain

$$\begin{aligned}
\|Q_h v\|_{H^1(\mathcal{T}_N)} &\le \|Q_h^1 v\|_{H^1(\mathcal{T}_N)} + \|Q_h v - Q_h^1 v\|_{H^1(\mathcal{T}_N)} \\
&\le \|v\|_{H^1(\mathcal{T}_N)} + c_I\, h^{-1}\, \|Q_h v - Q_h^1 v\|_{L_2(\mathcal{T}_N)} \\
&= \|v\|_{H^1(\mathcal{T}_N)} + c_I\, h^{-1}\, \|Q_h (u - Q_h^1 u)\|_{L_2(\mathcal{T}_N)}.
\end{aligned}$$

By applying the stability estimate (9.24) for $Q_h$ we further conclude

$$\|Q_h v\|_{H^1(\mathcal{T}_N)} \le \|v\|_{H^1(\mathcal{T}_N)} + c_I\, h^{-1}\, \|v - Q_h^1 v\|_{L_2(\mathcal{T}_N)}.$$

Now the stability estimate follows from the error estimate (9.35) for $Q_h^1$ and from the stability estimate (9.30). □

*Remark 9.14.* The $L_2$ projection $Q_h : H^1(\mathcal{T}_N) \to S_h^1(\mathcal{T}_N) \subset H^1(\mathcal{T}_N)$ is also bounded when the decomposition $\mathcal{T}_N$ is locally adaptive refined, if the ratio of local mesh sizes of neighboring elements does not vary too strongly [27].

In what follows we will always assume that the $L_2$ projection is stable in $H^1(\mathcal{T}_N)$. Then, by using an interpolation argument, it follows that the error estimate

$$\|u - Q_h u\|_{H^s(\mathcal{T}_N)} \le c\, h^{1-s}\, \|u\|_{H^1(\mathcal{T}_N)} \quad \text{for all } u \in H^1(\mathcal{T}_N) \tag{9.37}$$

is valid.

Based on the trial space of piecewise linear and globally continuous functions we can introduce trial spaces of locally higher polynomial degrees.

In the one–dimensional case $d = 1$ we can define quadratic basis functions locally by

$$\varphi_\ell^2(x) = 4\xi(1-\xi) \quad \text{for } x = x_{\ell_1} + \xi\, h_\ell \in \tau_\ell.$$

An arbitrary function $v_h \in S_h^2(\mathcal{T}_N)$ then can be written as

$$v_h(x) = \sum_{k=1}^{M} v_k \varphi_k^1(x) + \sum_{\ell=1}^{N} v_{M+\ell} \varphi_\ell^2(x).$$

Therefore we have $\dim S_h^2(\mathcal{T}_N) = M + N$. For both the two–dimensional case $d = 2$ and the three–dimensional case $d = 3$ we have to ensure the continuity of the quadratic basis functions $\varphi_\ell^2$. Since the quadratic form functions are defined locally with respect to the edges of the reference element, the support of a global quadratic basis function consists of those finite elements which share the corresponding edge. Denote by $K$ the number of all edges of the decomposition (9.1), then we can write for $d = 2, 3$ the global representation

$$v_h(x) = \sum_{k=1}^{M} v_k \varphi_k^1(x) + \sum_{j=1}^{K} v_{M+j} \varphi_j^2(x)$$

as well as the local representation

$$v_h(x) = \sum_{i=1}^{\frac{1}{2}(d+1)(d+2)} v_{\ell_i} \psi_i^2(\xi) \quad \text{for } x = x_{\ell_1} + J_\ell \xi,\ \xi \in \tau.$$

Here, $\ell_1, \ldots, \ell_{d+1}$ denote, as before, the indices of the associated global nodes, while $\ell_{d+2}, \ldots, \ell_{\frac{1}{2}(d+1)(d+2)}$ are the indices of the associated global edges, see also Fig. 9.3 for $d = 2$.

**Lemma 9.15.** *For $v_h \in S_h^2(\mathcal{T}_N)$ there hold the spectral equivalence inequalities*

$$c_1 \sum_{k=1}^{dim S_h^2(\mathcal{T}_N)} d_k v_k^2 \le \|v_h\|^2_{L_2(\mathcal{T}_N)} \le c_2 \sum_{k=1}^{dim S_h^2(\mathcal{T}_N)} d_k v_k^2$$

*with*

$$d_k := \begin{cases} \sum\limits_{\ell \in I(k)} \Delta_\ell & \text{for } k = 1, \ldots, M, \\ \Delta_{k-M} & \text{for } k = M+1, \ldots, M+N \end{cases}$$

*in the one–dimensional case $d = 1$, and*

$$d_k := \begin{cases} \sum\limits_{\ell \in I(k)} \Delta_\ell & \text{for } k = 1, \ldots, M, \\ \sum\limits_{\ell \in K(k-M)} \Delta_\ell & \text{for } k = M+1, \ldots, M+K \end{cases}$$

*when $d = 2, 3$.*

*Proof.* First we will use the local spectral equivalence inequalities (9.17). For $d = 1$ we have[1]

$$\|v_h\|^2_{L_2(\mathcal{T}_N)} = \sum_{\ell=1}^{N} \|v_h\|^2_{L_2(\tau_\ell)} \simeq \sum_{\ell=1}^{N} \Delta_\ell \sum_{i=1}^{3} v_{\ell_i}^2$$
$$= \sum_{k=1}^{M} \left( \sum_{\ell \in I(k)} \Delta_\ell \right) v_k^2 + \sum_{\ell=1}^{N} \Delta_\ell v_{M+\ell}^2.$$

[1] The equivalence $A \simeq B$ means that there are positive constants $c_1$ and $c_2$ such that $c_1 A \le B \le c_2 A$.

In the same way we have for $d = 2, 3$

$$\sum_{\ell=1}^{N} \|v_h\|^2_{L_2(\tau_\ell)} \simeq \sum_{\ell=1}^{N} \Delta_\ell \sum_{k=1}^{3} v^2_{\ell_k}$$
$$= \sum_{k=1}^{M} \left( \sum_{\ell \in I(k)} \Delta_\ell \right) v_k^2 + \sum_{j=1}^{K} \left( \sum_{\ell \in K(j)} \Delta_\ell \right) v^2_{M+j}. \qquad \square$$

By $I_h : C(\mathcal{T}_N) \to S_h^2(\mathcal{T}_N)$ we denote the interpolation operator into the trial space of locally quadratic functions. The interpolation nodes are hereby all $M$ nodes of the decomposition (9.1) and all $N$ element midpoints in the one–dimensional case $d = 1$ and all $K$ edge midpoints in the cases $d = 2, 3$. Note that the interpolation operator $I_h$ is exact for locally quadratic functions. Analogous to Lemma 9.9 as well as to the global error estimates (9.21) and (9.22) we can prove the following error estimates, when assuming $u \in H^3(\mathcal{T}_N)$,

$$\|u - I_h u\|_{L_2(\mathcal{T}_N)} \le c \sum_{\ell=1}^{N} h_\ell^6 \, |u|^2_{H^3(\tau_\ell)},$$

and

$$\|u - I_h u\|_{H^1(\mathcal{T}_N)} \le c \sum_{\ell=1}^{N} h_\ell^4 \, |u|^2_{H^3(\tau_\ell)}.$$

As in the case of piecewise linear basis functions we can show a global approximation property.

**Theorem 9.16.** *Let $u \in H^s(\mathcal{T}_N)$ with $s \in [\sigma, 3]$ and $\sigma = 0, 1$. Then there holds*

$$\inf_{v_h \in S_h^2(\mathcal{T}_N)} \|u - v_h\|_{H^\sigma(\mathcal{T}_N)} \le c\, h^{s-\sigma} \, |u|_{H^s(\mathcal{T}_N)}.$$

By $S_h^B(\mathcal{T}_N) = \text{span}\{\varphi_\ell^B\}_{\ell=1}^N$ we denote the global trial space of local bubble functions. For an arbitrary given $v_h \in S_h^B(\mathcal{T}_N)$ we can write

$$v_h(x) = \sum_{\ell=1}^{N} v_\ell^B \varphi_\ell^B(x).$$

If the decomposition $\mathcal{T}_N$ is globally quasi–uniform we can derive, by using the local inverse inequality (9.18), the global inverse inequality

$$\|\nabla v_h\|_{L_2(\mathcal{T}_N)} \le c_I \, h^{-1} \, \|v_h\|_{L_2(\mathcal{T}_N)} \quad \text{for all } v_h \in S_h^B(\mathcal{T}_N). \tag{9.38}$$

For a given $u \in L_2(\mathcal{T}_N)$ we denote by $Q_h^B : L_2(\mathcal{T}_N) \to S_h^B(\mathcal{T}_N)$ the projection into the trial space $S_h^B(\mathcal{T}_N)$ which is the unique solution of the variational problem

$$\int_{\tau_\ell} Q_h^B v(x) dx = \int_{\tau_\ell} v(x) dx \quad \text{for all } \ell = 1, \ldots, N. \tag{9.39}$$

**Lemma 9.17.** *For $v \in L_2(\mathcal{T}_N)$ let $Q_h^B v \in S_h^B(\mathcal{T}_N)$ be the projection as defined in* (9.39). *Then there holds the stability estimate*

$$\|Q_h^B v\|_{L_2(\mathcal{T}_N)} \le \sqrt{2}\,\|v\|_{L_2(\mathcal{T}_N)}.$$

*Proof.* From (9.39) we find for the coefficients of $v_h \in S_h^B(\mathcal{T}_N)$

$$v_\ell = \frac{c_d}{\Delta_\ell}\int_{\tau_\ell} v(x)dx, \quad c_d = \begin{cases} 6 & \text{for } d = 1, \\ 60 & \text{for } d = 2, \\ 840 & \text{for } d = 3. \end{cases}$$

Hence we have

$$\|Q_h^B v\|_{L_2(\tau_\ell)} = |v_\ell|^2\,\|\varphi_\ell^B\|^2_{L_2(\tau_\ell)} = \frac{c_d^2}{\Delta_\ell^2}\frac{\Delta_\ell}{c_d^B}\left[\int_{\tau_\ell} v(x)dx\right]^2$$

with

$$c_d^B = \begin{cases} 30 & \text{for } d = 1, \\ 2520 & \text{for } d = 2, \\ 415800 & \text{for } d = 3. \end{cases}$$

By using $c_d^2/c_d^B < 2$ for $d = 1,2,3$ and by applying the Cauchy–Schwarz inequality we therefore obtain

$$\|Q_h^B v\|_{L_2(\tau_\ell)} \le \frac{2}{\Delta_\ell}\left[\int_{\tau_\ell} v(x)dx\right]^2 \le \frac{2}{\Delta_\ell}\int_{\tau_\ell} dx \int_{\tau_\ell}[v(x)]^2 dx = 2\,\|v\|^2_{L_2(\tau_\ell)}.$$

Taking the sum over all elements this gives the desired stability estimate. □

## 9.4 Quasi Interpolation Operators

For a given function $v \in H^1(\mathcal{T}_N)$ we have considered the piecewise linear interpolation (9.20). By using Lemma 9.9 there holds for $v \in H^2(\mathcal{T}_N)$ the local error estimate

$$\|v - I_h v\|_{L_2(\tau_\ell)} \le c\,h_\ell^2\,|v|_{H^2(\tau_\ell)}$$

where we have to assume the continuity of the function to be interpolated. In particular, the interpolation operator $I_h$ is not defined for a general $v \in H^1(\mathcal{T}_N)$ and $d = 2,3$, and therefore $I_h$ is not a continuous operator in $H^1(\mathcal{T}_N)$.

On the other hand, the $L_2$ projection

$$Q_h : L_2(\mathcal{T}_N) \to S_h^1(\mathcal{T}_N) \subset L_2(\mathcal{T}_N)$$

as defined in (9.23) is bounded (see (9.24)), and there holds the global error estimate (9.27),

$$\|u - Q_h u\|_{L_2(\mathcal{T}_N)} \leq c\, h^2 \, |u|_{H^2(\mathcal{T}_N)}.$$

As already stated in Remark 9.14, the $L_2$ projection

$$Q_h : H^1(\mathcal{T}_N) \to S_h^1(\mathcal{T}_N) \subset H^1(\mathcal{T}_N)$$

is bounded, but it is not possible to derive a local error estimate. Hence we aim to construct a bounded projection operator

$$P_h : H^1(\mathcal{T}_N) \to S_h^1(\mathcal{T}_N) \subset H^1(\mathcal{T}_N),$$

which admits a local error estimate. This can be done by using quasi interpolation operators [42].

For any node $x_k$ of the locally uniform decomposition $\mathcal{T}_N$ we define

$$\overline{\omega}_k := \bigcup_{\ell \in I(k)} \overline{\tau}_\ell$$

to be the convex support of the associated piecewise linear basis function $\varphi_k^1 \in S_h^1(\mathcal{T}_N)$. By $\hat{h}_k$ we denote the averaged mesh size of $\omega_k$ which is equivalent to the local mesh sizes $h_\ell$ of all finite elements $\tau_\ell$ with $\ell \in I(k)$ when the decomposition is assumed to be locally quasi–uniform. Then we introduce $Q_h^k : L_2(\omega_k) \to S_h^1(\omega_k)$ as the local $L_2$ projection which is defined by the variational formulation

$$\langle Q_h^k u, v_h \rangle_{L_2(\omega_k)} = \langle u, v_h \rangle_{L_2(\omega_k)} \quad \text{for all } v_h \in S_h^1(\omega_k),$$

and by using (9.27) there holds the error estimate

$$\|u - Q_h^k u\|_{L_2(\omega_k)} \leq c\, \hat{h}_k \, |u|_{H^1(\omega_k)}$$

As in (9.24) we can prove the stability estimate

$$\|Q_h^k u\|_{L_2(\omega_k)} \leq \|u\|_{L_2(\omega_k)} \quad \text{for all } u \in L_2(\omega_k)$$

and by applying Lemma 9.13 there holds

$$\|Q_h^k u\|_{H^1(\omega_k)} \leq c\, \|u\|_{H^1(\omega_k)} \quad \text{for all } u \in H^1(\omega_k).$$

By using the local projection operators we can define the quasi interpolation operator or Clement operator

$$(P_h u)(x) = \sum_{k=1}^{M} (Q_h^k u)(x_k)\, \varphi_k^1(x).$$

It is easy to check that $P_h v_h = v_h \in S_h^1(\mathcal{T}_N)$.

**Theorem 9.18.** [27, 42, 139] *For* $u \in H^1(\mathcal{T}_N)$ *there holds the local error estimate*

$$\|u - P_h u\|_{L_2(\tau_\ell)} \leq c \sum_{k \in J(\ell)} \hat{h}_k \, |u|_{H^1(\omega_k)} \quad \textit{for all}\, \ell = 1, \ldots, N, \tag{9.40}$$

*and the global stability estimate*

$$\|P_h u\|_{H^1(\mathcal{T}_N)} \leq c \, \|u\|_{H^1(\mathcal{T}_N)}. \tag{9.41}$$

*Proof.* Let $\tau_\ell$ be an arbitrary but fixed finite element and let $\widetilde{k} \in J(\ell)$ be an arbitrary fixed index. For $x \in \tau_\ell$ we can write

$$(P_h)(x) = (Q_h^{\tilde{k}} u)(x) + \sum_{k \in J(\ell), k \neq \tilde{k}} [(Q_h^k u)(x_k) - (Q_h^{\tilde{k}})(x_k)] \varphi_k^1(x).$$

By using Lemma 9.4 we have

$$\|\varphi_k\|_{L_2(\tau_\ell)} \leq \frac{\Delta_\ell}{d+1},$$

and therefore

$$\|u - P_h u\|_{L_2(\tau_\ell)} \leq c_1 \, \hat{h}_{\tilde{k}} \, |u|_{H^1(\omega_{\tilde{k}})} + c_2 \, h_\ell^{d/2} \sum_{k \in J(\ell), k \neq \tilde{k}} |(Q_h^k u)(x_k) - (Q_h^{\tilde{k}} u)(x_k)| \, .$$

For an arbitrary $v_h \in S_h^1(\mathcal{T}_N)$ we conclude from Corollary 9.5

$$\|v_h\|_{L_\infty(\tau_\ell)} \leq c \, h_\ell^{-d/2} \, \|v_h\|_{L_2(\tau_\ell)}.$$

Therefore,

$$\begin{aligned}
|(Q_h^k u)(x_k) - (Q_h^{\tilde{k}} u)(x_k)| &\leq \|Q_h^k u - Q_h^{\tilde{k}} u\|_{L_\infty(\tau_\ell)} \\
&\leq c \, h_\ell^{-d/2} \, \|Q_h^k u - Q_h^{\tilde{k}} u\|_{L_2(\tau_\ell)} \\
&\leq c \, h_\ell^{-d/2} \left\{ \|Q_h^k u - u\|_{L_2(\tau_\ell)} + \|u - Q_h^{\tilde{k}}\|_{L_2(\tau_\ell)} \right\} \\
&\leq c \, h_\ell^{-d/2} \left\{ h_k \, |u|_{H^1(\omega_k)} + h_{\tilde{k}} \, |u|_{H^1(\omega_{\tilde{k}})} \right\},
\end{aligned}$$

from which the error estimate (9.40) follows. The stability estimate (9.41) can be shown in the same way. □

## 9.5 Exercises

**9.1** For an admissible decomposition of a bounded domain $\Omega \subset \mathbb{R}^2$ into triangular finite elements $\tau_\ell$ and for piecewise linear continuous basis functions $\varphi_k$ the mass matrix $M_h$ is defined by

$$M_h[j,k] = \int_\Omega \varphi_k(x)\varphi_j(x)dx, \quad j,k = 1,\ldots,M.$$

Find a diagonal matrix $D_h$ and positive constants $c_1$ and $c_2$ such that the spectral equivalence inequalities

$$c_1\,(D_h\underline{u},\underline{u}) \le (M_h\underline{u},\underline{u}) \le c_2\,(D_h\underline{u},\underline{u})$$

are satisfied for all $\underline{u} \in \mathbb{R}^M$.

**9.2** For the two–dimensional reference element $\tau \subset \mathbb{R}^2$ the local quadratic shape functions are given by

$$\begin{array}{lll} \psi_1^2(\xi) = 1 - \xi_1 - \xi_2, & \psi_2^2(\xi) = \xi_1, & \psi_3^2(\xi) = \xi_2, \\ \psi_4^2(\xi) = 4\xi_1(1 - \xi_1 - \xi_2), & \psi_5^2(\xi) = 4\xi_1\xi_2, & \psi_6^2(\xi) = 4\xi_2(1 - \xi_1 - \xi_2). \end{array}$$

Compute the local mass matrix $M_\ell$ as well as the minimal and maximal eigenvalues of $M_\ell$.

# 10

# Boundary Elements

For the approximate solution of the boundary integral equations as considered in Chapter 7 we introduce suitable finite–dimensional trial spaces. These are based on appropriate parametrizations of the boundary $\Gamma = \partial\Omega$ and on the use of finite elements in the parameter domain. In particular we can think of boundary elements as finite elements on the boundary.

## 10.1 Reference Elements

Let $\Gamma = \partial\Omega$ be a piecewise smooth Lipschitz boundary with $\overline{\Gamma} = \bigcup_{j=1}^{J} \overline{\Gamma}_j$ where any boundary part $\Gamma_j$ allows a local parametrization $\Gamma_j = \chi_j(\mathcal{Q})$ with respect to some parameter domain $\mathcal{Q} \subset \mathbb{R}^{d-1}$. We assume that

$$c_1^\chi \leq |\det \chi_j(\xi)| \leq c_2^\chi \quad \text{for all } \xi \in \mathcal{Q}, j = 1, \ldots, J. \tag{10.1}$$

Further we consider a sequence $\{\Gamma_N\}_{N \in \mathbb{N}}$ of decompositions (meshes)

$$\Gamma_N = \bigcup_{\ell=1}^{N} \overline{\tau}_\ell \tag{10.2}$$

with boundary elements $\tau_\ell$. We assume that for each boundary element $\tau_\ell$ there exists a unique index $j$ with $\tau_\ell \subset \Gamma_j$. A decomposition of the boundary part $\Gamma_j$ into boundary elements $\tau_\ell$ implies a decomposition of the parameter domain $\mathcal{Q}$ into finite elements $q_\ell^j$ with $\tau_\ell = \chi_j(q_\ell^j)$. In the simplest case the boundary elements $\tau_\ell$ are intervals in the two–dimensional case $d = 2$ or triangles in the three–dimensional case $d = 3$, see Fig. 10.1.

*Example 10.1.* The boundary of the two–dimensional L shaped domain as depicted in Fig. 10.1 can be described by using the following parametrization for $\xi \in \mathcal{Q} = (0, 1)$:

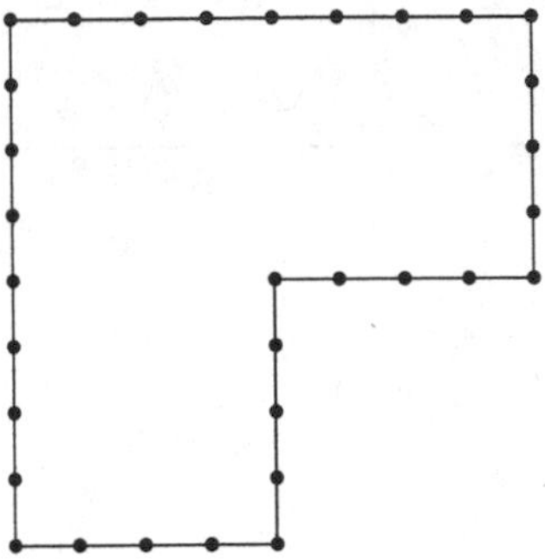

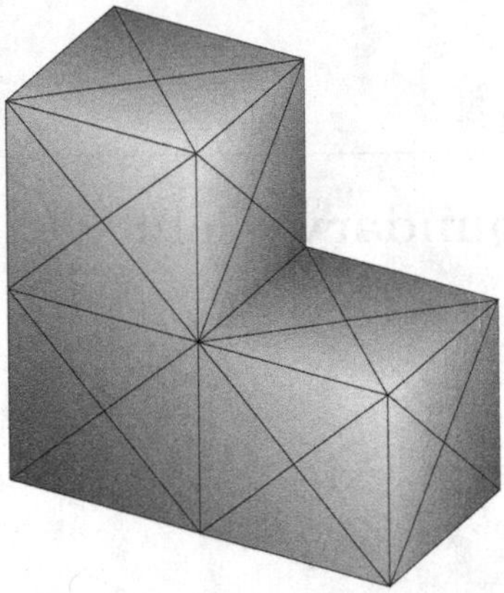

**Fig. 10.1.** Boundary discretization with 32 and 56 boundary elements.

$$\chi_1(\xi) = \begin{pmatrix} \frac{\xi}{4} \\ 0 \end{pmatrix}, \qquad \chi_2(\xi) = \begin{pmatrix} \frac{1}{4} \\ \frac{\xi}{4} \end{pmatrix}, \qquad \chi_3(\xi) = \begin{pmatrix} \frac{1-2\xi}{4} \\ \frac{1}{4} \end{pmatrix},$$

$$\chi_4(\xi) = \begin{pmatrix} -\frac{1}{4} \\ \frac{1-2\xi}{4} \end{pmatrix}, \chi_5(\xi) = \begin{pmatrix} \frac{\xi-1}{4} \\ -\frac{1}{4} \end{pmatrix}, \chi_6(\xi) = \begin{pmatrix} 0 \\ \frac{\xi-1}{4} \end{pmatrix}.$$

For $j = 1, 2, 5, 6$ the parameter domain $\mathcal{Q} = (0, 1)$ is decomposed into 4 equal sized elements $q_\ell^j$ while for $j = 3, 4$ we have 8 elements $q_\ell^j$ to be used.

By $\{x_k\}_{k=1}^M$ we denote the set of all nodes of the boundary decomposition $\Gamma_N$. The index set $I(k)$ describes all boundary elements $\tau_\ell$ where $x_k$ is a node, while $J(\ell)$ is the index set of all nodes $x_k$ describing the boundary element $\tau_\ell$. In the three–dimensional case $d = 3$ the boundary decomposition (10.2) is called admissible, if two neighboring boundary elements share either a node or an edge, see also Fig. 9.2. Analogous to (9.2) we compute by

$$\Delta_\ell := \int_{\tau_\ell} ds_x$$

the volume and by

$$h_\ell := \Delta_\ell^{1/(d-1)}$$

the local mesh size of the boundary element $\tau_\ell$. Then,

$$h := \max_{\ell=1,\ldots,N} h_\ell$$

is the global mesh size of the boundary decomposition (10.2). Moreover,

$$d_\ell := \sup_{x,y\in\tau_\ell} |x-y|$$

is the diameter of the boundary element $\tau_\ell$. Finally,

$$h_{\min} := \min_{\ell=1,\ldots,N} h_\ell.$$

is the minimal mesh size. The family of boundary decompositions (10.2) is called globally quasi–uniform if

$$\frac{h_{\max}}{h_{\min}} \le c_G$$

is satisfied with a global constant $c_G \ge 1$ which is independent of $N \in \mathbb{N}$. The family $\{\Gamma_N\}_{N\in\mathbb{N}}$ is called locally quasi–uniform if

$$\frac{h_\ell}{h_j} \le c_L \quad \text{for } \ell = 1,\ldots,N$$

is satisfied for all neighboring elements $\tau_j$ of $\tau_\ell$, i.e. $\tau_\ell$ and $\tau_j$ share either a node or an edge.

In the two–dimensional case $d = 2$ a boundary element $\tau_\ell$ with nodes $x_{\ell_1}$ and $x_{\ell_2}$ can be described via the parametrization

$$x(\xi) \;=\; x_{\ell_1} + \xi(x_{\ell_2} - x_{\ell_1}) \quad \text{for } \xi \in \tau = (0,1)$$

where $\tau = (0,1)$ is the reference element, and we have

$$d_\ell = h_\ell = \Delta_\ell \;=\; \int_{\tau_\ell} ds_x \;=\; \int_0^1 \sqrt{[x_1'(\xi)]^2 + [x_2'(\xi)]^2} d\xi \;=\; |x_{\ell_2} - x_{\ell_1}|\,.$$

In the three–dimensional case $d = 3$ we consider plane triangular boundary elements $\tau_\ell$ with nodes $x_{\ell_1}$, $x_{\ell_2}$ and $x_{\ell_3}$. The parametrization of $\tau_\ell$ with respect to the reference element

$$\tau := \left\{\xi \in \mathbb{R}^2 : 0 < \xi_1 < 1, 0 < \xi_2 < 1 - \xi_1\right\}$$

then reads

$$x(\xi) \;=\; x_{\ell_1} + \xi_1(x_{\ell_2} - x_{\ell_1}) + \xi_2(x_{\ell_3} - x_{\ell_1}) \quad \text{for } \xi \in \tau.$$

For the computation of the boundary element volume we obtain

$$\Delta_\ell \;=\; \int_{\tau_\ell} ds_x \;=\; \int_\tau \sqrt{EG - F^2} d\xi \;=\; \frac{1}{2}\sqrt{EG - F^2}$$

where

$$E = \sum_{i=1}^{3} \left[ \frac{\partial}{\partial \xi_1} x_i(\xi) \right]^2 = |x_{\ell_2} - x_{\ell_1}|^2,$$

$$G = \sum_{i=1}^{3} \left[ \frac{\partial}{\partial \xi_2} x_i(\xi) \right]^2 = |x_{\ell_3} - x_{\ell_1}|^2,$$

$$F = \sum_{i=1}^{3} \frac{\partial}{\partial \xi_1} x_i(\xi) \frac{\partial}{\partial \xi_2} x_i(\xi) = (x_{\ell_2} - x_{\ell_1}, x_{\ell_3} - x_{\ell_1}) .$$

In the three–dimensional case $d = 3$ we assume that all boundary elements $\tau_\ell$ are shape regular, i.e. there exists a constant $c_B$ independent of the boundary decomposition such that

$$d_\ell \leq c_B \, h_\ell \quad \text{for } \ell = 1, \ldots, N. \tag{10.3}$$

By using

$$J_\ell = \begin{pmatrix} x_{\ell_2,1} - x_{\ell_1,1} & x_{\ell_3,1} - x_{\ell_1,1} \\ x_{\ell_2,2} - x_{\ell_1,2} & x_{\ell_3,2} - x_{\ell_1,2} \\ x_{\ell_2,3} - x_{\ell_1,3} & x_{\ell_3,3} - x_{\ell_1,3} \end{pmatrix}$$

we can write a function $v(x)$ for $x \in \tau_\ell$ as

$$v(x) = v(x_{\ell_1} + J_\ell \xi) =: \widetilde{v}_\ell(\xi) \quad \text{for } \xi \in \tau.$$

Vice versa, for a function $\widetilde{v}(\xi)$ which is given for $\xi \in \tau$ we can define a function $v_\ell(x)$ for $x \in \tau_\ell$,

$$v_\ell(x) := v(x_{\ell_1} + J_\ell \xi) = \widetilde{v}(\xi) \quad \text{for } \xi \in \tau.$$

In the two–dimensional case $d = 2$ we have

$$\|v\|^2_{L_2(\tau_\ell)} = \int_{\tau_\ell} |v(x)|^2 ds_x = \int_{\tau} |\widetilde{v}_\ell(\xi)|^2 h_\ell d\xi = \Delta_\ell \, \|\widetilde{v}_\ell\|^2_{L_2(\tau)}$$

and for the three–dimensional case $d = 3$ it follows that

$$\|v\|^2_{L_2(\tau_\ell)} = \int_{\tau_\ell} |v(x)|^2 ds_x = 2\Delta_\ell \int_{\tau} |\widetilde{v}_\ell(\xi)|^2 d\xi = 2\Delta_\ell \, \|\widetilde{v}_\ell\|^2_{L_2(\tau)}.$$

To define Sobolev spaces $H^s(\Gamma)$ for $s \geq 1$ we have to use a parametrization $\Gamma_j = \chi_j(\mathcal{Q})$, see Section 2.5. In particular,

$$|v|^2_{H^1(\tau_\ell)} := \int_{q_\ell^j} |\nabla_\xi v(\chi_j(\xi))|^2 d\xi$$

and

$$\Delta_\ell = \int_{\tau_\ell} ds_x = \int_{q_\ell^j} \det|\chi_j(\xi)| \, d\xi.$$

## 10.2 Trial Spaces

With respect to the boundary decomposition (10.2) we now define trial spaces of local polynomials. In particular we will consider the trial space $S_h^0(\Gamma)$ of piecewise constant functions and the trial space $S_h^1(\Gamma)$ of piecewise linear continuous functions. By considering appropriate interpolation and projection operators we will prove certain approximation properties of these trial spaces.

Let

$$S_h^0(\Gamma) := \text{span}\{\varphi_k^0\}_{k=1}^N$$

be the space of functions which are piecewise constant with respect to the boundary decomposition (10.2). The basis functions $\varphi_k^0$ are given by

$$\varphi_k^0(x) = \begin{cases} 1 & \text{for } x \in \tau_k, \\ 0 & \text{elsewhere.} \end{cases}$$

If $u \in L_2(\Gamma)$ is a given function, the $L_2$ projection $Q_h u \in S_h^0(\Gamma)$ is defined as the unique solution of the variational problem

$$\langle Q_h u, v_h \rangle_{L_2(\Gamma)} = \langle u, v_h \rangle_{L_2(\Gamma)} \quad \text{for all } v_h \in S_h^0(\Gamma). \tag{10.4}$$

This is equivalent to finding the coefficient vector $\underline{u} \in \mathbb{R}^N$ as the solution of

$$\sum_{k=1}^N u_k \langle \varphi_k^0, \varphi_\ell^0 \rangle_{L_2(\Gamma)} = \langle u, \varphi_\ell^0 \rangle_{L_2(\Gamma)} \quad \text{for } \ell = 1, \ldots, N.$$

Due to

$$\langle \varphi_k^0, \varphi_\ell^0 \rangle_{L_2(\Gamma)} = \int_\Gamma \varphi_k^0(x)\varphi_\ell^0(x)ds_x = \begin{cases} \Delta_k & \text{for } k = \ell, \\ 0 & \text{for } k \neq \ell \end{cases}$$

we obtain

$$u_k = \frac{1}{\Delta_k} \int_{\tau_k} u(x)ds_x \quad \text{for } k = 1, \ldots, N.$$

**Theorem 10.2.** *Let $u \in H^s(\Gamma)$ be given for some $s \in [0,1]$, and let $Q_h u \in S_h^0(\Gamma)$ be the $L_2$ projection as defined by (10.4). Then there hold the error estimates*

$$\|u - Q_h u\|_{L_2(\Gamma)}^2 \leq c \sum_{k=1}^N h_k^{2s} \, |u|_{H^s(\tau_k)}^2 \tag{10.5}$$

*and*

$$\|u - Q_h u\|_{L_2(\Gamma)} \leq c\, h^s \, |u|_{H^s(\Gamma)}. \tag{10.6}$$

*Proof.* By using the Galerkin orthogonality

$$\langle u - Q_h u, v_h \rangle_{L_2(\Gamma)} = 0 \quad \text{for all } v_h \in S_h^0(\Gamma)$$

we obtain

$$\begin{aligned}\|u - Q_h u\|^2_{L_2(\Gamma)} &= \langle u - Q_h u, u - Q_h u \rangle_{L_2(\Gamma)} \\ &= \langle u - Q_h u, u \rangle_{L_2(\Gamma)} \le \|u - Q_h u\|_{L_2(\Gamma)} \|u\|_{L_2(\Gamma)}\end{aligned}$$

and therefore

$$\|u - Q_h u\|_{L_2(\Gamma)} \le \|u\|_{L_2(\Gamma)}$$

which is the error estimate for $s = 0$.

We now consider $s \in (0,1)$. For $x \in \tau_k$ we have $Q_h u(x) = u_k$ and therefore

$$u(x) - Q_h u(x) = \frac{1}{\Delta_k} \int_{\tau_k} [u(x) - u(y)] ds_y \quad \text{for } x \in \tau_k.$$

By taking the square and applying the Cauchy–Schwarz inequality we conclude

$$\begin{aligned}|u(x) - Q_h u(x)|^2 &= \frac{1}{\Delta_k^2} \left( \int_{\tau_k} [u(x) - u(y)] ds_y \right)^2 \\ &= \frac{1}{\Delta_k^2} \left( \int_{\tau_k} \frac{[u(x) - u(y)]}{|x-y|^{\frac{d-1}{2}+s}} |x-y|^{\frac{d-1}{2}+s} ds_y \right)^2 \\ &\le \frac{1}{\Delta_k^2} \int_{\tau_k} \frac{[u(x) - u(y)]^2}{|x-y|^{d-1+2s}} ds_y \int_{\tau_k} |x-y|^{d-1+2s} ds_y \\ &\le d_k^{d-1+2s} \frac{1}{\Delta_k} \int_{\tau_k} \frac{|u(x) - u(y)|^2}{|x-y|^{d-1+2s}} ds_y.\end{aligned}$$

By using the shape regularity (10.3) and $\Delta_k = h_k^{d-1}$ we can replace the diameter $d_k$ and the area $\Delta_k$ by the local mesh size $h_k$,

$$|u(x) - Q_h u(x)|^2 \le c_B^{d-1+2s} \, h_k^{2s} \int_{\tau_k} \frac{[u(x) - u(y)]^2}{|x-y|^{d-1+2s}} ds_y.$$

When integrating with respect to $x \in \tau_k$ this gives

$$\|u - Q_h u\|^2_{L_2(\tau_k)} \le c_B^{d-1+2s} \, h_k^{2s} \, |u|^2_{H^s(\tau_k)},$$

and by taking the sum over all boundary elements we obtain the error estimate for $s \in (0,1)$.

To prove the error estimate for $s = 1$ we first consider

$$u(x) - Q_h u(x) = \frac{1}{\Delta_k} \int\limits_{\tau_k} [u(x) - u(y)] ds_y \quad \text{for } x \in \tau_k.$$

By using the local parametrization $\tau_k = \chi_j(q_k^j)$ we further get

$$u(x) - Q_h u(x) = \frac{1}{\Delta_k} \int\limits_{q_k^j} [u(\chi_j(\xi)) - u(\chi_j(\eta))] |\det\chi_j(\eta)| \, d\eta. \qquad (10.7)$$

In the two–dimensional case $d = 2$ we have

$$u(x) - Q_h u(x) = \frac{1}{\Delta_k} \int\limits_{q_k^j} \int\limits_{\eta}^{\xi} \frac{d}{dt} u(\chi_j(t)) dt \, |\det\chi_j(\eta)| \, d\eta$$

and therefore

$$\begin{aligned} |u(x) - Q_h u(x)| &\leq \frac{1}{\Delta_k} \int\limits_{q_k^j} \int\limits_{q_k^j} \left| \frac{d}{dt} u(\chi_j(t)) |dt \right| \det\chi_j(\eta)| \, d\eta \\ &= \frac{1}{\Delta_k} \int\limits_{q_k^j} |\det\chi_j(\eta)| \, d\eta \int\limits_{q_k^j} |\nabla_\xi u(\chi_j(\xi))| d\xi \\ &= \int\limits_{q_k^j} |\nabla_\xi u(\chi_j(\xi))| d\xi. \end{aligned}$$

By taking the square and applying the Cauchy–Schwarz inequality we find by considering (10.1)

$$\begin{aligned} |u(x) - Q_h u(x)|^2 &= \left| \int\limits_{q_k^j} |\nabla_\xi u(\chi_j(\xi))| d\xi \right|^2 \leq \int\limits_{q_k^j} d\xi \int\limits_{q_k^j} |\nabla_\xi u(\chi_j(\xi))|^2 d\xi \\ &\leq \frac{1}{c_1^\chi} \int\limits_{q_k^j} |\det\chi_j(\xi)| d\xi \, |u|^2_{H^1(\tau_k)} = \frac{1}{c_1^\chi} \Delta_k \, |u|^2_{H^1(\tau_k)}. \end{aligned}$$

When integrating with respect to $x \in \tau_k$ and using $\Delta_k = h_k$ for $d = 2$ this gives

$$\int\limits_{\tau_k} [u(x) - Q_h u(x)]^2 ds_x \leq \frac{1}{c_1^\chi} h_k^2 \, |u|^2_{H^1(\tau_k)},$$

and by taking the sum over all boundary elements $\tau_k$ we finally obtain the error estimate for $s = 1$ and $d = 2$.

In the three–dimensional case $d = 3$ we have from the representation (10.7)

$$\begin{aligned}
u(x) - Q_h u(x) &= \frac{1}{\Delta_k} \int\limits_{q_k^j} [u(\chi_j(\xi)) - u(\chi_j(\eta))] |\det \chi_j(\eta)| \, d\eta \\
&= \frac{1}{\Delta_k} \int\limits_{q_k^j} \int\limits_0^1 \frac{d}{dt} u(\chi_j(\eta + t(\xi - \eta))) dt \, |\det \chi_j(\eta)| \, d\eta \\
&= \frac{1}{\Delta_k} \int\limits_{q_k^j} \int\limits_0^1 (\xi - \eta) \cdot \nabla_\eta u(\chi_j(\eta + t(\xi - \eta))) dt \, |\det \chi_j(\eta)| \, d\eta.
\end{aligned}$$

Taking the square

$$|u(x) - Q_h u(x)|^2 \leq \frac{1}{\Delta_k^2} \left( \int\limits_{q_k^j} \int\limits_0^1 (\xi - \eta) \cdot \nabla_\eta u(\chi_j(\eta + t(\xi - \eta))) dt \, |\det \chi_j(\eta)| \, d\eta \right)^2$$

and by applying the Cauchy–Schwarz inequality this gives

$$\begin{aligned}
|u(x) - Q_h u(x)|^2 &\leq \frac{1}{\Delta_k^2} \int\limits_{q_k^j} \left| \int\limits_0^1 \nabla_\eta u(\chi_j(\eta + t(\xi - \eta))) dt \right|^2 d\eta \\
&\qquad \int\limits_{q_k^j} |\xi - \eta|^2 |\det \chi_j(\eta)|^2 d\eta \\
&\leq c \int\limits_{q_k^j} \left| \int\limits_0^1 \nabla_\eta u(\chi_j(\eta + t(\xi - \eta))) dt \right|^2 d\eta.
\end{aligned}$$

When integrating over $\tau_k$ we obtain

$$\begin{aligned}
\int\limits_{\tau_k} [u(x) - Q_h u(x)]^2 ds_x &\leq c \int\limits_{q_k^j} \int\limits_{q_k^j} \left| \int\limits_0^1 \nabla_\eta u(\chi_j(\eta + t(\xi - \eta))) dt \right|^2 d\eta \, d\xi \\
&\leq c \int\limits_{q_k^j} \int\limits_{q_k^j} \int\limits_0^1 |\nabla_\eta u(\chi_j(\eta + t(\xi - \eta)))|^2 \, dt \, d\eta \, d\xi \\
&\leq c \, \Delta_k \int\limits_{q_k^j} |\nabla_\eta u(\chi_j(\eta))|^2 \, d\xi \; = \; c \, \Delta_k \, |u|^2_{H^1(\tau_k)}
\end{aligned}$$

and with $\Delta_k = h_k^2$ for $d = 3$ we find

$$\|u - Q_h u\|^2_{L_2(\tau_k)} \le c\, h_k^2\, |u|^2_{H^1(\tau_k)}.$$

By taking the sum over all boundary elements we finally get the error estimate for $s = 1$ and $d = 3$, □

**Corollary 10.3.** *Let $u \in H^s(\Gamma)$ be given for some $s \in [0,1]$. For $\sigma \in [-1,0)$ then there hold the error estimates*

$$\|u - Q_h u\|^2_{H^\sigma(\Gamma)} \le c\, h^{-2\sigma} \sum_{k=1}^N h_k^{2s}\, |u|^2_{H^s(\tau_k)}$$

*and*

$$\|u - Q_h u\|_{H^\sigma(\Gamma)} \le c\, h^{s-\sigma}\, |u|_{H^s(\Gamma)}. \tag{10.8}$$

*Proof.* For $\sigma \in [-1,0)$ we have by duality, by using the definition (10.4) of the $L_2$ projection, and by applying the Cauchy–Schwarz inequality

$$\begin{aligned}
\|u - Q_h u\|_{H^\sigma(\Gamma)} &= \sup_{0 \ne v \in H^{-\sigma}(\Gamma)} \frac{|\langle u - Q_h u, v\rangle_{L_2(\Gamma)}|}{\|v\|_{H^{-\sigma}(\Gamma)}} \\
&= \sup_{0 \ne v \in H^{-\sigma}(\Gamma)} \frac{|\langle u - Q_h u, v - Q_h v\rangle_{L_2(\Gamma)}|}{\|v\|_{H^{-\sigma}(\Gamma)}} \\
&\le \|u - Q_h u\|_{L_2(\Gamma)} \sup_{0 \ne v \in H^{-\sigma}(\Gamma)} \frac{\|v - Q_h v\|_{L_2(\Gamma)}}{\|v\|_{H^{-\sigma}(\Gamma)}}.
\end{aligned}$$

By using the error estimate (10.5) for $\|u - Q_h u\|_{L_2(\Gamma)}$ and the estimate (10.6) for $\|v - Q_h v\|_{L_2(\Gamma)}$ the assertion follows. □

Altogether we can formulate the approximation property of the trial space $S_h^0(\Gamma)$ of piecewise constant functions.

**Theorem 10.4.** *Let $\sigma \in [-1,0]$. For $u \in H^s(\Gamma)$ with some $s \in [\sigma,1]$ there holds the approximation property of $S_h^0(\Gamma)$*

$$\inf_{v_h \in S_h^0(\Gamma)} \|u - v_h\|_{H^\sigma(\Gamma)} \le c\, h^{s-\sigma}\, |u|_{H^s(\Gamma)}. \tag{10.9}$$

*Proof.* For $\sigma \in [-1,0]$ and $s \in [0,1]$ the approximation property is just the statement of Theorem 10.2 and Corollary 10.3. It remains to prove the approximation property for $\sigma \in [-1,0)$ and $s \in [\sigma,0)$.

For a given $u \in H^\sigma(\Gamma)$ let $Q_h^\sigma u \in S_h^0(\Gamma) \subset H^\sigma(\Gamma) \subset L_2(\Gamma)$ be the $H^\sigma(\Gamma)$ projection which is defined as the unique solution of the variational problem

$$\langle Q_h^\sigma u, v_h\rangle_{H^\sigma(\Gamma)} = \langle u, v_h\rangle_{H^\sigma(\Gamma)} \quad \text{for all } v_h \in S_h^0(\Gamma).$$

As for the $L_2$ projection there holds the error estimate for $s = \sigma$,

$$\|u - Q_h^\sigma u\|_{H^\sigma(\Gamma)} \le \|u\|_{H^\sigma(\Gamma)}.$$

Therefore, $I - Q_h^\sigma : H^\sigma(\Gamma) \to H^\sigma(\Gamma)$ is a bounded operator with norm

$$\|I - Q_h^\sigma\|_{H^\sigma(\Gamma)\to H^\sigma(\Gamma)} \le 1.$$

On the other hand, by using (10.8) and $s = 0$ we have

$$\|u - Q_h^\sigma u\|_{H^\sigma(\Gamma)} \le \|u - Q_h u\|_{H^\sigma(\Gamma)} \le c\,h^{-\sigma}\,\|u\|_{L_2(\Gamma)}.$$

Thus, $I - Q_h^\sigma : L_2(\Gamma) \to H^\sigma(\Gamma)$ is bounded with norm

$$\|I - Q_h^\sigma\|_{L_2(\Gamma)\to H^\sigma(\Gamma)} \le c\,h^{-\sigma}.$$

By applying the interpolation theorem (Theorem 2.18 and Remark 2.23) we conclude that the operator $I - Q_h^\sigma : H^s(\Gamma) \to H^\sigma(\Gamma)$ is bounded for all $s \in [\sigma, 0]$, and for the related operator norm it follows that

$$\begin{aligned} &\|I - Q_h^\sigma\|_{H^s(\Gamma)\to H^\sigma(\Gamma)} \\ &\quad\le \left(\|I - Q_h^\sigma\|_{H^\sigma(\Gamma)\to H^\sigma(\Gamma)}\right)^{\frac{s-0}{\sigma-0}} \left(\|I - Q_h^\sigma\|_{L_2(\Gamma)\to H^\sigma(\Gamma)}\right)^{\frac{s-\sigma}{-\sigma}} \\ &\quad\le \left(c\,h^{-\sigma}\right)^{\frac{s-\sigma}{-\sigma}} = c(s,\sigma)\,h^{s-\sigma}. \end{aligned}$$

This gives the approximation property for $\sigma \in [-1, 0)$ and $s \in [\sigma, 0)$. $\square$

Now we consider the case where $\Gamma_j \subset \Gamma$ is an open boundary part of $\Gamma = \partial\Omega$, and $S_h^0(\Gamma_j)$ is the associated trial space of piecewise constant basis functions. As in (10.6) there holds the error estimate

$$\|u - Q_h u\|_{L_2(\Gamma_j)} \le c\,h^s\,|u|_{H^s(\Gamma_j)}$$

for the $L_2$ projection $Q_h : L_2(\Gamma_j) \to S_h^0(\Gamma_j)$ which is defined accordingly. Analogous to Corollary 10.3 for $\sigma \in [-1, 0)$ we find the error estimate

$$\|u - Q_h u\|_{\widetilde{H}^\sigma(\Gamma_j)} \le c\,h^{s-\sigma}\,|u|_{H^s(\Gamma_j)}.$$

Hence we have the approximation property

$$\inf_{v_h \in S_h^0(\Gamma_j)} \|u - v_h\|_{\widetilde{H}^\sigma(\Gamma_j)} \le c\,h^{s-\tau}\,|u|_{H^s(\Gamma_j)} \qquad (10.10)$$

for $u \in H^s(\Gamma_j)$ and $-1 \le \sigma \le 0 \le s \le 1$.

In addition to the trial space $S_h^0(\Gamma)$ of piecewise constant basis functions $\varphi_k^0$ we next consider the trial space $S_h^1(\Gamma)$ of piecewise linear and globally continuous basis functions $\varphi_i^1$. If the boundary decomposition (10.2) is admissible, a function $v_h \in S_h^1(\Gamma)$ is determined by the nodal values which are described at the $M$ nodes $x_k$. Hence, a basis of $S_h^1(\Gamma)$ is given by

$$\varphi_i^1(x) = \begin{cases} 1 & \text{for } x = x_i, \\ 0 & \text{for } x = x_j \neq x_i, \\ \text{linear} & \text{elsewhere.} \end{cases}$$

If a piecewise linear function $v_h$ is considered in the boundary element $\tau_\ell$, this function is uniquely determined by the nodal values $v_h(x_k)$ for $k \in J(\ell)$. By using the parametrization $\tau_\ell = \chi_j(q_\ell^j)$ we can write

$$v_h(x) = v_h(\chi_j(\xi)) = \widetilde{v}_\ell^j(\xi) \quad \text{for } \xi \in q_\ell^j \subset \mathcal{Q} \subset \mathbb{R}^{d-1}.$$

Hence we can identify a boundary element $\tau_\ell \subset \Gamma$ where $\Gamma = \partial\Omega$ and $\Omega \subset \mathbb{R}^d$ with a finite element $q_\ell^j$ in the parameter domain $\mathcal{Q} \subset \mathbb{R}^{d-1}$. Thus we can transfer all local error estimates of piecewise linear basis functions, which were already proved in Chapter 9, to the finite element $q_\ell^j$ and therefore to the boundary element $\tau_\ell$.

**Lemma 10.5.** *For a function $v_h$ which is linear on $\tau_\ell$ there holds*

$$\frac{\Delta_\ell}{d(d+1)} \sum_{k=1}^{d} v_{\ell_k}^2 \leq \|v_h\|_{L_2(\tau_\ell)}^2 \leq \frac{\Delta_\ell}{d} \sum_{k=1}^{d} v_{\ell_k}^2.$$

*Proof.* By mapping the boundary element $\tau_\ell$ to the reference element $\tau$ we obtain

$$\begin{aligned}\|v_h\|_{L_2(\tau_\ell)}^2 &= \langle v_h, v_h\rangle_{L_2(\tau_\ell)} \\ &= \sum_{i=1}^{d}\sum_{j=1}^{d} v_i v_j \int_\tau \varphi_i^1(\xi)\varphi_j^1(\xi)|\det J_\ell|d\xi = (G_\ell \underline{v}^\ell, \underline{v}^\ell)\end{aligned}$$

where

$$G_\ell = \frac{\Delta_\ell}{d(d+1)}(I_d + \underline{e}_d \underline{e}_d^\top)$$

is the local mass matrix and $\underline{e}_d = \underline{1} \in \mathbb{R}^d$. The eigenvalues of the matrix $I_d + \underline{e}_d\underline{e}_d^\top$ are given by

$$\lambda_1 = d+1, \lambda_2 = \cdots = \lambda_d = 1$$

and therefore the assertion follows. □

**Corollary 10.6.** *For a function $v_h$ which is linear on $\tau_\ell$ there holds*

$$\frac{\Delta_\ell}{d(d+1)} \|v_h\|_{L_\infty(\tau_\ell)}^2 \leq \|v_h\|_{L_2(\tau_\ell)}^2 \leq \Delta_\ell \|v_h\|_{L_\infty(\tau_\ell)}^2.$$

*Proof.* Since the maximum of $|v_h|$ and therefore the $\|v_h\|_{L_\infty(\tau_\ell)}$ norm is equal to some nodal value $|v_h(x_{k^*})| = |v_{k^*}|$ for some $x_{k^*}$, the assertion follows immediately from Lemma 10.5. □

**Lemma 10.7.** *For a function $v_h$ which is linear on $\tau_\ell$ there holds the local inverse inequality*

$$|v_h|_{H^1(\tau_\ell)} \leq c_I\, h_\ell^{-1} \|v_h\|_{L_2(\tau_\ell)}.$$

*Proof.* First we have

$$|v_h|^2_{H^1(\tau_\ell)} = \int_{q^j_\ell} |\nabla_\xi v_h(\chi_j(\xi))|^2 |\det \chi_j(\xi)| d\xi \leq \Delta_\ell \, \|\nabla_\xi v_h(\chi_j(\cdot))\|^2_{L_\infty(q^j_\ell)} .$$

By mapping the finite element $q^j_\ell$ to the associated reference element we obtain

$$\|\nabla_\xi v_h(\chi_j(\cdot))\|^2_{L_\infty(q^j_\ell)} \leq c\, h_\ell^{-2} \, \|v_h(\chi_j(\cdot))\|^2_{L_\infty(q^j_\ell)} = c\, h_\ell^{-2} \, \|v_h\|^2_{L_\infty(\tau_\ell)},$$

and the inverse inequality follows from Corollary 10.6. □

Hence we also conclude the global inverse inequality

$$|v_h|^2_{H^1(\Gamma)} = \sum_{\ell=1}^N |v_h|^2_{H^1(\tau_\ell)} \leq c \sum_{\ell=1}^N h_\ell^{-2} \, \|v_h\|^2_{L_2(\tau_\ell)}$$

and, for a globally quasi–uniform boundary decomposition,

$$|v_h|_{H^1(\Gamma)} \leq c\, h^{-1} \, \|v_h\|_{L_2(\Gamma)}.$$

In particular for $h < 1$ we obtain

$$\|v_h\|_{H^1(\Gamma)} \leq c\, h^{-1} \, \|v_h\|_{L_2(\Gamma)},$$

and an interpolation argument gives

$$\|v_h\|_{H^s(\Gamma)} \leq c\, h^{-s} \, \|v_h\|_{L_2(\Gamma)} \quad \text{for } s \in [0,1].$$

Analogous to the error estimates (9.21) and (9.22) we can estimate the interpolation error of the piecewise linear interpolation operator $I_h : H^2(\Gamma) \to S^1_h(\Gamma)$ as follows.

**Lemma 10.8.** *Let $v \in H^2(\Gamma)$ be given. Assume that $\Gamma = \partial\Omega$ is sufficiently smooth where $\Omega \subset \mathbb{R}^d$. Let $I_h v$ be the piecewise linear interpolation satisfying $I_h v(x_k) = v(x_k)$ at all nodes $x_k$ of the admissible boundary decomposition* (10.2). *Then there hold the error estimates*

$$\|v - I_h v\|^2_{L_2(\Gamma)} \leq c \sum_{\ell=1}^N h_\ell^4 \, |v|^2_{H^2(\tau_\ell)} \leq c\, h^4 \, |v|^2_{H^2(\Gamma)}$$

*and*

$$\|v - I_h v\|^2_{H^1(\Gamma)} \leq c \sum_{\ell=1}^N h_\ell^2 \, |v|^2_{H^2(\tau_\ell)} \leq c\, h^2 \, |v|^2_{H^2(\Gamma)}.$$

By applying the interpolation theorem (Theorem 2.18, Remark 2.23) we can conclude the error estimate

$$\|v - I_h v\|_{H^\sigma(\Gamma)} \leq c\, h^{2-\sigma} \, |v|_{H^2(\Gamma)} \quad \text{for } \sigma \in [0,1].$$

The piecewise linear interpolation requires, as in the case of finite elements, the global continuity of the function to be interpolated. The function $v \in H^s(\Gamma)$ is continuous for $s \in (\frac{d-1}{2}, 2]$ and the following error estimate holds,

$$\|v - I_h v\|_{H^\sigma(\Gamma)} \leq c\, h^{s-\sigma}\, |v|_{H^s(\Gamma)}, \quad 0 \leq \sigma \leq \min\{1, s\}. \tag{10.11}$$

To prove more general error estimates we now consider projection operators which are defined by some variational problems. If $u \in L_2(\Gamma)$ is given the $L_2$ projection $Q_h u \in S_h^1(\Gamma)$ is defined as the unique solution of the variational problem

$$\langle Q_h u, v_h \rangle_{L_2(\Gamma)} = \langle u, v_h \rangle_{L_2(\Gamma)} \quad \text{for all } v_h \in S_h^1(\Gamma),$$

and there holds the error estimate

$$\|u - Q_h u\|_{L_2(\Gamma)} \leq \|u\|_{L_2(\Gamma)}.$$

On the other hand, by using Lemma 10.8 we also have the error estimate

$$\|u - Q_h u\|_{L_2(\Gamma)} \leq \|u - I_h u\|_{L_2(\Gamma)} \leq c \sum_{\ell=1}^{N} h_\ell^4\, |u|^2_{H^2(\Gamma)} \leq c\, h^2\, |u|^2_{H^2(\Gamma)}$$

and by applying the interpolation theorem (Theorem 2.18, Remark 2.23) we conclude the error estimate

$$\|u - Q_h u\|_{L_2(\Gamma)} \leq c\, h^s\, |u|_{H^s(\Gamma)} \quad \text{for } u \in H^s(\Gamma), \quad s \in [0, 2]. \tag{10.12}$$

Accordingly, for $u \in H^\sigma(\Gamma)$ and $\sigma \in (0, 1]$ we define the $H^\sigma$ projection $Q_h^\sigma : H^\sigma(\Gamma) \to S_h^1(\Gamma)$ as the unique solution of the variational problem

$$\langle Q_h^\sigma u, v_h \rangle_{H^\sigma(\Gamma)} = \langle u, v_h \rangle_{H^\sigma(\Gamma)} \quad \text{for all } v_h \in S_h^1(\Gamma)$$

satisfying the error estimate

$$\|u - Q_h^\sigma u\|_{H^\sigma(\Gamma)} \leq c\, h^{s-\sigma}\, |u|_{H^s(\Gamma)} \quad \text{for } u \in H^s(\Gamma), \quad s \in [\sigma, 2]. \tag{10.13}$$

**Theorem 10.9.** *Let $\Gamma = \partial\Omega$ be sufficiently smooth. For $\sigma \in [0, 1]$ and for some $s \in [\sigma, 2]$ we assume $u \in H^s(\Gamma)$. Then there holds the approximation property of $S_h^1(\Gamma)$,*

$$\inf_{v_h \in S_h^1(\Gamma)} \|u - v_h\|_{H^\sigma(\Gamma)} \leq c\, h^{s-\sigma}\, |u|_{H^s(\Gamma)}. \tag{10.14}$$

*Proof.* For $\sigma = 0$ and $\sigma \in (0, 1]$ as well as for $s \in [\sigma, 2]$ the approximation property is just the error estimate (10.12) and (10.13), respectively. □

As in Lemma 9.13 the $L_2$ projection $Q_h : H^{1/2}(\Gamma) \to S_h^1(\Gamma) \subset H^{1/2}(\Gamma)$ defines a bounded operator satisfying

$$\|Q_h v\|_{H^{1/2}(\Gamma)} \leq c\, \|v\|_{H^{1/2}(\Gamma)} \quad \text{for all } v \in H^{1/2}(\Gamma). \tag{10.15}$$

Note that if the boundary decomposition is locally adaptive, to ensure (10.15) we have to assume that the local mesh sizes of neighboring boundary elements do no vary too strongly [137].

It remains to prove the inverse inequality of the trial space $S_h^0(\Gamma)$ of piecewise constant basis functions. For this we first define the global trial space $S_h^B(\Gamma)$ of local bubble functions $\varphi_\ell^B$ which are defined on $\tau_\ell$. With respect to the reference element $\tau \subset \mathbb{R}^{d-1}$ the associated form functions are given by

$$\psi_B(\xi) = \begin{cases} \xi(1-\xi) & \text{for } d = 2, \\ \xi_1\xi_2(1-\xi_1-\xi_2) & \text{for } d = 3. \end{cases}$$

If the boundary decomposition is globally quasi–uniform there holds as in (9.38) a global inverse inequality

$$\|v_h\|_{H^1(\Gamma)} \leq c_I\, h^{-1}\, \|v_h\|_{L_2(\Gamma)} \quad \text{for all } v_h \in S_h^B(\Gamma).$$

By using an interpolation argument we then conclude

$$\|v_h\|_{H^{1/2}(\Gamma)} \leq c_I\, h^{-1/2}\, \|v_h\|_{L_2(\Gamma)} \quad \text{for all } v_h \in S_h^B(\Gamma). \tag{10.16}$$

For a given $u \in L_2(\Gamma)$ we define the projection operator $Q_h^B : L_2(\Gamma) \to S_h^B(\Gamma)$ as the unique solution of the variational problem

$$\int_\Gamma (Q_h^B u)(x) w_h(x) ds_x = \int_\Gamma u(x) w_h(x) ds_x \quad \text{for all } w_h \in S_h^0(\Gamma). \tag{10.17}$$

When considering piecewise constant test functions this is equivalent to

$$\int_{\tau_\ell} (Q_h^B u)(x) ds_x = \int_{\tau_\ell} u(x) ds_x \quad \text{for all } \ell = 1, \ldots, N.$$

As in Lemma 9.17 we can prove the stability estimate

$$\|Q_h^B u\|_{L_2(\Gamma)} \leq \sqrt{2}\, \|u\|_{L_2(\Gamma)} \quad \text{for all } u \in L_2(\Gamma). \tag{10.18}$$

Hence we can formulate an inverse inequality of the trial space $S_h^0(\Gamma)$ of piecewise constant basis functions.

**Lemma 10.10.** *Assume that the boundary decomposition* (10.2) *is globally quasi–uniform. Then there holds the global inverse inequality*

$$\|w_h\|_{L_2(\Gamma)} \leq c_I\, h^{-1/2}\, \|w_h\|_{H^{-1/2}(\Gamma)} \quad \textit{for all } w_h \in S_h^0(\Gamma).$$

*Proof.* For $w_h \in S_h^0(\Gamma)$ we have by using (10.17)

$$\begin{aligned} \|w_h\|_{L_2(\Gamma)} &= \sup_{0 \neq v \in L_2(\Gamma)} \frac{\langle w_h, v\rangle_{L_2(\Gamma)}}{\|v\|_{L_2(\Gamma)}} = \sup_{0 \neq v \in L_2(\Gamma)} \frac{\langle w_h, Q_h^B v\rangle_{L_2(\Gamma)}}{\|v\|_{L_2(\Gamma)}} \\ &\leq \|w_h\|_{H^{-1/2}(\Gamma)} \sup_{0 \neq v \in L_2(\Gamma)} \frac{\|Q_h^B v\|_{H^{1/2}(\Gamma)}}{\|v\|_{L_2(\Gamma)}}. \end{aligned}$$

By applying the inverse inequality (10.16) as well as the stability estimate (10.18) we finally obtain the assertion. □

# 11

# Finite Element Methods

For the approximate solution of the variational problems as described in Chapter 4 we will use the finite–dimensional trial spaces which were constructed in Chapter 9. Here we will just consider finite elements of lowest order, in particular we will use piecewise linear and continuous basis functions. The stability and error analysis is imbedded in the general theory as given in Chapter 8. Some numerical examples illustrate the theoretical results.

## 11.1 Dirichlet Boundary Value Problem

For the Poisson equation we consider the Dirichlet boundary value problem (1.10) and (1.11),

$$-\Delta u(x) \;=\; f(x) \quad \text{for } x \in \Omega, \quad \gamma_0^{\rm int} u(x) \;=\; g(x) \quad \text{for } x \in \Gamma = \partial\Omega. \qquad (11.1)$$

Let $u_g \in H^1(\Omega)$ be some bounded extension of the given Dirichlet datum $g \in H^{1/2}(\Gamma)$. Then the variational problem is to find $u_0 := u - u_g \in H^1_0(\Omega)$ such that

$$\int\limits_\Omega \nabla u_0(x) \nabla v(x) dx \;=\; \int\limits_\Omega f(x) v(x) dx - \int\limits_\Omega \nabla u_g(x) \nabla v(x) dx \qquad (11.2)$$

is satisfied for all $v \in H^1_0(\Omega)$. By using Theorem 4.3 we can state the unique solvability of the above variational formulation.

Let

$$X_h \;:=\; S^1_h(\Omega) \cap H^1_0(\Omega) \;=\; \text{span}\{\varphi^1_i\}_{i=1}^{\widetilde{M}}$$

be the conformal trial space of piecewise linear and globally continuous basis functions $\varphi^1_k$ which are zero on the boundary $\partial\Omega$. Note that the trial space is defined with respect to some admissible decomposition $\overline{\Omega} = \cup_{\ell=1}^N \overline{\tau}_\ell$ of $\Omega$ into finite elements $\tau_\ell$. Then the Galerkin variational problem of (11.2) is to find $u_{0,h} \in X_h$ such that

$$\int_\Omega \nabla u_{0,h}(x)\nabla v_h(x)dx = \int_\Omega f(x)v_h(x)dx - \int_\Omega \nabla u_g(x)\nabla v_h(x)dx \qquad (11.3)$$

is satisfied for all $v_h \in X_h$. By applying Theorem 8.1 (Cea's Lemma) there exists a unique solution $u_{0,h}$ of the Galerkin variational problem (11.3) satisfying the error estimate (8.7),

$$\|u_0 - u_{0,h}\|_{H^1(\Omega)} \le \frac{c_2^A}{c_1^A} \inf_{v_h \in X_h} \|u_0 - v_h\|_{H^1(\Omega)} .$$

If the solution $u$ of the Dirichlet boundary value problem (11.1) satisfies $u \in H^s(\Omega)$ for some $s \in [1,2]$, then we obtain by applying the trace theorem $g = \gamma_0^{\text{int}} u \in H^{s-1/2}(\Gamma)$, i.e. the extension $u_g$ of the given Dirichlet datum $g$ can be chosen such that $u_g \in H^s(\Omega)$ is satisfied. Therefore we have $u_0 = u - u_g \in H^s(\Omega)$ and from the approximation property (9.33) we conclude the error estimate

$$\|u_0 - u_{0,h}\|_{H^1(\Omega)} \le c\, h^{s-1} |u|_{H^s(\Omega)} \quad \text{for } u \in H^s(\Omega),\ s \in [1,2]. \qquad (11.4)$$

When assuming certain smoothness properties of the domain $\Omega$ we can prove, by using some duality arguments, error estimates which are valid in $L_2(\Omega)$.

**Theorem 11.1 (Aubin–Nitsche Trick).** *Suppose that $\Omega$ is convex or the boundary $\Gamma = \partial\Omega$ is smooth. For given $f \in L_2(\Omega)$ and $g = \gamma_0^{int} u_g$ with $u_g \in H^2(\Omega)$ let $u_0 \in H_0^1(\Omega)$ be the unique solution of the variational problem*

$$\int_\Omega \nabla u_0(x)\nabla v(x)dx = \int_\Omega f(x)v(x)dx - \int_\Omega \nabla u_g(x)\nabla v(x)dx$$

*to be satisfied for all $v \in H_0^1(\Omega)$. Assume that*

$$\|u_0\|_{H^2(\Omega)} \le c \left\{ \|f\|_{L_2(\Omega)} + \|u_g\|_{H^2(\Omega)} \right\} .$$

*The approximate solution $u_{0,h} \in X_h$ of the Galerkin variational problem* (11.3) *then satisfies the error estimate*

$$\|u_0 - u_{0,h}\|_{L_2(\Omega)} \le c\, h^2 \left[ \|f\|_{L_2(\Omega)} + \|u_g\|_{H^2(\Omega)} \right] .$$

*Proof.* By assumption we have $u_0 \in H^2(\Omega)$, and for the approximate solution $u_{0,h} \in X_h$ we get by using (11.4) the error estimate

$$\|u_0 - u_{0,h}\|_{H^1(\Omega)} \le c\, h\, |u_0|_{H^2(\Omega)} \le c\, h \left[ \|f\|_{L_2(\Omega)} + \|u_g\|_{H^2(\Omega)} \right] . \qquad (11.5)$$

Let $w \in H_0^1(\Omega)$ be the unique solution of the variational problem

$$\int_\Omega \nabla w(x)\nabla v(x)dx = \int_\Omega [u_0(x) - u_{0,h}(x)]v(x)dx$$

to be satisfied for all $v \in H_0^1(\Omega)$. Due to the assumptions made on $\Omega$ we conclude $w \in H^2(\Omega)$, and

$$\|w\|_{H^2(\Omega)} \leq c\,\|u_0 - u_{0,h}\|_{L_2(\Omega)}.$$

Due to $u_0 - u_{0,h} \in H_0^1(\Omega)$ and by using the Galerkin orthogonality

$$\int_\Omega \nabla[u_0(x) - u_{0,h}(x)]\nabla v_h(x)dx = 0 \quad \text{for all } v_h \in X_h$$

we get

$$\begin{aligned}\|u_0 - u_{0,h}\|^2_{L_2(\Omega)} &= \int_\Omega [u_0(x) - u_{0,h}(x)][u_0(x) - u_{0,h}(x)]dx \\ &= \int_\Omega \nabla w(x)\nabla[u_0(x) - u_{0,h}(x)]dx \\ &= \int_\Omega \nabla[w(x) - Q_h^1 w(x)]\nabla[u_0(x) - u_{0,h}(x)]dx \\ &\leq \|w - Q_h^1 w\|_{H^1(\Omega)}\|u_0 - u_{0,h}\|_{H^1(\Omega)}\end{aligned}$$

where $Q_h^1 : H_0^1(\Omega) \to X_h \subset H_0^1(\Omega)$ is the $H_0^1(\Omega)$ projection which is defined similar to (9.29). For $w \in H^2(\Omega)$ we obtain from the approximation property (9.32) of the trial space $X_h$ the error estimate

$$\|w - Q_h^1 w\|_{H^1(\Omega)} \leq c\,h\,\|w\|_{H^2(\Omega)} \leq c\,h\,\|u_0 - u_{0,h}\|_{L_2(\Omega)}.$$

Altogether we have

$$\|u_0 - u_{0,h}\|_{L_2(\Omega)} \leq c\,h\,\|u_0 - u_{0,h}\|_{H^1(\Omega)},$$

and by using (11.5) the assertion follows. □

To realize the Galerkin variational formulation (11.3) we need to know the bounded extension $u_g \in H^2(\Omega)$ of the given Dirichlet datum $g$. Formally, for $u_g \in H^2(\Omega)$ we denote by $I_h u_g \in S_h^1(\Omega)$ the piecewise linear interpolation which can be written as

$$I_h u_g(x) = \sum_{i=1}^M u_g(x_i)\varphi_i^1(x).$$

Instead of the exact Galerkin variational formulation (11.3) we now consider a perturbed variational problem to find $\widetilde{u}_{0,h} \in X_h$ such that

$$\int_\Omega \nabla\widetilde{u}_{0,h}(x)\nabla v_h(x)dx = \int_\Omega f(x)v_h(x)dx - \int_\Omega \nabla I_h u_g(x)\nabla v_h(x)dx \qquad (11.6)$$

is satisfied for all $v_h \in X_h$. For the unique solution $\widetilde{u}_{0,h} \in X_h$ we have, by using Theorem 8.2 (Strang Lemma), the error estimate

$$\|u_0 - \widetilde{u}_{0,h}\|_{H^1(\Omega)} \leq \frac{c_2^A}{c_1^A} \left[ \inf_{v_h \in X_h} \|u_0 - v_h\|_{H^1(\Omega)} + \|u_g - I_h u_g\|_{H^1(\Omega)} \right].$$

If we assume $u \in H^s(\Omega)$ for some $s \in [1,2]$ and $u_g \in H^2(\Omega)$, then by using the approximation property of the trial space $X_h$ as well as the interpolation estimate (9.22) we get

$$\begin{aligned}\|u_0 - \widetilde{u}_{0,h}\|_{H^1(\Omega)} &\leq c_1 h^{s-1} |u_0|_{H^s(\Omega)} + c_2 h |u_g|_{H^2(\Omega)} \\ &\leq c h^{s-1} \left[ |u_0|_{H^s(\Omega)} + \|g\|_{H^{3/2}(\Gamma)} \right].\end{aligned}$$

The piecewise linear interpolation $I_h u_g$ can be written, by considering the interior nodes $\{x_i\}_{i=1}^{\widetilde{M}}$, $x_i \in \Omega$ and the boundary nodes $\{x_i\}_{i=\widetilde{M}+1}^{M}$, $x_i \in \Gamma$, as

$$I_h u_g(x) = \sum_{i=1}^{\widetilde{M}} u_g(x_i) \varphi_i^1(x) + \sum_{i=\widetilde{M}+1}^{M} g(x_i) \varphi_i^1(x).$$

Hence the perturbed Galerkin variaitional formulation (11.6) is equivalent to finding

$$\bar{u}_{0,h}(x) := \widetilde{u}_{0,h}(x) + \sum_{i=1}^{\widetilde{M}} I_h u_g(x_i) \varphi_i^1(x) =: \sum_{i=1}^{\widetilde{M}} \bar{u}_i \varphi_i^1(x) \in X_h$$

as the unique solution of the variational problem

$$\int_\Omega \nabla \bar{u}_{0,h}(x) \nabla v_h(x) dx = \int_\Omega f(x) v_h(x) dx - \sum_{i=\widetilde{M}+1}^{M} g(x_i) \int_\Omega \nabla \varphi_i^1(x) \nabla v_h(x) dx \tag{11.7}$$

to be satisfied for all $v_h \in X_h$. The resulting approximate solution of the Dirichlet boundary value problem is then given by

$$u_h(x) := \sum_{i=1}^{\widetilde{M}} \bar{u}_k \varphi_i^1(x) + \sum_{i=\widetilde{M}+1}^{M} g(x_i) \varphi_i^1(x) \in S_h^1(\Omega). \tag{11.8}$$

**Theorem 11.2.** *Let $u \in H^s(\Omega)$ for some $s \in [1,2]$ be the unique solution of the Dirichlet boundary value problem* (11.1). *Let $u_g \in H^2(\Omega)$ be an appropriate chosen extension of the given Dirichlet datum $g$. Let $u_h$ be the approximate solution of* (11.1) *which is defined by* (11.8). *Then there holds the error estimate*

$$\|u - u_h\|_{H^1(\Omega)} \leq c h^{s-1} \left[ |u_0|_{H^s(\Omega)} + \|u_g\|_{H^2(\Omega)} \right] \tag{11.9}$$

*Proof.* By using the triangle inequality we have

$$\begin{aligned}\|u-u_h\|_{H^1(\Omega)} &= \|u_0+u_g-(\widetilde{u}_{0,h}+I_hu_g)\|_{H^1(\Omega)}\\ &\leq \|u_0-\widetilde{u}_{0,h}\|_{H^1(\Omega)}+\|u_g-I_hu_g\|_{H^1(\Omega)}.\end{aligned}$$

Therefore, the assertion follows from Theorem 8.2, by estimating the interpolation error, and by applying the inverse trace theorem. □

If the solution $u \in H^2(\Omega)$ is sufficiently regular, we therefore obtain the error estimate

$$\|u-u_h\|_{H^1(\Omega)} \leq c\,h\,|u|_{H^2(\Omega)}. \tag{11.10}$$

As in Theorem 11.1 (Aubin–Nitsche Trick) we can also prove an estimate for the error $\|u_0-\widetilde{u}_{0,h}\|_{L_2(\Omega)}$.

**Lemma 11.3.** *Let all assumptions of Theorem* 11.1 *be valid, in particular we assume* $u \in H^2(\Omega)$ *to be the unique solution of the Dirichlet boundary value problem* (11.1). *Then there holds the error estimate*

$$\|u_0-\widetilde{u}_{0,h}\|_{L_2(\Omega)} \leq c\,h^2\left[|u_0|_{H^2(\Omega)}+|u_g|_{H^2(\Omega)}\right]. \tag{11.11}$$

*Proof.* Let $w \in H_0^1(\Omega)$ be the unique solution of the variational problem

$$\int_\Omega \nabla w(x)\nabla v(x)dx \;=\; \int_\Omega [u_0(x)-\widetilde{u}_{0,h}(x)]v(x)dx$$

to be satisfied for all $v \in H_0^1(\Omega)$. For $u_0-\widetilde{u}_{0,h} \in L_2(\Omega)$ and due to the assumptions made on $\Omega$ we conclude $w \in H^2(\Omega)$, and therefore we have

$$\|w\|_{H^2(\Omega)} \leq c\|u_0-\widetilde{u}_{0,h}\|_{L_2(\Omega)}.$$

By subtracting the perturbed variational problem (11.6) from the variational problem (11.2) this gives

$$\int_\Omega \nabla[u_0(x)-\widetilde{u}_{0,h}(x)]\nabla v_h(x)dx \;=\; \int_\Omega \nabla[I_hu_g(x)-u_g(x)]\nabla v_h(x)dx$$

for all $v_h \in X_h$. Hence we have

$$\begin{aligned}\|u_0-\widetilde{u}_{0,h}\|^2_{L_2(\Omega)} &= \int_\Omega [u_0(x)-\widetilde{u}_{0,h}(x)][u_0(x)-\widetilde{u}_{0,h}(x)]dx\\ &= \int_\Omega \nabla w(x)\nabla[u_0(x)-\widetilde{u}_{0,h}(x)]dx\\ &= \int_\Omega \nabla[w(x)-Q_h^1w(x)]\nabla[u_0(x)-\widetilde{u}_{0,h}(x)]dx\\ &\quad+\int_\Omega \nabla Q_h^1w(x)\nabla[I_hu_g(x)-u_g(x)]dx.\end{aligned}$$

The first term can be estimated as in the proof of Theorem 11.1,

$$\int_\Omega \nabla[w(x) - Q_h^1 w(x)]\nabla[u_0(x) - \widetilde{u}_{0,h}(x)]dx \\ \leq c\,h\,\|u_0 - \widetilde{u}_{0,h}\|_{L_2(\Omega)}\|u_0 - \widetilde{u}_{0,h}\|_{H^1(\Omega)}.$$

For the second term we first have

$$\int_\Omega \nabla Q_h^1 w(x)\nabla[I_h u_g(x) - u_g(x)]dx \\ = \int_\Omega \nabla[Q_h^1 w(x) - w(x)]\nabla[I_h u_g(x) - u_g(x)]dx \\ + \int_\Omega \nabla w(x)\nabla[I_h u_g(x) - u_g(x)]dx.$$

The resulting first term can further be estimated by

$$\left|\int_\Omega \nabla[Q_h^1 w(x) - w(x)]\nabla[I_h u_g(x) - u_g(x)]dx\right| \\ \leq c_2^A\,\|w - Q_h^1 w\|_{H^1(\Omega)}\|u_g - I_h u_g\|_{H^1(\Omega)} \\ \leq c\,h\,\|w\|_{H^2(\Omega)}\|u_g - I_h u_g\|_{H^1(\Omega)} \\ \leq c\,h\,\|u_0 - \widetilde{u}_{0,h}\|_{L_2(\Omega)}\|u_g - I_h u_g\|_{H^1(\Omega)}.$$

For the remaining second term we have, by applying integration by parts, for $w \in H_0^1(\Omega) \cap H^2(\Omega)$,

$$\left|\int_\Omega \nabla w(x)\nabla[I_h u_g(x) - u_g(x)]dx\right| = \left|-\int_\Omega \Delta w(x)[I_h u_g(x) - u_g(x)]dx\right| \\ \leq \|w\|_{H^2(\Omega)}\|u_g - I_h u_g\|_{L_2(\Omega)} \\ \leq c\,\|u_0 - \widetilde{u}_{0,h}\|_{L_2(\Omega)}\|u_g - I_h u_g\|_{L_2(\Omega)}.$$

Altogether we have

$$\|u_0 - \widetilde{u}_{0,h}\|_{L_2(\Omega)} \\ \leq c_1\,h\,\left[\|u_0 - \widetilde{u}_{0,h}\|_{H^1(\Omega)} + \|u_g - I_h u_g\|_{H^1(\Omega)}\right] + \|u_g - I_h u_g\|_{L_2(\Omega)}.$$

For $u_0, u_g \in H^2(\Omega)$ the assertion now follows from the error estimates for $\|u_0 - \widetilde{u}_{0,h}\|_{H^1(\Omega)}$ as well as from the interpolation error estimates for $\|u_g - I_h u_g\|_{L_2(\Omega)}$ and $\|u_g - I_h u_g\|_{H^1(\Omega)}$. □

The Galerkin variational problem (11.7) to find the coefficients $\bar{u}_i$ for $i = 1, \ldots, \widetilde{M}$ is equivalent to an algebraic system of linear equations $A_h \underline{\bar{u}} = \underline{f}$ with the stiffness matrix $A_h$ defined by

$$A_h[j,i] = \int_\Omega \nabla\varphi_i^1(x) \nabla\varphi_j^1(x) dx$$

for $i, j = 1, \ldots, \widetilde{M}$, and with the right hand side vector $\underline{f}$ given by

$$f_j = \int_\Omega f(x)\varphi_j^1(x)dx - \sum_{i=\widetilde{M}+1}^{M} g(x_i) \int_\Omega \nabla\varphi_i^1(x)\nabla\varphi_j^1(x)dx$$

for $j = 1, \ldots, \widetilde{M}$. The stiffness matrix $A_h$ is symmetric and due to

$$(A_h \underline{v}, \underline{v}) = \int_\Omega \nabla v_h(x) \nabla v_h(x) dx \geq c_1^A \, \|v_h\|^2_{H^1(\Omega)}$$

for all $\underline{v} \in \mathbb{R}^{\widetilde{M}} \leftrightarrow v_h \in X_h \subset H_0^1(\Omega)$ positive definite. In particular we have the following result:

**Lemma 11.4.** *For all* $\underline{v} \in \mathbb{R}^{\widetilde{M}} \leftrightarrow v_h \in X_h \subset H_0^1(\Omega)$ *there hold the spectral equivalence inequalities*

$$c_1 \, h_{\min}^d \, \|\underline{v}\|_2^2 \leq (A_h \underline{v}, \underline{v}) \leq c_2 \, h_{\max}^{d-2} \, \|\underline{v}\|_2^2 \tag{11.12}$$

*Proof.* For $\underline{v} \in \mathbb{R}^{\widetilde{M}} \leftrightarrow v_h \in X_h \subset H_0^1(\Omega)$ it follows by localization and by applying the local inverse inequality (9.15)

$$\begin{aligned}
(A_h \underline{v}, \underline{v}) &= \sum_{i=1}^{\widetilde{M}} \sum_{j=1}^{\widetilde{M}} A_h[j,i] v_i v_j = \sum_{i=1}^{\widetilde{M}} \sum_{j=1}^{\widetilde{M}} a(\varphi_i^1, \varphi_j^1) v_i v_j \\
&= a\Big(\sum_{i=1}^{\widetilde{M}} v_i \varphi_i^1, \sum_{j=1}^{\widetilde{M}} v_j \varphi_j^1\Big) = a(v_h, v_h) \\
&= \int_\Omega |\nabla v_h(x)|^2 dx = \sum_{\ell=1}^{N} \int_{\tau_\ell} |\nabla v_h(x)|^2 dx \\
&= \sum_{\ell=1}^{N} \|\nabla v_h\|^2_{L_2(\tau_\ell)} \leq c_I \sum_{\ell=1}^{N} h_\ell^{-2} \|v_h\|^2_{L_2(\tau_\ell)} \\
&\leq c \sum_{\ell=1}^{N} h_\ell^{-2} \Delta_\ell \sum_{k \in J(\ell)} v_k^2 = c \sum_{k=1}^{\widetilde{M}} \left( \sum_{\ell \in I(k)} h_\ell^{d-2} \right) v_k^2
\end{aligned}$$

and therefore the upper estimate. On the other hand, by using the $H_0^1(\Omega)$ ellipticity of the bilinear form $a(\cdot,\cdot)$ and by changing to the $L_2(\Omega)$ norm we get

$$\begin{aligned}(A_h\underline{v},\underline{v}) &= a(v_h,v_h) \geq c_1^A \|v_h\|^2_{H^1(\Omega)} \geq c_1^A \|v_h\|^2_{L_2(\Omega)} \\ &= c_1^A \sum_{\ell=1}^{N} \|v_h\|^2_{L_2(\tau_\ell)} \geq c \sum_{\ell=1}^{N} \Delta_\ell \sum_{k\in J(\ell)} v_k^2 = c \sum_{k=1}^{\widetilde{M}} \left( \sum_{\ell\in I(k)} h_\ell^d \right) v_k^2\end{aligned}$$

and therefore the lower estimate. □

Note that the constants in the spectral equivalence inequalities (11.12) are sharp, i.e. the constants can not be improved. Hence we have for the spectral condition number of the stiffness matrix $A_h$ in the case of a globally quasi–uniform mesh

$$\kappa_2(A_h) \leq c\,h^{-2}, \tag{11.13}$$

in particular the spectral condition number increases when the mesh is refined. As an example we consider a Dirichlet boundary value problem where the domain is given by the square $\Omega = (0, 0.5)^2$. The initial mesh consists of four finite elements with five nodes, which are recursively refined by decomposing each finite element into four congruent elements, see Fig. 11.1 for the refinement levels $L = 0$ and $L = 3$.

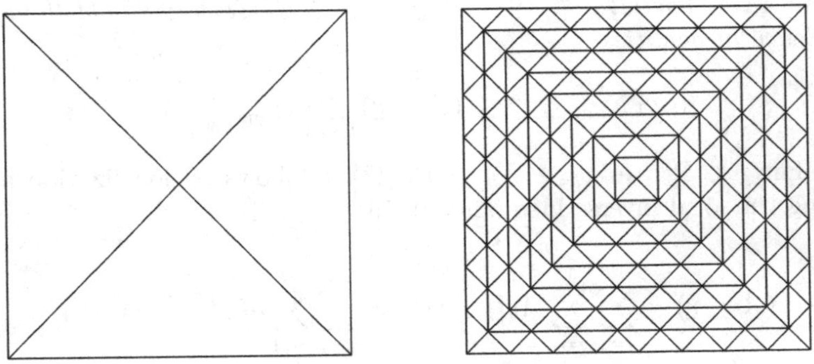

**Fig. 11.1.** Initial mesh ($L = 0$) and refined mesh $L = 3$.

The minimal and maximal eigenvalues and the resulting spectral condition numbers of the associated finite element stiffness matrices are given in Table 11.1. Note that when choosing $d = 2$ and $h = \mathcal{O}(N^{-2})$ the results of Lemma 11.4 are confirmed. Thus, when using a conjugate gradient scheme to solve the linear equation systems $A_h\underline{u} = \underline{f}$ with the symmetric and positive definite system matrices $A_h$ we need to have an appropriate preconditioner to bound the number of necessary iteration steps to reach a prescribed accuracy independent of the system size. Note that preconditioned iterative schemes are considered later in Chapter 13.

| $L$ | $N$ | $\lambda_{\min}(A_h)$ | $\lambda_{\max}(A_h)$ | $\kappa_2(A_h)$ |
|---|---|---|---|---|
| 2 | 64 | 5.86 −1 | 7.41 | 12.66 |
| 3 | 256 | 1.52 −1 | 7.85 | 51.55 |
| 4 | 1024 | 3.84 −2 | 7.96 | 207.17 |
| 5 | 4096 | 9.63 −3 | 7.99 | 829.69 |
| 6 | 16384 | 2.41 −3 | 8.00 | 3319.76 |
| 7 | 65536 | 6.02 −4 | 8.00 | 13280.04 |
| 8 | 262144 | 1.52 −4 | 8.00 | 52592.92 |
| Theory: | | $\mathcal{O}(h^2)$ | $\mathcal{O}(1)$ | $\mathcal{O}(h^{-2})$ |

**Table 11.1.** Spectral condition numbers of the stiffness matrices $A_h$.

Next we will discuss the computation of the load vector $\underline{f}$ and the realization of a matrix by vector multiplication with the stiffness matrix $A_h$ as needed in the application of an iterative solution scheme. For $j = 1, \ldots, \widetilde{M}$ we have by localization and parametrization, see Chapter 9,

$$\widetilde{f}_j = \int_\Omega f(x)\varphi_j^1(x)dx = \sum_{\ell\in I(j)} \int_{\tau_\ell} f(x)\varphi_j^1(x)dx$$

$$= \sum_{\ell\in\tau_j} |\det J_\ell| \int_\tau f(x_{\ell_1} + J_\ell\xi)\psi^1_{\ell_j}(\xi)d\xi$$

where $\ell_j$ is the local index of the global node $x_j$ with respect to the finite element $\tau_\ell$. Hence we can reduce the computation of the global load vector $\widetilde{\underline{f}}$ to the computation of local load vectors $\widetilde{\underline{f}}_\ell$. In particular, for each finite element $\tau_\ell$ we need to compute

$$\widetilde{f}_{\ell,\iota} = |\det J_\ell| \int_\tau f(x_{\ell_1} + J_\ell\xi)\psi_\iota^1(\xi)d\xi \quad \text{for } \iota = 1, \ldots, d+1.$$

For $\iota = 1, \ldots, d+1$ we denote by $\ell_\iota$ the corresponding global node index, then the global load vector $\widetilde{\underline{f}}$ is computed by assembling all local load vectors $\widetilde{\underline{f}}_\ell$, i.e.

$$\widetilde{f}_{\ell_\iota} := \widetilde{f}_{\ell_\iota} + \widetilde{f}_{\ell,\iota}.$$

If $\underline{u} \in \mathbb{R}^{\widetilde{M}} \leftrightarrow u_h \in X_h$ is given, the result $\underline{v} = A_h\underline{u}$ of a matrix by vector multiplication with the global stiffness matrix $A_h$ can be written as

$$v_j = \sum_{i=1}^{\widetilde{M}} A_h[j,i]u_i = a(u_h, \varphi_j^1) = \sum_{\ell\in I(j)} \int_{\tau_\ell} \nabla u_h(x)\nabla\varphi_j^1(x)dx$$

$$= \sum_{\ell\in I(j)} \sum_{i\in J(\ell)} u_i \int_{\tau_\ell} \nabla\varphi_i^1(x)\nabla\varphi_j^1(x)dx, \quad j = 1, \ldots, \widetilde{M}.$$

Therefore it is sufficient to compute the local stiffness matrices $A_h^\ell$ which are defined by

$$A_h^\ell[\iota', \iota] = |\det J_\ell| \int_T J_\ell^{-\top} \nabla_\xi \psi_\iota^1(\xi) J_\ell^{-\top} \nabla_\xi \psi_{\iota'}^1(\xi) d\xi$$

for $\iota, \iota' = 1, \ldots, d+1$. Hence we can reduce the matrix by vector multiplication with the global stiffness matrix $A_h$ to a localization of the global degrees of freedom $\underline{u} \in \mathbb{R}^{\widetilde{M}}$ to local degrees of freedom $\underline{u}_\ell \in \mathbb{R}^{d+1}$ where $u_{\ell,\iota} = u_{\ell_\iota}$, a multiplication with the local stiffness matrices $\underline{v}_\ell = A_h^\ell \underline{u}_\ell$, and the assembling of the global vector $\underline{v}$ from the local results $\underline{v}_\ell$, i.e.

$$v_{\ell_\iota} := v_{\ell_\iota} + v_{\ell,\iota}.$$

The incorporation of given Dirichlet boundary conditions to compute the global load vector $\underline{f}$ can be done in the same way. Hence, for a matrix by vector multiplication with the global stiffness matrix $A_h$ we only need to store the local stiffness matrices $A_h^\ell$ with an effort of $\mathcal{O}(N)$ essential operations. For alternative approaches to describe the sparse stiffness matrix $A_h$, see, for example, [85].

To check the theoretic error estimates (11.9) and (11.11) we now consider the Dirichlet boundary value problem (11.1) where $\Omega = (0, 0.5)^2$ and $f = 0$ are given, and where the Dirichlet boundary data are prescribed such that

$$u(x) = -\frac{1}{2} \log |x - x^*|, \quad x^* = (-0.1, -0.1)^\top \tag{11.14}$$

is the exact solution. In Table 11.2 we give the errors of the approximate Galerkin solutions $u_h$ for a sequence of uniformly refined meshes, where $L$ is the refinement level, $N$ is the number of finite elements, $M$ is the total number of nodes, and DoF is the number of degrees of freedom which coincides with the number of interior nodes.

| $L$ | $N$ | $M$ | DoF | $\lvert u - u_h \rvert_{H^1(\Omega)}$ | eoc | $\lVert u - u_h \rVert_{L_2(\Omega)}$ | eoc |
|---|---|---|---|---|---|---|---|
| 2 | 64 | 41 | 25 | 1.370 –1 | | 2.460 –3 | |
| 3 | 256 | 145 | 113 | 6.954 –2 | 0.98 | 5.717 –4 | 2.11 |
| 4 | 1024 | 545 | 481 | 3.494 –2 | 0.99 | 1.408 –4 | 2.02 |
| 5 | 4096 | 2113 | 1985 | 1.749 –2 | 1.00 | 3.511 –5 | 2.00 |
| 6 | 16384 | 8321 | 8065 | 8.748 –3 | 1.00 | 8.771 –6 | 2.00 |
| 7 | 65536 | 33025 | 32513 | 4.374 –3 | 1.00 | 2.192 –6 | 2.00 |
| 8 | 262144 | 131585 | 130561 | 2.187 –3 | 1.00 | 5.481 –7 | 2.00 |
| 9 | 1048576 | 525313 | 523265 | 1.094 –3 | 1.00 | 1.370 –7 | 2.00 |
| Theory: | | | | | 1 | | 2 |

**Table 11.2.** Errors and estimated order of convergence, Dirichlet problem.

Since the solution (11.14) of the Dirichlet boundary value problem is infinitely often differentiable, we can apply Theorem 11.2 and Lemma 11.3 for $s = 2$. Hence we obtain one as order of convergence when measuring the error in the energy norm $|u - u_h|_{H^1(\Omega)}$. Moreover, when applying the Aubin–Nitsche trick we get two as order of convergence when measuring the error in the $L_2$ norm $\|u - u_h\|_{L_2(\Omega)}$. By eoc we denote the estimated order of convergence which can be computed from

$$\text{eoc} := \frac{\log \|u - u_{h_\ell}\| - \log \|u - u_{h_{\ell+1}}\|}{\log h_\ell - \log h_{\ell+1}}.$$

Note that the theoretical error estimates are well confirmed by the numerical results as documented in Table 11.2.

## 11.2 Neumann Boundary Value Problem

We now consider the Neumann boundary value problem (1.10) and (1.12) for the Poisson equation,

$$-\Delta u(x) = f(x) \quad \text{for } x \in \Omega, \quad \gamma_1^{\text{int}} u(x) = g(x) \quad \text{for } x \in \Gamma = \partial\Omega \tag{11.15}$$

where we have to assume the solvability condition (1.17)

$$\int_\Omega f(x)\,dx + \int_\Gamma g(x)\,ds_x = 0. \tag{11.16}$$

For a finite element discretization we consider the modified variational problem (4.31) which admits, due to Section 4.1.3, a unique solution $u \in H^1(\Omega)$ such that

$$\int_\Omega \nabla u(x)\nabla v(x)dx + \int_\Omega u(x)dx \int_\Omega v(x)dx = \int_\Omega f(x)v(x)dx + \int_\Gamma g(x)\gamma_0^{\text{int}} v(x)ds_x \tag{11.17}$$

is satisfied for all $v \in H^1(\Omega)$. From the solvability condition (11.16) we then conclude the scaling condition $u \in H^1_*(\Omega)$, i.e. $u \in H^1(\Omega)$ satisfying $\langle u, 1\rangle_{L_2(\Omega)} = 0$.

Let

$$X_h := S_h^1(\Omega) = \text{span}\{\varphi_k^1\}_{k=1}^M \subset H^1(\Omega)$$

be the conforming trial space of piecewise linear and globally continuous basis functions $\varphi_k^1$ with respect to an admissible finite element mesh $\overline{\Omega} = \cup_{\ell=1}^N \overline{\tau}_\ell$. Note that the basis functions $\{\varphi_k^1\}_{k=1}^M$ build a partition of unity, i.e.

$$\sum_{k=1}^M \varphi_k^1(x) = 1 \quad \text{for all } x \in \Omega. \tag{11.18}$$

The Galerkin variational formulation of (11.17) is to find $u_h \in X_h$ such that

$$\int_\Omega \nabla u_h(x)\nabla v_h(x)dx + \int_\Omega u_h(x)dx \int_\Omega v_h(x)dx \tag{11.19}$$
$$= \int_\Omega f(x)v_h(x)dx + \int_\Gamma g(x)v_h(x)ds_x$$

is satisfied for all $v_h \in X_h$. By applying Theorem 8.1 (Cea's Lemma) we conclude that there exists a unique solution $u_h$ of the variational problem (11.19) satisfying the error estimate (8.7),

$$\|u - u_h\|_{H^1(\Omega)} \leq \frac{\hat{c}_1^A}{c_2^A} \inf_{v_h \in X_h} \|u - v_h\|_{H^1(\Omega)} \,.$$

If the solution $u$ of the Neumann boundary value problem (11.15) satisfies $u \in H^s(\Omega)$ for some $s \in [1,2]$, then we conclude, by using the approximation property (9.33) the error estimate

$$\|u - u_h\|_{H^1(\Omega)} \leq c\, h^{s-1}\, |u|_{H^s(\Omega)} \quad \text{for } u \in H^s(\Omega), \quad s \in [1,2].$$

Due to (11.18) we can choose $v_h \equiv 1$ as a test function of the Galerkin variational formulation (11.19). From the solvability condition (11.16) we then obtain

$$\int_\Omega u_h(x)dx \int_\Omega dx = 0$$

and therefore $u_h \in H^1_*(\Omega)$, i.e. the scaling condition is automatically satisfied for the Galerkin solution $u_h \in X_h$.

The Galerkin variational problem (11.19) to find the coefficient vector $\underline{u} \in \mathbb{R}^M$ is equivalent to the solution of the linear system of algebraic equations

$$\left[A_h + \underline{a}\,\underline{a}^\top\right]\underline{u} = \underline{f}$$

where the stiffness matrix $A_h$ defined by

$$A_h[j,i] = \int_\Omega \nabla\varphi_i^1(x)\nabla\varphi_j^1(x)dx$$

for $i,j = 1,\ldots,M$, and with the load vector given by

$$f_j := \int_\Omega f(x)\varphi_j^1(x)dx + \int_\Gamma g(x)\varphi_j^1(x)ds_x$$

for $j = 1,\ldots,M$. In addition, $\underline{a} \in \mathbb{R}^M$ is defined by

$$a_i = \int_\Omega \varphi_i^1(x)\,dx$$

for $i = 1, \ldots, M$. Note that the modified stiffness matrix $A_h + \underline{a}\,\underline{a}^\top$ is symmetric and positive definite. Moreover, the spectral equivalence inequalities (11.12) remain valid. Hence, when using the conjugate gradient scheme to solve the linear system iteratively we have to use again an appropriate preconditioner. Both the computation of the load vector $\underline{f}$ as well as an application of the matrix by vector product with the stiffness matrix $A_h$ can be realized as for the Dirichlet boundary value problem.

The approximate solution of boundary value problems with mixed boundary conditions as well as with Robin boundary conditions can be formulated and analyzed as for the Dirichlet or Neumann boundary value problem. Here, we will not discuss this in detail. The same is true for the approximate solution of boundary value problems in linear elasticity. However, when considering the Neumann boundary value problem we have to modify both the solvability conditions as well as the definition of the modified variational problem due to the rigid body motions.

## 11.3 Finite Element Methods with Lagrange Multipliers

For an alternative approximation of the Dirichlet boundary value problem (11.1) we consider the modified saddle point problem (4.22) and (4.23) to find $(u, \lambda) \in H^1(\Omega) \times H^{-1/2}(\Gamma)$ such that

$$\int\limits_\Gamma \gamma_0^{\text{int}} u(x) ds_x \int\limits_\Gamma \gamma_0^{\text{int}} v(x) ds_x + \int\limits_\Omega \nabla u(x) \nabla v(x) dx - \int\limits_\Gamma \gamma_0^{\text{int}} v(x) \lambda(x) ds_x$$
$$= \langle f, v\rangle_\Omega + \int\limits_\Gamma g(x) ds_x \int\limits_\Gamma \gamma_0^{\text{int}} v(x) ds_x \qquad (11.20)$$
$$\int\limits_\Gamma \gamma_0^{\text{int}} u(x) \mu(x) ds_x + \int\limits_\Gamma \lambda(x) ds_x \int\limits_\Gamma \mu(x) ds_x$$
$$= \langle g, \mu\rangle_\Gamma - \int\limits_\Omega f(x) dx \int\limits_\Gamma \mu(x) ds_x$$

is satisfied for all $(v, \mu) \in H^1(\Omega) \times H^{-1/2}(\Gamma)$.

Assume that there is given an admissible finite element mesh $\overline{\Omega} = \cup_{\ell=1}^{N_\Omega} \overline{\tau}_\ell$ of the polygonal or polyhedral bounded domain $\Omega \subset \mathbb{R}^d$. The restriction of the finite element mesh in $\Omega$ defines a boundary element mesh $\Gamma = \cup_{\ell=1}^{N_\Gamma} \overline{\Gamma}_\ell$ on $\Gamma = \partial\Omega$. Let

$$X_h(\Omega) := S_h^1(\Omega) = \text{span}\{\varphi_i^1\}_{i=1}^M \subset H^1(\Omega)$$

be the conforming finite element space of piecewise linear and globally continuous basis functions $\varphi_i^1$. The restriction of $X_h(\Omega)$ onto $\Gamma = \partial\Omega$ then defines a boundary element space

$$X_h(\Gamma) := S_h^1(\Gamma) = \operatorname{span}\{\phi_k^1\}_{i=1}^{M_\Gamma} \subset H^{1/2}(\Gamma)$$

of piecewise linear and continuous basis functions $\phi_k^1$. We denote by $M_\Omega$ the number of all interior nodes $x_i \in \Omega$, and we have $M = M_\Omega + M_\Gamma$ as well as

$$X_h(\Omega) = \operatorname{span}\{\varphi_i^1\}_{i=1}^{M_\Omega} \cup \operatorname{span}\{\varphi_i^1\}_{i=M_\Omega+1}^{M} \, .$$

In particular,

$$\phi_i^1 = \gamma_0^{\text{int}} \varphi_{M_\Omega+i}^1 \quad \text{for } i = 1, \ldots, M_\Gamma .$$

Moreover, let

$$\Pi_H := \operatorname{span}\{\psi_k\}_{k=1}^{N} \subset H^{-1/2}(\Gamma)$$

denote a suitable trial space to approximate the Lagrange multiplier $\lambda$.

The Galerkin discretization of the saddle point problem (11.20) is to find $(u_h, \lambda_H) \in X_h \times \Pi_H$ such that

$$\int_\Gamma \gamma_0^{\text{int}} u_h(x) ds_x \int_\Gamma \gamma_0^{\text{int}} v_h(x) ds_x + \int_\Omega \nabla u_h(x) \nabla v_h(x) dx - \int_\Gamma \gamma_0^{\text{int}} v_h(x) \lambda_H(x) ds_x$$
$$= \langle f, v_h \rangle_\Omega + \int_\Gamma g(x) ds_x \int_\Gamma \gamma_0^{\text{int}} v_h(x) ds_x \qquad (11.21)$$

$$\int_\Gamma \gamma_0^{\text{int}} u_h(x) \mu_H(x) ds_x + \int_\Gamma \lambda_H(x) ds_x \int_\Gamma \mu_H(x) ds_x$$
$$= \langle g, \mu_H \rangle_\Gamma - \int_\Omega f(x) dx \int_\Gamma \mu_H(x) ds_x$$

is satisfied for all $(v_h, \mu_H) \in X_h \times \Pi_H$. With

$$A_h[j,i] := \int_\Omega \nabla \varphi_i^1(x) \nabla \varphi_j^1(x) dx,$$

$$B_h[\ell, i] := \int_\Gamma \psi_\ell(x) \gamma_0^{\text{int}} \varphi_i^1(x) ds_x,$$

$$a_i := \int_\Gamma \gamma_0^{\text{int}} \varphi_i^1(x) ds_x,$$

$$b_\ell := \int_\Gamma \psi_\ell(x) ds_x$$

for $i, j = 1, \ldots, M$ and $\ell = 1, \ldots, N$, as well as with

$$f_j := \int_\Omega f(x) \varphi_j^1(x) dx + \int_\Gamma g(x) ds_x \int_\Gamma \gamma_0^{\text{int}} \varphi_j^1(x) ds_x,$$

$$g_\ell := \int_\Gamma g(x) \psi_\ell(x) ds_x - \int_\Omega f(x) dx \int_\Gamma \psi_\ell(x) ds_x$$

for $j = 1, \ldots, M$ and $\ell = 1, \ldots, N$ we conclude that the approximate saddle point problem (11.21) is equivalent to a linear system of algebraic equations,

$$\begin{pmatrix} \underline{a}\,\underline{a}^\top + A_h & -B_h^\top \\ B_h & \underline{b}\,\underline{b}^\top \end{pmatrix} \begin{pmatrix} \underline{u} \\ \underline{\lambda} \end{pmatrix} = \begin{pmatrix} \underline{f} \\ \underline{g} \end{pmatrix}. \tag{11.22}$$

Obviously,

$$B_h[\ell, i] = \begin{cases} 0 & \text{for } i = 1, \ldots, M_\Omega, \\ \bar{B}_h[\ell, i - M_\Omega] & \text{for } i = M_\Omega + 1, \ldots, M \end{cases}$$

with

$$\bar{B}_h[\ell, i] = \int_\Gamma \psi_\ell(x) \phi_i^1(x) ds_x$$

for $i = 1, \ldots, M_\Gamma$ and $\ell = 1, \ldots, N$.

The matrix $\underline{a}\,\underline{a}^\top + A_h$ is by construction symmetric and positive definite and therefore invertible. In particular, the first equation in (11.22) can be solved for $\underline{u}$ to obtain

$$\underline{u} = \left[\underline{a}\,\underline{a}^\top + A_h\right]^{-1} \left[\underline{f} + B_h^\top \underline{\lambda}\right].$$

Inserting this into the second equation of (11.22) we end up with the Schur complement system

$$\left[B_h \left[\underline{a}\,\underline{a}^\top + A_h\right]^{-1} B_h^\top + \underline{b}\,\underline{b}^\top\right] \underline{\lambda} = \underline{g} - B_h \left[\underline{a}\,\underline{a}^\top + A_h\right]^{-1} \underline{f}. \tag{11.23}$$

The unique solvability of the Schur complement system (11.23) and therefore of the linear system (11.22), and hence of the discrete saddle point problem (11.21), now follows from Lemma 8.6 where we have to ensure the discrete stability condition (8.25), i.e.

$$\widetilde{c}_S \, \|\mu_H\|_{H^{-1/2}(\Gamma)} \leq \sup_{0 \neq v_h \in X_h(\Omega)} \frac{\langle \mu_H, \gamma_0^{\text{int}} v_h \rangle_\Gamma}{\|v_h\|_{H^1(\Omega)}} \quad \text{for all } \mu_H \in \Pi_H. \tag{11.24}$$

This stability condition is first considered for the boundary element trial spaces $\Pi_H$ and $X_h(\Gamma)$.

**Theorem 11.5.** *The mesh size $h$ of the trial space $X_h(\Gamma)$ is assumed to be sufficiently small compared to the mesh size $H$ of $\Pi_H$, i.e. $h \leq c_0 H$. For the trial space $\Pi_H$ we assume a global inverse inequality. Then there holds the stability condition*

$$\bar{c}_S \, \|\mu_H\|_{H^{-1/2}(\Gamma)} \leq \sup_{0 \neq w_h \in X_h(\Gamma)} \frac{\langle \mu_H, w_h \rangle_\Gamma}{\|w_h\|_{H^{1/2}(\Gamma)}} \quad \textit{for all } \mu_H \in \Pi_H. \tag{11.25}$$

*Proof.* Let $\mu_H \in \Pi_H \subset H^{-1/2}(\Gamma)$ be arbitrary but fixed. By using the Riesz representation theorem (Theorem 3.3) there exists a unique $J\mu_H \in H^{1/2}(\Gamma)$ satisfying

$$\langle J\mu_H, w\rangle_{H^{1/2}(\Gamma)} = \langle \mu_H, w\rangle_\Gamma \quad \text{for all } w \in H^{1/2}(\Gamma)$$

and $\|J\mu_H\|_{H^{1/2}(\Gamma)} = \|\mu_H\|_{H^{-1/2}(\Gamma)}$. Let $Q_h^{1/2} J\mu_H \in X_h(\Gamma)$ be the unique solution of the variational problem

$$\langle Q_h^{1/2} J\mu_H, w_h\rangle_{H^{1/2}(\Gamma)} = \langle \mu_H, w_h\rangle_\Gamma \quad \text{for all } w_h \in X_h(\Gamma).$$

Then there holds the error estimate

$$\|(I - Q_h^{1/2})J\mu_H\|_{H^{1/2}(\Gamma)} \le \inf_{w_h \in X_h(\Gamma)} \|J\mu_H - w_h\|_{H^{1/2}(\Gamma)}.$$

Due to $\mu_H \in L_2(\Gamma)$ we obtain, by using duality and the definition of $J\mu_H$,

$$\|J\mu_H\|_{H^1(\Gamma)} = \sup_{0 \ne v \in L_2(\Gamma)} \frac{\langle J\mu_H, v\rangle_{H^{1/2}(\Gamma)}}{\|v\|_{L_2(\Gamma)}} = \sup_{0 \ne v \in L_2(\Gamma)} \frac{\langle \mu_H, v\rangle_\Gamma}{\|v\|_{L_2(\Gamma)}} \le \|\mu_H\|_{L_2(\Gamma)}$$

and therefore $J\mu_H \in H^1(\Gamma)$. From the approximation property of the trial space $X_h(\Gamma)$, see the error estimate (10.13), we then obtain

$$\|(I - Q_h^{1/2})J\mu_H\|_{H^{1/2}(\Gamma)} \le c_A\, h^{1/2}\, \|J\mu_H\|_{H^1(\Gamma)} \le c_A\, h^{1/2}\, \|\mu_H\|_{L_2(\Gamma)}.$$

By applying the inverse inequality in the trial space $\Pi_H$ this gives

$$\|(I - Q_h^{1/2})J\mu_H\|_{H^{1/2}(\Gamma)} \le c_A c_I \left(\frac{h}{H}\right)^{1/2} \|\mu_H\|_{H^{-1/2}(\Gamma)}.$$

Assume that the constant $c_0$ is chosen such that $h \le c_0 H$ and

$$\|(I - Q_h^{1/2})J\mu_H\|_{H^{1/2}(\Gamma)} \le \frac{1}{2}\, \|\mu_H\|_{H^{-1/2}(\Gamma)}$$

is satisfied. Then we have

$$\begin{aligned}
\|\mu_H\|_{H^{-1/2}(\Gamma)} &= \|J\mu_H\|_{H^{1/2}(\Gamma)} \\
&\le \|Q_h^{1/2} J\mu_H\|_{H^{1/2}(\Gamma)} + \|J\mu_H - Q_h^{1/2} J\mu_H\|_{H^{1/2}(\Gamma)} \\
&\le \|Q_h^{1/2} J\mu_H\|_{H^{1/2}(\Gamma)} + \frac{1}{2}\, \|\mu_H\|_{H^{-1/2}(\Gamma)}
\end{aligned}$$

and therefore

$$\|Q_h^{1/2} J\mu_H\|_{H^{1/2}(\Gamma)} \ge \frac{1}{2}\, \|\mu_H\|_{H^{-1/2}(\Gamma)}.$$

By using the definition of $Q_h^{1/2} J\mu_H$ we get

$$\begin{aligned}\langle \mu_H, Q_h^{1/2} J\mu_H\rangle_\Gamma &= \langle Q_h^{1/2} J\mu_H, Q_h^{1/2} J\mu_H\rangle_{H^{1/2}(\Gamma)}\\ &= \|Q_h^{1/2} J\mu_H\|^2_{H^{1/2}(\Gamma)}\\ &\geq \frac{1}{2}\,\|Q_h^{1/2} J\mu_H\|_{H^{1/2}(\Gamma)}\|\mu_H\|_{H^{-1/2}(\Gamma)}\end{aligned}$$

from which the stability estimate (11.25) follows. □

*Remark 11.6.* To establish the stability condition (11.25) one may also use Fortin's criterion (Lemma 8.9). Then we have to prove the boundedness of the projection operator $\widetilde{Q}_h : H^{1/2}(\Gamma) \to X_h(\Gamma) \subset H^{1/2}(\Gamma)$ which is defined via the variational formulation

$$\langle \widetilde{Q}_h u, \mu_H\rangle_\Gamma = \langle u, \mu_H\rangle_\Gamma \quad \text{for all } \mu_H \in \Pi_H.$$

If we assume a global inverse inequality in the trial space $X_h(\Gamma)$, then we can prove the $H^{1/2}(\Gamma)$–boundedness of $\widetilde{Q}_h$ by using the error estimates of $I - \widetilde{Q}_h$ and $I - Q_h^{1/2}$ in $L_2(\Gamma)$ as well as the stability of $Q_h^{1/2}$ in $H^{1/2}(\Gamma)$. For the case of trial spaces which are defined with respect to some adaptive boundary element mesh where we can not assume an inverse inequality globally, we refer to [137, 138].

By using the inverse trace theorem (Theorem 2.22) the stability condition (11.25) implies

$$\bar{c}_S\,\|\mu_H\|_{H^{-1/2}(\Gamma)} \leq c_{IT} \sup_{0\neq w_h\in X_h(\Gamma)} \frac{\langle \mu_H, w_h\rangle_\Gamma}{\|\mathcal{E}w_h\|_{H^1(\Omega)}} \quad \text{for all } \mu_H \in \Pi_H.$$

Finally, let $R_h : H^1(\Omega) \to X_h(\Omega) \subset H^1(\Omega)$ be some quasi interpolation operator [133] satisfying

$$\|R_h v\|_{H^1(\Omega)} \leq c_R\,\|v\|_{H^1(\Omega)}$$

and where Dirichlet boundary conditions are preserved. Then we obtain

$$\bar{c}_S\,\|\mu_H\|_{H^{-1/2}(\Gamma)} \leq c_{IT}c_R \sup_{0\neq w_h\in X_h(\Gamma)} \frac{\langle \mu_H, w_h\rangle_\Gamma}{\|R_h\mathcal{E}w_h\|_{H^1(\Omega)}} \quad \text{for all } \mu_H \in \Pi_H,$$

and by choosing $v_h = R_h\mathcal{E}w_h \in X_h(\Omega)$ we conclude the stability condition (11.24). This gives us the unique solvability of the Schur complement system (11.23) and therefore of the linear system (11.22). The application of Theorem 8.8 yields, when assuming $u \in H^2(\Omega)$ and $\lambda \in H^1_{\text{pw}}(\Gamma)$, the error estimate

$$\|u-u_h\|^2_{H^1(\Omega)}+\|\lambda-\lambda_H\|^2_{H^{-1/2}(\Gamma)} \leq c_1\,h^2\,|u|^2_{H^2(\Omega)}+c_2\,H^3\,\|\lambda\|^2_{H^1_{\text{pw}}(\Gamma)}. \quad (11.26)$$

To ensure the discrete stability condition (11.25) we need to assume $h \leq c_0 H$ where the constant $c_0 < 1$ is sufficiently small.

We now consider the numerical example of Section 11.1 with $h = \frac{1}{2}H$. For $L = 3$ the finite element mesh of $\Omega$ and the associated boundary element mesh are depicted in Fig. 11.2. In Table 11.3 the computed errors of the approximate solutions $(u_h, \lambda_H) \in S_h^1(\Omega) \times S_H^0(\Gamma)$ are given. The numerical results for the approximate solution $u_h$ of the primal variable $u$ confirm the theoretical error estimate (11.26).

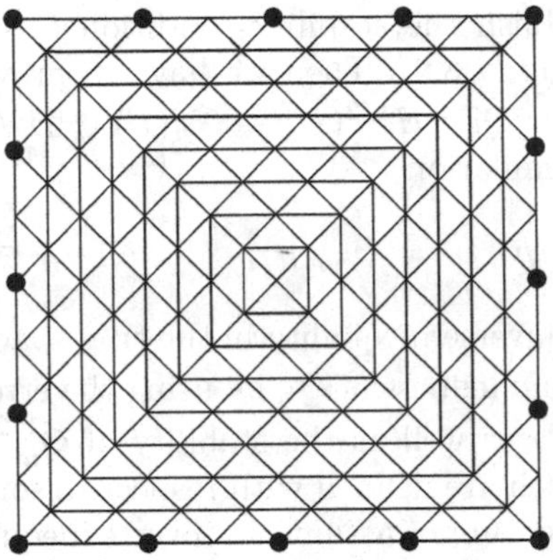

**Fig. 11.2.** Finite and boundary element meshes of $\Omega$ $(L = 3)$.

| L | $N_\Omega$ | $N_\Gamma$ | $\|u - u_h\|_{H^1(\Omega)}$ | eoc | $\|\lambda - \lambda_h\|_{L_2(\Gamma)}$ | eoc |
|---|---|---|---|---|---|---|
| 1 | 16 | 4 | 5.051 –1 | | 1.198 ±0 | |
| 2 | 64 | 8 | 2.248 –1 | 1.17 | 9.350 –1 | 0.36 |
| 3 | 256 | 16 | 9.897 –2 | 1.18 | 5.662 –1 | 0.72 |
| 4 | 1024 | 32 | 4.133 –2 | 1.26 | 2.970 –1 | 0.93 |
| 5 | 4096 | 64 | 1.851 –2 | 1.16 | 1.457 –1 | 1.03 |
| 6 | 16384 | 128 | 8.889 –3 | 1.06 | 7.093 –2 | 1.04 |
| 7 | 65536 | 256 | 4.393 –3 | 1.02 | 3.482 –2 | 1.03 |
| 8 | 262144 | 512 | 2.190 –3 | 1.00 | 1.723 –2 | 1.01 |
| 9 | 1048576 | 1024 | 1.094 –3 | 1.00 | 8.569 –3 | 1.01 |
| Theory: | | | | 1 | | 0.5 |

**Table 11.3.** Results for a Finite Element Method with Lagrange multipliers.

To describe the error of the approximation $\lambda_H$ of the Lagrange multiplier $\lambda$ we use the $L_2$ norm which is easier to compute.

**Lemma 11.7.** *For the trial space $S_H^0(\Gamma)$ we assume a global inverse inequality to be valid. If $u \in H^2(\Omega)$ and $\lambda \in H^1_{pw}(\Gamma)$ are satisfied, then there holds the error estimate*

$$\|\lambda - \lambda_H\|^2_{L_2(\Gamma)} \leq c_1\, h^2\, H^{-1}\, |u|^2_{H^2(\Omega)} + c_2\, H^2\, \|\lambda\|^2_{H^1_{pw}(\Gamma)}.$$

*Proof.* For $\lambda \in H^1_{\text{pw}}(\Gamma)$ we define $Q_H\lambda \in S^0_H(\Gamma)$ to be the $L_2$ projection as defined in (10.4). By using the triangle inequality and the inverse inequality we obtain

$$\begin{aligned}
\|\lambda - \lambda_H\|^2_{L_2(\Gamma)} &\leq 2\,\|\lambda - Q_H\lambda\|^2_{L_2(\Gamma)} + 2\,\|Q_H\lambda - \lambda_H\|^2_{L_2(\Gamma)} \\
&\leq 2\,\|\lambda - Q_H\lambda\|^2_{L_2(\Gamma)} + 2\,c_I^2\, H^{-1}\|Q_H\lambda - \lambda_H\|^2_{H^{-1/2}(\Gamma)} \\
&\leq 2\,\|\lambda - Q_H\lambda\|^2_{L_2(\Gamma)} + 4\,c_I^2\, H^{-1}\left[\|Q_H\lambda - \lambda\|^2_{H^{-1/2}(\Gamma)} + \|\lambda - \lambda_H\|^2_{H^{-1/2}(\Gamma)}\right].
\end{aligned}$$

The error estimate now follows from Theorem 10.2, Corollary 10.3, and by using the error estimate (11.26). □

When choosing $h = \frac{1}{2}H$ we conclude by applying Lemma 11.7 the error estimate

$$\|\lambda - \lambda_H\|^2_{L_2(\Gamma)} \leq \frac{1}{4}c_1\, H\, |u|^2_{H^2(\Omega)} + c_2\, H^2\, \|\lambda\|^2_{H^1_{\text{pw}}(\Gamma)}$$

and therefore an asymptotic order of convergence which is 0.5 when measuring the error in the $L_2$ norm. However, the numerical results in Table 11.3 indicate a higher order of convergence which is equal to 1. This preasymptotic behavior may be explained by different orders of magnitude in the constants $\frac{1}{4}c_1\|u\|^2_{H^1(\Omega)}$ and $c_2\|t\|^2_{H^1_{\text{pw}}(\Gamma)}$.

## 11.4 Exercises

**11.1** Consider the Dirichlet boundary value problem

$$-u''(x) = f(x) \quad \text{for } x \in (0,1), \quad u(0) = u(1) = 0.$$

Compute the finite element stiffness matrix when using piecewise linear basis functions with respect to a uniform decomposition of the interval $(0,1)$.

**11.2** Show that the eigenvectors of the finite element stiffness matrix as derived in Exercise 11.1 are given by the nodal interpolation of the eigenfunctions as obtained in Exercise 1.6. Compute the associated eigenvalues and discuss the behavior of the resulting spectral condition number.

**11.3** Derive a two–dimensional Gaussian quadrature formula which integrates cubic polynomials over the reference triangle

$$\tau = \{x \in \mathbb{R}^2 : x_1 \in (0,1), x_2 \in (0, 1-x_1)\}$$

exactly.

# 12

# Boundary Element Methods

The solution of the scalar homogeneous partial differential equation

$$Lu(x) = 0 \quad \text{for } x \in \Omega$$

is given by the representation formula

$$u(x) = (\widetilde{V}\gamma_1^{\text{int}}u)(x) - (W\gamma_0^{\text{int}}u)(x) \quad \text{for } x \in \Omega.$$

Since the Cauchy data $[\gamma_0^{\text{int}}u(x), \gamma_1^{\text{int}}u(x)]$ are given by the boundary conditions only partially for $x \in \Gamma$, we need to find the remaining data from the solution of suitable boundary integral equations. This chapter describes boundary element methods as numerical discretization schemes to solve these boundary integral equations approximately.

## 12.1 Dirichlet Boundary Value Problem

We first consider the Dirichlet boundary value problem (7.6),

$$Lu(x) = 0 \quad \text{for } x \in \Omega, \quad \gamma_0^{\text{int}}u(x) = g(x) \quad \text{for } x \in \Gamma. \tag{12.1}$$

The solution $u$ of (12.1) is given by the representation formula (6.1),

$$u(\widetilde{x}) = \int_\Gamma U^*(\widetilde{x}, y)\gamma_1^{\text{int}}u(y)ds_y - \int_\Gamma \gamma_{1,y}^{\text{int}}U^*(\widetilde{x}, y)g(y)ds_y, \quad \widetilde{x} \in \Omega. \tag{12.2}$$

We have to find the yet unknown conormal derivative $t := \gamma_1^{\text{int}}u \in H^{-1/2}(\Gamma)$ as the unique solution of the variational formulation (7.9),

$$\langle Vt, \tau\rangle_\Gamma = \langle(\frac{1}{2}I + K)g, \tau\rangle_\Gamma \quad \text{for all } \tau \in H^{-1/2}(\Gamma). \tag{12.3}$$

Let

$$S_h^0(\Gamma) = \operatorname{span}\{\varphi_k^0\}_{k=1}^N \subset H^{-1/2}(\Gamma)$$

be the trial space of piecewise constant basis functions $\varphi_k^0$. By using

$$t_h(x) = \sum_{k=1}^N t_k \, \varphi_k^0(x) \in S_h^0(\Gamma) \tag{12.4}$$

the Galerkin variational formulation of (12.3) reads to find $t_h \in S_h^0(\Gamma)$ such that

$$\langle V t_h, \tau_h \rangle_\Gamma = \langle (\frac{1}{2} I + K) g, \tau_h \rangle_\Gamma \tag{12.5}$$

is satisfied for all $\tau_h \in S_h^0(\Gamma)$.

Note that the single layer potential $V : H^{-1/2}(\Gamma) \to H^{1/2}(\Gamma)$ is bounded (see (6.8)) and $H^{-1/2}(\Gamma)$–elliptic (see Theorem 6.22 in the three–dimensional case $d = 3$, and Theorem 6.23 for $d = 2$ where we assume $\operatorname{diam} \Omega < 1$). Moreover, $S_h^0(\Gamma) \subset H^{-1/2}(\Gamma)$ is a conforming trial space. Therefore, all assumptions of Theorem 8.1 (Cea's Lemma) are satisfied, in particular there exists a unique solution $t_h \in S_h^0(\Gamma)$ of the Galerkin variational formulation (12.5) satisfying the stability estimate

$$\|t_h\|_{H^{-1/2}(\Gamma)} \leq \frac{1}{c_1^V} \, \|(\frac{1}{2} I + K) g\|_{H^{1/2}(\Gamma)} \leq \frac{c_2^W}{c_1^V} \, \|g\|_{H^{1/2}(\Gamma)},$$

as well as the error estimate

$$\|t - t_h\|_{H^{-1/2}(\Gamma)} \leq \frac{c_2^V}{c_1^V} \inf_{\tau_h \in S_h^0(\Gamma)} \|t - \tau_h\|_{H^{-1/2}(\Gamma)}. \tag{12.6}$$

In the case of a piecewise smooth Lipschitz boundary $\Gamma = \partial\Omega$ with the representation $\overline{\Gamma} = \cup_{j=1}^J \overline{\Gamma}_j$ we obtain, by applying Lemma 2.20 and by using the local definition of the trial space $S_h^0(\Gamma)$, the error estimate

$$\|t - t_h\|_{H^{-1/2}(\Gamma)} \leq \frac{c_2^V}{c_1^V} \sum_{j=1}^J \inf_{\tau_h^j \in S_h^0(\Gamma_j)} \|t_{|\Gamma_j} - \tau_h^j\|_{\widetilde{H}^{-1/2}(\Gamma_j)}. \tag{12.7}$$

From the approximation property (10.10) we then obtain the a priori error estimate

$$\|t - t_h\|_{H^{-1/2}(\Gamma)} \leq c \, h^{s+\frac{1}{2}} \, |t|_{H^s_{\text{pw}}(\Gamma)} \tag{12.8}$$

when assuming $t \in H^s_{\text{pw}}(\Gamma)$ for some $s \in [-\frac{1}{2}, 1]$. If the solution $t \in H^1_{\text{pw}}(\Gamma)$ is sufficiently smooth we then conclude the optimal error estimate

$$\|t - t_h\|_{H^{-1/2}(\Gamma)} \leq c \, h^{\frac{3}{2}} \, |t|_{H^1_{\text{pw}}(\Gamma)}$$

when using piecewise constant basis functions.

*Remark 12.1.* When considering a direct boundary integral approach for the Dirichlet boundary value problem of the Laplace equation, the solution $t \in H^{-1/2}(\Gamma)$ is the exterior normal derivative $t(x) = \underline{n}(x) \cdot \nabla u(x)$ for $x \in \Gamma$. When $\Gamma = \partial\Omega$ is the boundary of a domain with corners or edges, the exterior normal vector $\underline{n}(x)$ is discontinuous and therefore the exterior normal derivative $t(x)$ is also discontinuous. Hence we have $t \in H^{\frac{d-1}{2}-\varepsilon}(\Gamma)$ for any arbitrary small $\varepsilon > 0$. When using globally continuous trial spaces, e.g. the trial space $S_h^1(\Gamma)$ of piecewise linear basis functions, we can not apply the error estimate (12.7) due to the global definition of $S_h^1(\Gamma)$, but we can apply the global error estimate (12.6) for $s < \frac{1}{2}(d-1)$. Hence, when using higher order basis functions they have to be discontinuous globally.

Up to now we have only considered error estimates in the energy norm $\|t - t_h\|_{H^{-1/2}(\Gamma)}$. If the boundary mesh is globally quasi–uniform we can also derive an error estimate in $L_2(\Gamma)$.

**Lemma 12.2.** *For a globally quasi–uniform boundary mesh* (10.2) *let* $t_h \in S_h^0(\Gamma)$ *be the unique solution of the Galerkin variational problem* (12.5). *Let* $t \in H^s_{pw}(\Gamma)$ *for some* $s \in [0,1]$ *be satisfied. Then there holds the error estimate*

$$\|t - t_h\|_{L_2(\Gamma)} \le c\,h^s\,|t|_{H^s_{pw}(\Gamma)}. \tag{12.9}$$

*Proof.* For $t \in L_2(\Gamma)$ we define the $L_2$ projection $Q_h t \in S_h^0(\Gamma)$ as the unique solution of the variational problem (10.4). By applying the triangle inequality twice, and by using the global inverse inequality (see Lemma 10.10) we obtain

$$\begin{aligned}
\|t - t_h\|_{L_2(\Gamma)} &\le \|t - Q_h t\|_{L_2(\Gamma)} + \|Q_h t - t_h\|_{L_2(\Gamma)} \\
&\le \|t - Q_h t\|_{L_2(\Gamma)} + c\,h^{-1/2}\,\|Q_h t - t_h\|_{H^{-1/2}(\Gamma)} \\
&\le \|t - Q_h t\|_{L_2(\Gamma)} + c\,h^{-1/2}\,\left[\|t - Q_h t\|_{H^{-1/2}(\Gamma)} + \|t - t_h\|_{H^{-1/2}(\Gamma)}\right].
\end{aligned}$$

The assertion now follows from the error estimates (10.6), (10.8), and (12.8). □

After the approximate solution $t_h \in S_h^0(\Gamma)$ is determined as the unique solution of the Galerkin variational problem (12.5), we obtain from (12.2) an approximate solution of the Dirichlet boundary value problem (12.1), i.e. for $\widetilde{x} \in \Omega$ we have

$$\widetilde{u}(\widetilde{x}) = \int_\Gamma U^*(\widetilde{x}, y) t_h(y) ds_y - \int_\Gamma \gamma_{1,y}^{\text{int}} U^*(\widetilde{x}, y) g(y) ds_y. \tag{12.10}$$

For the related error we first obtain

$$|u(\widetilde{x}) - \widetilde{u}(\widetilde{x})| = \left| \int_\Gamma U^*(\widetilde{x}, y)[t(y) - t_h(y)] ds_y \right| \quad \text{for } \widetilde{x} \in \Omega.$$

Since for $\widetilde{x} \in \Omega$ and $y \in \Gamma$ the fundamental solution $U^*(\widetilde{x}, y)$ is a $C^\infty$ function, i.e. infinitely often differentiable, we conclude $U^*(\widetilde{x}, \cdot) \in H^{-\sigma}(\Gamma)$ for any $\sigma \in \mathbb{R}$. Hence we obtain

$$|u(\widetilde{x}) - \widetilde{u}(\widetilde{x})| \leq \|U^*(\widetilde{x}, \cdot)\|_{H^{-\sigma}(\Gamma)} \|t - t_h\|_{H^\sigma(\Gamma)}. \tag{12.11}$$

To derive an almost optimal error estimate for $|u(\widetilde{x}) - \widetilde{u}(\widetilde{x})|$, $\widetilde{x} \in \Omega$, we need to have an error estimate for $\|t - t_h\|_{H^\sigma(\Gamma)}$ where $\sigma \in \mathbb{R}$ is minimal.

**Theorem 12.3 (Aubin–Nitsche Trick [82]).** *For some $s \in [-\frac{1}{2}, 1]$ let $t \in H^s_{pw}(\Gamma)$ be the unique solution of the boundary integral equation* (7.8), *and let $t_h \in S^0_h(\Gamma)$ be the unique solution of the Galerkin variational problem* (12.5). *Assume that the single layer potential $V : H^{-1-\sigma}(\Gamma) \to H^{-\sigma}(\Gamma)$ is continuous and bijective for some $-2 \leq \sigma \leq -\frac{1}{2}$. Then there holds the error estimate*

$$\|t - t_h\|_{H^\sigma(\Gamma)} \leq c\, h^{s-\sigma}\, |t|_{H^s_{pw}(\Gamma)}. \tag{12.12}$$

*Proof.* For $\sigma < -\frac{1}{2}$ we have, by using the duality of Sobolev norms,

$$\|t - t_h\|_{H^\sigma(\Gamma)} = \sup_{0 \neq v \in H^{-\sigma}(\Gamma)} \frac{\langle t - t_h, v\rangle_\Gamma}{\|v\|_{H^{-\sigma}(\Gamma)}}.$$

By assumption, the single layer potential $V : H^{-1-\sigma}(\Gamma) \to H^{-\sigma}(\Gamma)$ is bijective. Hence, for any $v \in H^{-\sigma}(\Gamma)$ there exists a unique $w \in H^{-1-\sigma}(\Gamma)$ such that $v = Vw$ is satisfied. Therefore, by applying the Galerkin orthogonality

$$\langle V(t - t_h), \tau_h\rangle_\Gamma = 0 \quad \text{for all } \tau_h \in S^0_h(\Gamma)$$

we obtain

$$\begin{aligned}
\|t - t_h\|_{H^\sigma(\Gamma)} &= \sup_{0 \neq w \in H^{-1-\sigma}(\Gamma)} \frac{\langle t - t_h, Vw\rangle_\Gamma}{\|Vw\|_{H^{-\sigma}(\Gamma)}} \\
&= \sup_{0 \neq w \in H^{-1-\sigma}(\Gamma)} \frac{\langle V(t - t_h), w - Q_h w\rangle_\Gamma}{\|Vw\|_{H^{-\sigma}(\Gamma)}}.
\end{aligned}$$

Since the single layer potential $V : H^{-1-\sigma}(\Gamma) \to H^{-\sigma}(\Gamma)$ is assumed to be bijective, i.e. $\|Vw\|_{H^{-\sigma}(\Gamma)} \geq c\,\|w\|_{H^{-1-\sigma}(\Gamma)}$, we further conclude

$$\|t - t_h\|_{H^\sigma(\Gamma)} \leq \widetilde{c}\,\|t - t_h\|_{H^{-1/2}(\Gamma)} \sup_{0 \neq w \in H^{-1-\sigma}(\Gamma)} \frac{\|w - Q_h w\|_{H^{-1/2}(\Gamma)}}{\|w\|_{H^{-1-\sigma}(\Gamma)}}.$$

When considering $-1 - \sigma \leq 1$, i.e. $\sigma \geq -2$, we obtain from the error estimate (10.8)

$$\sup_{0 \neq w \in H^{-1-\sigma}(\Gamma)} \frac{\|w - Q_h w\|_{H^{-1/2}(\Gamma)}}{\|w\|_{H^{-1-\sigma}(\Gamma)}} \leq c\, h^{-1/2-\sigma}$$

and therefore

$$\|t - t_h\|_{H^\sigma(\Gamma)} \le \hat{c}\, h^{-1/2-\sigma}\, \|t - t_h\|_{H^{-1/2}(\Gamma)}.$$

The assertion now follows from the error estimate (12.8). □

If the solution $t$ of the boundary integral equation (7.8) is sufficiently regular, i.e. $t \in H^1_{\rm pw}(\Gamma)$, we obtain from (12.12) the optimal error estimate

$$\|t - t_h\|_{H^{-2}(\Gamma)} \le c\, h^3\, |t|_{H^1_{\rm pw}(\Gamma)}.$$

Moreover, when using (12.11) this gives the pointwise error estimate

$$|u(\widetilde{x}) - \widetilde{u}(\widetilde{x})| \le c\, h^3\, \|U^*(\widetilde{x},\cdot)\|_{H^2(\Gamma)} |t|_{H^1_{\rm pw}(\Gamma)}.$$

To obtain a global error estimate for $\|u - \widetilde{u}\|_{H^1(\Omega)}$ we first consider the trace $\widetilde{g}$ of the approximate solution $\widetilde{u} \in H^1(\Omega)$ which was defined via the representation formula (12.10). For $x \in \Gamma$ we have

$$\widetilde{g}(x) := (Vt_h)(x) + \frac{1}{2}g(x) - (Kg)(x).$$

On the other hand, the boundary integral equation (7.8) gives

$$g(x) = (Vt)(x) + \frac{1}{2}g(x) - (Kg)(x) \quad \text{for } x \in \Gamma.$$

Hence, by subtraction we get

$$g(x) - \widetilde{g}(x) = (V[t - t_h])(x) \quad \text{for } x \in \Gamma.$$

**Theorem 12.4.** *Let $u \in H^1(\Omega)$ be the weak solution of the Dirichlet boundary value problem* (12.1), *and let $\widetilde{u} \in H^1(\Omega)$ be the approximate solution as defined via the representation formula* (12.10). *Then there holds the global error estimate*

$$\|u - \widetilde{u}\|_{H^1(\Omega)} \le c\, \|t - t_h\|_{H^{-1/2}(\Gamma)}. \tag{12.13}$$

*Proof.* The weak solution $u = u_0 + \mathcal{E}g \in H^1(\Omega)$ of the Dirichlet boundary value problem (12.1) is given as the unique solution of the variational problem

$$a(u_0 + \mathcal{E}g, v) = 0 \quad \text{for all } v \in H^1_0(\Omega).$$

Correspondingly, the approximate solution $\widetilde{u} = \widetilde{u}_0 + \mathcal{E}\widetilde{g} \in H^1(\Omega)$ as defined via the representation formula (12.10) satisfies

$$a(\widetilde{u}_0 + \mathcal{E}\widetilde{g}, v) = 0 \quad \text{for all } v \in H^1_0(\Omega).$$

Hence we have

$$a(u_0 - \widetilde{u}_0, v) = a(\mathcal{E}(\widetilde{g} - g), v) \quad \text{for all } v \in H^1_0(\Omega).$$

By choosing $v := u_0 - \widetilde{u}_0 \in H^1_0(\Omega)$ we obtain, by using the $H^1_0(\Omega)$ ellipticity of the bilinear form $a(\cdot,\cdot)$,

$$c_1^A \, \|u_0 - \widetilde{u}_0\|^2_{H^1(\Omega)} \le a(u_0 - \widetilde{u}_0, u_0 - \widetilde{u}_0) \;=\; a(\mathcal{E}(\widetilde{g} - g), u_0 - \widetilde{u}_0)$$
$$\le c_2^A \, \|\mathcal{E}(\widetilde{g} - g)\|_{H^1(\Omega)} \|u_0 - \widetilde{u}_0\|_{H^1(\Omega)},$$

and therefore

$$\|u_0 - \widetilde{u}_0\|_{H^1(\Omega)} \;\le\; \frac{c_2^A}{c_1^A} \, \|\mathcal{E}(\widetilde{g} - g)\|_{H^1(\Omega)}.$$

By applying the triangle inequality, and the inverse trace theorem, we conclude

$$\begin{aligned}\|u - \widetilde{u}\|_{H^1(\Omega)} &\le \|u_0 - \widetilde{u}_0\|_{H^1(\Omega)} + \|\mathcal{E}(\widetilde{g} - g)\|_{H^1(\Omega)} \\ &\le \left(1 + \frac{c_2^A}{c_1^A}\right) \, \|\mathcal{E}(\widetilde{g} - g)\|_{H^1(\Omega)} \\ &\le c_{IT} \left(1 + \frac{c_2^A}{c_1^A}\right) \|\widetilde{g} - g\|_{H^{1/2}(\Gamma)} \\ &\le c_{IT} \left(1 + \frac{c_2^A}{c_1^A}\right) c_2^V \, \|t - t_h\|_{H^{-1/2}(\Gamma)}\end{aligned}$$

and therefore the claimed error estimate. □

If $t \in H^1_{\text{pw}}(\Gamma)$, i.e. $u \in H^{5/2}(\Omega)$, is satisfied, then from (12.13) we obtain for the approximate solution $\widetilde{u}$ the error estimate in the energy space $H^1$,

$$\|u - \widetilde{u}\|_{H^1(\Omega)} \;\le\; c\, h^{3/2} \, |t|_{H^1_{\text{pw}}(\Gamma)}. \tag{12.14}$$

*Remark 12.5.* When using lowest order, i.e. piecewise constant boundary elements, and when assuming $u \in H^{5/2}(\Omega)$ for the solution of the Dirichlet boundary value problem, then the approximate solution $\widetilde{u}$ as defined via the representation formula (12.10) converges with a convergence rate of 1.5. In contrast, a lowest order, i.e. piecewise linear finite element solution converges with a convergence rate of 1.0, see the error estimate (11.10), where we have to assume $u \in H^2(\Omega)$ only.

When inserting (12.4) into the Galerkin variational formulation (12.5), and when choosing $\tau_h = \varphi_\ell^0$ as a test function, this gives

$$\sum_{k=1}^N t_k \langle V\varphi_k^0, \varphi_\ell^0\rangle_\Gamma \;=\; \langle (\tfrac{1}{2}I + K)g, \varphi_\ell^0\rangle_\Gamma \quad \text{for } \ell = 1, \ldots, N.$$

By using

$$V_h[\ell, k] = \langle V\varphi_k^0, \varphi_\ell^0\rangle_\Gamma, \quad f_\ell = \langle (\tfrac{1}{2}I + K)g, \varphi_\ell^0\rangle_\Gamma$$

for $k, \ell = 1, \ldots, N$ the variational problem is equivalent to a linear system of algebraic equations,

$$V_h \underline{t} \;=\; \underline{f} \tag{12.15}$$

where the stiffness matrix $V_h$ is symmetric and positive definite, see Sec. 8.1.

**Lemma 12.6.** *Let the boundary mesh be globally quasi–uniform. Then, for all* $\underline{w} \in \mathbb{R}^N \leftrightarrow w_h \in S_h^0(\Gamma)$ *there hold the spectral equivalence inequalities*

$$c_1\, h^d\, \|\underline{w}\|_2^2 \leq (V_h \underline{w}, \underline{w}) \leq c_2\, h^{d-1}\, \|\underline{w}\|_2^2$$

*with some positive constants* $c_i$, $i = 1, 2$.

*Proof.* The upper estimate follows from

$$\begin{aligned}(V_h\underline{w}, \underline{w}) &= \langle V w_h, w_h\rangle_\Gamma \leq c_2^V\, \|w_h\|^2_{H^{-1/2}(\Gamma)} \\ &\leq c_2^V\, \|w_h\|^2_{L_2(\Gamma)} \;=\; c_2^V \sum_{\ell=1}^N w_\ell^2 \Delta_\ell\end{aligned}$$

by using $\Delta_\ell = h_\ell^{d-1}$.

To prove the lower estimate we consider an arbitrary but fixed $\underline{w} \in \mathbb{R}^N \leftrightarrow w_h \in S_h^0(\Gamma)$. The $L_2$ projection $Q_h^B w_h \in S_h^B(\Gamma)$ onto the space of local bubble functions is defined as the unique solution of the variational problem (10.17),

$$\langle Q_h^B w_h, \tau_h\rangle_{L_2(\Gamma)} = \langle w_h, \tau_h\rangle_{L_2(\Gamma)} \quad \text{for all } \tau_h \in S_h^0(\Gamma).$$

Since the single layer potential $V$ is $H^{-1/2}(\Gamma)$ elliptic, we first obtain

$$(V_h\underline{w}, \underline{w}) = \langle V w_h, w_h\rangle_\Gamma \geq c_1^V\, \|w_h\|^2_{H^{-1/2}(\Gamma)}\,.$$

By using duality, the inverse inequality (10.16) in $S_h^B(\Gamma)$, and the stability estimate (10.18) we further get

$$\begin{aligned}\|w_h\|_{H^{-1/2}(\Gamma)} &= \sup_{0\neq v\in H^{1/2}(\Gamma)} \frac{\langle w_h, v\rangle_\Gamma}{\|v\|_{H^{1/2}(\Gamma)}} \geq \frac{\langle w_h, Q_h^B w_h\rangle_{L_2(\Gamma)}}{\|Q_h^B w_h\|_{H^{1/2}(\Gamma)}} \\ &= \frac{\langle w_h, w_h\rangle_{L_2(\Gamma)}}{\|Q_h^B w\|_{H^{1/2}(\Gamma)}} = \frac{\|w_h\|^2_{L_2(\Gamma)}}{\|Q_h^B w_h\|_{H^{1/2}(\Gamma)}} \geq c\, h^{1/2}\, \|w_h\|_{L_2(\Gamma)}\end{aligned}$$

and therefore

$$(V_h\underline{w}, \underline{w}) \geq c\, h\, \|w_h\|^2_{L_2(\Gamma)} \;=\; c\, h \sum_{\ell=1}^N w_\ell^2 \Delta_\ell$$

which finally gives the lower estimate. □

To bound the spectral condition of the stiffness matrix $V_h$ of the single layer potential $V$ we now obtain the estimate

$$\kappa_2(V_h) \leq c\, h^{-1}\,.$$

As an example we consider a uniform boundary mesh of the boundary $\Gamma = \partial\Omega$ where $\Omega = (0, 0.5)^2$ is a square. By $N$ we denote the number of boundary

| $L$ | $N$ | $\lambda_{\min}(V_h)$ | $\lambda_{\max}(V_h)$ | $\kappa_2(V_h)$ |
|---|---|---|---|---|
| 2 | 16 | 2.04 –3 | 4.92 –2 | 24.14 |
| 3 | 32 | 5.14 –4 | 2.46 –2 | 47.86 |
| 4 | 64 | 1.29 –4 | 1.23 –2 | 95.64 |
| 5 | 128 | 3.22 –5 | 6.15 –3 | 191.01 |
| 6 | 256 | 8.06 –6 | 3.07 –3 | 381.32 |
| 7 | 512 | 2.02 –6 | 1.54 –3 | 760.73 |
| 8 | 1024 | 5.07 –7 | 7.68 –4 | 1516.02 |
| theory: | | $\mathcal{O}(h^2)$ | $\mathcal{O}(h)$ | $\mathcal{O}(h^{-1})$ |

**Table 12.1.** Spectral condition number of the stiffness matrix $V_h$.

elements $\tau_\ell$ while $h = \mathcal{O}(N^{-1})$ is the mesh size. In Table 12.1 there are given the minimal and maximal eigenvalues of the stiffness matrix $V_h$ as well as the resulting spectral condition number $\kappa_2(V_h)$.

The computation of the load vector $\underline{f}$ requires the evaluation of

$$f_\ell = \int_{\tau_\ell} (\frac{1}{2}I + K)g(x)ds_x \quad \text{for } \ell = 1, \ldots, N$$

where the double layer potential $K$ is applied to the given Dirichlet data $g$. If $g$ is replaced by some piecewise linear approximation $g_h \in S_h^1(\Gamma)$, we obtain

$$\widetilde{f}_\ell = \sum_{i=1}^{M} g_i \langle (\frac{1}{2}I + K)\varphi_i^1, \varphi_\ell^0 \rangle_\Gamma,$$

in particular

$$\underline{\widetilde{f}} = (\frac{1}{2}M_h + K_h)\underline{g}$$

where

$$M_h[\ell, i] = \langle \varphi_i^1, \varphi_\ell^0 \rangle_\Gamma, \quad K_h[\ell, i] = \langle K\varphi_i^1, \varphi_\ell^0 \rangle_\Gamma$$

for $i = 1, \ldots, M$ and $\ell = 1, \ldots, N$. Hence we have to find the solution vector $\underline{\widetilde{t}} \in \mathbb{R}^N$ of the linear system

$$V_h \underline{\widetilde{t}} = (\frac{1}{2}M_h + K_h)\underline{g}.$$

The associated approximate solution $\widetilde{t} \in S_h^0(\Gamma)$ is the unique solution of the perturbed variational problem

$$\langle V\widetilde{t}_h, \tau_h \rangle_\Gamma = \langle (\frac{1}{2}I + K)g_h, \tau_h \rangle_\Gamma \quad \text{for all } \tau_h \in S_h^0(\Gamma). \tag{12.16}$$

When applying Theorem 8.2 (Strang Lemma) we then obtain the error estimate

$$\|t - \widetilde{t}_h\|_{H^{-1/2}(\Gamma)} \leq \frac{1}{c_1^V} \left\{ c_2^V \inf_{\tau_h \in S_h^0(\Gamma)} \|t - \tau_h\|_{H^{-1/2}(\Gamma)} + c_2^W \|g - g_h\|_{H^{1/2}(\Gamma)} \right\}$$
$$\leq c_1\, h^{s+\frac{1}{2}}\, |t|_{H^s_{\rm pw}(\Gamma)} + c_2\, \|g - g_h\|_{H^{1/2}(\Gamma)} \qquad (12.17)$$

when assuming $t \in H^s_{\rm pw}(\Gamma)$ for some $s \in [-\frac{1}{2}, 1]$. It remains to estimate the error of the approximation $g_h$. When considering the piecewise linear interpolation

$$g_h(x) = I_h g(x) = \sum_{i=1}^{M} g(x_i)\varphi_i^1(x) \in S_h^1(\Gamma)$$

there holds the error estimate (10.11),

$$\|g - I_h g\|_{H^\sigma(\Gamma)} \leq c\, h^{s-\sigma}\, |g|_{H^s(\Gamma)} \qquad (12.18)$$

when assuming $g \in H^s(\Gamma)$ for some $s \in (\frac{d-1}{2}, 2]$ and $0 \leq \sigma \leq \min\{1, s\}$. Instead of an interpolation we may also consider the piecewise linear $L_2$ projection $Q_h g \in S_h^1(\Gamma)$ as the unique solution of the variational problem

$$\langle Q_h g, v_h \rangle_{L_2(\Gamma)} = \langle g, v_h \rangle_{L_2(\Gamma)} \quad \text{for all } v_h \in S_h^1(\Gamma).$$

In this case a similar error estimate as in (10.8) holds, i.e.

$$\|g - Q_h g\|_{H^\sigma(\Gamma)} \leq c\, h^{s-\sigma}\, |g|_{H^s(\Gamma)} \qquad (12.19)$$

when assuming $g \in H^s(\Gamma)$ for some $0 \leq s \leq 2$ and $-1 \leq \sigma \leq \min\{1, s\}$. In both cases we find from (12.17) the error estimate

$$\|t - \widetilde{t}_h\|_{H^{-1/2}(\Gamma)} \leq c_1\, h^{s+\frac{1}{2}}\, |t|_{H^s_{\rm pw}(\Gamma)} + c_2\, h^{\sigma-\frac{1}{2}}\, |g|_{H^\sigma(\Gamma)}$$

when assuming $t \in H^s_{\rm pw}(\Gamma)$ for some $s \in [-\frac{1}{2}, 1]$, and $g \in H^\sigma(\Gamma)$ for some $\sigma \in [\frac{1}{2}, 2]$ in the case of the $L_2$ projection, and $\sigma \in (\frac{d-1}{2}, 2]$ in the case of interpolation. In particular, when assuming $t \in H^1_{\rm pw}(\Gamma)$ and $g \in H^2(\Gamma)$ we conclude the error estimate

$$\|t - \widetilde{t}_h\|_{H^{-1/2}(\Gamma)} \leq c\, h^{3/2} \left\{ |t|_{H^1_{\rm pw}(\Gamma)} + |g|_{H^2(\Gamma)} \right\}.$$

For the error estimate $\|t - \widetilde{t}_h\|_{H^{-1/2}(\Gamma)}$ in the energy space it is not important whether the given data are replaced by some interpolation or by some projection. However, this changes when considering error estimates in lower Sobolev spaces, as they are used to obtain error estimates for the approximate solution as defined via the representation formula (12.10).

**Theorem 12.7.** *Let $t \in H^s_{pw}(\Gamma)$ for some $s \in [-\frac{1}{2}, 1]$ be the unique solution of the boundary integral equation (7.8), and let $\widetilde{t}_h \in S^0_h(\Gamma)$ be the unique solution of the perturbed variational problem (12.16). Assume that the single layer potential $V : H^{-1-\sigma}(\Gamma) \to H^{-\sigma}(\Gamma)$ is bounded and bijective for some $\sigma \in [-2, -\frac{1}{2}]$, and let the double layer potential $\frac{1}{2}I + K : H^{1+\sigma}(\Gamma) \to H^{1+\sigma}(\Gamma)$ be bounded. Moreover, let $g \in H^\varrho(\Gamma)$ for some $\varrho \in (1, 2]$. Then there holds the error estimate*

$$\|t - \widetilde{t}_h\|_{H^\sigma(\Gamma)} \leq c_1\, h^{s-\sigma}\, |t|_{H^s_{pw}(\Gamma)} + c_2\, h^{\varrho-\sigma-1}\, |g|_{H^\varrho(\Gamma)}$$

*where $\sigma \geq -1$ in the case of interpolation $g_h = I_h g$, and $\sigma \geq -2$ in the case of the $L_2$ projection $g_h = Q_h g$.*

*Proof.* As in the proof of Theorem 12.3 we have for $\sigma \in [-2, -\frac{1}{2})$

$$\|t - \widetilde{t}_h\|_{H^\sigma(\Gamma)} = \sup_{0 \neq w \in H^{-1-\sigma}(\Gamma)} \frac{\langle V(t - \widetilde{t}_h), w\rangle_\Gamma}{\|Vw\|_{H^{-\sigma}(\Gamma)}}.$$

By subtracting the perturbed variational formulation (12.16) from the exact formulation (12.3) we conclude the equality

$$\langle V(t - \widetilde{t}_h), \tau_h\rangle_\Gamma = \langle (\frac{1}{2}I + K)(g - g_h), \tau_h\rangle_\Gamma \quad \text{for all } \tau_h \in S^0_h(\Gamma).$$

Hence we obtain

$$\|t - \widetilde{t}_h\|_{H^\sigma(\Gamma)} \leq \sup_{0 \neq w \in H^{-1-\sigma}(\Gamma)} \frac{\langle V(t - \widetilde{t}_h), w - Q_h w\rangle_\Gamma}{\|Vw\|_{H^{-\sigma}(\Gamma)}} + \sup_{0 \neq w \in H^{-1-\sigma}(\Gamma)} \frac{\langle (\frac{1}{2}I + K)(g - g_h), Q_h w\rangle_\Gamma}{\|Vw\|_{H^{-\sigma}(\Gamma)}}.$$

As in the proof of Theorem 12.3 we get

$$\sup_{0 \neq w \in H^{-1-\sigma}(\Gamma)} \frac{\langle V(t - \widetilde{t}_h), w - Q_h w\rangle_\Gamma}{\|Vw\|_{H^{-\sigma}(\Gamma)}} \leq c\, h^{s-\sigma}\, |t|_{H^s_{\rm pw}(\Gamma)}$$

when assuming $\sigma \geq -2$. To estimate the second term we have

$$\begin{aligned}
|\langle (\frac{1}{2}I + K)(g - g_h), Q_h w\rangle_\Gamma| &= |\langle (\frac{1}{2}I + K)(g - g_h), w\rangle_\Gamma| \\
&\quad + |\langle (\frac{1}{2}I + K)(g - g_h), w - Q_h w\rangle_\Gamma| \\
&\leq c_1\, \|g - g_h\|_{H^{1+\sigma}(\Gamma)} \|w\|_{H^{-1-\sigma}(\Gamma)} + c_2\, \|g - g_h\|_{H^1(\Gamma)} \|w - Q_h w\|_{H^{-1}(\Gamma)} \\
&\leq c_1\, \|g - g_h\|_{H^{1+\sigma}(\Gamma)} \|w\|_{H^{-1-\sigma}(\Gamma)} + c_2\, h^{\varrho-1}\, |g|_{H^\varrho(\Gamma)} h^{-\sigma}\, \|w\|_{H^{-1-\sigma}(\Gamma)}
\end{aligned}$$

and therefore

$$\|t - \widetilde{t}_h\|_{H^\sigma(\Gamma)} \leq c_1 h^{s-\sigma} |t|_{H^s_{\rm pw}(\Gamma)} + c_2 h^{\varrho-\sigma-1} |g|_{H^\varrho(\Gamma)} + c_3 \|g - g_h\|_{H^{1+\sigma}(\Gamma)}.$$

Finally, for $\varrho \in (1,2]$ we have

$$\|g - g_h\|_{H^{1+\sigma}(\Gamma)} \leq c\, h^{\varrho-\sigma-1}\, |g|_{H^\varrho(\Gamma)}$$

for some $1+\sigma \geq 0$ in the case of interpolation $g_h = I_h g$, and $1+\sigma \geq -1$ in the case of the $L_2$ projection $g_h = Q_h g$, see the error estimates (12.18) and (12.19), respectively. This finishes the proof. □

When assuming $t \in H^1_{\rm pw}(\Gamma)$ and $g \in H^2(\Gamma)$ we therefore obtain the error estimate

$$\|t - \widetilde{t}_h\|_{H^\sigma(\Gamma)} \leq c\, h^{1-\sigma} \left\{ |t|_{H^1_{\rm pw}(\Gamma)} + |g|_{H^2(\Gamma)} \right\}.$$

In the case of an interpolation $g_h = I_h g$ we obtain by choosing $\sigma = -1$ the optimal error estimate

$$\|t - \widetilde{t}_h\|_{H^{-1}(\Gamma)} \leq c\, h^2 \left\{ |t|_{H^1_{\rm pw}(\Gamma)} + |g|_{H^2(\Gamma)} \right\},$$

while in the case of the $L_2$ projection $g_h = Q_h g$ we obtain by choosing $\sigma = -2$ the improved estimate

$$\|t - \widetilde{t}_h\|_{H^{-2}(\Gamma)} \leq c\, h^3 \left\{ |t|_{H^1_{\rm pw}(\Gamma)} + |g|_{H^2(\Gamma)} \right\}.$$

An approximate solution of the Dirichlet boundary value problem (12.1) is then given via the representation formula

$$\hat{u}(\widetilde{x}) = \int_\Gamma U^*(\widetilde{x}, y) \widetilde{t}_h(y) ds_y - \int_\Gamma \gamma^{\rm int}_{1,y} U^*(\widetilde{x}, y) g_h(y) ds_y \quad \text{for } \widetilde{x} \in \Omega \tag{12.20}$$

satisfying the error estimate

$$|u(\widetilde{x}) - \hat{u}(\widetilde{x})| \leq c\, h^2 \left\{ |t|_{H^1_{\rm pw}(\Gamma)} + |g|_{H^2(\Gamma)} \right\} \tag{12.21}$$

in the case of an interpolation $g_h = I_h g$, and

$$|u(\widetilde{x}) - \hat{u}(\widetilde{x})| \leq c\, h^3 \left\{ |t|_{H^1_{\rm pw}(\Gamma)} + |g|_{H^2(\Gamma)} \right\} \tag{12.22}$$

in the case of the $L_2$ projection $g_h = Q_h g$.

To check the theoretical convergence results, and to compare boundary and finite element methods we now consider a simple Dirichlet boundary value problem (12.1) where the domain $\Omega = (0, 0.5)^2$ is a square, and where the Dirichlet data are such that the solution is given by (11.14),

$$u(x) = -\frac{1}{2} \log |x - x^*|, \quad x^* = (-0.1, -0.1)^\top.$$

The boundary $\Gamma = \partial\Omega$ is discretized into $N$ boundary elements $\tau_\ell$ of mesh size $h$. The numerical results as given in Table 12.2 confirm the theoretical

| | FEM | interpolation | | $L_2$ projection | |
|---|---|---|---|---|---|
| $N$ | $\|\lambda - \lambda_h\|_{L_2(\Gamma)}$ | $\|t - t_h\|_{L_2(\Gamma)}$ | eoc | $\|t - t_h\|_{L_2(\Gamma)}$ | eoc |
| 8 | 9.350 -1 | 8.774 -1 | | 8.525 -1 | |
| 16 | 5.662 -1 | 5.217 -1 | 0.75 | 5.180 -1 | 0.72 |
| 32 | 2.970 -1 | 2.745 -1 | 0.93 | 2.751 -1 | 0.91 |
| 64 | 1.457 -1 | 1.384 -1 | 0.99 | 1.387 -1 | 0.99 |
| 128 | 7.093 -2 | 6.897 -2 | 1.00 | 6.905 -2 | 1.01 |
| 256 | 3.482 -2 | 3.433 -2 | 1.01 | 3.434 -2 | 1.01 |
| 512 | 1.723 -2 | 1.711 -2 | 1.00 | 1.711 -2 | 1.01 |
| 1024 | 8.569 -3 | 8.539 -3 | 1.00 | 8.539 -3 | 1.00 |
| 2048 | | 4.265 -3 | 1.00 | 4.265 -3 | 1.00 |
| 4096 | | 2.131 -3 | 1.00 | 2.131 -3 | 1.00 |
| theory: | | | 1 | | 1 |

**Table 12.2.** Error and order of convergence, Dirichlet boundary value problem.

estimate (12.9) which was obtained for the error $\|t - t_h\|_{L_2(\Gamma)}$. In addition we also give the errors $\|\lambda - \lambda_h\|_{L_2(\Gamma)}$ of the Lagrange multipliers as obtained by a mixed finite element method, see also Table 11.3.

Finally we also consider the error of the approximate solution (12.20). The results in the interior point $\hat{x} = (1/7, 2/7)^\top$ as listed in Table 12.3 again confirm the theoretical error estimates (12.21) and (12.22).

| | interpolation | | $L_2$ projection | |
|---|---|---|---|---|
| $N$ | $\|u(\hat{x}) - \hat{u}(\hat{x})\|$ | eoc | $\|u(\hat{x}) - \hat{u}(\hat{x})\|$ | eoc |
| 8 | 2.752 -2 | | 6.818 -3 | |
| 16 | 5.463 -3 | 2.33 | 5.233 -4 | 3.70 |
| 32 | 1.291 -3 | 2.08 | 6.197 -5 | 3.08 |
| 64 | 3.147 -4 | 2.04 | 6.753 -6 | 3.20 |
| 128 | 7.780 -5 | 2.02 | 7.587 -7 | 3.15 |
| 256 | 1.935 -5 | 2.01 | 8.878 -8 | 3.10 |
| 512 | 4.827 -6 | 2.00 | 1.070 -8 | 3.05 |
| 1024 | 1.205 -6 | 2.00 | 1.312 -9 | 3.03 |
| 2048 | 3.012 -7 | 2.00 | 1.620 -10 | 3.02 |
| 4096 | 7.528 -8 | 2.00 | 1.921 -11 | 3.08 |
| theory: | | 2 | | 3 |

**Table 12.3.** Error and order of convergence, interior point.

## 12.2 Neumann Boundary Value Problem

Next we consider the Neumann boundary value problem (7.16),

$$Lu(x) = 0 \quad \text{for } x \in \Omega, \quad \gamma_1^{\text{int}} u(x) = g(x) \quad \text{for } x \in \Gamma,$$

where we have to assume the solvability condition, see (1.17),

$$\int_\Gamma g(x) ds_x = 0.$$

The solution is given via the representation formula (6.1) for $\widetilde{x} \in \Omega$,

$$u(\widetilde{x}) = \int_\Gamma U^*(\widetilde{x}, y) g(y) ds_y - \int_\Gamma \gamma_{1,y}^{\text{int}} U^*(\widetilde{x}, y) \gamma_0^{\text{int}} u(y) ds. \qquad (12.23)$$

Hence we have to find the yet unknown Dirichlet datum $\gamma_0^{\text{int}} u \in H^{1/2}(\Gamma)$ as the unique solution of the stabilized variational problem (7.24) such that

$$\langle D\gamma_0^{\text{int}} u, v\rangle_\Gamma + \langle \gamma_0^{\text{int}} u, w_{\text{eq}}\rangle_\Gamma \langle v, w_{\text{eq}}\rangle_\Gamma = \langle (\frac{1}{2} I - K')g, v\rangle_\Gamma \qquad (12.24)$$

is satisfied for all $v \in H^{1/2}(\Gamma)$. Let

$$S_h^1(\Gamma) = \text{span}\{\varphi_i^1\}_{i=1}^M \subset H^{1/2}(\Gamma)$$

be the trial space of piecewise linear and continuous basis functions $\varphi_i^1$. By using

$$u_h(x) = \sum_{i=1}^M u_i \varphi_i^1(x) \in S_h^1(\Gamma)$$

the Galerkin variational formulation of (12.24) reads to find $u_h \in S_h^1(\Gamma)$ such that

$$\langle Du_h, v_h\rangle_\Gamma + \langle u_h, w_{\text{eq}}\rangle_\Gamma \langle v_h, w_{\text{eq}}\rangle_\Gamma = \langle (\frac{1}{2} I - K')g, v_h\rangle_\Gamma \qquad (12.25)$$

is satisfied for all $v_h \in S_h^1(\Gamma)$. Since all assumptions of Theorem 8.1 (Cea's Lemma) are satisfied, there exists a unique solution $u_h \in S_h^1(\Gamma)$ of the Galerkin variational problem (12.25) satisfying both the stability estimate

$$\|u_h\|_{H^{1/2}(\Gamma)} \leq c\,\|g\|_{H^{-1/2}(\Gamma)}$$

as well as the error estimate

$$\|\gamma_0^{\text{int}} u - u_h\|_{H^{1/2}(\Gamma)} \leq c \inf_{v_h \in S_h^1(\Gamma)} \|\gamma_0^{\text{int}} u - v_h\|_{H^{1/2}(\Gamma)}.$$

Due to

$$\ker D = \ker(\frac{1}{2} I + K) = \text{span}\{1\} \subset S_h^1(\Gamma)$$

we find from the solvability assumption (1.17) the orthogonality relation

$$\langle u_h, w_{\mathrm{eq}} \rangle_\Gamma = 0$$

and therefore $u_h \in H_*^{1/2}(\Gamma)$.

By using the approximation property (10.14) we further obtain the error estimate

$$\|\gamma_0^{\mathrm{int}} u - u_h\|_{H^{1/2}(\Gamma)} \leq c\, h^{s-1/2}\, |\gamma_0^{\mathrm{int}} u|_{H^s(\Gamma)} \tag{12.26}$$

when assuming $\gamma_0^{\mathrm{int}} u \in H^s(\Gamma)$ for some $s \in [\frac{1}{2}, 2]$. When $\gamma_0^{\mathrm{int}} u \in H^2(\Gamma)$ is sufficiently smooth, we then conclude the error estimate

$$\|\gamma_0^{\mathrm{int}} u - u_h\|_{H^{1/2}(\Gamma)} \leq c\, h^{3/2}\, |\gamma_0^{\mathrm{int}} u|_{H^2(\Gamma)}.$$

**Theorem 12.8 (Aubin–Nitsche Trick [82]).** *For some $s \in [\frac{1}{2}, 2]$ let $\gamma_0^{int} u \in H^s(\Gamma)$ be the unique solution of the variational problem (7.24), and let $u_h \in S_h^1(\Gamma)$ be the unique solution of the Galerkin variational problem (12.25). Assume that the stabilized hypersingular boundary integral operator $\widetilde{D} : H^{1-\sigma}(\Gamma) \to H^{-\sigma}(\Gamma)$ is bounded and bijective for some $\sigma \in [-1, \frac{1}{2}]$. Then there holds the error estimate*

$$\|\gamma_0^{int} u - u_h\|_{H^\sigma(\Gamma)} \leq c\, h^{s-\sigma}\, |\gamma_0^{int} u|_{H^s(\Gamma)}.$$

*Proof.* When considering $\sigma \leq 0$ we have, by using the duality of Sobolev norms,

$$\|\gamma_0^{\mathrm{int}} u - u_h\|_{H^\sigma(\Gamma)} = \sup_{0 \neq w \in H^{-\sigma}(\Gamma)} \frac{\langle \gamma_0^{\mathrm{int}} u - u_h, w \rangle_\Gamma}{\|w\|_{H^{-\sigma}(\Gamma)}}.$$

By assumption, the hypersingular integral operator $\widetilde{D} : H^{1-\sigma}(\Gamma) \to H^{-\sigma}(\Gamma)$ is bijective. Hence, for any $w \in H^{-\sigma}(\Gamma)$ there exists a unique $v \in H^{1-\sigma}(\Gamma)$ such that $w = \widetilde{D}v$ is satisfied. Therefore, by applying the Galerkin orthogonality

$$\langle \widetilde{D}(\gamma_0^{\mathrm{int}} u - u_h), v_h \rangle_\Gamma = 0 \quad \text{for all } v_h \in S_h^1(\Gamma)$$

we obtain

$$\begin{aligned} \|\gamma_0^{\mathrm{int}} u - u_h\|_{H^\sigma(\Gamma)} &= \sup_{0 \neq v \in H^{1-\sigma}(\Gamma)} \frac{\langle \gamma_0^{\mathrm{int}} u - u_h, \widetilde{D}v \rangle_\Gamma}{\|\widetilde{D}v\|_{H^{-\sigma}(\Gamma)}} \\ &= \sup_{0 \neq v \in H^{1-\sigma}(\Gamma)} \frac{\langle \widetilde{D}(\gamma_0^{\mathrm{int}} u - u_h), v - Q_h^{1/2} v \rangle_\Gamma}{\|\widetilde{D}v\|_{H^{-\sigma}(\Gamma)}} \end{aligned}$$

where $Q_h^{1/2} : H^{1/2}(\Gamma) \to S_h^1(\Gamma)$ is the projection which is defined via the variational formulation

$$\langle Q_h^{1/2} v, z_h \rangle_{H^{1/2}(\Gamma)} = \langle v, z_h \rangle_{H^{1/2}(\Gamma)} \quad \text{for all } z_h \in S_h^1(\Gamma).$$

Note that there holds the error estimate (10.13), i.e. for $s \in [1, 2]$ we have

$$\|v - Q_h^{1/2} v\|_{H^{1/2}(\Gamma)} \leq c\, h^{s-1/2}\, |v|_{H^s(\Gamma)}.$$

Since the stabilized hypersingular boundary integral operator $\widetilde{D} : H^{-\sigma}(\Gamma) \to H^{1-\sigma}(\Gamma)$ is assumed to be bijective, i.e. $\|\widetilde{D}v\|_{H^{-\sigma}(\Gamma)} \geq c\,\|v\|_{H^{1-\sigma}(\Gamma)}$, we further conclude

$$\|\gamma_0^{\text{int}} u - u_h\|_{H^\sigma(\Gamma)} \leq \frac{1}{c}\,\|\gamma_0^{\text{int}} u - u_h\|_{H^{1/2}(\Gamma)} \sup_{0 \neq v \in H^{1-\sigma}(\Gamma)} \frac{\|v - Q_h^{1/2} v\|_{H^{1/2}(\Gamma)}}{\|v\|_{H^{1-\sigma}(\Gamma)}}.$$

When considering $1 - \sigma \leq 2$, i.e. $\sigma \geq -1$, we further obtain

$$\sup_{0 \neq v \in H^{1-\sigma}(\Gamma)} \frac{\|v - Q_h^{1/2} v\|_{H^{1/2}(\Gamma)}}{\|v\|_{H^{1-\sigma}(\Gamma)}} \leq c\, h^{1/2-\sigma}$$

and therefore

$$\|\gamma_0^{\text{int}} u - u_h\|_{H^\sigma(\Gamma)} \leq c\, h^{1/2-\sigma}\, \|\gamma_0^{\text{int}} u - u_h\|_{H^{1/2}(\Gamma)}.$$

Now, the assertion follows from (12.26). Finally, for $\sigma \in (0, \frac{1}{2})$ we can apply an interpolation argument. □

When assuming $\gamma_0^{\text{int}} u \in H^2(\Gamma)$ we then find the optimal error estimate

$$\|\gamma_0^{\text{int}} u - u_h\|_{H^{-1}(\Gamma)} \leq c\, h^3\, |\gamma_0^{\text{int}} u|_{H^2(\Gamma)}.$$

Inserting the unique solution $u_h$ into the representation formula (12.23) this defines an approximate solution of the Neumann boundary value problem (7.16),

$$\widetilde{u}(\widetilde{x}) = \int_\Gamma U^*(\widetilde{x}, y) g(y) ds_y - \int_\Gamma \gamma_{1,y}^{\text{int}} U^*(\widetilde{x}, y) u_h(y) ds_y \quad \text{for } \widetilde{x} \in \Omega,$$

which satisfies, when assuming $\gamma_0^{\text{int}} u \in H^2(\Gamma)$, the error estimate

$$\begin{aligned} |u(\widetilde{x}) - \widetilde{u}(\widetilde{x})| &= \left| \int_\Gamma \gamma_1^{\text{int}} U^*(\widetilde{x}, y) [\gamma_0^{\text{int}} u(y) - u_h(y)] ds_y \right| \\ &\leq \|\gamma_1^{\text{int}} U^*(\widetilde{x}, \cdot)\|_{H^1(\Gamma)} \|\gamma_0^{\text{int}} u - u_h\|_{H^{-1}(\Gamma)} \\ &\leq c\, h^3\, \|\gamma_1^{\text{int}} U^*(\widetilde{x}, \cdot)\|_{H^1(\Gamma)}\, |\gamma_0^{\text{int}} u|_{H^2(\Gamma)}. \end{aligned}$$

The Galerkin variational problem (12.25) is equivalent to a linear system of algebraic equations,

$$(D_h + \alpha\, \underline{a}\, \underline{a}^\top) \underline{u} = \underline{f} \tag{12.27}$$

where

$$D_h[j,i] = \langle D\varphi_i^1, \varphi_j^1 \rangle_\Gamma, \quad a_j = \langle \varphi_j^1, w_{\text{eq}} \rangle_\Gamma, \quad f_j = \langle (\frac{1}{2} I - K') g, \varphi_j^1 \rangle_\Gamma$$

for $i, j = 1, \ldots, M$. The stabilized stiffness matrix $\widetilde{D}_h := D_h + \alpha\, \underline{a}\, \underline{a}^\top$ is symmetric and positive definite, see also the general discussion in Sec. 8.1.

**Lemma 12.9.** *Let the boundary mesh be globally quasi–uniform. Then, for all* $\underline{v} \in \mathbb{R}^M \leftrightarrow v_h \in S_h^1(\Gamma)$ *there hold the spectral equivalence inequalities*

$$c_1 \, h^{d-1} \, \|\underline{v}\|_2^2 \le (\widetilde{D}_h \underline{v}, \underline{v}) \le c_2 \, h^{d-2} \, \|\underline{v}\|_2^2$$

*with some positive constants* $c_i$, $i = 1, 2$.

*Proof.* Since the stabilized hypersingular boundary integral operator $\widetilde{D} : H^{1/2}(\Gamma) \to H^{-1/2}(\Gamma)$ is bounded and $H^{1/2}(\Gamma)$ elliptic, we first obtain the spectral equivalence inequalities

$$c_1^D \, \|v_h\|^2_{H^{1/2}(\Gamma)} \le (\widetilde{D}_h \underline{v}, \underline{v}) \le c_2^D \, \|v_h\|^2_{H^{1/2}(\Gamma)} \quad \text{for all } v_h \in S_h^1(\Gamma).$$

By using the inverse inequality in $S_h^1(\Gamma)$, and by applying Lemma 10.5 we further have

$$\|v_h\|^2_{H^{1/2}(\Gamma)} \le c_I \, h^{-1} \, \|v_h\|^2_{L_2(\Gamma)} = c \, h^{-1} \sum_{\ell=1}^N \Delta_\ell \sum_{k \in J(\ell)} v_k^2$$

and with $\Delta_\ell = h_\ell^{d-1}$ the upper estimate follows. The lower estimate finally results from

$$\|v_h\|^2_{H^{1/2}(\Gamma)} \ge \|v_h\|^2_{L_2(\Gamma)} \ge c \sum_{\ell=1}^N \Delta_\ell \sum_{k \in J(\ell)} v_k^2 \, . \qquad \square$$

To bound the spectral condition number of the stiffness matrix $\widetilde{D}_h$ of the stabilized hypersingular boundary integral operator we now obtain the estimate

$$\kappa_2(\widetilde{D}_h) \le c \, h^{-1} \, .$$

For the computation of the load vector $\underline{f}$ again we can introduce, as for the Dirichlet boundary value problem, a projection of the given Neumann data $g \in H^{-1/2}(\Gamma)$ onto the space of piecewise constant basis functions. When assuming $g \in L_2(\Gamma)$ the $L_2$ projection $Q_h g \in S_h^0(\Gamma)$ is defined as the unique solution of the variational problem

$$\langle Q_h g, \tau_h \rangle_{L_2(\Gamma)} = \langle g, \tau_h \rangle_{L_2(\Gamma)} \quad \text{for all } \tau_h \in S_h^0(\Gamma).$$

Note that there holds the error estimate

$$\|g - Q_h g\|_{H^\sigma(\Gamma)} \le c \, h^{s-\sigma} \, |g|_{H^s_{\text{pw}}(\Gamma)}$$

when assuming $g \in H^s_{\text{pw}}(\Gamma)$ for some $\sigma \in [-1, 0]$, if $s \in [0, 1]$ is satisfied. By using $g_h = Q_h g$ the perturbed variational problem is to find $\widetilde{u}_h \in S_h^1(\Gamma)$ such that

$$\langle D\widetilde{u}_h, v_h \rangle_\Gamma + \alpha \, \langle \widetilde{u}_h, w_{\text{eq}} \rangle_\Gamma \langle v_h, w_{\text{eq}} \rangle_\Gamma = \langle (\tfrac{1}{2} I - K') g_h, v_h \rangle_\Gamma \qquad (12.28)$$

is satisfied for all $v_h \in S_h^1(\Gamma)$. This is equivalent to a linear system of equations,

$$(D_h + \alpha \underline{a}\,\underline{a}^\top)\underline{\widetilde{u}} = (\frac{1}{2}M_h^\top - K_h^\top)\underline{g}\,.$$

If $\mathcal{R} \subset S_h^0(\Gamma)$ is satisfied, then we have

$$\langle Q_h g, v_k\rangle_{L_2(\Gamma)} = \langle g, v_k\rangle_{L_2(\Gamma)} = 0 \quad \text{for all } v_k \in \mathcal{R}$$

and therefore the solvability condition

$$\int_\Gamma Q_h g(x) v_k(x) ds_x = 0 \quad \text{for all } v_k \in \mathcal{R}$$

is satisfied, i.e. $\widetilde{u}_h \in H_*^{1/2}(\Gamma)$.

The application of Theorem 8.2 (Strang Lemma) gives, when assuming $\gamma_0^{\text{int}} u \in H^s(\Gamma)$ and $g \in H^\varrho_{\text{pw}}(\Gamma)$ for some $s \in [\frac{1}{2}, 2]$ and $\varrho \in [0,1]$, respectively, the error estimate

$$\begin{aligned}\|\gamma_0^{\text{int}} u - \widetilde{u}_h\|_{H^{1/2}(\Gamma)} &\leq \frac{c_2^{\widetilde{D}}}{c_1^{\widetilde{D}}} \inf_{v_h \in S_h^1(\Gamma)} \|\gamma_0^{\text{int}} u - v_h\|_{H^{1/2}(\Gamma)} + \frac{c_2^W}{c_1^{\widetilde{D}}} \|g - g_h\|_{H^{-1/2}(\Gamma)} \\ &\leq c_1\, h^{s-1/2}\, |\gamma_0^{\text{int}} u|_{H^s(\Gamma)} + c_2\, h^{\varrho + \frac{1}{2}}\, |g|_{H^\varrho(\Gamma)}. \qquad (12.29)\end{aligned}$$

In particular, when assuming $\gamma_0^{\text{int}} u \in H^2(\Gamma)$ and $g \in H^1_{\text{pw}}(\Gamma)$ we then obtain the error estimate

$$\|\gamma_0^{\text{int}} u - \widetilde{u}_h\|_{H^{1/2}(\Gamma)} \leq c\, h^{3/2} \left\{ |\gamma_0^{\text{int}} u|_{H^2(\Gamma)} + |g|_{H^1(\Gamma)} \right\}.$$

**Theorem 12.10.** *For some $s \in [\frac{1}{2}, 2]$ let $\gamma_0^{int} u \in H^s(\Gamma)$ be the unique solution of the variational problem (7.24), and let $\widetilde{u}_h \in S_h^1(\Gamma)$ be the unique solution of the perturbed Galerkin variational problem (12.28) where $g \in H^\varrho_{pw}(\Gamma)$ for some $\varrho \in [-\frac{1}{2}, 1]$. Assume that the stabilized hypersingular boundary integral operator $\widetilde{D} : H^{1-\sigma}(\Gamma) \to H^{-\sigma}(\Gamma)$ is bounded and bijective for some $\sigma \in [-1, \frac{1}{2}]$, and let the adjoint double layer potential $\frac{1}{2}I - K' : H^{-1+\sigma}(\Gamma) \to H^{-1+\sigma}(\Gamma)$ be bounded. Then there holds the error estimate*

$$\|\gamma_0^{int} u - \widetilde{u}_h\|_{H^\sigma(\Gamma)} \leq c_1\, h^{s-\sigma}\, |\gamma_0^{int} u|_{H^s(\Gamma)} + c_2\, h^{\varrho+1-\sigma}\, |g|_{H^\varrho_{pw}(\Gamma)}$$

*where $\sigma \in [0, \frac{1}{2}]$.*

*Proof.* As in the proof of Theorem 12.8 we have for some $\sigma \in [-1, \frac{1}{2}]$

$$\|\gamma_0^{\text{int}} u - \widetilde{u}_h\|_{H^\sigma(\Gamma)} = \sup_{0 \neq v \in H^{1-\sigma}(\Gamma)} \frac{\langle \widetilde{D}(\gamma_0^{\text{int}} u - \widetilde{u}_h), v\rangle_\Gamma}{\|\widetilde{D} v\|_{H^{-\sigma}(\Gamma)}}.$$

By subtracting the perturbed variational formulation (12.28) from the exact formulation (12.25) we conclude the equality

$$\langle \widetilde{D}(\gamma_0^{\text{int}} u - \widetilde{u}_h), v_h \rangle_\Gamma = \langle (\frac{1}{2} I - K')(g - g_h), v_h \rangle_\Gamma \quad \text{for all } v_h \in S_h^1(\Gamma).$$

Therefore,

$$\begin{aligned} \|\gamma_0^{\text{int}} u - \widetilde{u}_h\|_{H^\sigma(\Gamma)} \leq & \sup_{0 \neq v \in H^{1-\sigma}(\Gamma)} \frac{\langle \widetilde{D}(\gamma_0^{\text{int}} u - \widetilde{u}_h), v - Q_h^{1/2} v \rangle_\Gamma}{\|\widetilde{D} v\|_{H^{-\sigma}(\Gamma)}} \\ & + \sup_{0 \neq v \in H^{1-\sigma}(\Gamma)} \frac{\langle (\frac{1}{2} I - K')(g - g_h), Q_h^{1/2} v \rangle_\Gamma}{\|\widetilde{D} v\|_{H^{-\sigma}(\Gamma)}} \end{aligned}$$

As in the proof of Theorem 12.8 we have

$$\sup_{0 \neq v \in H^{1-\sigma}(\Gamma)} \frac{\langle \widetilde{D}(\gamma_0^{\text{int}} u - \widetilde{u}_h), v - Q_h^{1/2} v \rangle_\Gamma}{\|\widetilde{D} v\|_{H^{-\sigma}(\Gamma)}} \leq c_1 \, h^{s-\sigma} \, |\gamma_0^{\text{int}} u|_{H^s(\Gamma)}$$

for some $\sigma \geq -1$. The second term can be estimated by

$$\begin{aligned} |\langle (\frac{1}{2} I - K')(g - g_h), Q_h^{1/2} v \rangle_\Gamma| \leq & \, |\langle (\frac{1}{2} I - K')(g - g_h), v \rangle_\Gamma| \\ & + |\langle (\frac{1}{2} I - K')(g - g_h), v - Q_h^{1/2} v \rangle_\Gamma| \\ \leq & \, c_2 \|g - g_h\|_{H^{-1+\sigma}(\Gamma)} \|v\|_{H^{1-\sigma}(\Gamma)} + c_3 \|g - g_h\|_{L_2(\Gamma)} \|v - Q_h^{1/2} v\|_{L_2(\Gamma)} \\ \leq & \, c_2 \|g - g_h\|_{H^{-1+\sigma}(\Gamma)} \|v\|_{H^{1-\sigma}(\Gamma)} + c_3 h^\varrho \, |g|_{H^\varrho_{\text{pw}}(\Gamma)} h^{1-\sigma} \, \|v\|_{H^{1-\sigma}(\Gamma)} \end{aligned}$$

and therefore we get

$$\begin{aligned} \|\gamma_0^{\text{int}} u - \widetilde{u}_h\|_{H^\sigma(\Gamma)} \leq & \, c_1 h^{s-\sigma} |\gamma_0^{\text{int}} u|_{H^s(\Gamma)} + c_2 h^{\varrho+1-\sigma} \, |g|_{H^\varrho_{\text{pw}}(\Gamma)} \\ & + c_3 \|g - g_h\|_{H^{-1+\sigma}(\Gamma)}. \end{aligned}$$

Finally, for the $L_2$ projection $g_h = Q_h g$ there holds the error estimate

$$\|g - Q_h g\|_{H^{-1+\sigma}(\Gamma)} \leq c \, h^{\varrho+1-\sigma} \, |g|_{H^\varrho_{\text{pw}}(\Gamma)}$$

for $-1 + \sigma \geq -1$, i.e. $\sigma \geq 0$, and when assuming $g \in H^\varrho_{\text{pw}}(\Gamma)$ for some $\varrho \in [0, 1]$. □

When assuming $\gamma_0^{\text{int}} u \in H^2(\Gamma)$ and $g \in H^1_{\text{pw}}(\Gamma)$ we therefore obtain the error estimate

$$\|\gamma_0^{\text{int}} u - \widetilde{u}_h\|_{L_2(\Gamma)} \leq c \, h^2 \left\{ |\gamma_0^{\text{int}} u|_{H^2(\Gamma)} + |g|_{H^1_{\text{pw}}(\Gamma)} \right\},$$

and in the sequel we can compute the approximate solution

$$\hat{u}(\widetilde{x}) = \int_\Gamma U^*(\widetilde{x}, y) g_h(y) ds_y - \int_\Gamma \gamma_{1,y}^{\text{int}} U^*(\widetilde{x}, y) u_h(y) ds_y \quad \text{for } x \in \Omega$$

which satisfies the pointwise error estimate

$$|u(\widetilde{x}) - \hat{u}(\widetilde{x})| \le c\, h^2 \left\{ |\gamma_0^{\rm int} u|_{H^2(\Gamma)} + |g|_{H^1_{\rm pw}(\Gamma)} \right\}.$$

Hence, when using approximate boundary conditions, in particular when using a piecewise constant $L_2$ projection $Q_h g \in S_h^0(\Gamma)$ of the given Neumann data $g$ this results in non–optimal error estimates, i.e. the order of convergence is reduced in comparison to the case when using the exact boundary data.

Instead of the stabilized variational formulation (7.24) we may also consider the modified variational problem (7.25). Then, the associated Galerkin variational formulation is to find $u_h \in S_h^1(\Gamma)$ such that

$$\langle D u_h, v_h \rangle_\Gamma + \bar{\alpha}\, \langle u_h, 1 \rangle_\Gamma \langle v_h, 1 \rangle_\Gamma = \langle (\frac{1}{2} I - K') g, v_h \rangle_\Gamma \tag{12.30}$$

is satisfied for all $v_h \in S_h^1(\Gamma)$. From the solvability assumption (1.17) we then find $u_h \in H^{1/2}_{**}(\Gamma)$. All error estimates for the approximate solution $u_h \in S_h^1(\Gamma)$ follow as in Theorem 12.8. The variational problem (12.30) is equivalent to a linear system of algebraic equations,

$$(D_h + \bar{\alpha}\, \underline{\bar{a}}\, \underline{\bar{a}}^\top)\, \underline{u} = \underline{f},$$

where

$$\bar{a}_j = \langle \varphi_j^1, 1 \rangle_\Gamma \quad \text{for } j = 1, \ldots, M.$$

## 12.3 Mixed Boundary Conditions

The solution of the mixed boundary value problem (7.35),

$$\begin{aligned} Lu(x) &= 0 && \text{for } x \in \Omega \\ \gamma_0^{\rm int} u(x) &= g_D(x) && \text{for } x \in \Gamma_D, \\ \gamma_1^{\rm int} u(x) &= g_N(x) && \text{for } x \in \Gamma_N \end{aligned}$$

is given via the representation formula (7.36).

By $\widetilde{g}_D \in H^{1/2}(\Gamma)$ and $\widetilde{g}_N \in H^{-1/2}(\Gamma)$ we denote some appropriate extensions of the given data $g_D \in H^{1/2}(\Gamma_D)$ and $g_N \in H^{-1/2}(\Gamma_N)$, respectively. Then, the symmetric formulation of boundary integral equations reads to find the yet unknown Cauchy data $(\widetilde{t}, \widetilde{u}) \in \widetilde{H}^{-1/2}(\Gamma_D) \times \widetilde{H}^{1/2}(\Gamma_N)$ as the unique solution of the variational problem (7.39) such that

$$a(\widetilde{t}, \widetilde{u}; \tau, v) = F(\tau, v)$$

is satisfied for all $(\tau, v) \in \widetilde{H}^{-1/2}(\Gamma_D) \times \widetilde{H}^{1/2}(\Gamma_N)$. By using

$$\widetilde{t}_h(x) = \sum_{k=1}^{N_D} \widetilde{t}_k \varphi_k^0(x) \in S_h^0(\Gamma_D),$$

$$\widetilde{u}_h(x) = \sum_{i=1}^{M_N} \widetilde{u}_i \varphi_i^1(x) \in S_h^1(\Gamma_N) \cap \widetilde{H}^{1/2}(\Gamma_N)$$

the approximate solution $(\widetilde{t}_h, \widetilde{u}_h)$ is given as the unique solution of the Galerkin variational formulation such that

$$a(\widetilde{t}_h, \widetilde{u}_h; \tau_h, v_h) = F(\tau_h, v_h) \tag{12.31}$$

is satisfied for all $(\tau_h, v_h) \in S_h^0(\Gamma_D) \times S_h^1(\Gamma_N) \cap \widetilde{H}^{1/2}(\Gamma_N)$. Based on Lemma 7.4 all assumptions of Theorem 8.1 (Cea's Lemma) are satisfied. Therefore we have the unique solvability of the Galerkin variational problem (12.31) as well as the a priori error estimate

$$\begin{aligned} &\|\widetilde{u} - \widetilde{u}_h\|^2_{H^{1/2}(\Gamma)} + \|\widetilde{t} - \widetilde{t}_h\|^2_{H^{-1/2}(\Gamma)} \qquad (12.32) \\ &\le c \left\{ \inf_{v_h \in S_h^1(\Gamma_N) \cap \widetilde{H}^{1/2}(\Gamma_N)} \|\widetilde{u} - v_h\|^2_{H^{1/2}(\Gamma)} + \inf_{\tau_h \in S_h^0(\Gamma_D)} \|\widetilde{t} - \tau_h\|^2_{H^{-1/2}(\Gamma)} \right\}. \end{aligned}$$

Assume that the solution $u$ of the mixed boundary value problem (7.35) satisfies $u \in H^s(\Omega)$ for some $s > \frac{3}{2}$. Applying the trace theorem we obtain $\gamma_0^{\text{int}} u \in H^{s-1/2}(\Gamma)$ as well as $\gamma_1^{\text{int}} u \in H_{\text{pw}}^{s-3/2}(\Gamma)$. When assuming that the extensions $\widetilde{g}_D \in H^{s-1/2}(\Gamma)$ and $\widetilde{g}_N \in H_{\text{pw}}^{s-3/2}(\Gamma)$ are sufficiently regular, and using the approximation properties of the trial spaces $S_h^0(\Gamma_D)$ and $S_h^1(\Gamma_N)$ we obtain from (12.32) the error estimate

$$\|\widetilde{u} - \widetilde{u}_h\|^2_{H^{1/2}(\Gamma)} + \|\widetilde{t} - \widetilde{t}_h\|^2_{H^{-1/2}(\Gamma)} \le c_1 h^{2\sigma_1 - 1} |\widetilde{u}|^2_{H^{\sigma_1}(\Gamma)} + c_2 h^{2\sigma_2 + 1} |\widetilde{t}|^2_{H_{\text{pw}}^{\sigma_2}(\Gamma)}$$

where

$$\frac{1}{2} \le \sigma_1 \le \min\left\{2, s - \frac{1}{2}\right\}, \quad -\frac{1}{2} \le \sigma_2 \le \min\left\{1, s - \frac{3}{2}\right\}.$$

In particular for $\sigma_1 = \sigma$ and $\sigma_2 = \sigma - 1$ we have the error estimate

$$\|\widetilde{u} - \widetilde{u}_h\|^2_{H^{1/2}(\Gamma)} + \|\widetilde{t} - \widetilde{t}_h\|^2_{H^{-1/2}(\Gamma)} \le c\, h^{2\sigma - 1} \left\{ |\widetilde{u}|^2_{H^{\sigma}(\Gamma)} + |\widetilde{t}|^2_{H_{\text{pw}}^{\sigma-1}(\Gamma)} \right\}$$

for all $\frac{1}{2} \le \sigma \le \min\{2, s - \frac{1}{2}\}$. If $u \in H^{5/2}(\Omega)$ is sufficiently smooth, we then have

$$\|\widetilde{u} - \widetilde{u}_h\|^2_{H^{1/2}(\Gamma)} + \|\widetilde{t} - \widetilde{t}_h\|^2_{H^{-1/2}(\Gamma)} \le c\, h^3 \left\{ |\widetilde{u}|^2_{H^2(\Gamma)} + |\widetilde{t}|^2_{H_{\text{pw}}^1(\Gamma)} \right\},$$

i.e. the order of convergence for approximate solutions of the mixed boundary value problem corresponds to those of the pure Dirichlet or Neumann boundary value problem. As for the pure Dirichlet or Neumann boundary value problem we may also obtain error estimates in other Sobolev norms. In particular, as in Theorem 12.10 (Aubin–Nitsche Trick) we obtain, when assuming $\widetilde{u} \in H^2(\Gamma)$, the error estimate

$$\|\widetilde{u} - \widetilde{u}_h\|_{L_2(\Gamma)} \leq c\,h^2 \left[|\widetilde{u}|^2_{H^2(\Gamma)} + |\widetilde{t}|^2_{H^1_{\mathrm{pw}}(\Gamma)}\right]^{1/2}. \tag{12.33}$$

In analogy to Lemma 12.2 we finally have

$$\|\widetilde{t} - \widetilde{t}_h\|_{L_2(\Gamma)} \leq c\,h \left[|\widetilde{u}|^2_{H^2(\Gamma)} + |\widetilde{t}|^2_{H^1_{\mathrm{pw}}(\Gamma)}\right]^{1/2}. \tag{12.34}$$

The Galerkin variational formulation (12.31) is equivalent to a linear system of algebraic equations,

$$\begin{pmatrix} V_h & -K_h \\ K_h^\top & D_h \end{pmatrix} \begin{pmatrix} \underline{\widetilde{t}} \\ \underline{\widetilde{u}} \end{pmatrix} = \begin{pmatrix} \underline{f}_1 \\ \underline{f}_2 \end{pmatrix} \tag{12.35}$$

where the block matrices are given by

$$V_h[\ell,k] = \langle V\varphi^0_k, \varphi^0_\ell\rangle_{\Gamma_D},\; D_h[j,i] = \langle D\varphi^1_i, \varphi^1_j\rangle_{\Gamma_N},\; K_h[\ell,i] = \langle K\varphi^1_i, \varphi^0_\ell\rangle_{\Gamma_D}$$

and where the load vectors are defined via

$$f_{1,\ell} = \langle (\frac{1}{2}I + K)\widetilde{g}_D - V\widetilde{g}_N, \varphi^0_\ell\rangle_{\Gamma_D},\; f_{2,j} = \langle (\frac{1}{2}I - K')\widetilde{g}_N - D\widetilde{g}_D, \varphi^1_j\rangle_{\Gamma_N}$$

for $k,\ell = 1,\ldots,N_D$; $i,j = 1,\ldots,M_N$.

The stiffness matrix in (12.35) is block skew–symmetric and positive definite, or symmetric and indefinite. From the $H^{-1/2}(\Gamma_D)$–ellipticity of the single layer potential $V$ we conclude the positive definiteness of the Galerkin matrix $V_h$. Hence we can solve the first equation in (12.35) for the first unknown $\underline{\widetilde{t}}$, i.e.

$$\underline{\widetilde{t}} = V_h^{-1} K_h \underline{\widetilde{u}} + V_h^{-1} \underline{f}_1.$$

Inserting this into the second equation of (12.35) this gives the Schur complement system

$$\left[D_h + K_h^\top V_h^{-1} K_h\right] \underline{\widetilde{u}} = \underline{f}_2 - K_h^\top V_h^{-1} \underline{f}_1$$

where the Schur complement matrix

$$\widetilde{S}_h = D_h + K_h^\top V_h^{-1} K_h \tag{12.36}$$

is symmetric and positive definite.

If we use appropriate approximations of the given boundary data to evaluate the load vectors in (12.35), we can apply Theorem 8.2 to obtain the corresponding error estimates.

In what follows we will describe an alternative approach to solve the mixed boundary value problem (7.35). For this we consider the variational formulation (7.40) to find $\widetilde{u} \in \widetilde{H}^{1/2}(\Gamma_N)$ such that

$$\langle S\widetilde{u}, v\rangle_\Gamma \;=\; \langle g_N - S\widetilde{g}_D, v\rangle_{\Gamma_N}$$

is satisfied for all $v \in \widetilde{H}^{1/2}(\Gamma_N)$. Recall that

$$S := D + (\frac{1}{2}I + K')V^{-1}(\frac{1}{2}I + K) : H^{1/2}(\Gamma) \to H^{-1/2}(\Gamma)$$

is the Steklov–Poincaré operator. By using

$$\widetilde{u}_h(x) = \sum_{i=1}^{M_N} \widetilde{u}_i \varphi_i^1(x) \in S_h^1(\Gamma) \cap \widetilde{H}^{1/2}(\Gamma_N)$$

we obtain an associated Galerkin variational formulation and the corresponding linear system of algebraic equations, $S_h\underline{\widetilde{u}} = \underline{f}$, where

$$S_h[j,i] \;=\; \langle S\varphi_i^1, \varphi_j^1\rangle_{\Gamma_N}, \quad f_j \;=\; \langle g_N - S\widetilde{g}_D, \varphi_j\rangle_{\Gamma_N}$$

for $i, j = 1, \ldots, M_N$. However, due to the inverse single layer potential $V^{-1}$, the symmetric representation of the Steklov–Poincaré operator $S$ does not allow a direct computation of the stiffness matrix $S_h$. Hence we first have to define a symmetric approximation of the continuous Steklov–Poincaré operators $S$.

Let $v \in H^{1/2}(\Gamma)$ be some given function. Then, the application of the Steklov–Poincaré operator by using the symmetric representation (6.43) reads

$$Sv = Dv + (\frac{1}{2}I + K')V^{-1}(\frac{1}{2}I + K)v \;=\; Dv + (\frac{1}{2}I + K')w$$

where $w = V^{-1}(\frac{1}{2}I + K)v \in H^{-1/2}(\Gamma)$, i.e. $w \in H^{-1/2}(\Gamma)$ is the unique solution of the variational problem

$$\langle Vw, \tau\rangle_\Gamma \;=\; \langle(\frac{1}{2}I + K)v, \tau\rangle_\Gamma \quad \text{for all } \tau \in H^{-1/2}(\Gamma).$$

Let $S_h^0(\Gamma) \;=\; \operatorname{span}\{\varphi_k^0\}_{k=1}^N \subset H^{-1/2}(\Gamma)$ be the conforming trial space of piecewise constant basis functions, then the associated Galerkin variational formulation is to find $w_h \in S_h^0(\Gamma)$ such that

$$\langle Vw_h, \tau_h\rangle_\Gamma = \langle(\frac{1}{2}I + K)v, \tau_h\rangle_\Gamma \tag{12.37}$$

is satisfied for all $\tau_h \in S_h^0(\Gamma)$. By

$$\widetilde{S}v \;=\; Dv + (\frac{1}{2}I + K')w_h \tag{12.38}$$

we then define an approximation $\widetilde{S}$ of the Steklov–Poincaré operator $S$.

**Lemma 12.11.** *The approximate Steklov–Poincareé operator $\widetilde{S}$ as defined in (12.38) is bounded, i.e. $\widetilde{S} : H^{1/2}(\Gamma) \to H^{-1/2}(\Gamma)$ satisfying*

$$\|\widetilde{S}v\|_{H^{-1/2}(\Gamma)} \leq c_2^{\widetilde{S}}\, \|v\|_{H^{1/2}(\Gamma)} \quad \textit{for all } v \in H^{1/2}(\Gamma).$$

*Moreover, $\widetilde{S}$ is $\widetilde{H}^{1/2}(\Gamma_N)$–elliptic,*

$$\langle \widetilde{S}v, v\rangle_\Gamma \geq c_1^D\, \|v\|^2_{H^{1/2}(\Gamma)} \quad \textit{for all } v \in \widetilde{H}^{1/2}(\Gamma_N),$$

*and satisfies the error estimate*

$$\|(S-\widetilde{S})v\|_{H^{-1/2}(\Gamma)} \leq c \inf_{\tau_h \in S_h^0(\Gamma)} \|Sv - \tau_h\|_{H^{-1/2}(\Gamma)}.$$

*Proof.* For the solution $w_h \in S_h^0(\Gamma)$ of the Galerkin variational problem (12.37) we have, by using the $H^{-1/2}(\Gamma)$–ellipticity of the single layer potential $V$,

$$\begin{aligned} c_1^V\, \|w_h\|^2_{H^{-1/2}(\Gamma)} \leq \langle Vw_h, w_h\rangle_\Gamma &= \langle (\frac{1}{2}I + K)v, w_h\rangle_\Gamma \\ &\leq (\frac{1}{2} + c_2^K)\, \|v\|_{H^{1/2}(\Gamma)} \|w_h\|_{H^{-1/2}(\Gamma)} \end{aligned}$$

and therefore

$$\|w_h\|_{H^{-1/2}(\Gamma)} \leq \frac{1}{c_1^V}(\frac{1}{2} + c_2^K)\, \|v\|_{H^{1/2}(\Gamma)}.$$

Then we can conclude the boundedness of the approximate Steklov–Poincaré operator $\widetilde{S}$ by

$$\begin{aligned} \|\widetilde{S}v\|_{H^{-1/2}(\Gamma)} &= \|Dv + (\frac{1}{2}I + K')w_h\|_{H^{-1/2}(\Gamma)} \\ &\leq c_2^D\, \|v\|_{H^{1/2}(\Gamma)} + (\frac{1}{2} + c_2^K)\, \|w_h\|_{H^{-1/2}(\Gamma)} \\ &\leq \left[c_2^D + \frac{1}{c_1^V}(\frac{1}{2} + c_2^K)^2\right] \|v\|_{H^{1/2}(\Gamma)}. \end{aligned}$$

The $\widetilde{H}^{1/2}(\Gamma_N)$ ellipticity of $\widetilde{S}$ follows due to

$$\begin{aligned} \langle \widetilde{S}v, v\rangle_\Gamma &= \langle Dv, v\rangle_\Gamma + \langle (\frac{1}{2}I + K')w_h, v\rangle_\Gamma \\ &= \langle Dv, v\rangle_\Gamma + \langle w_h, (\frac{1}{2}I + K)v\rangle_\Gamma \\ &= \langle Dv, v\rangle_\Gamma + \langle Vw_h, w_h\rangle_\Gamma \geq c_1^D\, \|v\|^2_{H^{1/2}(\Gamma)} \end{aligned}$$

from the $H^{-1/2}(\Gamma)$ ellipticity of the single layer potential $V$ and from the $\widetilde{H}^{1/2}(\Gamma_N)$ ellipticity of the hypersingular boundary integral operator $D$.

By taking the difference

$$(S - \widetilde{S})v = (\frac{1}{2}I + K')(w - w_h)$$

we finally obtain

$$\begin{aligned}\|(S - \widetilde{S})v\|_{H^{-1/2}(\Gamma)} &= \|(\frac{1}{2}I + K')(w - w_h)\|_{H^{-1/2}(\Gamma)} \\ &\leq (\frac{1}{2} + c_2^K)\|w - w_h\|_{H^{-1/2}(\Gamma)},\end{aligned}$$

and by using the error estimate (12.6) for the Dirichlet boundary value problem we complete the proof. □

We now consider a modified Galerkin variational formulation to find an approximate solution $\hat{u}_h \in S_h^1(\Gamma) \cap \widetilde{H}^{1/2}(\Gamma_N)$ such that

$$\langle \widetilde{S}\hat{u}_h, \varphi_j^1 \rangle_{\Gamma_N} = \langle g_N - \widetilde{S}\widetilde{g}_D, \varphi_j^1 \rangle_{\Gamma_N} \tag{12.39}$$

is satisfied for all $j = 1, \ldots, M_N$. By using Theorem 8.3 (Strang Lemma) we conclude that the perturbed variational problem (12.39) is unique solvable, and there holds the error estimate

$$\begin{aligned}&\|\widetilde{u} - \hat{u}_h\|_{H^{1/2}(\Gamma)} \\ &\leq c \left\{ \inf_{v_h \in S_h^1(\Gamma)\cap\widetilde{H}^{1/2}(\Gamma_N)} \|\widetilde{u} - v_h\|_{H^{1/2}(\Gamma)} + \inf_{\tau_h \in S_h^0(\Gamma)} \|Su - \tau_h\|_{H^{-1/2}(\Gamma)} \right\}.\end{aligned}$$

When assuming $u \in H^2(\Gamma)$ and $Su \in H^1_{\text{pw}}(\Gamma)$ we then obtain the error estimate

$$\|\widetilde{u} - \hat{u}_h\|_{H^{1/2}(\Gamma)} \leq c\, h^{3/2} \left\{ \|u\|_{H^2(\Gamma)} + \|Su\|_{H^1_{\text{pw}}(\Gamma)} \right\}.$$

As in Theorem 12.10 we also find the error estimate

$$\|\widetilde{u} - \hat{u}_h\|_{L_2(\Gamma)} \leq c\, h^2 \left\{ \|u\|_{H^2(\Gamma)} + \|Su\|_{H^1_{\text{pw}}(\Gamma)} \right\}. \tag{12.40}$$

If the complete Dirichlet datum $u_h := \hat{u}_h + \widetilde{g}_D$ is known we can solve a Dirichlet boundary value problem to find the approximate Neumann datum $t_h \in S_h^0(\Gamma)$ satisfying the error estimate

$$\|t - t_h\|_{L_2(\Gamma)} \leq c\, h \left\{ \|u\|_{H^2(\Gamma)} + \|Su\|_{H^1_{\text{pw}}(\Gamma)} \right\}. \tag{12.41}$$

The Galerkin variational formulation (12.39) is equivalent to a linear system of algebraic equations, $\widetilde{S}_h \underline{\hat{u}} = \underline{f}$, where

$$\widetilde{S}_h = D_h + (\frac{1}{2}M_h^\top + K_h^\top)V_h^{-1}(\frac{1}{2}M_h + K_h) \tag{12.42}$$

is the discrete Steklov–Poincaré operator which is defined by using the matrices

$$D_h[j,i] = \langle D\varphi_i^1, \varphi_j^1\rangle_{\Gamma_N}, \qquad K_h[\ell,i] = \langle K\varphi_i^1, \varphi_\ell^0\rangle_\Gamma,$$
$$V_h[\ell,k] = \langle V\varphi_k^0, \varphi_\ell^0\rangle_\Gamma, \qquad M_h[\ell,i] = \langle \varphi_i^1, \varphi_\ell^0\rangle_\Gamma$$

for $i,j = 1,\ldots,M_N$ and $k,\ell = 1,\ldots,N$. By using

$$\underline{w} = V_h^{-1}(\frac{1}{2}M_h + K_h)\underline{\hat{u}}$$

the linear system $\widetilde{S}_h\underline{\hat{u}} = \underline{f}$ is equivalent to

$$\begin{pmatrix} V_h & -\frac{1}{2}M_h - K_h \\ \frac{1}{2}M_h^\top + K_h^\top & D_h \end{pmatrix} \begin{pmatrix} \underline{w} \\ \underline{\hat{u}} \end{pmatrix} = \begin{pmatrix} \underline{0} \\ \underline{f} \end{pmatrix}. \tag{12.43}$$

For a comparison of the symmetric boundary integral formulation (12.31) with the symmetric boundary element approximation (12.38) of the Steklov–Poincaré operator, and to check the theoretical error estimates we now consider a simple model problem. Let the domain $\Omega = (0,0.5)^2$ be a square with the boundary $\Gamma = \partial\Omega$ which is decomposed into a Dirichlet part

$$\Gamma_D = \Gamma \cap \{x \in \mathbb{R}^2 : x_2 = 0\},$$

and the remaining Neumann part $\Gamma_N = \Gamma\backslash\Gamma_D$. The boundary data $g_D$ and $g_N$ are given in such a way such that the solution of the mixed boundary value problem (7.35) is

$$u(x) = -\frac{1}{2}\log|x - x^*|, \quad x^* = (-0.1,-0.1)^\top.$$

The boundary $\Gamma = \partial\Omega$ is decomposed uniformly into $N$ boundary elements $\tau_\ell$ of the same mesh width $h$. The boundary conditions on $\Gamma_D$ and $\Gamma_N$ are interpolated by using piecewise linear and piecewise constant basis functions, respectively. In Table 12.4 we give the errors and the related order of convergence for the symmetric formulation (12.31) which confirm the theoretical error estimates (12.33) and (12.34). The same holds true for the error estimates (12.40) and (12.41) in the case of the symmetric approximation, for which the numerical results are given in Table 12.5.

## 12.4 Robin Boundary Conditions

Finally we consider the variational problem (7.41) of the homogeneous Robin boundary value problem (1.10) and (1.13) to find $\gamma_0^{\text{int}}u \in H^{1/2}(\Gamma)$ such that

$$\langle S\gamma_0^{\text{int}}u, v\rangle_\Gamma + \langle \kappa\gamma_0^{\text{int}}u, v\rangle_\Gamma = \langle g, v\rangle_\Gamma$$

| $N$ | $\|t - t_h\|_{L_2(\Gamma)}$ | eoc | $\|u - u_h\|_{L_2(\Gamma)}$ | eoc |
|---|---|---|---|---|
| 8 | 9.151 –1 | | 7.623 –2 | |
| 16 | 5.304 –1 | 0.79 | 2.305 –2 | 1.73 |
| 32 | 2.767 –1 | 0.94 | 6.166 –3 | 1.90 |
| 64 | 1.389 –1 | 0.99 | 1.544 –3 | 2.00 |
| 128 | 6.913 –2 | 1.01 | 3.776 –4 | 2.03 |
| 256 | 3.438 –2 | 1.01 | 9.180 –5 | 2.04 |
| 512 | 1.713 –2 | 1.01 | 2.234 –5 | 2.04 |
| 1024 | 8.544 –3 | 1.00 | 5.451 –6 | 2.04 |
| 2048 | 4.266 –3 | 1.00 | 1.328 –6 | 2.04 |
| theory: | | 1 | | 2 |

**Table 12.4.** Error and order of convergence, symmetric formulation.

| $N$ | $\|t - t_h\|_{L_2(\Gamma)}$ | eoc | $\|u - u_h\|_{L_2(\Gamma)}$ | eoc |
|---|---|---|---|---|
| 8 | 8.725 –1 | | 6.437 –2 | |
| 16 | 5.140 –1 | 0.76 | 1.832 –2 | 1.81 |
| 32 | 2.710 –1 | 0.92 | 4.851 –3 | 1.92 |
| 64 | 1.370 –1 | 0.98 | 1.244 –3 | 1.96 |
| 128 | 6.855 –2 | 1.00 | 3.143 –4 | 1.98 |
| 256 | 3.422 –2 | 1.00 | 7.893 –5 | 1.99 |
| 512 | 1.708 –2 | 1.00 | 1.977 –5 | 2.00 |
| 1024 | 8.531 –3 | 1.00 | 4.948 –6 | 2.00 |
| 2048 | 4.263 –3 | 1.00 | 1.239 –6 | 2.00 |
| theory: | | 1 | | 2 |

**Table 12.5.** Error and order of convergence, symmetric approximation.

is satisfied for all $v \in H^{1/2}(\Gamma)$. By using the ansatz

$$u_h(x) = \sum_{i=1}^{M} u_i \varphi_i^1(x) \in S_h^1(\Gamma)$$

the corresponding Galerkin variational formulation is equivalent to a linear system of algebraic equations,

$$\left[\bar{M}_h^\kappa + S_h\right] \underline{u} = \underline{f}, \tag{12.44}$$

where

$$S_h[j,i] = \langle S\varphi_i^1, \varphi_j^1\rangle_\Gamma, \; \bar{M}_h^\kappa[j,i] = \langle \kappa\varphi_i^1, \varphi_j^1\rangle_\Gamma, \; f_j = \langle g, \varphi_j^1\rangle_\Gamma$$

and $i, j = 1, \ldots, M$. As in the case when considering a mixed boundary value problem we need to introduce a suitable approximation of the Galerkin stiffness matrix $S_h$ of the Steklov Poincaré operator $S$. Then, instead of (12.44) we have to solve

$$\left[\bar{M}_h^\kappa + D_h + (\frac{1}{2}M_h^\top + K_h^\top)V_h^{-1}(\frac{1}{2}M_h + K_h)\right]\underline{\hat{u}} = \underline{f},$$

or the equivalent system

$$\begin{pmatrix} V_h & -\frac{1}{2}M_h - K_h \\ \frac{1}{2}M_h^\top + K_h^\top & \bar{M}_h^\kappa + D_h \end{pmatrix} \begin{pmatrix} \underline{w} \\ \underline{\hat{u}} \end{pmatrix} = \begin{pmatrix} \underline{0} \\ \underline{f} \end{pmatrix}.$$

## 12.5 Exercises

**12.1** Let $\tau_k$ be a one–dimensional boundary element of length $h_k$ with mid-point $x_k^*$. Compute the collocation matrix element

$$V_h^C[k,k] = -\frac{1}{2\pi}\int_{\tau_k} \log|y - x_k^*|ds_y.$$

**12.2** Let $\tau_k$ be a one–dimensional boundary element of length $h_k$. Compute the Galerkin matrix element

$$V_h^G[k,k] = -\frac{1}{2\pi}\int_{\tau_k}\int_{\tau_k} \log|x - y|ds_y ds_x.$$

**12.3** Using a simple midpoint rule the Galerkin matrix element of Exercise 12.2 may be approximated by

$$V_h^G[k,k] \approx h_k\, V_h^C[k,k].$$

Compute the relative error

$$\frac{|V_h^G[k,k] - h_k V_h^C[k,k]|}{V_h^G[k,k]}$$

and investigate the behavior as $h_k \to 0$.

**12.4** Give an explicite integration formula for the pointwise evaluation of the double layer potential of the Laplace operator when using piecewise linear basis functions.

**12.5** Let $\tau_k$ be some plane boundary element in $\mathbb{R}^3$. For a function $u(x)$ defined for $x \in \tau_k$ the extension $\widetilde{u}$ is defined to be constant along the normal vector $n(x)$. The surface curl is then given as

$$\underline{\text{curl}}_{\tau_k} u(x) = n(x) \times \nabla_x \widetilde{u}(x).$$

Compute the surface curl when $u(x)$ is linear on $\tau_k$.

**12.6** Let $\Gamma = \partial\Omega$ be the boundary of the circle $\Omega = B_r(0) \subset \mathbb{R}^2$ which can be described by using polar coordinates. Describe the Galerkin discretization of the Laplace single layer potential (cf. Exercise 6.1) when using piecewise constant basis functions with respect to a uniform decomposition of the parameter domain $[0, 1]$. In particular prove that $V_h$ is a circulant matrix, i.e.

$$V_h[\ell + 1, k + 1] = V_h[\ell, k] \quad \text{for } k, \ell = 1, \ldots, N - 1$$

and

$$V_h[\ell + 1, 1] = V_h[\ell, N] \quad \text{for } \ell = 1, \ldots, N - 1.$$

**12.7** Compute all eigenvalues and eigenvectors of the matrix

$$J = \begin{pmatrix} 0 & 1 & 0 & \cdots & \cdots & 0 \\ 0 & 0 & 1 & 0 & \cdots & 0 \\ \vdots & & \ddots & \ddots & \ddots & \vdots \\ 0 & \cdots & \cdots & 0 & 1 & 0 \\ 0 & \cdots & \cdots & \cdots & 0 & 1 \\ 1 & 0 & \cdots & \cdots & \cdots & 0 \end{pmatrix} \in \mathbb{R}^{N \times N}$$

as well as of $J^\ell$, $\ell = 2, \ldots, N$.

**12.8** Compute all eigenvalues and eigenvectors of a circulant matrix $A$. Describe the inverse matrix $A^{-1}$.

# 13

# Iterative Solution Methods

The Galerkin discretization of variational problems as described in Chapter 8 leads to large linear systems of algebraic equations. In the case of an elliptic and self adjoint partial differential operator the system matrix is symmetric and positive definite. Therefore we may use the method of conjugate gradients to solve the resulting system iteratively. Instead, the Galerkin discretization of a saddle point problem, e.g. when considering a mixed finite element scheme or the symmetric formulation of boundary integral equations, leads to a linear system where the system matrix is positive definite but block skew symmetric. By applying an appropriate transformation this system can be solved again by using a conjugate gradient method. Since we are interested in iterative solution algorithms where the convergence behavior is independent of the problem size, i.e. which is robust with respect to the mesh size, we need to use appropriate preconditioning strategies. For this we describe and analyze first a quite general approach which is based on the use of operators of the opposite order, and give later two examples for both finite and boundary element methods. For a more detailed theory of general iterative methods we refer to [4, 11, 70, 143].

## 13.1 The Method of Conjugate Gradients

We need to compute the solution vectors $\underline{u} \in \mathbb{R}^M$ of a sequence of linear systems of algebraic equations (8.5), $A_M\underline{u} = \underline{f}$, where the system matrix $A_M \in \mathbb{R}^{M\times M}$ is symmetric and positive definite, and where $M \in \mathbb{N}$ is the dimension of the trial space to be used for the discretization of the underlying elliptic variational problem (8.1).

To derive the method of conjugate gradients we start with a system of conjugate or $A_M$–orthogonal vectors $\{\underline{p}^k\}_{k=0}^{M-1}$ satisfying

$$(A_M\underline{p}^k, \underline{p}^\ell) = 0 \quad \text{for } k, \ell = 0, \ldots, M-1, k \neq \ell.$$

Since the system matrix $A_M$ is supposed to be positive definite, we have

$$(A_M\underline{p}^k, \underline{p}^k) > 0 \quad \text{for } k = 0, 1, \ldots, M-1.$$

For an arbitrary given initial guess $\underline{u}^0 \in \mathbb{R}^M$ we can write the unique solution $\underline{u} \in \mathbb{R}^M$ of the linear system $A_M\underline{u} = \underline{f}$ as a linear combination of conjugate vectors as

$$\underline{u} = \underline{u}^0 - \sum_{\ell=0}^{M-1} \alpha_\ell \underline{p}^\ell.$$

Hence we have

$$A_M\underline{u} = A_M\underline{u}^0 - \sum_{\ell=0}^{M-1} \alpha_\ell A_M\underline{p}^\ell = \underline{f},$$

and from the $A_M$–orthogonality of the basis vectors $\underline{p}^\ell$ we can compute the yet unknown coefficients from

$$\alpha_\ell = \frac{(A_M\underline{u}^0 - \underline{f}, \underline{p}^\ell)}{(A_M\underline{p}^\ell, \underline{p}^\ell)} \quad \text{for } \ell = 0, 1, \ldots, M-1.$$

For some $k = 0, 1, \ldots, M$ we may define an approximate solution

$$\underline{u}^k := \underline{u}^0 - \sum_{\ell=0}^{k-1} \alpha_\ell \underline{p}^\ell \in \mathbb{R}^M$$

of the linear system $A_M\underline{u} = \underline{f}$. Obviously, $\underline{u}^M = \underline{u}$ is just the exact solution. By construction we have

$$\underline{u}^{k+1} := \underline{u}^k - \alpha_k \underline{p}^k \quad \text{for } k = 0, 1, \ldots, M-1,$$

and from the $A_M$–orthogonality of the vectors $\{\underline{p}^\ell\}_{\ell=0}^{M-1}$ we obtain

$$\alpha_k = \frac{(A_M\underline{u}^0 - \underline{f}, \underline{p}^k)}{(A_M\underline{p}^k, \underline{p}^k)} = \frac{\left(A_M\underline{u}^0 - \sum\limits_{\ell=0}^{k-1} \alpha_\ell A_M\underline{p}^\ell - \underline{f}, \underline{p}^k\right)}{(A_M\underline{p}^k, \underline{p}^k)} = \frac{(A_M\underline{u}^k - \underline{f}, \underline{p}^k)}{(A_M\underline{p}^k, \underline{p}^k)}.$$

If we denote by

$$\underline{r}^k := A_M\underline{u}^k - \underline{f}$$

the residual of the approximate solution $\underline{u}^k$ we finally have

$$\alpha_k = \frac{(\underline{r}^k, \underline{p}^k)}{(A_M\underline{p}^k, \underline{p}^k)}. \tag{13.1}$$

On the other hand, for $k = 0, 1, \ldots, M-1$ we can compute the residual $\underline{r}^{k+1}$ recursively by

$$\underline{r}^{k+1} = A_M\underline{u}^{k+1} - \underline{f} = A_M(\underline{u}^k - \alpha_k\underline{p}^k) - \underline{f} = \underline{r}^k - \alpha_k A_M\underline{p}^k .$$

The above approach is based on the use of $A_M$–orthogonal vectors $\{\underline{p}^\ell\}_{\ell=0}^{M_1}$. Such a vector system can be constructed by applying the Gram–Schmidt orthogonalization algorithm which is applied to some given system $\{\underline{w}^\ell\}_{\ell=0}^{M-1}$ of vectors which are linear independent, see Algorithm 13.1.

Initialize for $k = 0$:
$\underline{p}^0 := \underline{w}^0$
Compute for $k = 0, 1, \ldots, M-2$:
$$\underline{p}^{k+1} := \underline{w}^{k+1} - \sum_{\ell=0}^{k} \beta_{k\ell}\underline{p}^\ell, \quad \beta_{k\ell} = \frac{(A_M\underline{w}^{k+1}, \underline{p}^\ell)}{(A\underline{p}^\ell, \underline{p}^\ell)}$$

Algorithm 13.1: Gram–Schmidt orthogonalization.

By construction we have for $k = 0, 1, \ldots, M-1$

$$\text{span}\,\{\underline{p}^\ell\}_{\ell=0}^{k} = \text{span}\,\{\underline{w}^\ell\}_{\ell=0}^{k} .$$

It remains to define the initial vector system $\{\underline{w}^\ell\}_{\ell=0}^{M-1}$. One possibility is to choose the unit basis vectors $\underline{w}^k := \underline{e}^k = (\delta_{k+1,\ell})_{\ell=1}^{M}$ [59]. Alternatively we may find the basis vector $\underline{w}^{k+1}$ from the properties of the already constructed vector systems $\{\underline{p}^\ell\}_{\ell=0}^{k}$ and $\{\underline{r}^\ell\}_{\ell=0}^{k}$.

**Lemma 13.1.** *For $k = 0, 1, \ldots, M-2$ we have*

$$(\underline{r}^{k+1}, \underline{p}^\ell) = 0 \quad \textit{for}\, \ell = 0, 1, \ldots, k.$$

*Proof.* For $\ell = k = 1, \ldots, M-1$ we have by using (13.1) to define the coefficients $\alpha_k$ and by using the recursion of the residual $\underline{r}^{k+1}$ the orthogonality

$$(\underline{r}^{k+1}, \underline{p}^k) = (\underline{r}^k, \underline{p}^k) - \alpha_k(A_M\underline{p}^k, \underline{p}^k) = 0 .$$

For $\ell = k-1$ we then obtain

$$(\underline{r}^{k+1}, \underline{p}^{k-1}) = (\underline{r}^k, \underline{p}^{k-1}) - \alpha_k(A_M\underline{p}^k, \underline{p}^{k-1}) = 0$$

by applying the $A_M$–orthogonality of $\underline{p}^k$ and $\underline{p}^{k-1}$. Now the assertion follows by induction. □

From the orthogonality relation between the residual $\underline{r}^{k+1}$ and the search directions $\underline{p}^\ell$ we can immediately conclude a orthogonality of the residual $\underline{r}^{k+1}$ with the initial vectors $\underline{w}^\ell$.

**Corollary 13.2.** *For $k = 0, 1, \ldots, M-2$ we have*

$$(\underline{r}^{k+1}, \underline{w}^\ell) = 0 \quad \textit{for}\, \ell = 0, 1, \ldots, k.$$

*Proof.* By construction of the search directions $\underline{p}^\ell$ from the Gram–Schmidt orthogonalization we first have the representation

$$\underline{w}^\ell = \underline{p}^\ell + \sum_{j=0}^{\ell-1} \beta_{\ell-1,j}\underline{p}^j .$$

Hence we obtain

$$(\underline{r}^{k+1}, \underline{w}^\ell) = (\underline{r}^{k+1}, \underline{p}^\ell) + \sum_{j=0}^{\ell-1} \beta_{\ell-1,j}(\underline{r}^{k+1}, \underline{p}^j),$$

and the orthogonality relation follows from Lemma 13.1. □

Hence we have that all vectors

$$\underline{w}^0, \underline{w}^1, \ldots, \underline{w}^k, \underline{r}^{k+1}$$

are orthogonal to each other, and therefore linear independent. Since we need only to know the search directions $\underline{p}^0, \ldots, \underline{p}^k$ and therefore the initial vectors $\underline{w}^0, \ldots, \underline{w}^k$ to construct the approximate solution $\underline{u}^{k+1}$ and therefore the residual $\underline{r}^{k+1}$, we can define the new initial vector as

$$\underline{w}^{k+1} := \underline{r}^{k+1} \quad \text{for } k = 0, \ldots, M-2$$

where $\underline{w}^0 := \underline{r}^0$. By Corollary 13.2 we then have the orthogonality

$$(\underline{r}^{k+1}, \underline{r}^\ell) = 0 \quad \text{for } \ell = 0, \ldots, k; k = 0, \ldots, M-2.$$

Moreover, for the numerator of the coefficient $\alpha_k$ we obtain

$$(\underline{r}^k, \underline{p}^k) = (\underline{r}^k, \underline{r}^k) + \sum_{\ell=0}^{k-1} \beta_{k-1,\ell}(\underline{r}^k, \underline{p}^\ell) = (\underline{r}^k, \underline{r}^k),$$

and therefore, instead of (13.1),

$$\alpha_k = \frac{(\underline{r}^k, \underline{r}^k)}{(A_M \underline{p}^k, \underline{p}^k)} \quad \text{for } k = 0, \ldots, M-1.$$

In what follows we can assume

$$\alpha_\ell > 0 \quad \text{for } \ell = 0, \ldots, k.$$

Otherwise we would have

$$(\underline{r}^{\ell+1}, \underline{r}^{\ell+1}) = (\underline{r}^\ell - \alpha_\ell A \underline{p}^\ell, \underline{r}^{\ell+1}) = (\underline{r}^\ell, \underline{r}^{\ell+1}) = 0$$

implying $\underline{r}^{\ell+1} = \underline{0}$ and therefore $\underline{u}^{\ell+1} = \underline{u}$ would be the exact solution of the linear system $A_M \underline{u} = \underline{f}$.

From the recursion of the residual $\underline{r}^{\ell+1}$ we then obtain

$$A_M \underline{p}^\ell = \frac{1}{\alpha_\ell}\left(\underline{r}^\ell - \underline{r}^{\ell+1}\right) \quad \text{for } \ell = 0,\ldots,k.$$

Now, by using $\underline{w}^{k+1} = \underline{r}^{k+1}$ and by using the symmetry of the system matrix $A_M = A_M^\top$ we can compute the nominator of the coefficient $\beta_{k\ell}$ as

$$(A_M \underline{w}^{k+1}, \underline{p}^\ell) = (\underline{r}^{k+1}, A_M \underline{p}^\ell) = \frac{1}{\alpha_\ell}(\underline{r}^{k+1}, \underline{r}^\ell - \underline{r}^{\ell+1}) = \begin{cases} 0 & \text{for } \ell < k, \\ -\dfrac{(\underline{r}^{k+1}, \underline{r}^{k+1})}{\alpha_k} & \text{for } \ell = k. \end{cases}$$

The recursion of the Gram–Schmidt orthogonalization algorithm now reduces to

$$\underline{p}^{k+1} = \underline{r}^{k+1} - \beta_{kk}\underline{p}^k \quad \text{where } \beta_{kk} = -\frac{1}{\alpha_k}\frac{(\underline{r}^{k+1}, \underline{r}^{k+1})}{(A_M \underline{p}^k, \underline{p}^k)}.$$

On the other hand we have

$$\alpha_k (A_M \underline{p}^k, \underline{p}^k) = (\underline{r}^k - \underline{r}^{k+1}, \underline{p}^k) = (\underline{r}^k, \underline{p}^k) = (\underline{r}^k, \underline{r}^k - \beta_{k-1,k-1}\underline{p}^{k-1}) = (\underline{r}^k, \underline{r}^k)$$

and therefore

$$\underline{p}^{k+1} = \underline{r}^{k+1} + \beta_k \underline{p}^k \quad \text{where } \beta_k = \frac{(\underline{r}^{k+1}, \underline{r}^{k+1})}{(\underline{r}^k, \underline{r}^k)}.$$

Summarizing the above we obtain the iterative method of conjugate gradients [78] as described in Algorithm 13.2.

For an arbitrary initial guess $\underline{u}^0$ compute

$$\underline{r}^0 := A_M \underline{u}^0 - \underline{f},\ \underline{p}^0 := \underline{r}^0,\ \varrho_0 := (\underline{r}^0, \underline{r}^0).$$

For $k = 0, 1, 2, \ldots, M-2$:

$$\underline{s}^k := A_M \underline{p}^k,\ \sigma_k := (\underline{s}^k, \underline{p}^k),\ \alpha_k := \varrho_k / \sigma_k;$$
$$\underline{u}^{k+1} := \underline{u}^k - \alpha_k \underline{p}^k,\ \underline{r}^{k+1} := \underline{r}^k - \alpha_k \underline{s}^k;$$
$$\varrho_{k+1} := (\underline{r}^{k+1}, \underline{r}^{k+1}).$$

Stop, if $\varrho_{k+1} \leq \varepsilon \varrho_0$ is satisfied for some given accuracy $\varepsilon$.
Otherwise, compute the new search direction

$$\beta_k := \varrho_{k+1}/\varrho_k,\ \underline{p}^{k+1} := \underline{r}^{k+1} + \beta_k \underline{p}^k.$$

Algorithm 13.2: Method of conjugate gradients.

If the matrix $A_M$ is symmetric and positive definite we may define

$$\|\cdot\|_{A_M} := \sqrt{(A_M\cdot,\cdot)}$$

to be an equivalent norm in $\mathbb{R}^M$. Moreover,

$$\kappa_2(A_M) := \|A_M\|_2 \|A_M^{-1}\|_2 = \frac{\lambda_{\max}(A_M)}{\lambda_{\min}(A_M)}$$

is the spectral condition number of the positive definite and symmetric matrix $A_M$ where $\|\cdot\|_2$ is the matrix norm which is induced by the Euclidean inner product. Then one can prove the following estimate for the approximate solution $\underline{u}^k$, see for example [70, 143].

**Theorem 13.3.** *Let $A_M = A_M^\top > 0$ be symmetric and positive definite, and let $\underline{u} \in \mathbb{R}^M$ be the unique solution of the linear system $A_M\underline{u} = \underline{f}$. Then the method of conjugate gradients as described in Algorithm 13.2 is convergent for any initial guess $\underline{u}^0 \in \mathbb{R}^M$, and there holds the error estimate*

$$\|\underline{u}^k - \underline{u}\|_A \leq \frac{2q^k}{1+q^{2k}} \|\underline{u}^0 - \underline{u}\|_A \quad \textit{where } q := \frac{\sqrt{\kappa_2(A_M)}+1}{\sqrt{\kappa_2(A_M)}-1}.$$

To ensure a certain given relative accuracy $\varepsilon \in (0,1)$ we find the number $k_\varepsilon \in \mathbb{N}$ of required iteration steps from

$$\frac{\|\underline{u}^k - \underline{u}\|_A}{\|\underline{u}^0 - \underline{u}\|_A} \leq \frac{2q^k}{1+q^{2k}} \leq \varepsilon$$

and therefore

$$k_\varepsilon > \frac{\ln[1-\sqrt{1-\varepsilon^2}] - \ln\varepsilon}{\ln q}.$$

The number $k_\varepsilon$ obviously depends on $q$ and therefore on the spectral condition number $\kappa_2(A_M)$ of $A_M$. When considering the discretization of elliptic variational problems by using either finite or boundary elements the spectral condition number $\kappa_2(A_M)$ depends on the dimension $M \in \mathbb{N}$ of the used finite dimensional trial space, or on the underlying mesh size $h$.

In the case of a finite element discretization we have for the spectral condition number, by using the estimate (11.13),

$$\kappa_2(A_h^{\text{FEM}}) = \mathcal{O}(h^{-2}), \qquad \text{i.e.} \quad \kappa_2(A_{h/2}^{\text{FEM}}) \approx 4\,\kappa_2(A_h^{\text{FEM}})$$

when considering a globally quasi–uniform mesh refinement strategy. Asymptotically, this gives

$$\begin{aligned} \ln q_{h/2} &= \ln \frac{\sqrt{\kappa_2(A_{h/2}^{\text{FEM}})}+1}{\sqrt{\kappa_2(A_{h/2}^{\text{FEM}})}-1} \approx \ln \frac{2\sqrt{\kappa_2(A_h^{\text{FEM}})}+1}{2\sqrt{\kappa_2(A_h^{\text{FEM}})}-1} \\ &\approx \frac{1}{2} \ln \frac{\sqrt{\kappa_2(A_h^{\text{FEM}})}+1}{\sqrt{\kappa_2(A_h^{\text{FEM}})}-1} = \frac{1}{2} \ln q_h\,. \end{aligned}$$

Therefore, in the case of an uniform refinement step, i.e. halving the mesh size $h$, the number of required iterations is doubled to reach the same relative accuracy $\varepsilon$. As an example we choose $\varepsilon = 10^{-10}$. In Table 13.1 we give the number of iterations of the conjugate gradient method to obtain the results which were already presented in Table 11.2.

| | FEM | | | BEM | | |
|---|---|---|---|---|---|---|
| L | $N$ | $\kappa_2(A_h)$ | Iter | $N$ | $\kappa_2(V_h)$ | Iter |
| 2 | 64 | 12.66 | 13 | 16 | 24.14 | 8 |
| 3 | 256 | 51.55 | 38 | 32 | 47.86 | 18 |
| 4 | 1024 | 207.17 | 79 | 64 | 95.64 | 28 |
| 5 | 4096 | 829.69 | 157 | 128 | 191.01 | 39 |
| 6 | 16384 | 3319.76 | 309 | 256 | 381.32 | 52 |
| 7 | 65536 | 13280.04 | 607 | 512 | 760.73 | 69 |
| 8 | 262144 | 52592.92 | 1191 | 1024 | 1516.02 | 91 |
| Theory: | | $\mathcal{O}(h^{-2})$ | $\mathcal{O}(h^{-1})$ | | $\mathcal{O}(h^{-1})$ | $\mathcal{O}(h^{-1/2})$ |

**Table 13.1.** Number of CG iteration steps when $\varepsilon = 10^{-10}$.

When considering a comparable discretization by using boundary elements as already discussed in Table 12.2 we obtain for the spectral condition number of the system matrix

$$\kappa_2(A_h^{\mathrm{BEM}}) = \mathcal{O}(h^{-1}) \qquad \text{i.e.} \quad \kappa_2(A_{h/2}^{\mathrm{BEM}}) \approx 2\,\kappa_2(A_h^{\mathrm{BEM}}).$$

The number of required iterations to reach a certain relative accuracy $\varepsilon$ then grows with a factor of $\sqrt{2}$, see Table 13.1.

Hence there is a serious need to construct iterative algorithms which are almost robust with respect to all discretization parameters, i.e. with respect to the mesh size $h$. In general this can be done by introducing the concept of preconditioning the linear system $A_M\underline{u} = \underline{f}$.

Let $C_A \in \mathbb{R}^{M\times M}$ be a symmetric and positive definite matrix which can be factorized as

$$C_A = JD_{C_A}J^\top, \quad D_{C_A} = \mathrm{diag}(\lambda_k(C_A)), \quad \lambda_k(C_A) > 0$$

where $J \in \mathbb{R}^{M\times M}$ contains all eigenvectors of $C_A$ which are assumed to be orthonormal. Hence we can define

$$C_A^{1/2} = JD_{C_A}^{1/2}J^\top, \quad D_{C_A} = \mathrm{diag}(\sqrt{\lambda_k(C_A)})$$

satisfying

$$C_A = C_A^{1/2}C_A^{1/2}, \qquad C_A^{-1/2} := (C_A^{1/2})^{-1}.$$

Instead of the linear system $A_M\underline{u} = \underline{f}$ we now consider the equivalent system

$$\widetilde{A}\widetilde{\underline{u}} := C_A^{-1/2} A_M C_A^{-1/2} C_A^{1/2} \underline{u} = C_A^{-1/2} \underline{f} =: \widetilde{\underline{f}}$$

where the transformed system matrix

$$\widetilde{A} := C_A^{-1/2} A_M C_A^{-1/2}$$

is again symmetric and positive definite. Hence we can apply the method of conjugate gradients as described in Algorithm 13.2 to compute the transformed solution vector $\widetilde{\underline{u}} = C_A^{1/2} \underline{u}$. Inserting all the transformations we finally obtain the preconditioned method of conjugate gradients, see Algorithm 13.3.

For an arbitrary initial guess $\underline{x}^0$ compute

$$\underline{r}^0 := A_M \underline{x}^0 - \underline{f},\ \underline{v}^0 := C_A^{-1} \underline{r}^0,\ \underline{p}^0 := \underline{v}^0,\ \varrho_0 := (\underline{v}^0, \underline{r}^0).$$

For $k = 0, 1, 2, \ldots, M-2$:

$$\underline{s}^k := A_M \underline{p}^k,\ \sigma_k := (\underline{s}^k, \underline{p}^k),\ \alpha_k := \varrho_k / \sigma_k;$$

$$\underline{x}^{k+1} := \underline{x}^k - \alpha_k \underline{p}^k,\ \underline{r}^{k+1} := \underline{r}^k - \alpha_k \underline{s}^k;$$

$$\underline{v}^{k+1} := C_A^{-1} \underline{r}^{k+1},\ \varrho_{k+1} := (\underline{v}^{k+1}, \underline{r}^{k+1}) \,.$$

Stop, if $\varrho_{k+1} \leq \varepsilon \varrho_0$ is satisfied for some given accuracy $\varepsilon$. Otherwise compute the new search direction

$$\beta_k := \varrho_{k+1}/\varrho_k,\ \underline{p}^{k+1} := \underline{v}^{k+1} + \beta_k \underline{p}^k \,.$$

Algorithm 13.3: Preconditioned method of conjugate gradients.

The Algorithm 13.3 of the preconditioned method of conjugate gradients requires one matrix by vector product per iteration step, $\underline{s}^k = A_M \underline{p}^k$, and one application of the inverse preconditioning matrix, $\underline{v}^{k+1} = C_A^{-1} \underline{r}^{k+1}$. From Theorem 13.3 we obtain an error estimate for the approximate solution $\widetilde{\underline{u}}^k$,

$$\|\widetilde{\underline{u}}^k - \widetilde{\underline{u}}\|_{\widetilde{A}} \leq \frac{2\widetilde{q}}{1+\widetilde{q}^{2k}} \|\widetilde{\underline{u}}^0 - \widetilde{\underline{u}}\|_{\widetilde{A}} \quad \text{where } \widetilde{q} = \frac{\sqrt{\kappa_2(\widetilde{A})}+1}{\sqrt{\kappa_2(\widetilde{A})}-1}.$$

Note that for $\widetilde{\underline{z}} = C_A^{1/2} \underline{z}$ we have

$$\|\widetilde{\underline{z}}\|^2_{\widetilde{A}} = (\widetilde{A} C_A^{1/2} \underline{z}, C_A^{1/2} \underline{z}) = (A_M \underline{z}, \underline{z}) = \|\underline{z}\|^2_{A_M}.$$

Hence, for the approximate solution $\underline{u}^k = C_A^{-1/2} \widetilde{\underline{u}}^k$ we find the error estimate

$$\|\underline{u}^k - \underline{u}\|_{A_M} \leq \frac{2\widetilde{q}}{1+\widetilde{q}^{2k}} \|\underline{u}^0 - \underline{u}\|_{A_M}.$$

To bound the extremal eigenvalues of the transformed system matrix $\widetilde{A}$ we get from the Rayleigh quotient

$$\lambda_{\min}(\widetilde{A}) = \min_{\widetilde{\underline{z}}\in\mathbb{R}^M} \frac{(\widetilde{A}\widetilde{\underline{z}},\widetilde{\underline{z}})}{(\widetilde{\underline{z}},\widetilde{\underline{z}})} \leq \max_{\widetilde{\underline{z}}\in\mathbb{R}^M} \frac{(\widetilde{A}\widetilde{\underline{z}},\widetilde{\underline{z}})}{(\widetilde{\underline{z}},\widetilde{\underline{z}})} = \lambda_{\max}(\widetilde{A}).$$

When inserting the transformations $\widetilde{A} = C_A^{-1/2} A_M C_A^{-1/2}$ and $\widetilde{\underline{z}} = C_A^{1/2}\underline{z}$ this gives

$$\lambda_{\min}(\widetilde{A}) = \min_{\underline{z}\in\mathbb{R}^M} \frac{(A_M\underline{z},\underline{z})}{(C_A\underline{z},\underline{z})} \leq \max_{\underline{z}\in\mathbb{R}^M} \frac{(A_M\underline{z},\underline{z})}{(C_A\underline{z},\underline{z})} = \lambda_{\max}(\widetilde{A}).$$

Hence we have to assume that the preconditioning matrix $C_A$ satisfies the spectral equivalence inequalities

$$c_1^A\,(C_A\underline{z},\underline{z}) \leq (A_M\underline{z},\underline{z}) \leq c_2^A\,(C_A\underline{z},\underline{z}) \quad \text{for } \underline{z}\in\mathbb{R}^M \tag{13.2}$$

independent of $M$. Then we can bound the spectral condition number of the transformed system matrix as

$$\kappa_2(C_A^{-1/2} A C_A^{-1/2}) = \kappa_2(C_A^{-1}A_M) \leq \frac{c_2^A}{c_1^A}.$$

If the spectral condition number $\kappa_2(C_A^{-1}A_M)$ of the preconditioned system matrix can be bounded independent of the dimension $M$, i.e. independent of the mesh size $h$, then there is a fixed number $k_\varepsilon$ of required iterations to reach a certain given relative accuracy $\varepsilon$.

## 13.2 A General Preconditioning Strategy

We need to construct a matrix $C_A$ as a preconditioner for a given matrix $A_M$ such that the spectral equivalence inequalities (13.2) are satisfied, and an efficient realization of the preconditioning $\underline{v}^k = C_A^{-1}\underline{r}^k$ is possible. Here we consider the case where the matrix $A_M$ represents a Galerkin discretization of a bounded, $X$–elliptic, and self–adjoint operator $A : X \to X'$ satisfying

$$\langle Av, v\rangle \geq c_1^A\,\|v\|_X^2, \quad \|Av\|_{X'} \leq c_2^A\,\|v\|_X \quad \text{for } v\in X. \tag{13.3}$$

In particular, the matrix $A_M$ is given by

$$A_M[\ell,k] = \langle A\varphi_k, \varphi_\ell\rangle \quad \text{for } k,\ell = 1,\ldots,M$$

where $X_M := \text{span}\{\varphi_k\}_{k=1}^M \subset X$ is some conforming trial space.

Let $B : X' \to X$ be some bounded, $X'$–elliptic, and self–adjoint operator, i.e. for $f\in X'$ we assume

$$\langle Bf, f\rangle \geq c_1^B\,\|f\|_{X'}^2, \quad \|Bf\|_X \leq c_2^B\,\|f\|_{X'}.$$

By applying Theorem 3.4 there exists the inverse operator $B^{-1} : X \to X'$. In particular, by using (3.13) and Lemma 3.5 we have

$$\|B^{-1}v\|_{X'} \leq \frac{1}{c_1^B} \|v\|_X, \quad \langle B^{-1}v, v\rangle \geq \frac{1}{c_2^B} \|v\|_X^2 \quad \text{for } v \in X. \tag{13.4}$$

From the assumptions (13.3) and (13.4) we immediately conclude:

**Corollary 13.4.** *For $v \in X$ there hold the spectral equivalence inequalities*

$$c_1^A c_1^B \langle B^{-1}v, v\rangle \leq \langle Av, v\rangle \leq c_2^A c_2^B \langle B^{-1}v, v\rangle.$$

Then, by defining the preconditioning matrix

$$C_A[\ell, k] = \langle B^{-1}\varphi_k, \varphi_\ell\rangle \quad \text{for } k, \ell = 1, \ldots, M \tag{13.5}$$

we obtain from Corollary 13.4 by using the isomorphism

$$\underline{v} \in \mathbb{R}^M \quad \leftrightarrow \quad v_M = \sum_{k=1}^M v_k \varphi_k \in X_M \subset X$$

the required spectral equivalence inequalities

$$c_1^A c_1^B (C_A \underline{v}, \underline{v}) \leq (A_M \underline{v}, \underline{v}) \leq c_2^A c_2^B (C_A \underline{v}, \underline{v}) \quad \text{for } \underline{v} \in \mathbb{R}^M. \tag{13.6}$$

Although the constants in (13.6) only express the continuous mapping properties of the operators $A$ and $B$, and therefore they are independent of the discretization to be used, the above approach seems on a first glance useless, since in general only the operator $B$ is given explicitly. Moreover, neither can the preconditioning matrix $C_A$ be computed nor can the inverse $C_A^{-1}$ be applied efficiently. Hence we introduce a conforming trial space in the dual space $X'$,

$$X'_M := \text{span}\{\psi_k\}_{k=1}^M \subset X',$$

and define

$$B_M[\ell, k] = \langle B\psi_k, \psi_\ell\rangle, \quad M_M[\ell, k] = \langle \varphi_k, \psi_\ell\rangle \quad \text{for } k, \ell = 1, \ldots, M.$$

Note that the Galerkin matrix $B_M$ is symmetric and positive definite, and therefore invertible. Therefore we can define an approximation of the preconditioning matrix $C_A$ by

$$\widetilde{C}_A := M_M^\top B_M^{-1} M_M \,. \tag{13.7}$$

We need to prove that the approximated preconditioning matrix $\widetilde{C}_A$ is spectrally equivalent to $C_A$, and therefore to $A_M$.

**Lemma 13.5.** *Let $C_A$ be the Galerkin matrix of $B^{-1}$ as defined in* (13.5), *and let $\widetilde{C}_A$ be the approximation as given in* (13.7). *Then there holds*

$$(\widetilde{C}_A \underline{v}, \underline{v}) \leq (C_A \underline{v}, \underline{v}) \quad \textit{for } \underline{v} \in \mathbb{R}^M.$$

*Proof.* Let $\underline{v} \in \mathbb{R}^M \leftrightarrow v_M \in X_M \subset X$ be arbitrary but fixed. Then, $w = B^{-1}v_M \in X'$ is the unique solution of the variational problem

$$\langle Bw, z\rangle \;=\; \langle v_M, z\rangle \quad \text{for } z \in X'.$$

Note that

$$(C_A\underline{v}, \underline{v}) = \langle B^{-1}v_M, v_M\rangle = \langle w, v_M\rangle = \langle Bw, w\rangle. \tag{13.8}$$

In the same way we define $\underline{w} = B_M^{-1}M_M\underline{v} \leftrightarrow w_M \in X'_M$ as the unique solution of the Galerkin variational problem

$$\langle Bw_M, z_M\rangle \;=\; \langle v_M, z_M\rangle \quad \text{for } z_M \in X'_M,$$

Again,

$$(\widetilde{C}_A\underline{v}, \underline{v}) = (B_M^{-1}M_M\underline{v}, M_M\underline{v}) = (\underline{w}, M_M\underline{v}) = \langle w_M, v_M\rangle = \langle Bw_M, w_M\rangle. \tag{13.9}$$

Moreover we have the Galerkin orthogonality

$$\langle B(w - w_M), z_M\rangle \;=\; 0 \quad \text{for } z_M \in X'_M.$$

By using the $X'$–ellipticity of $B$ we now have

$$\begin{aligned} 0 \le c_1^B \,\|w - w_M\|^2_{X'} &\le \langle B(w - w_M), w - w_M\rangle \\ &= \langle B(w - w_M), w\rangle \;=\; \langle Bw, w\rangle - \langle Bw_M, w_M\rangle \end{aligned}$$

and therefore

$$\langle Bw_M, w_M\rangle \;\le\; \langle Bw, w\rangle.$$

By using (13.8) and (13.9) this finally gives the assertion. □

Note that Lemma 13.5 holds for any arbitrary conforming trial spaces $X_M \subset X$ and $X'_M \subset X'$. However, to prove the reverse estimate we need to assume a certain stability condition of the trial space $X'_M \subset X'$.

**Lemma 13.6.** *In addition to the assumptions of Lemma* 13.5 *we assume the stability condition*

$$c_S\,\|v_M\|_X \;\le\; \sup_{0 \ne z_M \in X'_M} \frac{\langle v_M, z_M\rangle}{\|z_M\|_{X'}} \quad \textit{for all } v_M \in X_M. \tag{13.10}$$

*Then,*

$$\left(c_S\,\frac{c_1^B}{c_2^B}\right)^2 \;(C_A\underline{v}, \underline{v}) \;\le\; (\widetilde{C}_A\underline{v}, \underline{v}) \quad \textit{for all } \underline{v} \in \mathbb{R}^M.$$

*Proof.* Let $\underline{v} \in \mathbb{R}^M \leftrightarrow v_M \in X_M$ be arbitrary but fixed. From the properties (13.4) we then obtain

$$(C_A\underline{v}, \underline{v}) \;=\; \langle B^{-1}v_M, v_M\rangle \;\le\; \|B^{-1}v_M\|_{X'}\|v_M\|_X \;\le\; \frac{1}{c_1^B}\,\|v_M\|^2_X.$$

As in the proof of Lemma 13.5 let $\underline{w} = B_M^{-1} M_M \underline{v} \leftrightarrow w_M \in X'_M$. Then, by using the stability assumption (13.10),

$$c_S \, \|v_M\|_X \leq \sup_{0 \neq z_M \in X'_M} \frac{\langle v_M, z_M \rangle}{\|z_M\|_{X'}} = \sup_{0 \neq z_M \in X'_M} \frac{\langle B w_M, z_M \rangle}{\|z_M\|_{X'}} \leq c_2^B \, \|w_M\|_{X'},$$

and hence

$$(C_A \underline{v}, \underline{v}) \leq \frac{1}{c_1^B} \left( \frac{c_2^B}{c_S} \right)^2 \|w_M\|^2_{X'} \leq \left( \frac{1}{c_S} \frac{c_2^B}{c_1^B} \right)^2 \langle B w_M, w_M \rangle.$$

By using (13.9) this gives the assertion. □

Together with (13.6) we now conclude the spectral equivalence inequalities of $\widetilde{C}_A$ and $A_M$.

**Corollary 13.7.** *Let all assumptions of Lemma* 13.6 *be satisfied, in particular we assume the stability condition* (13.10). *Then there hold the spectral equivalence inequalities*

$$c_1^A c_1^B \, (\widetilde{C}_A \underline{v}, \underline{v}) \leq (A_M \underline{v}, \underline{v}) \leq c_2^A c_2^B \left( \frac{1}{c_S} \frac{c_2^B}{c_1^B} \right)^2 (\widetilde{C}_A \underline{v}, \underline{v}) \quad \textit{for all } \underline{v} \in \mathbb{R}^M .$$

Due to $\dim X_M = \dim X'_M$ the discrete stability condition (13.10) also ensures the invertibility of the matrix $M_M$. Hence for the inverse of the approximated preconditioning matrix $\widetilde{C}_A$ we obtain

$$\widetilde{C}_A^{-1} = M_M^{-1} B_M M_M^{-\top},$$

in particular we need to invert sparse matrices $M_M$ and $M_M^\top$, and in addition we have to perform one matrix by vector multiplication with $B_M$.

### 13.2.1 An Application in Boundary Element Methods

The general approach of preconditioning as described in Section 13.2 is now applied to construct some preconditioners to be used in boundary element methods. By considering the single layer potential $V : H^{-1/2}(\Gamma) \to H^{1/2}(\Gamma)$ and the hypersingular boundary integral operator $D : H^{1/2}(\Gamma) \to H^{-1/2}(\Gamma)$ there is given a suitable pair of boundary integral operators of opposite order [104, 105, 144]. However, the hypersingular boundary integral operator $D$ is only semi–elliptic, hence we have to use appropriate factor spaces $H_*^{\pm 1/2}(\Gamma)$ as already considered in Section 6.6.1.

**Lemma 13.8.** *For the single layer potential $V$ and for the hypersingular boundary integral operator $D$ there hold the spectral equivalence inequalities*

$$c_1^V c_1^D \, \langle V^{-1} \widetilde{v}, \widetilde{v} \rangle_\Gamma \leq \langle D \widetilde{v}, \widetilde{v} \rangle_\Gamma \leq \frac{1}{4} \, \langle V^{-1} \widetilde{v}, \widetilde{v} \rangle_\Gamma$$

*for all* $\widetilde{v} \in H_*^{1/2}(\Gamma) = \{ v \in H^{1/2}(\Gamma) : \langle v, w_{eq} \rangle_\Gamma = 0 \}$.

*Proof.* The single layer potential $V : H_*^{-1/2}(\Gamma) \to H_*^{1/2}(\Gamma)$ defines an isomorphism. Hence, for an arbitrary given $\widetilde{v} \in H_*^{1/2}(\Gamma)$ there exists a unique $\widetilde{w} \in H_*^{-1/2}(\Gamma)$ such that $\widetilde{v} = V\widetilde{w}$. By using the symmetry relations (6.25) and (6.26) of all boundary integral operators we obtain the upper estimate,

$$\begin{aligned}
\langle D\widetilde{v}, \widetilde{v}\rangle_\Gamma &= \langle DV\widetilde{w}, V\widetilde{w}\rangle_\Gamma \\
&= \langle (\tfrac{1}{2}I - K')(\tfrac{1}{2}I + K')\widetilde{w}, V\widetilde{w}\rangle_\Gamma \\
&= \langle (\tfrac{1}{2}I + K')\widetilde{w}, (\tfrac{1}{2}I - K)V\widetilde{w}\rangle_\Gamma \\
&= \langle (\tfrac{1}{2}I + K')\widetilde{w}, V(\tfrac{1}{2}I - K')\widetilde{w}\rangle_\Gamma \\
&= \frac{1}{4}\langle V\widetilde{w}, \widetilde{w}\rangle_\Gamma - \langle VK'\widetilde{w}, K'\widetilde{w}\rangle_\Gamma \\
&\le \frac{1}{4}\langle V\widetilde{w}, \widetilde{w}\rangle_\Gamma = \frac{1}{4}\langle V^{-1}\widetilde{v}, \widetilde{v}\rangle_\Gamma .
\end{aligned}$$

From the $H^{-1/2}(\Gamma)$–ellipticity of the single layer potential $V$ (see Theorem 6.22 in the three–dimensional case $d = 3$, and Theorem 6.23 in the two–dimensional case $d = 2$) we conclude, by using the estimate (3.13), the boundedness of the inverse single layer potential,

$$\langle V^{-1}v, v\rangle_\Gamma \le \frac{1}{c_1^V} \|v\|^2_{H^{1/2}(\Gamma)} \quad \text{for all } v \in H^{1/2}(\Gamma).$$

By using the $H_*^{1/2}(\Gamma)$–ellipticity of the hypersingular boundary integral operator $D$ (see Theorem 6.24) we then obtain the lower estimate

$$\langle D\widetilde{v}, \widetilde{v}\rangle_\Gamma \ge \widetilde{c}_1^D \|\widetilde{v}\|^2_{H^{1/2}(\Gamma)} \ge c_1^D c_1^V \langle V^{-1}\widetilde{v}, \widetilde{v}\rangle_\Gamma$$

for all $\widetilde{v} \in H_*^{1/2}(\Gamma)$. □

By the bilinear form

$$\langle \widetilde{D}u, v\rangle_\Gamma := \langle Du, v\rangle_\Gamma + \alpha\,\langle u, w_{\text{eq}}\rangle_\Gamma \langle v, w_{\text{eq}}\rangle_\Gamma$$

for $u, v \in H^{1/2}(\Gamma)$ we may define the modified hypersingular boundary integral operator $\widetilde{D} : H^{1/2}(\Gamma) \to H^{-1/2}(\Gamma)$ where $\alpha \in \mathbb{R}_+$ is some parameter to be chosen appropriately, and $w_{\text{eq}} = V^{-1}1 \in H^{-1/2}(\Gamma)$ is the natural density.

**Theorem 13.9.** *For the single layer potential $V$ and for the modified hypersingular boundary integral operator $\widetilde{D}$ there hold the spectral equivalence inequalities*

$$\gamma_1 \langle V^{-1}v, v\rangle_\Gamma \le \langle \widetilde{D}v, v\rangle_\Gamma \le \gamma_2 \langle V^{-1}v, v\rangle_\Gamma \tag{13.11}$$

*for all $v \in H^{1/2}(\Gamma)$ where*

$$\gamma_1 := \min\left\{c_1^V c_1^D, \alpha\langle 1, w_{eq}\rangle_\Gamma\right\}, \quad \gamma_2 := \max\left\{\frac{1}{4}, \alpha\langle 1, w_{eq}\rangle_\Gamma\right\}.$$

*Proof.* For any $v \in H^{1/2}(\Gamma)$ we consider the orthogonal decomposition

$$v = \widetilde{v} + \gamma, \quad \gamma := \frac{\langle v, w_{\rm eq}\rangle_\Gamma}{\langle 1, w_{\rm eq}\rangle_\Gamma}, \quad \widetilde{v} \in H^{1/2}_*(\Gamma).$$

The bilinear form of the inverse single layer potential can then be written as

$$\langle V^{-1}v, v\rangle_\Gamma = \langle V^{-1}\widetilde{v}, \widetilde{v}\rangle_\Gamma + \frac{[\langle v, w_{\rm eq}\rangle_\Gamma]^2}{\langle 1, w_{\rm eq}\rangle_\Gamma}.$$

By using Lemma 13.8 we now obtain

$$\begin{aligned}\langle \widetilde{D}v, v\rangle_\Gamma &= \langle D\widetilde{v}, \widetilde{v}\rangle_\Gamma + \alpha\,[\langle v, w_{\rm eq}\rangle_\Gamma]^2 \\ &\le \frac{1}{4}\,\langle V^{-1}\widetilde{v}, \widetilde{v}\rangle_\Gamma + \alpha\,\langle 1, w_{\rm eq}\rangle_\Gamma\,\frac{[\langle v, w_{\rm eq}\rangle_\Gamma]^2}{\langle 1, w_{\rm eq}\rangle_\Gamma} \\ &\le \max\left\{\frac{1}{4}, \alpha\langle 1, w_{\rm eq}\rangle_\Gamma\right\}\langle V^{-1}v, v\rangle_\Gamma.\end{aligned}$$

The lower estimate follows in the same way. □

From the previous theorem we can find an optimal choice of the positive parameter $\alpha \in \mathbb{R}_+$.

**Corollary 13.10.** *When choosing*

$$\alpha := \frac{1}{4\langle 1, w_{eq}\rangle_\Gamma}$$

*we obtain the spectral equivalence inequalities*

$$c_1^V c_1^D\,\langle V^{-1}v, v\rangle_\Gamma \le \langle \widetilde{D}v, v\rangle_\Gamma \le \frac{1}{4}\,\langle V^{-1}v, v\rangle_\Gamma$$

*for all $v \in H^{1/2}(\Gamma)$.*

By using Corollary 13.10 we now can define a preconditioner for the linear system (12.15) of the Dirichlet boundary value problem, and for the system (12.27) of the Neumann boundary value problem. The system matrix in (12.27) is $\widetilde{D}_h := D_h + \alpha\,\underline{a}\,\underline{a}^\top$ where

$$D_h[j,i] = \langle D\varphi_i^1, \varphi_j^1\rangle_\Gamma, \quad a_j = \langle \varphi_j^1, w_{\rm eq}\rangle_\Gamma$$

for $i,j = 1,\ldots,M$ and $\varphi_i^1 \in S_h^1(\Gamma)$ are piecewise linear continuous basis functions. In addition we define

$$\bar{V}_h[j,i] = \langle V\varphi_i^1, \varphi_j^1\rangle_\Gamma, \quad \bar{M}_h[j,i] = \langle \varphi_i^1, \varphi_j^1\rangle_\Gamma$$

for $i,j = 1,\ldots,M$.

**Lemma 13.11.** *Let the $L_2$ projection $Q_h : H^{1/2}(\Gamma) \to S_h^1(\Gamma) \subset H^{1/2}(\Gamma)$ be bounded. Then there holds the stability condition*

$$\frac{1}{c_Q} \|v_h\|_{H^{1/2}(\Gamma)} \leq \sup_{0 \neq w_h \in S_h^1(\Gamma)} \frac{\langle v_h, w_h \rangle_\Gamma}{\|w_h\|_{H^{-1/2}(\Gamma)}} \quad \textit{for all } v_h \in S_h^1(\Gamma).$$

*Proof.* The $L_2$ projection $Q_h : H^{1/2}(\Gamma) \to S_h^1(\Gamma) \subset H^{1/2}(\Gamma)$ is bounded, i.e.

$$\|Q_h v\|_{H^{1/2}(\Gamma)} \leq c_Q \|v\|_{H^{1/2}(\Gamma)} \quad \text{for all } v \in H^{1/2}(\Gamma).$$

For any $w \in H^{-1/2}(\Gamma)$ the $L_2$ projection $Q_h w \in S_h^1(\Gamma)$ is defined as the unique solution of the variational problem

$$\langle Q_h w, v_h \rangle_{L_2(\Gamma)} = \langle w, v_h \rangle_\Gamma \quad \text{for all } v_h \in S_h^1(\Gamma).$$

Then,

$$\begin{aligned}\|Q_h w\|_{H^{-1/2}(\Gamma)} &= \sup_{0 \neq v \in H^{1/2}(\Gamma)} \frac{\langle Q_h w, v \rangle_\Gamma}{\|v\|_{H^{1/2}(\Gamma)}} = \sup_{0 \neq v \in H^{1/2}(\Gamma)} \frac{\langle Q_h w, Q_h v \rangle_{L_2(\Gamma)}}{\|v\|_{H^{1/2}(\Gamma)}} \\ &= \sup_{0 \neq v \in H^{1/2}(\Gamma)} \frac{\langle w, Q_h v \rangle_\Gamma}{\|v\|_{H^{1/2}(\Gamma)}} \leq \|w\|_{H^{-1/2}(\Gamma)} \sup_{0 \neq v \in H^{1/2}(\Gamma)} \frac{\|Q_h v\|_{H^{1/2}(\Gamma)}}{\|v\|_{H^{1/2}(\Gamma)}} \\ &\leq c_Q \|w\|_{H^{-1/2}(\Gamma)},\end{aligned}$$

which implies the boundedness of $Q_h : H^{-1/2}(\Gamma) \to S_h^1(\Gamma) \subset H^{-1/2}(\Gamma)$. Now the stability estimate follows by applying Lemma 8.5. □

By using Lemma 13.11 all assumptions of Lemma 13.6 are satisfied, i.e.

$$C_{\widetilde{D}} := \bar{M}_h \bar{V}_h^{-1} \bar{M}_h$$

defines a preconditioning matrix which is spectrally equivalent to $\widetilde{D}_h$. In particular there hold the spectral equivalence inequalities

$$c_1^V c_1^D (C_{\widetilde{D}} \underline{v}, \underline{v}) \leq (\widetilde{D}_h \underline{v}, \underline{v}) \leq \frac{1}{4} \left( c_Q \frac{c_2^V}{c_1^V} \right)^2 (C_{\widetilde{D}} \underline{v}, \underline{v}) \quad \text{for all } \underline{v} \in \mathbb{R}^M.$$

In Table 13.2 the extremal eigenvalues and the resulting spectral condition numbers of the preconditioned system matrix $C_{\widetilde{D}}^{-1} \widetilde{D}_h$ are listed for the L–shaped domain as given in Fig. 10.1. For comparison we also give the corresponding values in the case of a simple diagonal preconditioning which show a linear dependency on the inverse mesh parameter $h^{-1}$.

By applying Corollary 13.10 we can use the Galerkin discretization of the modified hypersingular boundary integral operator $\widetilde{D}$ as a preconditioner for the discrete single layer potential $V_h$ in (12.15). However, when using piecewise constant basis functions to discretize the single layer potential, for the Galerkin discretization of the hypersingular boundary integral operator

| | | $C_{\widetilde{D}} = \operatorname{diag} \widetilde{D}_h$ | | | $C_{\widetilde{D}} = \bar{M}_h \bar{V}_h^{-1} \bar{M}_h$ | | |
|---|---|---|---|---|---|---|---|
| $L$ | $N$ | $\lambda_{\min}$ | $\lambda_{\max}$ | $\kappa(C_{\widetilde{D}}^{-1}\widetilde{D}_h)$ | $\lambda_{\min}$ | $\lambda_{\max}$ | $\kappa(C_{\widetilde{D}}^{-1}\widetilde{D}_h)$ |
| 0 | 28 | 9.05 –3 | 2.88 –2 | 3.18 | 1.02 –1 | 2.56 –1 | 2.50 |
| 1 | 112 | 4.07 –3 | 2.82 –2 | 6.94 | 9.24 –2 | 2.66 –1 | 2.88 |
| 2 | 448 | 1.98 –3 | 2.87 –2 | 14.47 | 8.96 –2 | 2.82 –1 | 3.14 |
| 3 | 1792 | 9.84 –3 | 2.90 –2 | 29.52 | 8.86 –2 | 2.89 –1 | 3.26 |
| 4 | 7168 | 4.91 –3 | 2.91 –2 | 59.35 | 8.80 –2 | 2.92 –1 | 3.31 |
| 5 | 28672 | 2.46 –4 | 2.92 –2 | 118.72 | 8.79 –2 | 2.92 –1 | 3.32 |
| 6 | 114688 | 1.23 –4 | 2.92 –2 | 237.66 | 8.78 –2 | 2.92 –1 | 3.33 |
| Theory: | | | | $\mathcal{O}(h^{-1})$ | | | $\mathcal{O}(1)$ |

**Table 13.2.** Extremal eigenvalues and spectral condition number (BEM).

$D$ requires the use of globally continuous basis functions. Moreover, as an assumption of Lemma 13.6 we need to guarantee a related stability condition, too. One possibility is to use locally quadratic basis functions [144]. For the analysis of boundary integral preconditioners in the case of open curves, see [104].

### 13.2.2 A Multilevel Preconditioner in Finite Element Methods

For $u, v \in H^1(\Omega)$ we consider the bilinear form

$$a(u,v) = \int_\Gamma \gamma_0^{\text{int}} u(x) ds_x \int_\Gamma \gamma_0^{\text{int}} v(x) ds_x + \int_\Omega \nabla u(x) \nabla v(x) dx$$

which induces a bounded and $H^1(\Omega)$–elliptic operator $A : H^1(\Omega) \to \widetilde{H}^{-1}(\Omega)$. This bilinear form is either related to the stabilized variational formulation (4.31) of the Neumann boundary value problem, or to the variational formulation of the Robin boundary value problem, or to the modified saddle point formulation (4.22) when using Lagrange multipliers.

Let us assume that there is given a sequence $\{\mathcal{T}_{N_j}\}_{j\in\mathbb{N}_0}$ of globally quasi–uniform decompositions of a bounded domain $\Omega \subset \mathbb{R}^d$ where the global mesh size $h_j$ of a decomposition $\mathcal{T}_{N_j}$ satisfies

$$c_1 \, 2^{-j} \le h_j \le c_2 \, 2^{-j} \tag{13.12}$$

for all $j = 0, 1, 2, \ldots$ with some global constants $c_1$ and $c_2$. In particular, this condition is satisfied when applying a globally uniform refinement strategy to a given uniform coarse decomposition $\mathcal{T}_{N_0}$.

For each decomposition $\mathcal{T}_{N_j}$ the associated trial space of piecewise linear continuous basis functions is given by

$$V_j := S^1_{h_j}(\Omega) = \operatorname{span}\{\varphi^j_k\}_{k=1}^{M_j} \subset H^1(\Omega), \quad j \in \mathbb{N}_0.$$

By construction we have

$$V_0 \subset V_1 \subset \cdots \subset V_L = X_h = S^1_{h_L}(\Omega) \subset V_{L+1} \subset \cdots \subset H^1(\Omega)$$

where $X_h = S^1_h(\Omega) \subset H^1(\Omega)$ is the trial space to be used for the Galerkin discretization of the operator $A : H^1(\Omega) \to \widetilde{H}^{-1}(\Omega)$ which is induced by the bilinear form $a(\cdot,\cdot)$, i.e.

$$A_{h_L}[\ell, k] = a(\varphi^L_k, \varphi^L_\ell) \quad \text{for } k, \ell = 1, \ldots, M_L.$$

It remains to construct a preconditioning matrix $\widetilde{C}_A$ which is spectrally equivalent to $A_h$. For this we need to have a preconditioning operator $B : \widetilde{H}^{-1}(\Omega) \to H^1(\Omega)$ which satisfies the spectral equivalence inequalities

$$c^B_1 \, \|f\|^2_{\widetilde{H}^{-1}(\Omega)} \leq \langle Bf, f\rangle_\Omega \leq c^B_2 \, \|f\|^2_{\widetilde{H}^{-1}(\Omega)} \tag{13.13}$$

for all $f \in \widetilde{H}^{-1}(\Omega)$ with some positive constants $c^B_1$ and $c^B_2$. Such an operator can be constructed when using an appropriately weighted multilevel representation of $L_2$ projection operators, see [28, 162].

For any trial space $V_j \subset H^1(\Omega)$ let $Q_j : L_2(\Omega) \to V_j$ be the $L_2$ projection operator as defined in (9.23), i.e. $Q_j u \in V_j$ is the unique solution of the variational problem

$$\langle Q_j u, v_j\rangle_{L_2(\Omega)} = \langle u, v_j\rangle_{L_2(\Omega)} \quad \text{for all } v_j \in V_j.$$

Note that there holds the error estimate (9.28),

$$\|(I - Q_j)u\|_{L_2(\Omega)} \leq c\, h_j \, |u|_{H^1(\Omega)} \quad \text{for all } u \in H^1(\Omega). \tag{13.14}$$

In addition we assume an inverse inequality (9.19) to hold uniformly for all trial spaces $V_j$, i.e.,

$$\|v_j\|_{H^1(\Omega)} \leq c_I \, h_j^{-1} \, \|v_j\|_{L_2(\Omega)} \quad \text{for all } v_j \in V_j. \tag{13.15}$$

Finally, for $j = -1$ we define $Q_{-1} := 0$.

**Lemma 13.12.** *For the sequence $\{Q_j\}_{j\in\mathbb{N}_0}$ of $L_2$ projection operators $Q_j$ we have the following properties:*

1. $Q_k Q_j = Q_{\min\{k,j\}}$,
2. $(Q_k - Q_{k-1})(Q_j - Q_{j-1}) = 0$ *for* $k \neq j$,
3. $(Q_j - Q_{j-1})^2 = Q_j - Q_{j-1}$.

*Proof.* For $u_j \in V_j$ we have $Q_j v_j = v_j \in V_j$ and therefore $Q_j Q_j v = Q_j v$ for all $v \in L_2(\Omega)$. In the case $j < k$ we find $V_j \subset V_k$. Then, $Q_j v \in V_j \subset V_k$ and thus $Q_k Q_j v = Q_j v$. Finally, for $j > k$ we obtain

$$\langle Q_j v, v_j \rangle_{L_2(\Omega)} = \langle v, v_j \rangle_{L_2(\Omega)} \quad \text{for all } v_j \in V_j$$

and therefore

$$\langle Q_k Q_j v, v_k \rangle_{L_2(\Omega)} = \langle Q_j v, v_k \rangle_{L_2(\Omega)} = \langle v, v_k \rangle_{L_2(\Omega)} = \langle Q_k v, v_k \rangle_{L_2(\Omega)}$$

is satisfied for all $v_k \in V_k \subset V_j$. This concludes the proof of 1. To show 2. we assume $j < k$ and therefore $j \leq k-1$. Then, by using 1. we obtain

$$\begin{aligned}(Q_k - Q_{k-1})(Q_j - Q_{j-1}) &= Q_k Q_j - Q_{k-1} Q_j - Q_k Q_{j-1} + Q_{k-1} Q_{j-1} \\ &= Q_j - Q_j - Q_{j-1} + Q_{j-1} = 0.\end{aligned}$$

By using 1. we finally get

$$\begin{aligned}(Q_j - Q_{j-1})^2 &= Q_j Q_j - Q_j Q_{j-1} - Q_{j-1} Q_j + Q_{j-1} Q_{j-1} \\ &= Q_j - Q_{j-1} - Q_{j-1} + Q_{j-1} = Q_j - Q_{j-1}. \quad \square\end{aligned}$$

By considering a weighted linear combination of $L_2$ projection operators $Q_k$ we define the multilevel operator

$$B^1 := \sum_{k=0}^{\infty} h_k^{-2} (Q_k - Q_{k-1}) \tag{13.16}$$

which induces an equivalent norm in the Sobolev space $H^1(\Omega)$.

**Theorem 13.13.** *For the multilevel operator $B^1$ as defined in* (13.16) *there hold the spectral equivalence inequalities*

$$c_1^B \, \|v\|^2_{H^1(\Omega)} \leq \langle B^1 v, v \rangle_{L_2(\Omega)} \leq c_2^B \, \|v\|^2_{H^1(\Omega)}$$

*for all $v \in H^1(\Omega)$.*

The proof of Theorem 13.13 is based on several results. First we consider a consequence of Lemma 13.12:

**Corollary 13.14.** *For $v \in H^1(\Omega)$ we have the representation*

$$\langle B^1 v, v \rangle_{L_2(\Omega)} = \sum_{k=0}^{\infty} h_k^{-2} \, \|(Q_k - Q_{k-1}) v\|^2_{L_2(\Omega)}.$$

*Proof.* By the definition of $B^1$ and by using Lemma 13.12, 3., we have

$$\begin{aligned}
\langle B^1 v, v\rangle_{L_2(\Omega)} &= \sum_{k=0}^{\infty} h_k^{-2} \langle (Q_k - Q_{k-1})v, v\rangle_{L_2(\Omega)} \\
&= \sum_{k=0}^{\infty} h_k^{-2} \langle (Q_k - Q_{k-1})^2 v, v\rangle_{L_2(\Omega)} \\
&= \sum_{k=0}^{\infty} h_k^{-2} \langle (Q_k - Q_{k-1})v, (Q_k - Q_{k-1})v\rangle_{L_2(\Omega)} \\
&= \sum_{k=0}^{\infty} h_k^{-2} \|(Q_k - Q_{k-1})v\|^2_{L_2(\Omega)}. \qquad \square
\end{aligned}$$

From the inverse inequalities of the trial spaces $V_k$ and by using the error estimates of the $L_2$ projection operators $Q_k$ we further obtain from Corollary 13.14:

**Lemma 13.15.** *For all $v \in H^1(\Omega)$ there hold the spectral equivalence inequalities*

$$c_1 \sum_{k=0}^{\infty} \|(Q_k - Q_{k-1})v\|^2_{H^1(\Omega)} \leq \langle B^1 v, v\rangle_{L_2(\Omega)} \leq c_2 \sum_{k=0}^{\infty} \|(Q_k - Q_{k-1})v\|^2_{H^1(\Omega)} .$$

*Proof.* By using Lemma 13.12, 3., the triangle inequality, the error estimate (13.14), and assumption (13.12) we have

$$\begin{aligned}
\langle B^1 v, v\rangle_{L_2(\Omega)} &= \sum_{k=0}^{\infty} h_k^{-2} \|(Q_k - Q_{k-1})v\|^2_{L_2(\Omega)} \\
&= \sum_{k=0}^{\infty} h_k^{-2} \|(Q_k - Q_{k-1})(Q_k - Q_{k-1})v\|^2_{L_2(\Omega)} \\
&\leq 2\sum_{k=0}^{\infty} h_k^{-2} \Big\{ \|(Q_k - I)(Q_k - Q_{k-1})v\|^2_{L_2(\Omega)} \\
&\qquad\qquad + \|(I - Q_{k-1})(Q_k - Q_{k-1})v\|^2_{L_2(\Omega)} \Big\} \\
&\leq 2c\sum_{k=0}^{\infty} h_k^{-2} \Big\{ h_k^2 \|(Q_k - Q_{k-1})v\|^2_{H^1(\Omega)} + h_{k-1}^2 \|(Q_k - Q_{k-1})v\|^2_{H^1(\Omega)} \Big\} \\
&\leq c_2 \sum_{k=0}^{\infty} \|(Q_k - Q_{k-1})v\|^2_{H^1(\Omega)}
\end{aligned}$$

and therefore the upper estimate. To prove the lower estimate we get from the global inverse inequality (13.15) for $(Q_k - Q_{k-1})v \in V_{k-1}$, and by using assumption (13.12),

$$\begin{aligned}\sum_{k=0}^{\infty} \|(Q_k - Q_{k-1})v\|^2_{H^1(\Omega)} &\leq c_I^2 \sum_{k=0}^{\infty} h_{k-1}^{-2} \|(Q_k - Q_{k-1})v\|^2_{L_2(\Omega)} \\ &\leq c \sum_{k=0}^{\infty} h_k^{-2} \|(Q_k - Q_{k-1})v\|^2_{L_2(\Omega)} \\ &= c\,\langle B^1 v, v\rangle_{L_2(\Omega)}. \qquad \square\end{aligned}$$

The statement of Theorem 13.13 now follows from Lemma 13.15 and from the following spectral equivalence inequalities.

**Lemma 13.16.** *For all $v \in H^1(\Omega)$ there hold the spectral equivalence inequalities*

$$\bar{c}_1 \|v\|^2_{H^1(\Omega)} \leq \sum_{k=0}^{\infty} \|(Q_k - Q_{k-1})v\|^2_{H^1(\Omega)} \leq \bar{c}_2 \|v\|^2_{H^1(\Omega)}.$$

To prove Lemma 13.16 we first need a tool to estimate some matrix norms.

**Lemma 13.17 (Schur Lemma).** *For a countable index set $I$ we consider the matrix $A = (A[\ell,k])_{k,\ell \in I}$ and the vector $\underline{u} = (u_k)_{k\in I}$. For an arbitrary $\alpha \in \mathbb{R}$ we then have*

$$\|A\underline{u}\|_2^2 \leq \left[\sup_{\ell\in I} \sum_{k\in I} |A[\ell,k]|\, 2^{\alpha(k-\ell)}\right] \left[\sup_{k\in I} \sum_{\ell\in I} |A[\ell,k]|\, 2^{\alpha(\ell-k)}\right] \|\underline{u}\|_2^2.$$

*Proof.* Let $\underline{v} = A\underline{u}$. For an arbitrary $\ell \in I$ we first have

$$\begin{aligned}|v_\ell| &= \left|\sum_{k\in I} A[\ell,k] u_k\right| \leq \sum_{k\in I} |A[\ell,k]| \cdot |u_k| \\ &= \sum_{k\in I} \sqrt{|A[\ell,k]|}\, 2^{\alpha(k-\ell)/2} \sqrt{|A[\ell,k]|}\, 2^{\alpha(\ell-k)/2} |u_k|.\end{aligned}$$

By applying the Cauchy–Schwarz inequality this gives

$$|v_\ell|^2 \leq \left[\sum_{k\in I} |A[\ell,k]|\, 2^{\alpha(k-\ell)}\right] \left[\sum_{k\in I} |A[\ell,k]|\, 2^{\alpha(\ell-k)}\, u_k^2\right].$$

Hence we have

$$\begin{aligned}\sum_{\ell\in I}|v_\ell|^2 &\leq \sum_{\ell\in I}\left[\sum_{k\in I}|A[\ell,k]|\,2^{\alpha(k-\ell)}\right]\left[\sum_{k\in I}|A[\ell,k]|\,2^{\alpha(\ell-k)}\,u_k^2\right]\\ &\leq \sup_{\ell\in I}\left[\sum_{k\in I}|A[\ell,k]|\,2^{\alpha(k-\ell)}\right]\sum_{\ell\in I}\left[\sum_{k\in I}|A[\ell,k]|\,2^{\alpha(\ell-k)}\,u_k^2\right]\\ &= \sup_{\ell\in I}\left[\sum_{k\in I}|A[\ell,k]|\,2^{\alpha(k-\ell)}\right]\sum_{k\in I}\left[\sum_{\ell\in I}|A[\ell,k]|\,2^{\alpha(\ell-k)}\right]u_k^2\\ &\leq \sup_{\ell\in I}\left[\sum_{k\in I}|A[\ell,k]|\,2^{\alpha(k-\ell)}\right]\sup_{k\in I}\left[\sum_{\ell\in I}|A[\ell,k]|\,2^{\alpha(\ell-k)}\right]\sum_{k\in I}u_k^2\end{aligned}$$

which concludes the proof. □

As a consequence of Lemma 13.17 we immediately obtain the norm estimate

$$\|A\|_2 \leq \left[\sup_{\ell\in I}\sum_{k\in I}|A[\ell,k]|\,2^{\alpha(k-\ell)}\right]^{1/2}\left[\sup_{k\in I}\sum_{\ell\in I}|A[\ell,k]|\,2^{\alpha(\ell-k)}\right]^{1/2} \tag{13.17}$$

where $\alpha\in\mathbb{R}$ is arbitrary. In particular for a symmetric matrix $A$ and when considering $\alpha=0$ the estimate

$$\|A\|_2 \leq \sup_{\ell\in I}\sum_{k\in I}|A[\ell,k]| \tag{13.18}$$

follows. To prove the lower estimate in the spectral equivalence inequalities of Lemma 13.16 we need to have a strengthened Cauchy–Schwarz inequality.

**Lemma 13.18 (Strengthened Cauchy–Schwarz Inequality).** *Let assumption* (13.12) *be satisfied. Then there exists a* $q<1$ *such that*

$$\begin{aligned}&\left|\langle(Q_i-Q_{i-1})v,(Q_j-Q_{j-1})v\rangle_{H^1(\Omega)}\right|\\ &\qquad\qquad \leq c\,q^{|i-j|}\,\|(Q_i-Q_{i-1})v\|_{H^1(\Omega)}\|(Q_j-Q_{j-1})v\|_{H^1(\Omega)}\end{aligned}$$

*holds for all* $v\in H^1(\Omega)$.

*Proof.* Without loss of generality we may assume $j<i$. For $v_j\in V_j$ we have for the $H^1$ projection $Q_j^1 v_j = v_j\in V_j$ and therefore

$$\begin{aligned}\langle(Q_i-Q_{i-1})v,(Q_j-Q_{j-1})v\rangle_{H^1(\Omega)} &= \langle(Q_i-Q_{i-1})v,Q_j^1(Q_j-Q_{j-1})v\rangle_{H^1(\Omega)}\\ &= \langle Q_j^1(Q_i-Q_{i-1})v,(Q_j-Q_{j-1})v\rangle_{H^1(\Omega)}\\ &\leq \|Q_j^1(Q_i-Q_{i-1})v\|_{H^1(\Omega)}\|(Q_j-Q_{j-1})v\|_{H^1(\Omega)}.\end{aligned}$$

Due to $V_j=S^1_{h_j}(\Omega)\subset H^{1+\sigma}(\Omega)$ the $H^1$ projection as given in (9.29) is well defined for $u\in H^{1-\sigma}(\Omega)$ and for $\sigma\in(0,\frac{1}{2})$. Dependent on the regularity of the computational domain $\Omega$ there exists an index $s \in (0,\sigma]$ such that $Q_j^1 : H^{1-s}(\Omega) \to V_j \subset H^{1-s}(\Omega)$ is bounded, see Lemma 9.11. By using the inverse inequality in $V_j$ and by using the error estimate (9.37) of the $L_2$ projection $Q_j$ we then have

$$\begin{aligned}
\|Q_j^1(Q_i - Q_{i-1})v\|_{H^1(\Omega)} &\le c_I\, h_j^{-s}\, \|(Q_i - Q_{i-1})v\|_{H^{1-s}(\Omega)} \\
&= c_I\, h_j^{-s}\, \|(Q_i - Q_{i-1})(Q_i - Q_{i-1})v\|_{H^{1-s}(\Omega)} \\
&\le c_I\, h_j^{-s}\, \Big[\|(Q_i - I)(Q_i - Q_{i-1})v\|_{H^{1-s}(\Omega)} \\
&\qquad + \|(I - Q_{i-1})(Q_i - Q_{i-1})v\|_{H^{1-s}(\Omega)}\Big] \\
&\le c\, h_j^{-s}\, \left[h_i^s + h_{i-1}^s\right] \|(Q_i - Q_{i-1})v\|_{H^1(\Omega)} \\
&\le \widetilde{c}\, 2^{s(j-i)}\, \|(Q_i - Q_{i-1})v\|_{H^1(\Omega)}.
\end{aligned}$$

With $q := 2^{-s}$ we obtain the strengthened Cauchy–Schwarz inequality. □

**Proof of Lemma 13.16:** Let $Q_j^1 : H^1(\Omega) \to S_{h_j}^1(\Omega) \subset H^1(\Omega)$ be the $H^1$ projection as defined by the variational problem (9.29), in particular for a given $u \in H^1(\Omega)$ the projection $Q_j^1 u \in V_j$ is the unique solution of

$$\langle Q_j^1 u, v_j\rangle_{H^1(\Omega)} = \langle u, v_j\rangle_{H^1(\Omega)} \quad \text{for all } v_j \in V_j.$$

Then, dependent on the regularity of the computational domain $\Omega$, and by applying Lemma 9.11, there exists an index $s \in (0,1]$, such that the following error estimate holds,

$$\|(I - Q_h^1)u\|_{H^{1-s}(\Omega)} \le c\,h^s\, \|u\|_{H^1(\Omega)}.$$

As in Lemma 13.12 we also have

$$(Q_j^1 - Q_{j-1}^1)(Q_j^1 - Q_{j-1}^1) = Q_j^1 - Q_{j-1}^1.$$

Therefore, for $v \in H^1(\Omega)$ we obtain the representation

$$v = \sum_{i=0}^{\infty}(Q_i^1 - Q_{i-1}^1)v = \sum_{i=0}^{\infty} v_i \quad \text{where } v_i := (Q_i^1 - Q_{i-1}^1)v.$$

For $i < k$ we therefore have $v_i = (Q_i^1 - Q_{i-1}^1)v \in V_{i-1} \subset V_{k-1}$, and thus $(Q_k - Q_{k-1})v_i = 0$. Hence, by interchanging the order of summation,

$$\begin{aligned}
\sum_{k=0}^{\infty} \|(Q_k - Q_{k-1})v\|^2_{H^1(\Omega)} &= \sum_{k=0}^{\infty}\sum_{i,j=0}^{\infty} \langle (Q_k - Q_{k-1})v_i, (Q_k - Q_{k-1})v_j\rangle_{H^1(\Omega)} \\
&= \sum_{i,j=0}^{\infty}\sum_{k=0}^{\min\{i,j\}} \langle (Q_k - Q_{k-1})v_i, (Q_k - Q_{k-1})v_j\rangle_{H^1(\Omega)} \\
&\le \sum_{i,j=0}^{\infty}\sum_{k=0}^{\min\{i,j\}} \|(Q_k - Q_{k-1})v_i\|_{H^1(\Omega)} \|(Q_k - Q_{k-1})v_j\|_{H^1(\Omega)}.
\end{aligned}$$

By using the global inverse inequality (13.15), the stability of the $L_2$ projection (see Remark 9.14), and applying some interpolation argument, we obtain from assumption (13.12) for the already fixed parameter $s \in (0,1]$ the estimate

$$\begin{aligned}\|(Q_k - Q_{k-1})v_i\|_{H^1(\Omega)} &\le c\, h_k^{-s}\, \|(Q_k - Q_{k-1})v_i\|_{H^{1-s}(\Omega)} \\ &\le c\, h_k^{-s}\, \|v_i\|_{H^{1-s}(\Omega)}.\end{aligned}$$

Moreover,

$$\begin{aligned}\|v_i\|_{H^{1-s}(\Omega)} &= \|(Q_i^1 - Q_{i-1}^1)v\|_{H^{1-s}(\Omega)} \\ &= \|(Q_i^1 - Q_{i-1}^1)(Q_i^1 - Q_{i-1}^1)v\|_{H^{1-s}(\Omega)} \\ &\le \|(Q_i^1 - I)v_i\|_{H^{1-s}(\Omega)} + \|(I - Q_{i-1}^1)v_i\|_{H^{1-s}(\Omega)} \\ &\le c\, h_i^s\, \|v_i\|_{H^1(\Omega)}.\end{aligned}$$

Hence we obtain

$$\sum_{k=0}^{\infty} \|(Q_k - Q_{k-1})v\|^2_{H^1(\Omega)} \le c \sum_{i,j=0}^{\infty} \sum_{k=0}^{\min\{i,j\}} h_k^{-2s}\, h_i^s\, h_j^s\, \|v_i\|_{H^1(\Omega)} \|v_j\|_{H^1(\Omega)}.$$

By using assumption (13.12) we further have

$$h_k^{-2s} \le c\left(2^{-k}\right)^{-2s} = c\left(2^{\min\{i,j\}-k}\right)^{-2s} 2^{2s\min\{i,j\}}.$$

Then, for the already fixed parameter $s \in (0,1]$ it follows that

$$\sum_{k=0}^{\min\{i,j\}} h_k^{-2s} \le c\, 2^{2s\min\{i,j\}} \sum_{k=0}^{\min\{i,j\}} \left(2^{-2s}\right)^{\min\{i,j\}-k} \le \widetilde{c}\, 2^{2s\min\{i,j\}}.$$

By using assumption (13.12) this gives

$$\begin{aligned}\sum_{k=0}^{\infty} \|(Q_k - Q_{k-1})v\|^2_{H^1(\Omega)} &\le c \sum_{i,j=0}^{\infty} 2^{2s\min\{i,j\}}\, 2^{-s(i+j)}\, \|v_i\|_{H^1(\Omega)} \|v_j\|_{H^1(\Omega)} \\ &= c \sum_{i,j=0}^{\infty} 2^{-s|i-j|}\, \|v_i\|_{H^1(\Omega)} \|v_j\|_{H^1(\Omega)}.\end{aligned}$$

If we define a symmetric matrix $A$ by its entries $A[j,i] = 2^{-s|i-j|}$, we then get

$$\sum_{k=0}^{\infty} \|(Q_k - Q_{k-1})v\|^2_{H^1(\Omega)} \le c\, \|A\|_2 \sum_{i=0}^{\infty} \|v_i\|^2_{H^1(\Omega)}.$$

By using the estimate (13.18) of the Schur lemma we obtain

$$\|A\|_2 \leq \sup_{j\in\mathbb{N}_0} \sum_{i=0}^{\infty} 2^{-s|i-j|}\,.$$

For $q := 2^{-s} < 1$ and for some $j \in \mathbb{N}_0$ this norm is bounded by

$$\sum_{i=0}^{\infty} q^{|i-j|} = \sum_{i=0}^{j-1} q^{j-i} + \sum_{i=j}^{\infty} q^{i-j} = \sum_{i=1}^{j} q^i + \sum_{i=0}^{\infty} q^i \leq 2\sum_{i=0}^{\infty} q^i = \frac{2}{1-q},$$

and therefore it follows that

$$\sum_{k=0}^{\infty} \|(Q_k - Q_{k-1})v\|^2_{H^1(\Omega)} \leq \bar{c}\sum_{i=0}^{\infty} \|v_i\|^2_{H^1(\Omega)}\,.$$

Finally,

$$\begin{aligned}
\sum_{i=0}^{\infty} \|v_i\|^2_{H^1(\Omega)} &= \sum_{i=0}^{\infty} \langle (Q_i^1 - Q_{i-1}^1)v, (Q_i^1 - Q_{i-1}^1)v\rangle_{H^1(\Omega)} \\
&= \sum_{i=0}^{\infty} \langle (Q_i^1 - Q_{i-1}^1)(Q_i^1 - Q_{i-1}^1)v, v\rangle_{H^1(\Omega)} \\
&= \sum_{i=0}^{\infty} \langle (Q_i^1 - Q_{i-1}^1)v, v\rangle_{H^1(\Omega)} = \langle v, v\rangle_{H^1(\Omega)} = \|v\|^2_{H^1(\Omega)},
\end{aligned}$$

which gives the upper estimate.

To prove the lower estimate we use the strengenthed Cauchy–Schwarz inequality (Lemma 13.18) for some $q < 1$ to obtain

$$\begin{aligned}
\|v\|^2_{H^1(\Omega)} &= \sum_{i,j=0}^{\infty} \langle (Q_i - Q_{i-1})v, (Q_j - Q_{j-1})v\rangle_{H^1(\Omega)} \\
&\leq \sum_{i,j=0}^{\infty} q^{|i-j|}\, \|(Q_i - Q_{i-1})v\|_{H^1(\Omega)} \|(Q_j - Q_{j-1})v\|_{H^1(\Omega)}.
\end{aligned}$$

Now the assertion follows as above by applying the Schur lemma. □

*Remark 13.19.* For $s \in [0, \frac{3}{2})$ we may define the more general multilevel operator

$$B^s := \sum_{k=0}^{\infty} h_k^{-2s}(Q_k - Q_{k-1})$$

which satisfies, as in the special case $s = 1$, the spectral equivalence inequalities

$$c_1\,\|v\|^2_{H^s(\Omega)} \leq \langle B^s v, v\rangle_{L_2(\Omega)} \leq c_2\,\|v\|^2_{H^s(\Omega)} \quad \text{for all } v \in H^s(\Omega).$$

Although the following considerations are only done for the special case $s = 1$, these investigations can be extended to the more general case $s \in [0, \frac{3}{2})$, too.

By using Theorem 13.13 the multilevel operator $B^1 : H^1(\Omega) \to \widetilde{H}^{-1}(\Omega)$ is bounded and $H^1(\Omega)$–elliptic. The inverse operator $(B^1)^{-1} : \widetilde{H}^{-1}(\Omega) \to H^1(\Omega)$ is then bounded and $\widetilde{H}^{-1}(\Omega)$–elliptic, in particular the spectral equivalence inequalities (13.13) are valid. For the inverse operator $(B^1)^{-1}$ again a multilevel representation can be given.

**Lemma 13.20.** *The inverse operator* $(B^1)^{-1}$ *allows the representation*

$$B^{-1} := (B^1)^{-1} = \sum_{k=0}^{\infty} h_k^2 (Q_k - Q_{k-1}).$$

*Proof.* The assertion follows directly from

$$\begin{aligned} B^{-1}B^1 &= \sum_{k=0}^{\infty}\sum_{j=0}^{\infty} h_k^{-2} h_j^2 (Q_k - Q_{k-1})(Q_j - Q_{j-1}) \\ &= \sum_{k=0}^{\infty} (Q_k - Q_{k-1}) = I. \qquad \square \end{aligned}$$

*Remark 13.21.* If we define the $L_2$ projection operators $Q_j : L_2(\Omega) \to S^0_{h_j}(\Omega)$ onto the space of piecewise constant basis functions $\varphi_k^{0,j}$, then the related multilevel operator $B^s$ satisfies the spectral equivalence inequalities

$$c_1 \, \|v\|^2_{H^s(\Omega)} \leq \langle B^s v, v\rangle_{L_2(\Omega)} \leq c_2 \, \|v\|^2_{H^s(\Omega)} \quad \text{for all } v \in H^s(\Omega)$$

where $s \in (-\frac{1}{2}, \frac{1}{2})$.

By using corollary 13.7 we can now establish the spectral equivalence of the system matrix $A_{h_L}$ with the discrete preconditioning matrix

$$\widetilde{C}_A = \bar{M}_{h_L} B_{h_L}^{-1} \bar{M}_{h_L}$$

where

$$B_{h_L}[\ell, k] = \langle B^{-1}\varphi_k^L, \varphi_\ell^L\rangle_{L_2(\Omega)}, \quad \bar{M}_{h_L}[\ell, k] = \langle \varphi_k^L, \varphi_\ell^L\rangle_{L_2(\Omega)}$$

for all $\varphi_k^L, \varphi_\ell^L \in V_L = S^1_{h_L}(\Omega)$.

It remains to describe the application $\underline{v} = \widetilde{C}_A^{-1}\underline{r}$ inside the algorithm of the preconditioned method of conjugate gradients, see Algorithm 13.3. There we have to compute

$$\underline{v} := \widetilde{C}_A^{-1}\underline{r} = M_{h_L}^{-1} B_{h_L} M_{h_L}^{-1}\underline{r},$$

or,

$$\underline{u} := M_{h_L}^{-1}\underline{r}, \quad \underline{w} := B_{h_L}\underline{u}, \quad \underline{v} := M_{h_L}^{-1}\underline{w}.$$

By using the isomorphism $\underline{u} \in \mathbb{R}^{M_L} \leftrightarrow u_{h_L} \in V_L$ we obtain for the components of $\underline{w} = B_{h_L}\underline{u}$

$$w_\ell := \sum_{k=1}^{M_L} B_{h_L}[\ell,k]u_k = \sum_{k=1}^{M_L} \langle B^{-1}\varphi_k^L, \varphi_\ell^L\rangle_{L_2(\Omega)} u_k = \langle B^{-1}u_{h_L}, \varphi_\ell^L\rangle_{L_2(\Omega)}.$$

Hence, for $u_{h_L} \in V_L$ we need to evaluate

$$z_{h_L} := B^{-1}u_{h_L} = \sum_{k=0}^{\infty} h_k^2 (Q_k - Q_{k-1})u_{h_L} = \sum_{k=0}^{L} h_k^2 (Q_k - Q_{k-1})u_{h_L} \in V_L$$

which is a finite sum due to $Q_k u_{h_L} = u_{h_L}$ for $k \geq L$. For the components of $\underline{w} = B_{h_L}\underline{u}$ we then obtain

$$w_\ell = \langle B^{-1}u_{h_L}, \varphi_\ell^L\rangle_{L_2(\Omega)} = \langle z_{h_L}, \varphi_\ell^L\rangle_{L_2(\Omega)} = \sum_{k=1}^{M_L} z_k \langle \varphi_k^L, \varphi_\ell^L\rangle_{L_2(\Omega)}.$$

This is equivalent to

$$\underline{w} = M_{h_L}\underline{z},$$

and therefore there is no need to invert the inverse mass matrix $M_{h_L}$ when computing the preconditioned residual,

$$\underline{v} = M_{h_L}^{-1}\underline{w} = M_{h_L}^{-1}M_{h_L}\underline{z} = \underline{z}.$$

It remains to compute the coefficients of $\underline{z} \in \mathbb{R}^{M_L} \leftrightarrow z_{h_L} \in V_L$. For this we have the representation

$$\begin{aligned} z_{h_L} &= \sum_{k=0}^{L} h_k^2 (Q_k - Q_{k-1})u_{h_L} \\ &= h_L^2 Q_L u_{h_L} + \sum_{k=0}^{L-1} (h_k^2 - h_{k+1}^2) Q_k u_{h_L} \\ &= h_L^2 \, \bar{u}_{h_L} + \sum_{k=0}^{L-1} (h_k^2 - h_{k+1}^2) \bar{u}_{h_k} \end{aligned}$$

where

$$\bar{u}_{h_k} = Q_k u_{h_L} = \sum_{\ell=1}^{M_k} \bar{u}_\ell^k \varphi_\ell^k \in V_k$$

is the $L_2$ projection of $u_{h_L}$ into the trial space $V_k$, $k = 0, 1, \ldots, L$. Due to

$$c\,h_k^2 \le h_k^2 - h_{k+1}^2 \le h_k^2$$

we can define

$$\bar{z}_{h_L} := \sum_{k=0}^{L} h_k^2\, \bar{u}_{h_k}$$

to be spectrally equivalent to $\widetilde{C}_A^{-1}$. The evaluation of $\bar{z}_{h_L}$ can be done recursively. Starting from $\bar{z}_{h_0} = h_0^2 \bar{u}_{h_0} \in V_0$ we have

$$\bar{z}_{h_k} := \bar{z}_{h_{k-1}} + h_k^2 \bar{u}_{h_k} = \sum_{\ell=1}^{M_{k-1}} \bar{z}_\ell^{k-1} \varphi_\ell^{k-1} + \sum_{\ell=1}^{M_k} h_k^2 \bar{u}_\ell^k \varphi_\ell^k .$$

Due to the inclusion $V_{k-1} \subset V_k$ we can write each basis function $\varphi_\ell^{k-1} \in V_{k-1}$ as a linear combination of basis functions $\varphi_j^k \in V_k$,

$$\varphi_\ell^{k-1} = \sum_{j=1}^{M_k} r_{\ell,j}^k \varphi_j^k \quad \text{for all } \ell = 1, \ldots, M_{k-1}.$$

Hence we can write

$$\sum_{\ell=1}^{M_{k-1}} \bar{z}_\ell^{k-1} \varphi_\ell^{k-1} = \sum_{\ell=1}^{M_{k-1}} \bar{z}_\ell^{k-1} \sum_{j=1}^{M_k} r_{\ell,j}^k \varphi_j^k = \sum_{j=1}^{M_k} \sum_{\ell=1}^{M_{k-1}} \bar{z}_\ell^{k-1} r_{\ell,j}^k \varphi_j^k .$$

By introducing the matrices

$$R_{k-1,k}[j,\ell] = r_{\ell,j}^k \quad \text{for } j = 1, \ldots, M_k, \ell = 1, \ldots, M_{k-1}$$

we obtain for the coefficient vector

$$\underline{\bar{z}}^k := R_{k-1,k} \underline{\bar{z}}^{k-1} + h_k^2 \underline{\bar{u}}^k .$$

When considering a uniform mesh refinement strategy the coefficients $r_{\ell,j}^k$ are given by the nodal interpolation of the basis functions $\varphi_\ell^{k-1} \in V_{k-1}$ at the nodes $x_j$ of the decomposition $\mathcal{T}_{N_k}$, see Fig. 13.1.
By using the matrices

$$R_k := R_{L-1,L} \ldots R_{k,k+1} \quad \text{for } k = 0, \ldots, L-1, \quad R_L := I$$

we obtain by induction

$$\underline{\bar{z}}^L = \sum_{k=0}^{L} h_k^2 R_k \underline{\bar{u}}^k .$$

It remains to compute the $L_2$ projections $\bar{u}_{h_k} = Q_k u_{h_L}$ as the unique solutions of the variational problems

$$\langle \bar{u}_{h_k}, \varphi_\ell^k \rangle_{L_2(\Omega)} = \langle u_{h_L}, \varphi_\ell^k \rangle_{L_2(\Omega)} \quad \text{for all } \varphi_\ell^k \in V_k .$$

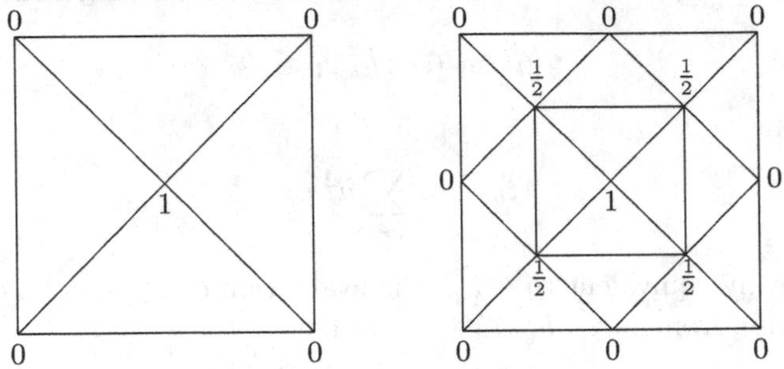


**Fig. 13.1.** Basis functions $\varphi_\ell^{k-1}$ and coefficients $r_{\ell,j}^k$ $(d = 2)$.

This is equivalent to a linear system of algebraic equations,

$$M_{h_k} \underline{\bar{u}}^k = \underline{f}^k$$

where

$$M_{h_k}[\ell, j] = \langle \varphi_j^k, \varphi_\ell^k \rangle_{L_2(\Omega)}, \quad f_\ell^k = \langle u_{h_L}, \varphi_\ell^k \rangle_{L_2(\Omega)}.$$

In particular for $k = L$ we have

$$\underline{f}^L = M_{h_L} \underline{u} = M_{h_L} M_{h_L}^{-1} \underline{r} = \underline{r}$$

and therefore

$$\underline{\bar{u}}^L = M_{h_L}^{-1} \underline{r}.$$

Due to

$$f_\ell^{k-1} = \langle u_{h_L}, \varphi_\ell^{k-1} \rangle_{L_2(\Omega)} = \sum_{j=1}^{M_k} r_{\ell,j}^k \langle u_{h_L}, \varphi_j^k \rangle_{L_2(\Omega)} = \sum_{j=1}^{M} r_{\ell,j}^k f_j^k$$

we get

$$\underline{f}^{k-1} = R_{k-1,k}^\top \underline{f}^k = R_{k-1}^\top \underline{r}.$$

By recursion we therefore have

$$\underline{\bar{u}}^k = M_{h_k}^{-1} R_k^\top \underline{r},$$

and the application of the preconditioner reads

$$\underline{v} = \sum_{k=0}^{L} h_k^2 R_k M_{h_k}^{-1} R_k^\top \underline{r}.$$

Taking into account the spectral equivalence of the mass matrices with the diagonal matrices $h_k^d I$, see Lemma 9.7, we then obtain for the application of the multilevel preconditioner

$$\underline{v} = \sum_{k=0}^{L} h_k^{2-d} R_k R_k^\top \underline{r}. \tag{13.19}$$

The realization of the multilevel preconditioner (13.19) therefore requires the restriction of a residual vector $\underline{r}$ which is given on the computational level $L$, and a weighted summation of prolongated coarse grid vectors. Thus, an application of the multilevel preconditioner requires $\mathcal{O}(M)$ operations only.

| | | $C_A = I$ | | | $C_A = \bar{M}_{h_L} B_{h_L}^{-1} \bar{M}_{h_L}$ | | |
|---|---|---|---|---|---|---|---|
| $L$ | $M$ | $\lambda_{\min}$ | $\lambda_{\max}$ | $\kappa(C_A^{-1}A_h)$ | $\lambda_{\min}$ | $\lambda_{\max}$ | $\kappa(C_A^{-1}A_h)$ |
| 1 | 13 | 2.88 –1 | 6.65 | 23.13 | 16.34 | 130.33 | 7.98 |
| 2 | 41 | 8.79 –2 | 7.54 | 85.71 | 16.69 | 160.04 | 9.59 |
| 3 | 145 | 2.42 –2 | 7.87 | 324.90 | 16.32 | 179.78 | 11.02 |
| 4 | 545 | 6.34 –3 | 7.96 | 1255.75 | 15.47 | 193.36 | 12.50 |
| 5 | 2113 | 1.62 –3 | 7.99 | 4925.47 | 15.48 | 202.94 | 13.11 |
| 6 | 8321 | 4.10 –4 | 8.00 | 19496.15 | 15.58 | 209.85 | 13.47 |
| 7 | 33025 | | | ≈80000 | 15.76 | 214.87 | 13.63 |
| 8 | 131585 | | | ≈320000 | 15.87 | 218.78 | 13.79 |
| 9 | 525313 | | | ≈1280000 | 15.96 | 221.65 | 13.89 |
| Theory: | | | | $\mathcal{O}(h^{-2})$ | | | $\mathcal{O}(1)$ |

**Table 13.3.** Extremal eigenvalues and spectral condition number (FEM).

In Table 13.3 we give the extremal eigenvalue and the resulting spectral condition numbers of the preconditioned finite element stiffness matrix $C_A^{-1}[\underline{a}\,\underline{a}^\top + A_{h_L}]$. This preconditioner is also needed for an efficient solve of the linear system (11.22), as it will be considered in the next section. The results for the non–preconditioned system ($C_A = I$) confirm the statement of Lemma 11.4, while the boundedness of the spectral condition of the preconditioned systems coincides with the results of this section.

## 13.3 Solution Methods for Saddle Point Problems

The boundary element discretization of the symmetric formulation of boundary integral equations to solve mixed boundary value problems, as well as the finite element discretization of saddle point problems, both lead to linear systems of algebraic equations of the form

$$\begin{pmatrix} A & -B \\ B^\top & D \end{pmatrix} \begin{pmatrix} \underline{u}_1 \\ \underline{u}_2 \end{pmatrix} = \begin{pmatrix} \underline{f}_1 \\ \underline{f}_2 \end{pmatrix} \tag{13.20}$$

where the block $A \in \mathbb{R}^{M_1 \times M_1}$ is symmetric and positive definite, and where $D \in \mathbb{R}^{M_2 \times M_2}$ is symmetric but positive semi–definite. Accordingly, $B \in \mathbb{R}^{M_1 \times M_2}$. Since the matrix $A$ is assumed to be positive definite, we can solve the first equation in (13.20) for $\underline{u}_1$ to obtain

$$\underline{u}_1 = A^{-1}B\underline{u}_2 + A^{-1}\underline{f}_1.$$

Inserting this into the second equation of (13.20) this results in the Schur complement system

$$\left[D + B^\top A^{-1}B\right]\underline{u}_2 = \underline{f}_2 - B^\top A^{-1}\underline{f}_1 \tag{13.21}$$

where

$$S = D + B^\top A^{-1}B \in \mathbb{R}^{M_2 \times M_2}. \tag{13.22}$$

is the Schur complement. From the symmetry properties of the block matrices $A$, $B$ and $D$ we conclude the symmetry of $S$, while, at this point, we assume the positive definiteness of $S$.

We assume that for the symmetric and positive definite matrices $A$ and $S = D + B^\top A^{-1}B$ there are given some positive definite and symmetric preconditioning matrices $C_A$ and $C_S$ satisfying the spectral equivalence inequalities

$$c_1^A\,(C_A\underline{x}_1, \underline{x}_1) \le (A\underline{x}_1, \underline{x}_1) \le c_2^A\,(C_A\underline{x}_1, \underline{x}_1) \tag{13.23}$$

for all $\underline{x}_1 \in \mathbb{R}^{M_1}$ as well as

$$c_1^S\,(C_S\underline{x}_2, \underline{x}_2) \le (S\underline{x}_2, \underline{x}_2) \le c_2^S\,(C_S\underline{x}_2, \underline{x}_2) \tag{13.24}$$

for all $\underline{x}_2 \in \mathbb{R}^{M_2}$. Hence, to solve the Schur complement system (13.21) we can apply the $C_S$ preconditioned method of conjugate gradients (Algorithm 13.3). There, the matrix by vector multiplication $\underline{s}^k = S\underline{p}^k$ for the Schur complement (13.22) reads

$$\underline{s}^k := D\underline{p}^k + B^\top A^{-1}B\underline{p}^k = D\underline{p}^k + B^\top\underline{w}^k,$$

where $\underline{w}^k$ is the unique solution of the linear system

$$A\underline{w}^k = B\underline{p}^k.$$

This system can be solved either by a direct method, for example by the Cholesky approach, or again by using a $C_A$ preconditioned method of conjugate gradients (Algorithm 13.3). Depending on the application under consideration the Schur complement approach can be disadvantageous. Then, an iterative solution strategy for the system (13.20) should be used. Possible iterative solution methods for general non–symmetric linear systems of the form (13.20) are the method of the generalized minimal residual (GMRES, [120]), or the stabilized method of biorthogonal search directions (BiCGStab, [155]).

Here, following [26], we will describe a transformation of the block–skew symmetric but positive definite system (13.20) leading to a symmetric and positive definite system for which a preconditioned conjugate gradient approach can be used.
For the preconditioning matrix $C_A$ we need to assume that the spectral equivalence inequalities (13.23) hold where

$$c_1^A > 1 \tag{13.25}$$

is satisfied. This can always be guaranteed by using an appropriate scaling, i.e. for a given preconditioning matrix $C_A$ we need to compute the minimal eigenvalue of the preconditioned system $C_A^{-1}A$. From the assumption (13.25) we find that the matrix $A - C_A$ is positive definite,

$$((A - C_A)\underline{x}_1, \underline{x}_1) \geq (c_1^A - 1)\,(C_A\underline{x}_1, \underline{x}_1) \quad \text{for all } \underline{x}_1 \in \mathbb{R}^{M_1},$$

and hence invertible. Thus, also the matrix

$$AC_A^{-1} - I = (A - C_A)C_A^{-1}$$

is invertible, and

$$T = \begin{pmatrix} AC_A^{-1} - I & 0 \\ -B^\top C_A^{-1} & I \end{pmatrix}$$

defines a invertible matrix. By multiplying the linear system (13.20) with the transformation matrix $T$ this gives

$$\begin{pmatrix} AC_A^{-1} - I & 0 \\ -B^\top C_A^{-1} & I \end{pmatrix} \begin{pmatrix} A & -B \\ B^\top & D \end{pmatrix} \begin{pmatrix} \underline{u}_1 \\ \underline{u}_2 \end{pmatrix} = \begin{pmatrix} AC_A^{-1} - I & 0 \\ -B^\top C_A^{-1} & I \end{pmatrix} \begin{pmatrix} \underline{f}_1 \\ \underline{f}_2 \end{pmatrix} \tag{13.26}$$

where the system matrix

$$\begin{aligned} M &= \begin{pmatrix} AC_A^{-1} - I & 0 \\ -B^\top C_A^{-1} & I \end{pmatrix} \begin{pmatrix} A & -B \\ B^\top & D \end{pmatrix} \\ &= \begin{pmatrix} AC_A^{-1}A - A & (I - AC_A^{-1})B \\ B^\top(I - C_A^{-1}A) & D + B^\top C_A^{-1}B \end{pmatrix} \end{aligned} \tag{13.27}$$

is symmetric. From the spectral equivalence inequalities of the transformed system matrix $M$ with the preconditioning matrix

$$C_M := \begin{pmatrix} A - C_A & 0 \\ 0 & C_S \end{pmatrix} \tag{13.28}$$

then the positive definiteness of $M$ follows. Hence we can use a preconditioned conjugate gradient scheme to solve the transformed linear system (13.26).

**Theorem 13.22.** *For the preconditioning matrix $C_M$ as defined in* (13.28) *there hold the spectral equivalence inequalities*

$$c_1^M (C_M\underline{x}, \underline{x}) \leq (M\underline{x}, \underline{x}) \leq c_2^M (C_M\underline{x}, \underline{x}) \quad \textit{for all } \underline{x} \in \mathbb{R}^{M_1+M_2}$$

*where*

$$c_1^M = \frac{1}{2}c_2^A[1+c_1^S] - \sqrt{\frac{1}{4}[c_2^A(1+c_1^S)]^2 - c_1^S c_2^A}\,,$$
$$c_2^M = \frac{1}{2}c_2^A[1+c_2^S] + \sqrt{\frac{1}{4}[c_2^A(1+c_2^S)]^2 - c_2^S c_2^A}\,.$$

*Proof.* We need to estimate the extremal eigenvalues of the preconditioned system matrix $C_M^{-1}M$, in particular we have to consider the eigenvalue problem

$$\begin{pmatrix} AC_A^{-1}A - A & (I - AC_A^{-1})B \\ B^\top(I - C_A^{-1}A) & D + B^\top C_A^{-1}B \end{pmatrix} \begin{pmatrix} \underline{x}_1 \\ \underline{x}_2 \end{pmatrix} = \lambda \begin{pmatrix} A - C_A & 0 \\ 0 & C_S \end{pmatrix} \begin{pmatrix} \underline{x}_1 \\ \underline{x}_2 \end{pmatrix}.$$

Let $\lambda_i$ be an eigenvalue with associated eigenvectors $\underline{x}_1^i$ and $\underline{x}_2^i$. From the first equation,

$$(AC_A^{-1}A - A)\underline{x}_1^i + (I - AC_A^{-1})B\underline{x}_2^i = \lambda_i(A - C_A)\underline{x}_1^i,$$

we find by some simple manipulations

$$-B\underline{x}_2^i = (\lambda_i C_A - A)\underline{x}_1^i\,.$$

For $\lambda_i \in [1, c_2^A]$ nothing is to be shown. Hence we only consider $\lambda_i \notin [1, c_2^A]$ were $\lambda_i C_A - A$ is invertible. Thus,

$$\underline{x}_1^i = -(\lambda_i C_A - A)^{-1}B\underline{x}_2^i.$$

Inserting this result into the second equation of the eigenvalue problem,

$$B^\top(I - C_A^{-1}A)\underline{x}_1^i + [D + B^\top C_A^{-1}B]\underline{x}_2^i = \lambda_i C_S \underline{x}_2^i,$$

this gives

$$-B^\top C_A^{-1}(C_A - A)(\lambda_i C_A - A)^{-1}B\underline{x}_2^i + [D + B^\top C_A^{-1}B]\underline{x}_2^i = \lambda_i C_S \underline{x}_2^i\,.$$

Due to

$$\begin{aligned} -C_A^{-1}(C_A - A)(\lambda_i C_A - A)^{-1} &= -C_A^{-1}[\lambda_i C_A - A + (1-\lambda_i)C_A](\lambda_i C_A - A)^{-1} \\ &= (\lambda_i - 1)(\lambda_i C_A - A)^{-1} - C_A^{-1} \end{aligned}$$

this is equivalent to

$$(\lambda_i - 1)B^\top(\lambda_i C_A - A)^{-1}B\underline{x}_2^i + D\underline{x}_2^i = \lambda_i C_S \underline{x}_2^i\,.$$

When $\lambda_i > c_2^A$ is satisfied we have that $\lambda_i C_A - A$ is positive definite. By using the spectral equivalence inequalities (13.23) we then obtain

$$\frac{\lambda_i - c_2^A}{c_2^A}(A\underline{x}_1, \underline{x}_1) \le ((\lambda_i C_A - A)\underline{x}_1, \underline{x}_1)$$

for all $\underline{x}_1 \in \mathbb{R}^{M_1}$, and therefore

$$((\lambda_i C_A - A)^{-1}\underline{x}_1, \underline{x}_1) \leq \frac{c_2^A}{\lambda_i - c_2^A}(A^{-1}\underline{x}_1, \underline{x}_1) \quad \text{for all } \underline{x}_1 \in \mathbb{R}^{n_1}.$$

From the spectral equivalence inequalities (13.24) we conclude

$$\begin{aligned}
\frac{\lambda_i}{c_2^S}(S\underline{x}_2^i, \underline{x}_2^i) &\leq \lambda_i (C_S \underline{x}_2^i, \underline{x}_2^i) \\
&= (D\underline{x}_2^i, \underline{x}_2^i) + (\lambda_i - 1)((\lambda_i C_A - A)^{-1} B\underline{x}_2^i, B\underline{x}_2^i) \\
&\leq (D\underline{x}_2^i, \underline{x}_2^i) + (\lambda_i - 1)\frac{c_2^A}{\lambda_i - c_2^A}(A^{-1}B\underline{x}_2^i, B\underline{x}_2^i) \\
&\leq c_2^A \frac{\lambda_i - 1}{\lambda_i - c_2^A}(S\underline{x}_2^i, \underline{x}_2^i)
\end{aligned}$$

and therefore

$$\frac{\lambda_i}{c_2^S} \leq c_2^A \frac{\lambda_i - 1}{\lambda_i - c_2^A},$$

i.e.

$$\lambda_i^2 - c_2^A[1 + c_2^S]\lambda_i + c_2^S c_2^A \leq 0.$$

From this we obtain

$$\lambda_- \leq \lambda_i \leq \lambda_+$$

where

$$\lambda_\pm = \frac{1}{2}c_2^A[1 + c_2^S] \pm \sqrt{\frac{1}{4}[c_2^A(1 + c_2^S)]^2 - c_2^A c_2^S}.$$

Altogether we therefore have

$$c_2^A < \lambda_i \leq \frac{1}{2}c_2^A[1 + c_2^S] + \sqrt{\frac{1}{4}[c_2^A(1 + c_2^S)]^2 - c_2^A c_2^S} = c_2^M.$$

It remains to consider the case $\lambda_i < 1$ where $A - \lambda_i C_A$ is positive definite. By using the spectral equivalence inequalities (13.23) we get

$$((A - \lambda_i C_A)\underline{x}_1, \underline{x}_1) \leq \frac{c_2^A - \lambda_i}{c_2^A}(A\underline{x}_1, \underline{x}_1)$$

and therefore

$$((A - \lambda_i C_A)^{-1}\underline{x}_1, \underline{x}_1) \geq \frac{c_2^A}{c_2^A - \lambda_i}(A^{-1}\underline{x}_1, \underline{x}_1)$$

for all $\underline{x}_1 \in \mathbb{R}^{M_1}$. Again, by using the spectral equivalence inequalities (13.24) we conclude

$$\begin{aligned}\frac{\lambda_i}{c_1^S}(S\underline{x}_2^i,\underline{x}_2^i) &\geq \lambda_i(C_S\underline{x}_2^i,\underline{x}_2^i)\\ &= (D\underline{x}_2^i,\underline{x}_2^i) + (1-\lambda_i)((A-\lambda_i C_A)^{-1}B\underline{x}_2^i, B\underline{x}_2^i)\\ &\geq (D\underline{x}_2^i,\underline{x}_2^i) + (1-\lambda_i)\frac{c_2^A}{c_2^A-\lambda_i}(A^{-1}B\underline{x}_2^i, B\underline{x}_2^i)\\ &\geq (1-\lambda_i)\frac{c_2^A}{c_2^A-\lambda_i}(S\underline{x}_2^i,\underline{x}_2^i)\end{aligned}$$

and therefore

$$c_2^A\frac{1-\lambda_i}{c_2^A-\lambda_i} \leq \frac{\lambda_i}{c_1^S}\,.$$

This is equivalent to

$$\lambda_i^2 - c_2^A[c_1^S+1]\lambda_i + c_1^S c_2^A \leq 0,$$

i.e.

$$\lambda_- \leq \lambda_i \leq \lambda_+$$

where

$$\lambda_\pm = \frac{1}{2}c_2^A[1+c_1^S] \pm \sqrt{\frac{1}{4}[c_2^A(1+c_1^S)]^2 - c_1^S c_2^A}.$$

Summarizing we have

$$1 > \lambda_i \geq \frac{1}{2}c_2^A[1+c_1^S] - \sqrt{\frac{1}{4}[c_2^A(1+c_1^S)]^2 - c_1^S c_2^A} = c_1^M\,.$$

This completes the proof. □

For the solution of the transformed linear system (13.26) Algorithm 13.3 of the preconditioned conjugate gradient approach can be applied. On a first glance the multiplication with the inverse preconditioning matrix $\underline{v}^{k+1} = C_M^{-1}\underline{r}^{k+1}$, in particular the evaluation of $\underline{v}_1^{k+1} = (A-C_A)^{-1}\underline{r}^{k+1}$ seems to be difficult. However, from the recursion of the residual, $\underline{r}^{k+1} = \underline{r}^k - \alpha_k M\underline{p}^k$, we find the representation

$$\underline{r}_1^{k+1} := \underline{r}_1^k - \alpha_k(AC_A^{-1} - I)(A\underline{p}_1^k - B\underline{p}_2^k).$$

Hence we can write the preconditioned residual $\underline{v}_1^k$ recursively as

$$\underline{v}_1^{k+1} := \underline{v}_1^k - \alpha_k C_A^{-1}(A\underline{p}_1^k - B\underline{p}_2^k).$$

In particular for $k=0$ we have

$$\underline{v}_1^0 := C_A^{-1}\left[A\underline{x}_1^0 - B\underline{x}_2^0 - \underline{f}_1\right].$$

The resulting preconditioned iterative scheme is summarized in Algorithm 13.4.

For an arbitrary initial guess $\underline{u}^0 \in \mathbb{R}^{M_1+M_2}$ compute the residual

$\underline{\bar{r}}_1^0 := A\underline{u}_1^0 - B\underline{u}_2^0 - \underline{f}_1$, $\underline{\bar{r}}_2^0 := B^\top \underline{u}_1^0 + D\underline{u}_2^0 - \underline{f}_2$.

Compute the transformed residual

$\underline{w}_1^0 := C_A^{-1}\underline{\bar{r}}_1^0$, $\underline{r}_1^0 := A\underline{w}_1^0 - \underline{\bar{r}}_1^0$, $\underline{r}_2^0 := \underline{\bar{r}}_2^0 - B^\top \underline{w}_1^0$.

Initialize the method of conjugate gradients:

$\underline{v}_1^0 := \underline{w}_1^0$, $\underline{v}_2^0 := C_S^{-1}\underline{r}_2^0$, $\underline{p}^0 := \underline{v}^0$, $\varrho_0 := (\underline{v}^0, \underline{r}^0)$.

For $k = 0, 1, 2, \ldots, n-1$:

Realize the matrix by vector multiplication

$\underline{\widetilde{s}}_1^k := A\underline{p}_1^k - B\underline{p}_2^k$, $\underline{\widetilde{s}}_2^k := B^\top \underline{p}_1^k + D\underline{p}_2^k$.

Compute the transformation

$\underline{w}_1^k := C_A^{-1}\underline{\widetilde{s}}_1^k$, $\underline{s}_1^k := A\underline{w}_1^k - \underline{\widetilde{s}}_1^k$, $\underline{s}_2^k := \underline{\widetilde{s}}_2^k - B^\top \underline{w}_1^k$.

Compute the new iterates

$\sigma_k := (\underline{s}^k, \underline{p}^k)$, $\alpha_k := \varrho_k/\sigma_k$;

$\underline{u}^{k+1} := \underline{u}^k - \alpha_k \underline{p}^k$, $\underline{r}^{k+1} := \underline{r}^k - \alpha_k \underline{s}^k$;

$\underline{v}_1^{k+1} := \underline{v}_1^k - \alpha_k \underline{w}_1^{k+1}$, $\underline{v}_2^{k+1} := C_S^{-1}\underline{r}_2^{k+1}$, $\varrho_{k+1} := (\underline{v}^{k+1}, \underline{r}^{k+1})$.

Stop, if $\varrho_{k+1} \le \varepsilon \varrho_0$ is satisfied for some given accuracy $\varepsilon$.

Otherwise compute the new search direction

$\beta_k := \varrho_{k+1}/\varrho_k$, $\underline{p}^{k+1} := \underline{v}^{k+1} + \beta_k \underline{p}^k$.

Algorithm 13.4: Conjugate gradient method with Bramble/Pasciak transformation.

| $L$ | $N$ | $M$ | Schur CG | BP CG |
|---|---|---|---|---|
| 2 | 16 | 11 | 11 | 16 |
| 3 | 32 | 23 | 13 | 19 |
| 4 | 64 | 47 | 14 | 21 |
| 5 | 128 | 95 | 14 | 21 |
| 6 | 256 | 191 | 15 | 23 |
| 7 | 512 | 383 | 16 | 23 |
| 8 | 1024 | 767 | 16 | 23 |
| 9 | 2048 | 1535 | 16 | 24 |

**Table 13.4.** Comparison of Schur CG and Bramble/Pasciak CG.

As an example we consider the solution of the linear system (12.43) which results from a Galerkin boundary element approximation, see Section 12.3,

$$\begin{pmatrix} V_h & -\frac{1}{2}M_h - K_h \\ \frac{1}{2}M_h^\top + K_h^\top & D_h \end{pmatrix} \begin{pmatrix} \underline{w} \\ \underline{\hat{u}} \end{pmatrix} = \begin{pmatrix} \underline{0} \\ \underline{f} \end{pmatrix}. \tag{13.29}$$

The associated Schur complement system reads

$$S_h\underline{\hat{u}} = \left[D_h + (\frac{1}{2}M_h^\top + K_h^\top)V_h^{-1}(\frac{1}{2}M_h + K_h)\right]\underline{\hat{u}} = \underline{f}. \qquad (13.30)$$

As preconditioner for the Schur complement matrix $S_h$ we can apply the preconditioning strategy as described in section 13.2.1. But in this case the spectral equivalence inequalities (13.11) of the hypersingular boundary integral operator $D : \widetilde{H}^{1/2}(\Gamma_N) \to H^{-1/2}(\Gamma)$ and of the inverse single layer potential $V : \widetilde{H}^{-1/2}(\Gamma_N) \to H^{1/2}(\Gamma_N)$ are not satisfied due to the different function spaces to be used. But for finite dimensional conformal trial spaces $S_h^1(\Gamma_N) \subset \widetilde{H}^{1/2}(\Gamma_N)$ one can prove related estimates [104], i.e. for all $v_h \in S_h^1(\Gamma_N)$ there hold the spectral equivalence inequalities

$$\gamma_1 \langle V^{-1}v_h, v_h\rangle_\Gamma \le \langle Dv_h, v_h\rangle_\Gamma \le \gamma_2 [1 + \log|h|]^2 \langle V^{-1}v_h, v_h\rangle_\Gamma.$$

When using the preconditioning matrix $C_D = \bar{M}_h \bar{V}_h^{-1} \bar{M}_h$ we then obtain the estimate for the spectral condition number,

$$\kappa_2(C_D^{-1}D_h) \le c\,[1 + \log|h|]^2.$$

As described in section 13.2.1 we can also define a preconditioning matrix $C_V$ for the discrete single layer potential $V_h$ which is based on the modified hypersingular boundary integral operator $\hat{D} : H^{1/2}(\Gamma) \to H^{-1/2}(\Gamma)$, see [144]. In Table 13.4 we give the number of iterations of the preconditioned conjugate gradient approach for the solution of the Schur complement system (13.30), and of the conjugate gradient approach with the Bramble/Pasciak transformation to solve the system (13.29). As relative accuracy we have considered $\varepsilon = 10^{-8}$, and the scaling of the preconditioning matrix $C_V$ of the discrete single layer potential was chosen such that the spectral equivalence inequalities (13.23) are satisfied with $c_1^A = 1.2$.

# 14

# Fast Boundary Element Methods

Boundary element methods as described in Chapter 12 result in dense stiffness matrices. In particular, both the storage requirements and the numerical amount of work to compute all entries of a boundary element stiffness matrix is quadratic in the number of degrees of freedom. Hence there is a serious need to derive and to describe fast boundary element methods which exhibit an almost linear, up to some polylogarithmic factors, behavior in the number of degrees of freedom. Here we constrict our considerations to the case of a two–dimensional model problem, for the three–dimensional case, see, for example, [117].

As a model problem we consider the Dirichlet boundary value problem

$$-\Delta u(x) = 0 \quad \text{for } x \in \Omega \subset \mathbb{R}^2, \quad \gamma_0^{\text{int}} u(x) = g(x) \quad \text{for } x \in \Gamma = \partial\Omega$$

where we assume $\operatorname{diam} \Omega < 1$ to ensure the invertibility of the single layer potential $V$. When using an indirect single layer potential (7.4) the solution of the above problem is given by

$$u(\widetilde{x}) = -\frac{1}{2\pi} \int_\Gamma \log |\widetilde{x} - y| w(y) ds_y \quad \text{for } \widetilde{x} \in \Omega.$$

The yet unknown density $w \in H^{-1/2}(\Gamma)$ is then given as the unique solution of the boundary integral equation (7.12),

$$(Vw)(x) = -\frac{1}{2\pi} \int_\Gamma \log |x - y| w(y) ds_y = g(x) \quad \text{for } x \in \Gamma.$$

Let $S_h^0(\Gamma) = \operatorname{span}\{\varphi_k^0\}_{k=1}^N$ be the trial space of piecewise constant basis functions $\varphi_k^0$ which are defined with respect to a globally quasi–uniform boundary mesh $\{\tau_k\}_{k=1}^N$ with a global mesh size $h$. Then we can find an approximate solution $w_h \in S_h^0(\Gamma)$ as the unique solution of the Galerkin variational problem

$$\langle V w_h, \tau_h \rangle_\Gamma = \langle g, \tau_h \rangle_\Gamma \quad \text{for all } \tau_h \in S_h^0(\Gamma). \tag{14.1}$$

This variational problem is equivalent to a linear system $V_h \underline{w} = \underline{f}$ of algebraic equations where the stiffness matrix $V_h$ is defined as

$$V_h[\ell, k] = \langle V \varphi_k^0, \varphi_\ell^0 \rangle_\Gamma = -\frac{1}{2\pi} \int_{\tau_\ell} \int_{\tau_k} \log|x - y| ds_y ds_x \tag{14.2}$$

for all $k, \ell = 1, \ldots, N$. Note that for the approximate solution $w_h \in S_h^0(\Gamma)$ there holds the error estimate ,

$$\|w - w_h\|_{H^{-1/2}(\Gamma)} \le c h^{3/2} |w|_{H^1_{pw}(\Gamma)}. \tag{14.3}$$

when assuming $w \in H^1_{\text{pw}}(\Gamma)$. Due to the nonlocal definition of the fundamental solution the stiffness matrix $V_h$ is dense, i.e. to describe the symmetric matrix $V_h$ we need to store $\frac{1}{2}N(N+1)$ matrix entries. Moreover, a realization of a matrix by vector product within the use of a preconditioned conjugate gradient scheme (Algorithm 13.3) requires $N^2$ multiplications. Hence we have a quadratic amount of work in both storage of the matrix and in a matrix by vector product with respect to the number $N$ of degrees of freedom. In contrast to standard boundary element methods we are interested in the design of fast boundary element methods where the numerical amount of work will be of the order $\mathcal{O}(N(\log_2 N)^\alpha)$ where we have to ensure an error estimate as given in (14.3) for a standard boundary element method.

In this chapter we will consider two different approaches to derive fast boundary element methods. The use of wavelets [46, 47, 94, 127] leads to dense stiffness matrices $V_h$, but since most of the matrix entries can be neglected this results in a sparse approximation $\widetilde{V}_h$ of the stiffness matrix. A second approach is based on a hierarchical clustering of boundary elements [14, 62, 65, 73, 122] which defines a block partitioning of the stiffness matrix $V_h$. If two clusters are well separated the related block can be approximated by some low rank matrices.

## 14.1 Hierarchical Cluster Methods

Since the fundamental solution $U^*(x,y) = -\frac{1}{2\pi} \log|x-y|$ is only a function of the distance $|x-y|$, all matrix entries $V_h[\ell, k]$ of the discrete single layer potential $V_h$ as defined in (14.2) only depend on the distance, on the size, and on the shape of the boundary elements $\tau_k$ and $\tau_\ell$. Hence we can cluster all boundary elements when taking into account their size and the distances to each other. The ratio of the cluster size and the distance between two clusters will then serve as an admissibility criterion to define an approximation of the fundamental solution. A larger distance between two clusters then also allows to consider larger clusters of boundary elements. For this we will consider an appropriate hierarchy of clusters. The interaction between two boundary elements $\tau_k$ and $\tau_\ell$ to compute the matrix entry $V_h[\ell, k]$ is then replaced by the interaction between the associated clusters.

For a given globally quasi–uniform boundary decomposition $\Gamma = \bigcup_{\ell=1}^{N} \tau_\ell$ we first have to construct a suitable cluster hierarchy. Due to the general assumption diam $\Omega < 1$ we may assume $\Omega \subset (0,1)^2$. Hence, all boundary elements $\tau_\ell$ are contained in the surrounding square box $\Omega_1^0 = (0,1)^2$. Applying a recursive decomposition of $\Omega_j^{\lambda-1}$ into four congruent boxes $\Omega_i^\lambda$ as depicted in Fig. 14.1 this first defines a hierarchy of boxes $\Omega_i^\lambda$, and from this we easily find a hierarchical clustering of the boundary elements $\tau_\ell$. The number $\lambda$ of recursive refinement steps is called the level of the hierarchy.

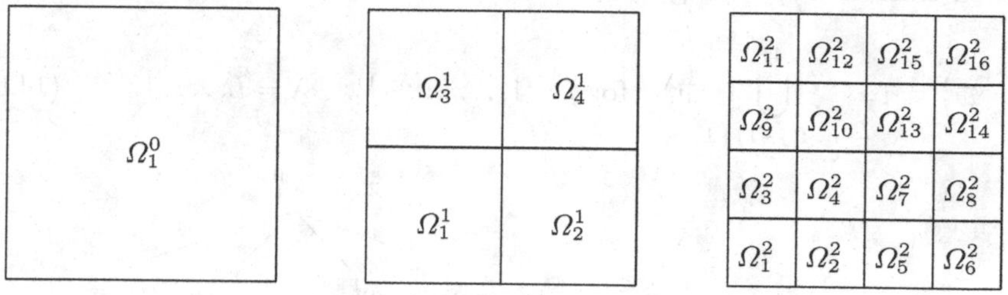


**Fig. 14.1.** Hierarchy of boxes $\Omega_j^\lambda$ for $\lambda = 0, 1, 2$.

For $\lambda = 1, \ldots, L$ we therefore have the representation

$$\overline{\Omega}_j^{\lambda-1} = \bigcup_{i=4(j-1)+1}^{4j} \overline{\Omega}_i^\lambda, \quad j = 1, \ldots, 4^{(\lambda-1)} \tag{14.4}$$

where the length $d_j^\lambda$ of an edge of the box $\Omega_j^\lambda$ is given by

$$d_j^\lambda = 2^{-\lambda}, \quad j = 1, \ldots, 4^\lambda.$$

The refinement strategy (14.4) is applied recursively until the edge length $d_j^L$ of a box $\Omega_j^L$ on the finest level $L$ is proportional to the mesh size $h$ of the globally quasi–uniform boundary mesh $\{\tau_\ell\}_{\ell=1}^N$, i.e.

$$d_j^L = 2^{-L} \leq c_L \, h$$

induces a maximal number of boundary elements $\tau_\ell$ which are contained in the box $\Omega_j^L$. Then we find for the maximal level of the cluster tree

$$L \geq \frac{c_L \ln(1/h)}{\ln 2}.$$

Since the boundary decomposition is assumed to be globally quasi–uniform, this implies that the surface measure $|\Gamma|$ is proportional to $Nh$ and therefore we obtain

$$L = \mathcal{O}(\ln N)\,. \tag{14.5}$$

To describe the clustering of the boundary elements $\{\tau_\ell\}_{\ell=1}^N$ we may consider the clustering of the associated element midpoints $\hat{x}_\ell \in \tau_\ell$ for $\ell = 1, \ldots, N$. For $j = 1, \ldots, 4^L$ we first collect all boundary elements $\tau_\ell$ where the midpoint $\hat{x}_\ell$ is in the box $\Omega_j^L$ in a cluster $\omega_j^L$,

$$\overline{\omega}_j^L \;:=\; \bigcup_{\hat{x}_\ell \in \Omega_j^L} \overline{\tau}_\ell .$$

The hierarchy (14.4) of boxes $\Omega_j^\lambda$ now transfers directly to a hierarchy of the related clusters $\omega_j^\lambda$, see Fig. 14.2,

$$\overline{\omega}_j^{\lambda-1} \;:=\; \bigcup_{i=4(j-1)+1}^{4j} \overline{\omega}_i^\lambda \quad \text{for } j = 1, \ldots, 4^{(\lambda-1)}, \quad \lambda = L, \ldots, 1. \tag{14.6}$$

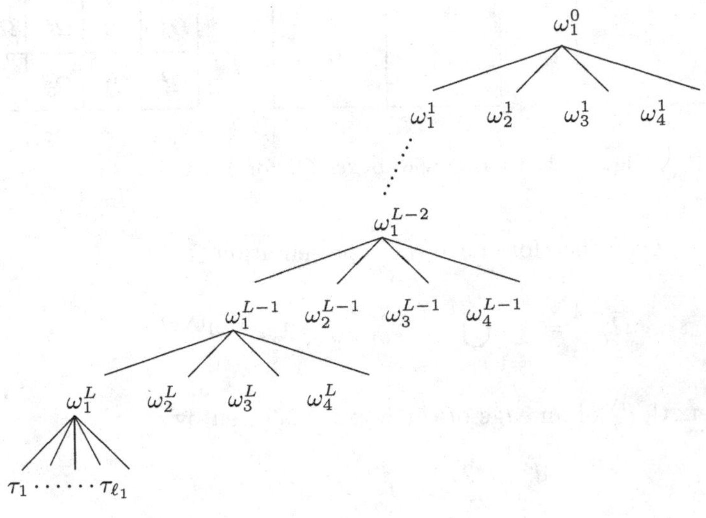


**Fig. 14.2.** Cluster tree $\omega_j^\lambda$.

For each cluster $\omega_j^\lambda$ we define

$$I_j^\lambda \;:=\; \left\{\ell \in \mathbb{N} \,:\, \tau_\ell \subset \omega_j^\lambda\right\}$$

as the index set of all associated boundary elements $\tau_\ell$ where

$$P_j^\lambda \,:\, I_1^0 = \{1, 2, \ldots, N\} \to I_j^\lambda$$

describes the assignment of the boundary elements $\{\tau_\ell\}_{\ell=1}^N$ to the associated cluster $\omega_j^\lambda$. Finally,

$$N_j^\lambda \;:=\; \dim \omega_j^\lambda$$

denotes the number of boundary elements $\tau_\ell$ inside the cluster $\omega_j^\lambda$. By construction we have for each $\lambda = 0, 1, 2, \ldots, L$

$$\sum_{j=1}^{4^\lambda} N_j^\lambda = N. \qquad (14.7)$$

By

$$\operatorname{diam} \omega_j^\lambda := \sup_{x,y \in \omega_j^\lambda} |x-y|$$

we denote the diameter of the cluster $\omega_j^\lambda$, and by

$$\operatorname{dist}(\omega_i^\kappa, \omega_j^\lambda) := \inf_{(x,y) \in \omega_i^\kappa \times \omega_j^\lambda} |x-y|$$

we define the distance between the clusters $\omega_i^\kappa$ and $\omega_j^\lambda$. A pair of clusters $\omega_i^\kappa$ and $\omega_j^\lambda$ is called admissible, if

$$\operatorname{dist}(\omega_i^\kappa, \omega_j^\lambda) \geq \eta \max\left\{\operatorname{diam} \omega_i^\kappa, \operatorname{diam} \omega_j^\lambda\right\} \qquad (14.8)$$

is satisfied where $\eta > 1$ is a prescribed parameter. For simplicity we only consider admissible clusters $\omega_i^\kappa$ and $\omega_j^\lambda$ which are defined for the same level $\kappa = \lambda$. If there are given two admissible clusters $\omega_i^\lambda$ and $\omega_j^\lambda$, then also all subsets $\omega_{i'}^{\lambda+1} \subset \omega_i^\lambda$ and $\omega_{j'}^{\lambda+1} \subset \omega_j^\lambda$ are admissible. Therefore, a pair of clusters $\omega_i^\lambda$ and $\omega_j^\lambda$ is called maximally admissible if there exist inadmissible clusters $\omega_{i'}^{\lambda-1}$ and $\omega_{j'}^{\lambda-1}$ where $\omega_i^\lambda \subset \omega_{i'}^{\lambda-1}$ and $\omega_j^\lambda \subset \omega_{j'}^{\lambda-1}$ are satisfied.

For the stiffness matrix $V_h$ as defined in (14.2) the hierarchical clustering (14.6) allows the representation

$$V_h = \sum_{\lambda=0}^{L} \underbrace{\sum_{j=1}^{4^\lambda}\sum_{i=1}^{4^\lambda}}_{\omega_i^\lambda, \omega_j^\lambda \text{ maximally admissible}} (P_j^\lambda)^\top V_h^{\lambda,ij} P_i^\lambda + \underbrace{\sum_{j=1}^{4^L}\sum_{i=1}^{4^L}}_{\omega_i^L, \omega_j^L \text{ inadmissible}} (P_j^L)^\top V_h^{L,ij} P_i^L \qquad (14.9)$$

where the block matrices $V_h^{\lambda,ij} \in \mathbb{R}^{N_j^\lambda \times N_i^\lambda}$, are defined by

$$V_h^{\lambda,ij}[\ell,k] = -\frac{1}{2\pi} \int_{\tau_\ell} \int_{\tau_k} \log|x-y| ds_y ds_x \quad \text{for } \tau_k \in \omega_i^\lambda,\ \tau_\ell \in \omega_j^\lambda, \qquad (14.10)$$

see also Fig. 14.3. The sum is to be taken over all inadmissible clusters $\omega_i^L$ and $\omega_j^L$ includes in particular the interaction of a cluster with itself and with all neighboring clusters. Hence we denote this part as the near field of the stiffness matrix $V_h$ while the remainder, i.e. the sum over all maximally admissible clusters is called the far field. A box $\Omega_i^L$ has maximal 8 direct neighbors $\Omega_j^L$, the associated cluster $\omega_i^L$ therefore has a certain number of inadmissible

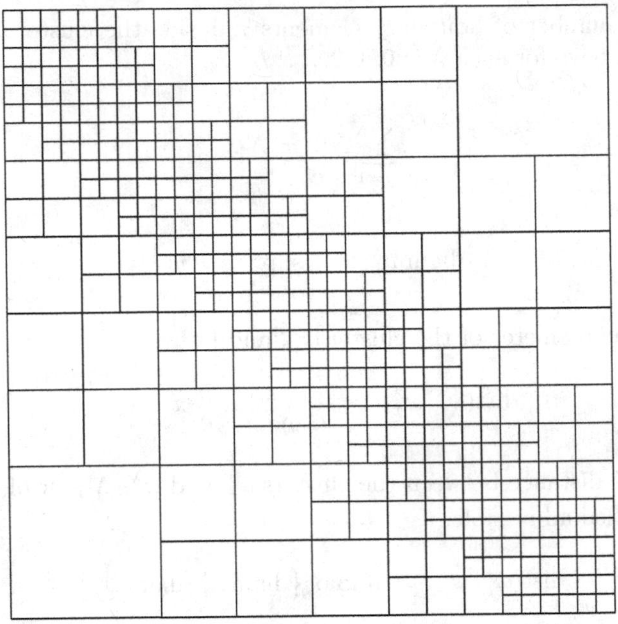

**Fig. 14.3.** Hierarchical partitioning of the stiffness matrix $V_h$.

clusters $\omega_j^L$ where the number only depends on the parameter $\eta > 1$. All other clusters are thus admissible and therefore they are included in the far field.

The near field part therefore contains only $\mathcal{O}(\eta^2 4^L) = \mathcal{O}(\eta^2 N)$ summands. While the block matrices of the near field part can be evaluated directly as in a standard boundary element method, the block matrices of the far field part can be approximated by using low rank matrices which allow for a more efficient application. The resulting matrices are called hierarchical matrices, or $\mathcal{H}$ matrices. [72].

## 14.2 Approximation of the Stiffness Matrix

For a maximally admissible pair of clusters $\omega_i^\lambda$ and $\omega_j^\lambda$ we have to compute all entries of the block matrix $V_h^{\lambda,ij}$,

$$V_h^{\lambda,ij}[\ell,k] = \int_{\tau_\ell}\int_{\tau_k} U^*(x,y)ds_y ds_x \quad \text{for } \tau_k \in \omega_i^\lambda,\ \tau_\ell \in \omega_j^\lambda .$$

The basic idea for the derivation of fast boundary element methods is an approximate splitting of the fundamental solution $U^*(x,y) = -\frac{1}{2\pi}\log|x-y|$ into functions which only depend on the integration point $y \in \omega_i^\lambda$, and on the observation point $x \in \omega_j^\lambda$,

$$U_\varrho^*(x,y) = \sum_{m=0}^{\varrho} f_m^{\lambda,j}(x) g_m^{\lambda,i}(y) \quad \text{for } (x,y) \in \omega_j^\lambda \times \omega_i^\lambda. \tag{14.11}$$

We assume that there is given an error estimate

$$|U^*(x,y) - U_\varrho^*(x,y)| \le c(\eta,\varrho) \quad \text{for } (x,y) \in \omega_j^\lambda \times \omega_i^\lambda, \tag{14.12}$$

where $c(\eta,\varrho)$ is a constant which only depends on the admissibility parameter $\eta$ and on the approximation order $\varrho$. By using the decomposition (14.11) we can approximate the entries of the block matrix $V_h^{\lambda,ij}$ by computing

$$\widetilde{V}_h^{\lambda,ij}[\ell,k] := \int_{\tau_\ell} \int_{\tau_k} U_\varrho^*(x,y) ds_y ds_x \quad \text{for } \tau_k \in \omega_i^\lambda, \tau_\ell \in \omega_j^\lambda. \tag{14.13}$$

From the error estimate (14.12) we then obtain

$$|V_h^{\lambda,ij}[\ell,k] - \widetilde{V}_h^{\lambda,ij}[\ell,k]| \le c(\eta,\varrho)\, \Delta_k \Delta_\ell \quad \text{for } \tau_k \in \omega_i^\lambda, \tau_\ell \in \omega_j^\lambda. \tag{14.14}$$

Inserting the series expansion (14.11) into (14.13) this gives

$$\widetilde{V}_h^{\lambda,ij}[\ell,k] = \sum_{m=0}^{\varrho} \int_{\tau_\ell} f_m^{\lambda,j}(x) ds_x \int_{\tau_k} g_m^{\lambda,i}(y) ds_y \quad \text{for all } \tau_k \in \omega_i^\lambda, \tau_\ell \in \omega_j^\lambda.$$

Hence, for all boundary elements $\tau_k \in \omega_i^\lambda$ and $\tau_\ell \in \omega_j^\lambda$ we need to compute vectors defined by the entries

$$a_{m,\ell}^{\lambda,j} := \int_{\tau_\ell} f_m^{\lambda,j}(x) ds_x, \quad b_{m,k}^{\lambda,i} := \int_{\tau_k} g_m^{\lambda,i}(y) ds_y,$$

where $\ell = 1, \ldots, N_j^\lambda, k = 1, \ldots, N_i^\lambda$ and $m = 0, \ldots, \varrho$. The numerical amount of work to store and to apply the approximate block matrix

$$\widetilde{V}_h^{\lambda,ij} = \sum_{m=0}^{\varrho} \underline{a}_m^{\lambda,j} (\underline{b}_m^{\lambda,i})^\top,$$

which is a matrix of rank $\varrho + 1$, is therefore

$$(\varrho + 1)(N_i^\lambda + N_j^\lambda).$$

As in (14.9) we can now define an approximation $\widetilde{V}_h$ of the global stiffness matrix $V_h$,

$$\widetilde{V}_h = \sum_{\lambda=0}^{L} \underbrace{\sum_{j=1}^{4^\lambda} \sum_{i=1}^{4^\lambda}}_{\omega_i^\lambda, \omega_j^\lambda \text{maximally admissible}} (P_j^\lambda)^\top \widetilde{V}_h^{\lambda,ij} P_i^\lambda + \underbrace{\sum_{j=1}^{4^L} \sum_{i=1}^{4^L}}_{\omega_i^L, \omega_j^L \text{inadmissible}} (P_j^L)^\top V_h^{L,ij} P_i^L. \tag{14.15}$$

Due to (14.7) the total amount of work to store and to apply the approximate stiffness matrix $\widetilde{V}_h$ is proportional to, by taking into account the near field part,

$$(\varrho+1)(\eta+1)^2(L+1)N+\eta^2 N. \tag{14.16}$$

Instead of the original linear system $V_h \underline{w} = \underline{f}$ we now have to solve the perturbed system $\widetilde{V}_h \underline{\widetilde{w}} = \underline{f}$ with an associated approximate solution $\widetilde{w}_h \in S_h^0(\Gamma)$. The stability and error analysis of the perturbed problem is based on the Strang lemma (Theorem 8.3). Hence we need to prove the positive definiteness of the approximate stiffness matrix $\widetilde{V}_h$, which will follow from an estimate of the approximation error $V_h - \widetilde{V}_h$.

**Lemma 14.1.** *For a pair of maximally admissible clusters $\omega_i^\lambda$ and $\omega_j^\lambda$ the error estimate* (14.12) *is assumed. Let $\widetilde{V}_h$ be the approximate stiffness matrix as defined in* (14.15). *Then there holds the error estimate*

$$|((V_h - \widetilde{V}_h)\underline{w}, \underline{v})| \leq c(\eta, \varrho)\, |\Gamma|\, \|w_h\|_{L_2(\Gamma)} \|v_h\|_{L_2(\Gamma)}$$

*for all $\underline{w}, \underline{v} \in \mathbb{R}^N \leftrightarrow w_h, v_h \in S_h^0(\Gamma)$.*

*Proof.* By using the error estimate (14.14) we first have

$$\begin{aligned}
|((V_h - \widetilde{V}_h)\underline{w}, \underline{v})| &\leq \sum_{\lambda=0}^{L} \underbrace{\sum_{j=1}^{4^\lambda}\sum_{i=1}^{4^\lambda}}_{\omega_i^\lambda,\omega_j^\lambda \text{maximally admissible}} \sum_{\tau_k \in \omega_i^\lambda} \sum_{\tau_\ell \in \omega_j^\lambda} |V_h^{\lambda,ij}[\ell,k] - \widetilde{V}_h^{\lambda,ij}[\ell,k]|\, |w_k|\, |v_\ell| \\
&\leq c(\eta,\varrho) \sum_{\lambda=0}^{L} \underbrace{\sum_{j=1}^{4^\lambda}\sum_{i=1}^{4^\lambda}}_{\omega_i^\lambda,\omega_j^\lambda \text{maximally admissible}} \sum_{\tau_k \in \omega_i^\lambda} \sum_{\tau_\ell \in \omega_j^\lambda} \Delta_k\, |w_k|\, \Delta_\ell\, |v_\ell| .
\end{aligned}$$

Due to the assumption of the maximal admissibility of clusters $\omega_i^\lambda$ and $\omega_j^\lambda$ each pair of boundary elements $\tau_k$ and $\tau_\ell$ appears maximal only once. Hence we have

$$|((V_h - \widetilde{V}_h)\underline{u}, \underline{v})| \leq c(\eta,\varrho) \sum_{k=1}^{N}\sum_{\ell=1}^{N} \Delta_k\, |w_k|\, \Delta_\ell\, |v_\ell| .$$

By applying the Hölder inequality we finally obtain

$$\sum_{k=1}^{N} \Delta_k\, |w_k| \leq \left(\sum_{k=1}^{N} \Delta_k\right)^{1/2} \left(\sum_{k=1}^{N} \Delta_k\, w_k^2\right)^{1/2} = |\Gamma|^{1/2}\, \|w_h\|_{L_2(\Gamma)}$$

from which the assertion follows. □

By using the error estimate of Lemma 14.1, by applying the inverse inequality in $S_h^0(\Gamma)$, and by an appropriate choice of the parameter $\eta$ for the definition of the near field and of the approximation order $\varrho$ we now can prove the positive definiteness of the approximate stiffness matrix $\widetilde{V}_h$.

**Theorem 14.2.** *Let all assumptions of Lemma 14.1 be satisfied. For an appropriate choice of the parameter $\eta$ and $\varrho$ the approximate stiffness matrix $\widetilde{V}_h$ is positive definite, i.e.*

$$(\widetilde{V}_h\underline{w},\underline{w}) \geq \frac{1}{2}c_1^V\,\|w_h\|^2_{H^{-1/2}(\Gamma)} \quad \text{for all } w_h \in S_h^0(\Gamma).$$

*Proof.* By using Lemma 14.1, the $H^{-1/2}(\Gamma)$–ellipticity of the single layer potential $V$, and the inverse inequality in $S_h^0(\Gamma)$ we obtain

$$\begin{aligned}
(\widetilde{V}_h\underline{w},\underline{w}) &= (V_h\underline{w},\underline{w}) + ((\widetilde{V}_h - V_h)\underline{w},\underline{w})\\
&\geq \langle Vw_h, w_h\rangle_\Gamma - |((V_h-\widetilde{V}_h)\underline{w},\underline{w})|\\
&\geq c_1^V\,\|w_h\|^2_{H^{-1/2}(\Gamma)} - c(\eta,\varrho)\,|\Gamma|\,\|w_h\|^2_{L_2(\Gamma)}\\
&\geq \left[c_1^V - c(\eta,\varrho)\,|\Gamma|\,c_I^2\,h^{-1}\right]\,\|w_h\|^2_{H^{-1/2}(\Gamma)}\\
&\geq \frac{1}{2}c_1^V\,\|w_h\|^2_{H^{-1/2}(\Gamma)},
\end{aligned}$$

if

$$c(\eta,\varrho) \leq \frac{c_1^V}{2|\Gamma|c_I^2}\,h \tag{14.17}$$

is satisfied. □

In the same way as in the proof of Theorem 14.2 the boundedness of the approximate single layer potential $\widetilde{V}_h$ follows. Due to the positive definiteness of the approximate stiffness matrix $\widetilde{V}_h$ we then obtain the unique solvability of the perturbed linear system $\widetilde{V}_h\underline{\widetilde{w}} = \underline{f}$. Moreover, as in Theorem 8.3 we can also estimate the error $\|w-\widetilde{w}_h\|_{H^{-1/2}(\Gamma)}$ of the computed approximate solution $\widetilde{w}_h \in S_h^0(\Gamma)$.

**Theorem 14.3.** *Let the parameter $\eta$ and $\varrho$ be chosen such that the approximate stiffness matrix $\widetilde{V}_h$ as defined in (14.15) is positive definite. The uniquely determined solution $\underline{\widetilde{w}} \in \mathbb{R}^N \leftrightarrow \widetilde{w}_h \in S_h^0(\Gamma)$ of the perturbed linear system $\widetilde{V}_h\underline{\widetilde{w}} = \underline{f}$ then satisfies the error estimate*

$$\|w-\widetilde{w}_h\|_{H^{-1/2}(\Gamma)} \leq \|w-w_h\|_{H^{-1/2}(\Gamma)} + \widetilde{c}(\eta,\varrho)\,h^{-1/2}\,\|w\|_{L_2(\Gamma)}.$$

*Proof.* Let $\underline{w},\underline{\widetilde{w}} \in \mathbb{R}^N$ be the uniquely determined solutions of the linear systems $V_h\underline{w} = \underline{f}$ and $\widetilde{V}_h\underline{\widetilde{w}} = \underline{f}$, respectively. Then there holds the orthogonality relation

$$(V_h\underline{w} - \widetilde{V}_h\underline{\widetilde{w}}, \underline{v}) = 0 \quad \text{for all } \underline{v} \in \mathbb{R}^N.$$

Since the approximated stiffness matrix $\widetilde{V}_h$ is positive definite, we obtain, by using Lemma 14.1

$$\begin{aligned}\frac{1}{2}c_1^V \, \|w_h - \widetilde{w}_h\|^2_{H^{-1/2}(\Gamma)} &\leq (\widetilde{V}_h(\underline{w} - \underline{\widetilde{w}}), \underline{w} - \underline{\widetilde{w}}) \\ &= ((\widetilde{V}_h - V_h)\underline{w}, \underline{w} - \underline{\widetilde{w}}) \\ &\leq c(\eta, \varrho)\, \|w_h\|_{L_2(\Gamma)} \|w_h - \widetilde{w}_h\|_{L_2(\Gamma)}.\end{aligned}$$

Applying Lemma 12.2 this gives the stability estimate

$$\|w_h\|_{L_2(\Gamma)} \leq \|w\|_{L_2(\Gamma)} + \|w - w_h\|_{L_2(\Gamma)} \leq c\,\|w\|_{L_2(\Gamma)}.$$

Then, using the inverse inequality in $S_h^0(\Gamma)$ we conclude

$$\|w_h - \widetilde{w}_h\|_{H^{-1/2}(\Gamma)} \leq \widetilde{c}(\eta, \varrho)\, h^{-1/2}\, \|w\|_{L_2(\Gamma)}.$$

The assigned error estimate now follows from the triangle inequality. □

From the error estimate (14.3) we conclude, when assuming $w \in H^1_{\mathrm{pw}}(\Gamma)$,

$$\|w - \widetilde{w}_h\|_{H^{-1/2}(\Gamma)} \leq c_1\, h^{3/2}\, |w|_{H^1_{\mathrm{pw}}(\Gamma)} + \widetilde{c}(\eta, \varrho)\, h^{-1/2}\, \|w\|_{L_2(\Gamma)}.$$

To ensure an asymptotically optimal order of convergence we therefore need to satisfy the condition

$$\widetilde{c}(\eta, \varrho) \leq c_2\, h^2. \tag{14.18}$$

In this case, the error estimate

$$\|w - \widetilde{w}_h\|_{H^{-1/2}(\Gamma)} \leq c_1\, h^{3/2}\, |w|_{H^1_{\mathrm{pw}}(\Gamma)} + c_2\, h^{3/2}\, \|w\|_{L_2(\Gamma)}$$

is asymptotically of the same order of convergence as the corresponding error estimate (14.3) of a standard Galerkin boundary element method. A comparison with the condition (14.17) which was needed to ensure the positive definiteness of the approximate stiffness matrix $\widetilde{V}_h$ shows, that asymptotically condition (14.17) follows from (14.18).

It remains to find suitable representations (14.11) of the fundamental solution $U^*(x,y) = -\frac{1}{2\pi}\log|x-y|$ satisfying the error estimate (14.12). Then, the condition (14.18) also implies a suitable choice of the parameters $\eta$ and $\varrho$.

### 14.2.1 Taylor Series Representations

A first possibility to derive a representation (14.11) is to consider a Taylor expansion of the fundamental solution $U^*(x,y)$ with respect to the integration variable $y \in \omega_i^\lambda$ [73]. First we consider the Taylor expansion of a scalar function $f(t)$. For $p \in \mathbb{N}$ we have

$$f(1) = f(0) + \sum_{n=1}^{p} \frac{1}{n!}\frac{d^n}{dt^n} f(t)_{|t=0} + \frac{1}{p!}\int_0^1 (1-s)^p \frac{d^{p+1}}{ds^{p+1}} f(s)ds.$$

Let $y_i^\lambda$ be the center of the cluster $\omega_i^\lambda$. For an arbitrary $y \in \omega_i^\lambda$ and $t \in [0,1]$ let

$$f(t) := U^*(x, y_i^\lambda + t(y - y_i^\lambda)).$$

Then we have

$$\frac{d}{dt} f(t) = \sum_{j=1}^{2} (y_j - y_{i,j}^\lambda) \frac{\partial}{\partial z_j} U^*(x,z)_{|z=y_i^\lambda + t(y-y_i^\lambda)},$$

and by applying this recursively, we obtain for $1 \leq n \leq p$

$$\frac{d^n}{dt^n} f(t) = \sum_{|\alpha|=n} \frac{n!}{\alpha!} (y - y_i^\lambda)^\alpha D_z^\alpha U^*(x,z)_{|z=y_i^\lambda + t(y-y_i^\lambda)}.$$

Thus, the Taylor expansion of the fundamental solution $U^*(x,y)$ with respect to the cluster center $y_i^\lambda$ gives the representation

$$U^*(x,y) = U_\varrho^*(x,y) + R_p(x,y)$$

where

$$U_\varrho^*(x,y) = U^*(x, y_i^\lambda) + \sum_{n=1}^{p} \sum_{|\alpha|=n} \frac{1}{\alpha!} (y - y_i^\lambda)^\alpha D_z^\alpha U^*(x,z)_{|z=y_i^\lambda} \tag{14.19}$$

defines an approximation of the fundamental solution. By setting

$$f_0^{\lambda,j}(x) := U^*(x, y_i^\lambda), \quad g_0^{\lambda,i}(y) := 1$$

and

$$f_{n,\alpha}^{\lambda,j}(x) := D_z^\alpha U^*(x,z)_{|z=y_i^\lambda}, \quad g_{n,\alpha}^{\lambda,i}(y) := \frac{1}{\alpha!}(y - y_i^\lambda)^\alpha$$

for $n = 1, \ldots, p$ and $|\alpha| = n$, we then obtain the representation (14.11). For any $n \in [1,p]$ there exist $n+1$ multi–indices $\alpha \in \mathbb{N}_0^2$ satisfying $|\alpha| = n$. Then, the number $\varrho$ of terms in the series representation (14.11) is

$$\varrho = 1 + \sum_{n=1}^{p} (n+1) = \frac{1}{2}(p+1)(p+2).$$

To derive the error estimate (14.12) we have to consider the remainder

$$R_p(y, y_i^\lambda) = \frac{1}{p!} \int_0^1 (1-s)^p \sum_{|\alpha|=p+1} (y - y_i^\lambda)^\alpha D_z^\alpha U^*(x,z)_{|z=y_i^\lambda + s(y-y_i^\lambda)} ds.$$

For this we first need to estimate certain derivatives of the fundamental solution $U^*(x,y)$.

**Lemma 14.4.** *Let $\omega_i^\lambda$ and $\omega_j^\lambda$ be a pair of maximally admissible clusters. For $|\alpha| = p \in \mathbb{N}$ there holds the estimate*

$$\left|D_y^\alpha U^*(x,y)\right| \leq \frac{1}{2\pi}\frac{3^{p-1}(p-1)!}{|x-y|^p} \quad \textit{for all } (x,y) \in \omega_j^\lambda \times \omega_i^\lambda.$$

*Proof.* For the derivatives of the function $f(x,y) = \log|x-y|$ we first have

$$\frac{\partial}{\partial y_i} f(x,y) \;=\; \frac{y_i - x_i}{|x-y|^2} \quad \text{for } i = 1,2.$$

For the second order derivatives we further obtain

$$\frac{\partial^2}{\partial y_i^2} f(x,y) \;=\; \frac{1}{|x-y|^2} - 2\,\frac{(y_i-x_i)^2}{|x-y|^4} \quad \text{for } i = 1,2,$$

and

$$\frac{\partial^2}{\partial y_1 \partial y_2} f(x,y) \;=\; -2\,\frac{(y_1-x_1)(y_2-x_2)}{|x-y|^4}.$$

In general we find for $|\alpha| = \varrho \in \mathbb{N}$ a representation of the form

$$D_y^\alpha f(x,y) \;=\; \sum_{|\beta| \leq \varrho} a_\beta^\varrho \frac{(y-x)^\beta}{|x-y|^{|\beta|+\varrho}} \tag{14.20}$$

where $a_\beta^\varrho$ are some coefficients to be characterized. Hence we conclude

$$\left|D_y^\alpha f(x,y)\right| \leq \sum_{|\beta| \leq \varrho} \left|a_\beta^\varrho\right| \frac{1}{|x-y|^\varrho} \;=\; \frac{c_\varrho}{|x-y|^\varrho}.$$

A comparison with the first and second order derivatives of $f(x,y)$ gives $c_1 = 1$ and $c_2 = 3$. Now, a general estimate of the constant $c_\varrho$ for $\varrho \geq 2$ follows by induction. From (14.20) we obtain for $i = 1,2$ and $j \neq i$

$$\begin{aligned}
\frac{\partial}{\partial y_i} D_y^\alpha f(x,y) &= \sum_{|\beta| \leq \varrho} a_\beta^\varrho \frac{\partial}{\partial y_i} \frac{(y-x)^\beta}{|x-y|^{|\beta|+\varrho}} \\
&= \sum_{|\beta| \leq \varrho} a_\beta^\varrho \Bigg[\beta_i \frac{(y_i-x_i)^{\beta_i-1}(y_j-x_j)^{\beta_j}}{|x-y|^{|\beta|+\varrho}} \\
&\qquad\qquad - (|\beta|+\varrho)\frac{(y_i-x_i)^{\beta_i+1}(y_j-x_j)^{\beta_j}}{|x-y|^{|\beta|+\varrho+2}}\Bigg].
\end{aligned}$$

By using

$$|y_i - x_i| \leq |x-y|, \quad \beta_i \leq |\beta| \leq \varrho, \quad \beta_i + \beta_j \leq |\beta| \quad \text{for } i \neq j$$

we then obtain

$$\left|\frac{\partial}{\partial y_i} D_y^\alpha f(x,y)\right| \leq \frac{3\varrho}{|x-y|^{\varrho+1}} \sum_{|\beta|\leq\varrho} \left|a_\beta^\varrho\right| = \frac{3\varrho c_\varrho}{|x-y|^{\varrho+1}} = \frac{c_{\varrho+1}}{|x-y|^{\varrho+1}}$$

and therefore

$$c_{\varrho+1} = 3\varrho c_\varrho = 3^\varrho \varrho!\,.$$

In particular for $\varrho = p$ this gives the assertion. □

By using Lemma 14.4 we now can derive an estimate of the remainder $R_p(x,y)$ of the Taylor series approximation (14.19).

**Lemma 14.5.** *Let $\omega_i^\lambda$ and $\omega_j^\lambda$ be a pair of maximally admissible clusters. For the approximation $U_\varrho^*(x,y)$ of the fundamental solution $U^*(x,y)$ as defined in (14.19) there holds the error estimate*

$$\left|U^*(x,y) - U_\varrho^*(x,y)\right| \leq 3^p \left(\frac{1}{\eta}\right)^{p+1} \quad \textit{for all } (x,y) \in \omega_j^\lambda \times \omega_i^\lambda.$$

*Proof.* By applying Lemma 14.4 for $(x,y) \in \omega_j^\lambda \times \omega_i^\lambda$ and by using the admissibility condition (14.8) we obtain

$$\begin{aligned}
\left|U^*(x,y) - U_\varrho^*(x,y)\right| &= \left|R_p(y,y_i^\lambda)\right| \\
&\leq \frac{1}{p!}\,|y-y_i^\lambda|^{p+1} \max_{\bar{y}\in\omega_i^\lambda} \sum_{|\alpha|=p+1} |D_z^\alpha U^*(x,z)_{z=\bar{y}}| \int_0^1 (1-s)^p ds \\
&= \frac{1}{(p+1)!}|y-y_i^\lambda|^{p+1} \max_{\bar{y}\in\omega_i^\lambda} \sum_{|\alpha|=p+1} \frac{3^p p!}{|x-\bar{y}|^{p+1}} \\
&\leq 3^p\,\frac{[\operatorname{diam}\omega_i^\lambda]^{p+1}}{[\operatorname{dist}(\omega_i^\lambda,\omega_j^\lambda)]^{p+1}} \leq 3^p \left(\frac{1}{\eta}\right)^{p+1}. \quad \square
\end{aligned}$$

The decomposition (14.19) of the fundamental solution defines via (14.15) an approximated stiffness matrix $\widetilde{V}_h$. To ensure the asymptotically optimal error estimate (14.3) the related condition (14.18) reads

$$3^p \left(\frac{1}{\eta}\right)^{p+1} \leq c\,h^2\,.$$

If we choose a fixed admissibility parameter $\eta > 3$, then due to $h = \mathcal{O}(1/N)$ we therefore obtain the estimate

$$\frac{1}{3}\left(\frac{3}{\eta}\right)^{p+1} \leq \widetilde{c}\,\frac{1}{N^2}$$

from which we finally conclude

$$p = \mathcal{O}(\log_2 N)\,.$$

The total amount of work (14.16) to store $\widetilde{V}_h$ and to realize a matrix by vector product with the approximated stiffness matrix $\widetilde{V}_h$ is then proportional to

$$N\,(\log_2 N)^3\,.$$

### 14.2.2 Series Representations of the Fundamental Solution

Instead of a Taylor expansion of the fundamental solution $U^*(x,y)$ one may derive alternative series expansions which are valid in the far field for $(x,y) \in \omega_j^\lambda \times \omega_i^\lambda$ where $\omega_i^\lambda$ and $\omega_j^\lambda$ is a pair of admissible clusters. In particular, these series expansions are the starting point to derive the Fast Multipole Method [64, 65] which combines the series expansions on different levels. However, the resulting hierarchical algorithms will not discussed here, see, e.g. [48, 62, 109, 110].

Let $y_i^\lambda$ be the center of the cluster $\omega_i^\lambda$. By considering the fundamental solution in the complex plane we have

$$-\frac{1}{2\pi}\log|x-y| \,=\, -\frac{1}{2\pi}\log|(x-y_i^\lambda)-(y-y_i^\lambda)| \,=\, \mathrm{Re}\left(-\frac{1}{2\pi}\log(z-z_0)\right)$$

where

$$z := (x_1-y_{i,1}^\lambda)+i(x_2-y_{i,2}^\lambda) \,=\, |x-y_i^\lambda|\,e^{i\varphi(x-y_i^\lambda)},$$
$$z_0 := (y_1-y_{i,1}^\lambda)+i(y_2-y_{i,2}^\lambda) \,=\, |y-y_i^\lambda|\,e^{i\varphi(y-y_i^\lambda)}.$$

Since the clusters $\omega_i^\lambda$ and $\omega_j^\lambda$ are assumed to be admissible, it follows that

$$\frac{|z_0|}{|z|} \,=\, \frac{|y-y_i^\lambda|}{|x-y_i^\lambda|} \,\leq\, \frac{\operatorname{diam}\omega_i^\lambda}{\operatorname{dist}(\omega_i^\lambda,\omega_j^\lambda)} \,\leq\, \frac{1}{\eta} < 1.$$

Hence we can apply the series representation of the logarithm,

$$\begin{aligned}-\frac{1}{2\pi}\log(z-z_0) &= -\frac{1}{2\pi}\log z - \frac{1}{2\pi}\log\left(1-\frac{z_0}{z}\right)\\ &= -\frac{1}{2\pi}\log z + \frac{1}{2\pi}\sum_{n=1}^{\infty}\frac{1}{n}\left(\frac{z_0}{z}\right)^n,\end{aligned}$$

and for $p \in \mathbb{N}$ we can define an approximation

$$U_\varrho^*(x,y) \,:=\, \mathrm{Re}\left(-\frac{1}{2\pi}\log z + \frac{1}{2\pi}\sum_{n=1}^{p}\frac{1}{n}\left(\frac{z_0}{z}\right)^n\right) \tag{14.21}$$

of the fundamental solution $U^*(x,y)$. By using

$$\left(\frac{z_0}{z}\right)^n \,=\, z_0^n z^{-n} \,=\, \frac{|y-y_i^\lambda|^n}{|x-y_i^\lambda|^n}\,e^{in\varphi(y-y_i^\lambda)}e^{-in\varphi(x-y_i^\lambda)}$$

we then obtain the representation

$$\begin{aligned}U_\varrho^*(x,y) &= -\frac{1}{2\pi}\log|x-y_i^\lambda|\\ &\quad+\frac{1}{2\pi}\sum_{n=1}^{p}\frac{1}{n}|y-y_i^\lambda|^n\cos n\varphi(y-y_i^\lambda)\frac{\cos n\varphi(x-y_i^\lambda)}{|x-y_i^\lambda|^n}\\ &\quad+\frac{1}{2\pi}\sum_{n=1}^{p}\frac{1}{n}|y-y_i^\lambda|^n\sin n\varphi(y-y_i^\lambda)\frac{\sin n\varphi(x-y_i^\lambda)}{|x-y_i^\lambda|^n}.\end{aligned}$$

By introducing

$$f_0^{\lambda,j}(x) := U^*(x, y_i^\lambda), \quad g_0^{\lambda,i}(y) := 1$$

and

$$f_{2n-1}^{\lambda,j}(x) := \frac{1}{2\pi} \frac{\cos n\varphi(x - y_i^\lambda)}{|x - y_i^\lambda|^n}, \quad g_{2n-1}^{\lambda,i}(y) := \frac{1}{n}|y - y_i^\lambda|^n \cos n\varphi(y - y_i^\lambda),$$

as well as

$$f_{2n}^{\lambda,j}(x) := \frac{1}{2\pi} \frac{\sin n\varphi(x - y_i^\lambda)}{|x - y_i^\lambda|^n}, \quad g_{2n}^{\lambda,i}(y) := \frac{1}{n}|y - y_i^\lambda|^n \sin n\varphi(y - y_i^\lambda)$$

for $n = 1, \ldots, p$ we finally obtain the representation (14.11) where $\varrho = 2p+1$.

**Lemma 14.6.** *Let $\omega_i^\lambda$ and $\omega_j^\lambda$ be a pair of admissible clusters. For the approximation $U_\varrho(x,y)$ of the fundamental solution $U^*(x,y)$ as defined in (14.21) there holds the error estimate*

$$|U^*(x,y) - U_\varrho^*(x,y)| \le \frac{1}{2\pi} \frac{1}{p+1} \frac{1}{\eta - 1} \left(\frac{1}{\eta}\right)^p \text{ for all } (x,y) \in \omega_j^\lambda \times \omega_i^\lambda.$$

*Proof.* By using the series expansion

$$-\frac{1}{2\pi} \log(z - z_0) = -\frac{1}{2\pi} \log z + \frac{1}{2\pi} \sum_{n=1}^\infty \frac{1}{n} \left(\frac{z_0}{z}\right)^n$$

we conclude from the admissibility condition (14.8)

$$\begin{aligned} |U^*(x,y) - U_\varrho^*(x,y)| &= \left| \text{Re} \left( \frac{1}{2\pi} \sum_{n=p+1}^\infty \frac{1}{n} \left(\frac{z_0}{z}\right)^n \right) \right| \\ &\le \frac{1}{2\pi} \sum_{n=p+1}^\infty \frac{1}{n} \left(\frac{1}{\eta}\right)^n \\ &\le \frac{1}{2\pi} \frac{1}{p+1} \left(\frac{1}{\eta}\right)^{p+1} \sum_{n=0}^\infty \left(\frac{1}{\eta}\right)^n . \quad \square \end{aligned}$$

To ensure the asymptotically optimal error estimate (14.3) the related condition (14.18) now reads

$$\frac{1}{2\pi} \frac{1}{p+1} \frac{1}{\eta - 1} \left(\frac{1}{\eta}\right)^p \le c h^2 .$$

If we choose a fixed admissibility parameter $\eta > 1$ this finally gives

$$p = \mathcal{O}(\log_2 N) .$$

The total amount of work (14.16) to store $\widetilde{V}_h$ and to realize a matrix by vector multiplication with the approximated stiffness matrix $\widetilde{V}_h$ is then proportional to

$$N\,(\log_2 N)^2\,.$$

Note that both the approximation (14.19) based on the Taylor expansion as well as the series expansion (14.21) define nonsymmetric approximations $U^*_\varrho(x,y)$ of the fundamental solution $U^*(x,y)$, and therefore this results in a nonsymmetric approximated stiffness matrix $\widetilde{V}_h$. Hence we aim to derive a symmetric approximation $U^*_\varrho(x,y)$. For this we first consider the representation (14.21),

$$U^*_\varrho(x,y) = \text{Re}\left(-\frac{1}{2\pi}\log z + \frac{1}{2\pi}\sum_{n=1}^{p}\frac{1}{n}\left(\frac{z_0}{z}\right)^n\right)$$

where

$$z = |x-y_i^\lambda|e^{i\varphi(x-y_i^\lambda)}, \quad z_0 = |y-y_i^\lambda|e^{i\varphi(y-y_i^\lambda)}.$$

For the center $y_j^\lambda$ of the cluster $\omega_j^\lambda$ we consider $z = w - z_1$ where

$$w := |y_j^\lambda - y_i^\lambda|e^{i\varphi(y_j^\lambda - y_i^\lambda)}, \quad z_1 := |y_j^\lambda - x|e^{i\varphi(y_j^\lambda - x)}.$$

By using the admissibility condition (14.8) we have

$$\frac{|z_1|}{|w|} = \frac{|y_j^\lambda - x|}{|y_j^\lambda - y_i^\lambda|} \le \frac{\text{diam}\,\omega_j^\lambda}{\text{dist}(\omega_i^\lambda,\omega_j^\lambda)} \le \frac{1}{\eta} < 1,$$

and therefore we can write

$$-\frac{1}{2\pi}\log z = -\frac{1}{2\pi}\log w + \frac{1}{2\pi}\sum_{n=1}^{\infty}\frac{1}{n}\left(\frac{z_1}{w}\right)^n.$$

**Lemma 14.7.** *Let $w, z_1 \in \mathbb{C}$ satisfying $|z_1| < |w|$. For $n \in \mathbb{N}$ then there holds*

$$\frac{1}{(w-z_1)^n} = \sum_{m=0}^{\infty}\binom{m+n-1}{n-1}\frac{z_1^m}{w^{m+n}}.$$

*Proof.* For $|z_1| < |w|$ we first have

$$\frac{1}{w-z_1} = \frac{1}{w}\frac{1}{1-\frac{z_1}{w}} = \frac{1}{w}\sum_{m=0}^{\infty}\left(\frac{z_1}{w}\right)^m$$

and therefore the assertion in the case $n = 1$. For $n > 1$ we have

$$\begin{aligned}
\frac{1}{(w-z_1)^n} &= \frac{1}{(n-1)!}\frac{d^{n-1}}{dz_1^{n-1}}\frac{1}{w-z_1} \\
&= \frac{1}{(n-1)!}\frac{d^{n-1}}{dz_1^{n-1}}\left[\frac{1}{w}\sum_{m=0}^{\infty}\left(\frac{z_1}{w}\right)^m\right] \\
&= \frac{1}{(n-1)!}\frac{1}{w}\sum_{m=n-1}^{\infty}\frac{m!}{(m-n+1)!}\frac{z_1^{m-n+1}}{w^m} \\
&= \frac{1}{(n-1)!}\frac{1}{w}\sum_{m=0}^{\infty}\frac{(m+n-1)!}{m!}\frac{z_1^m}{w^{m+n-1}} \\
&= \sum_{m=0}^{\infty}\binom{m+n-1}{n-1}\frac{z_1^m}{w^{m+n}}. \quad \square
\end{aligned}$$

By using (14.21), $z = w - z_1$, and by applying Lemma 14.7 for $p \in \mathbb{N}$ we can define a symmetric approximation of the fundamental solution $U^*(x,y)$ as

$$\begin{aligned}
\widetilde{U}_\varrho^*(x,y) = \operatorname{Re}\Bigg( &-\frac{1}{2\pi}\log w + \frac{1}{2\pi}\sum_{m=1}^{p}\frac{1}{m}\left(\frac{z_1}{w}\right)^m + \frac{1}{2\pi}\sum_{n=1}^{p}\frac{1}{n}\left(\frac{z_0}{w}\right)^n \\
&+ \frac{1}{2\pi}\sum_{n=1}^{p}\sum_{m=1}^{p}\frac{1}{n}\binom{m+n-1}{n-1}\frac{z_0^n z_1^m}{w^{m+n}}\Bigg).
\end{aligned} \tag{14.22}$$

**Lemma 14.8.** *Let $\omega_i^\lambda$ and $\omega_j^\lambda$ be a pair of admissible clusters. For the approximation $\widetilde{U}_\varrho^*(x,y)$ of the fundamental solution $U^*(x,y)$ as defined in (14.22) there holds the error estimate*

$$\left|U^*(x,y) - \widetilde{U}_\varrho^*(x,y)\right| \le \frac{1}{2\pi}\frac{1}{p+1}\frac{1}{\eta-1}\left[2 + \frac{1}{\eta-1}\right]\left(\frac{1}{\eta}\right)^p$$

*for all $(x,y) \in \omega_j^\lambda \times \omega_i^\lambda$.*

*Proof.* By applying the triangle inequality and Lemma 14.6 we have

$$\begin{aligned}
&\left|U^*(x,y) - \widetilde{U}_\varrho^*(x,y)\right| \le \left|U^*(x,y) - U_\varrho^*(x,y)\right| + \left|U_\varrho^*(x,y) - \widetilde{U}_\varrho^*(x,y)\right| \\
&\le \frac{1}{2\pi}\frac{1}{p+1}\frac{\eta}{\eta-1}\left(\frac{1}{\eta}\right)^{p+1} \\
&\quad + \left|\operatorname{Re}\left(\frac{1}{2\pi}\sum_{m=p+1}^{\infty}\frac{1}{m}\left(\frac{z_1}{w}\right)^m + \frac{1}{2\pi}\sum_{n=1}^{p}\sum_{m=p+1}^{\infty}\frac{1}{n}\binom{m-n-1}{n-1}\frac{z_0^n z_1^m}{w^{m+n}}\right)\right| \\
&\le \frac{1}{\pi}\frac{1}{p+1}\frac{\eta}{\eta-1}\left(\frac{1}{\eta}\right)^{p+1} + \frac{1}{2\pi}\sum_{n=1}^{p}\sum_{m=p+1}^{\infty}\frac{1}{n}\binom{m-n-1}{n-1}\left(\frac{1}{\eta}\right)^{m+n}
\end{aligned}$$

where we have used the admissibility condition (14.8). With

$$\frac{1}{n}\binom{m-n-1}{n-1} = \frac{1}{n}\frac{(m-n-1)!}{(n-1)!m!} = \frac{1}{m}\frac{(m-n-1)!}{n!(m-1)!} \leq \frac{1}{m}$$

we obtain for the remaining term

$$\begin{aligned}\frac{1}{2\pi}\sum_{n=1}^{p}\sum_{m=p+1}^{\infty}\frac{1}{n}\binom{m-n-1}{n-1}\left(\frac{1}{\eta}\right)^{m+n} &\leq \frac{1}{2\pi}\sum_{n=1}^{p}\sum_{m=p+1}^{\infty}\frac{1}{m}\left(\frac{1}{\eta}\right)^{m+n}\\ &\leq \frac{1}{2\pi}\frac{1}{p+1}\frac{\eta}{(\eta-1)^2}\left(\frac{1}{\eta}\right)^{p+1}\end{aligned}$$

which concludes the proof. □

### 14.2.3 Adaptive Cross Approximation

All approximations $U^*_\varrho(x,y)$ of the fundamental solution $U^*(x,y)$ as described before require either the knowledge of a suitable series expansion, or the computation of higher order derivatives of the fundamental solution. Hence we want to define approximations which only require the evaluation of the fundamental solution in appropriate interpolation nodes. One possibility is to consider the Tschebyscheff interpolation of the fundamental solution. Here we describe an alternative interpolation algorithm which was first given by Tyrtyshnikov in [151], see also [12, 13].

Let $\omega_i^\lambda$ and $\omega_j^\lambda$ be a pair of admissible clusters. To define an approximation of the fundamental solution $U^*(x,y)$ for arguments $(x,y) \in \omega_j^\lambda \times \omega_i^\lambda$ we consider two sequences of functions $s_k(x,y)$ and $r_k(x,y)$. In particular, $r_k(x,y)$ describes the residual of the associated approximation $s_k(x,y)$. To initialize this construction we first define

$$s_0(x,y) := 0, \quad r_0(x,y) := U^*(x,y).$$

For $k = 1,2,\ldots,\varrho$ let $(x_k,y_k) \in \omega_j^\lambda \times \omega_i^\lambda$ be a pair of interpolation nodes with a nonzero residual, i.e. $\alpha_k := r_{k-1}(x_k,y_k) \neq 0$. Then we define the recursion as

$$s_k(x,y) := s_{k-1}(x,y) + \frac{1}{\alpha_k} r_{k-1}(x,y_k) r_{k-1}(x_k,y), \tag{14.23}$$

$$r_k(x,y) := r_{k-1}(x,y) - \frac{1}{\alpha_k} r_{k-1}(x,y_k) r_{k-1}(x_k,y). \tag{14.24}$$

For $\varrho \in \mathbb{N}_0$ and for $(x,y) \in \omega_j^\lambda \times \omega_i^\lambda$ we finally define the approximation

$$U^*_\varrho(x,y) = s_\varrho(x,y) = \sum_{k=1}^{\varrho} \frac{r_{k-1}(x,y_k) r_{k-1}(x_k,y)}{r_{k-1}(x_k,y_k)} \tag{14.25}$$

of the fundamental solution $U^*(x,y)$ with respect to an admissible pair of clusters $(\omega_j^\lambda, \omega_i^\lambda)$. The recursion as defined in (14.23) and (14.24) admits the following properties.

**Lemma 14.9.** *For $0 \leq k \leq \varrho$ and for all $(x, y) \in \omega_j^\lambda \times \omega_i^\lambda$ there holds*

$$U^*(x,y) \;=\; s_k(x,y) + r_k(x,y). \tag{14.26}$$

*Moreover, there holds the interpolation property*

$$r_k(x,y_i) \;=\; 0 \quad \textit{for all}\, 1 \leq i \leq k \tag{14.27}$$

*as well as*

$$r_k(x_j,y) \;=\; 0 \quad \textit{for all}\, 1 \leq j \leq k. \tag{14.28}$$

*Proof.* By taking the sum of the recursions (14.23) and (14.24) we first have

$$r_k(x,y) + s_k(x,y) \;=\; r_{k-1}(x,y) + s_{k-1}(x,y)$$

for all $k = 1, \ldots, p$, and therefore

$$r_k(x,y) + s_k(x,y) = r_0(x,y) + s_0(x,y) = U^*(x,y).$$

The interpolation properties follow by induction with respect to $j, k$. For $k = 1$ we have

$$r_1(x,y_1) \;=\; r_0(x,y_1) - \frac{1}{r_0(x_1,y_1)} r_0(x,y_1) r_0(x_1,y_1) \;=\; 0.$$

Hence we have $r_k(x,y_i) = 0$ for $k = 1, 2, \ldots, p$ and $i = 1, \ldots, k$. Then we conclude

$$r_{k+1}(x,y_i) \;=\; r_k(x,y_i) - \frac{1}{\alpha_{k+1}} r_k(x,y_{k+1}) r_k(x_{k+1},y_i) \;=\; 0,$$

i.e. $r_{k+1}(x,y_i) = 0$ for all $i = 1, \ldots, k$. By using (14.24) we finally obtain

$$r_{k+1}(x,y_{k+1}) = r_k(x,y_{k+1}) - \frac{1}{r_k(x_{k+1},y_{k+1})} r_k(x,y_{k+1}) r_k(x_{k+1},y_{k+1}) = 0.$$

The other interpolation property follows in the same way. □

When inserting in (14.27) $x = x_j$ for $j = 1, \ldots, k$ this gives

$$r_k(x_j,y_i) \;=\; 0 \quad \text{for all } i,j = 1, \ldots, k,$$

and due to (14.26) we conclude

$$s_k(x_j,y_i) \;=\; U^*(x_j,y_i) \quad \text{for all } i,j = 1, \ldots, k; k = 1, \ldots, \varrho.$$

This means that the approximations $s_k(x,y)$ interpolate the fundamental solution $U^*(x,y)$ at the interpolation nodes $(x_j,y_i)$ for $i,j = 1, \ldots, k$.

To analyze the approximations $U_\varrho^*(x,y)$ as defined via (14.23) and (14.24) we consider a sequence of matrices

$$M_k[j,i] \;=\; U^*(x_j,y_i) \quad \text{for } i,j = 1, \ldots, k; k = 1, \ldots, \varrho \tag{14.29}$$

where we compute the determinants as follows.

**Lemma 14.10.** *Let $M_k$, $k = 1, \ldots, \varrho$, be the sequence of matrices as defined in (14.29). Then there holds*

$$det M_k \; = \; r_0(x_1, y_1) \cdots r_{k-1}(x_k, y_k) \; = \; \prod_{i=1}^{k} r_{i-1}(x_i, y_i) \; = \; \prod_{i=1}^{k} \alpha_i \, . \tag{14.30}$$

*Proof.* To prove (14.30) we consider an induction with respect to $k = 1, \ldots, \varrho$. For $k = 1$ we first have

$$\det M_1 \; = \; M_1[1,1] \; = \; U^*(x_1, y_1) \; = \; r_0(x_1, y_1) \; = \; \alpha_1 .$$

Assume that (14.30) holds for $M_k$. The matrix $M_{k+1}$ allows the representation

$$M_{k+1} \; = \; \begin{pmatrix} U^*(x_1, y_1) & \cdots & U^*(x_1, y_k) & U^*(x_1, y_{k+1}) \\ \vdots & & \vdots & \vdots \\ U^*(x_k, y_1) & \cdots & U^*(x_k, y_k) & U^*(x_k, y_{k+1}) \\ U^*(x_{k+1}, y_1) & \cdots & U^*(x_{k+1}, y_k) & U^*(x_{k+1}, y_{k+1}) \end{pmatrix} .$$

For any $(x, y_i) \in \omega_j^\lambda \times \omega_i^\lambda$ we have by the recursion (14.24)

$$\begin{aligned} r_0(x, y_i) &= U^*(x, y_i), \\ r_1(x, y_i) &= r_0(x, y_i) - \frac{r_0(x, y_1) r_0(x_1, y_i)}{r_0(x_1, y_1)} \\ &= U^*(x, y_i) - \frac{r_0(x_1, y_i)}{r_0(x_1, y_1)} U^*(x, y_1) \end{aligned}$$

and therefore

$$r_1(x, y_i) \; = \; U^*(x, y_i) - \alpha_1^1(y_i) \, U^*(x, y_1), \quad \alpha_1^1(y_i) \; := \; \frac{r_0(x_1, y_i)}{r_0(x_1, y_1)} .$$

By induction, this representation can be generalized to all residuals $r_k(x, y_i)$ for all $k = 1, \ldots, \varrho$ and for all $(x, y_i) \in \omega_j^\lambda \times \omega_i^\lambda$. Note that all residuals satisfy

$$r_k(x, y_i) \; = \; U^*(x, y_i) - \sum_{j=1}^{k} \alpha_j^k(\underline{y}) U^*(x, y_j) \tag{14.31}$$

By using the recursion (14.24) and inserting twice the assumption of the induction this gives

$$
\begin{aligned}
r_{k+1}(x,y_i) &= r_k(x,y_i) - \frac{r_k(x,y_{k+1})r_k(x_{k+1},y_i)}{r_k(x_{k+1},y_{k+1})} \\
&= U^*(x,y_i) - \sum_{j=1}^{k} \alpha_j^k(\underline{y})U^*(x,y_j) \\
&\quad - \frac{r_k(x_{k+1},y_i)}{r_k(x_{k+1},y_{k+1})}\left[U^*(x,y_{k+1}) - \sum_{j=1}^{k} \alpha_j^k(\underline{y})U^*(x,y_j)\right] \\
&= U^*(x,y_i) - \sum_{j=1}^{k+1} \alpha_j^{k+1}(\underline{y})U^*(x,y_j)\,.
\end{aligned}
$$

By

$$\widetilde{M}_{k+1}[j,i] \;:=\; M_{k+1}[j,i] \quad \text{for } i=1,\ldots,k, j=1,\ldots,k+1$$

and

$$\widetilde{M}_{k+1}[j,k+1] \;=\; M_{k+1}[j,k+1] \;-\; \sum_{\ell=1}^{k} \alpha_\ell^k(\underline{y})M_{k+1}[j,\ell] \quad \text{for } j=1,\ldots,k+1$$

we define a transformed matrix $\widetilde{M}_{k+1}$ satisfying $\det \widetilde{M}_{k+1} = \det M_{k+1}$. Inserting the definition of $M_{k+1}[j,\cdot]$ this gives for $j=1,\ldots,k+1$, due to (14.31),

$$\widetilde{M}_{k+1}[j,k+1] \;=\; U^*(x_j,y_{k+1}) \;-\; \sum_{\ell=1}^{k} \alpha_\ell^k(\underline{y})U^*(x_j,y_\ell) \;=\; r_k(x_j,y_{k+1}).$$

Note that, see Lemma 14.9,

$$r_k(x_j,y_{k+1}) \;=\; 0 \quad \text{for all } j=1,\ldots,k,$$

In particular, the determinant of the matrix $M_{k+1}$ remains unchanged when a row $M_k[j,\cdot]$ multiplied by $\alpha_j^k$ is subtracted from the last row of $M_k$. By using (14.31) we then conclude

$$\widetilde{M}_{k+1} = \begin{pmatrix} U^*(x_1,y_1) & \cdots & U^*(x_1,y_k) & 0 \\ \vdots & & \vdots & \vdots \\ U^*(x_k,y_1) & \cdots & U^*(x_k,y_k) & 0 \\ U^*(x_{k+1},y_1) & \cdots & U^*(x_{k+1},y_k) & r_k(x_{k+1},y_{k+1}) \end{pmatrix}.$$

The computation of $\det \widetilde{M}_{k+1}$ via an expansion with respect to the last column of $\widetilde{M}_{k+1}$ now gives

$$\det M_{k+1} \;=\; \det \widetilde{M}_{k+1} \;=\; r_k(x_{k+1},y_{k+1}) \det M_k$$

which concludes the induction. □

By using the matrices $M_k$ now we can represent the approximations $s_k(x,y)$ as follows.

**Lemma 14.11.** *For $k = 1, \ldots, \varrho$ the approximations $s_k(x,y)$ allow the representations*

$$s_k(x,y) = (U^*(x,y_i))^\top_{i=1,\ldots,k} M_k^{-1} (U^*(x_i,y))_{i=1,\ldots,k} \,.$$

*Proof.* By using the recursion (14.23) the approximation $s_k(x,y)$ is defined as

$$s_k(x,y) = \sum_{i=1}^{k} \frac{1}{\alpha_i} r_{i-1}(x,y_i) r_{i-1}(x_i,y) \,.$$

From (14.31) we find for the residual $r_{i-1}(x,y_i)$ the representation

$$r_{i-1}(x,y_i) = U^*(x,y_i) - \sum_{\ell=1}^{i-1} \alpha_\ell^{i-1}(\underline{y}) U^*(x,y_\ell).$$

Analogously we also have

$$r_{i-1}(x_i,y) = U^*(x_i,y) - \sum_{\ell=1}^{i-1} \beta_\ell^{i-1}(\underline{x}) U^*(x_\ell,y).$$

Hence there exists a matrix $A_k \in \mathbb{R}^{k\times k}$ satisfying

$$s_k(x,y) = (U^*(x,y_\ell))^\top_{\ell=1,\ldots,k} A_k U^*(x_\ell,y)_{\ell=1,\ldots,k} \,.$$

In particular for $(x,y) = (x_j,y_i)$ we have $r_k(x_j,y_i) = 0$ and therefore

$$U^*(x_j,y_i) = s_k(x_j,y_i) = (U^*(x_j,y_\ell))^\top_{\ell=1,\ldots,k} A_k U^*(x_\ell,y_i)_{\ell=1,\ldots,k}$$

for $i,j = 1,\ldots,k$. This is equivalent to

$$M_k = M_k A_k M_k$$

and since $M_k$ is invertible this gives $A_k = M_k^{-1}$. □

To estimate the residual $r_\varrho(x,y)$ of the approximation $U_\varrho^*(x,y)$ as defined in (14.25) we will consider a relation of the above approach with an interpolation by using Lagrange polynomials. For $p \in \mathbb{N}_0$ let $P_p(\mathbb{R}^2)$ denote the space of polynomials $y^\alpha$ of degree $|\alpha| \le p$ where $y \in \mathbb{R}^2$. We assume that the number of interpolation nodes $(x_k, y_k)$ to define (14.25) corresponds with the dimension of $P_p(\mathbb{R}^2)$, i.e.

$$\varrho := \dim P_p(\mathbb{R}^2) = \frac{1}{2} p(p+1).$$

For $k \neq \ell$ let $y_k \neq y_\ell$, then the Lagrange polynomials $L_k \in P_p(\mathbb{R}^2)$ are well defined for $k = 1,\ldots,\varrho$, and we have

$$L_k(y_\ell) = \delta_{k\ell} \quad \text{for } k,\ell = 1,\ldots,\varrho.$$

The Lagrange interpolation of the fundamental solution $U^*(x,y)$ is then given as

$$U^*_{L,\varrho}(x,y) = \sum_{k=1}^{\varrho} U^*(x,y_k)L_k(y) = (U^*(x,y_k))^\top_{k=1,\ldots,\varrho}\,(L_k(y))_{k=1,\ldots,\varrho}$$

where the associated residual is

$$E_\varrho(x,y) := U^*(x,y) - (U^*(x,y_k))^\top_{k=1,\ldots,\varrho}\,(L_k(y))_{k=1,\ldots,\varrho}\,.$$

Let

$$M_{\varrho,\ell}(x) := \begin{pmatrix} U^*(x_1,y_1) & \cdots\cdots & U^*(x_1,y_\varrho) \\ \vdots & & \vdots \\ U^*(x_{\ell-1},y_1) & \cdots\cdots & U^*(x_{\ell-1},y_\varrho) \\ U^*(x,y_1) & \cdots\cdots & U^*(x,y_\varrho) \\ U^*(x_{\ell+1},y_1) & \cdots\cdots & U^*(x_{\ell+1},y_\varrho) \\ \vdots & & \vdots \\ U^*(x_\varrho,y_1) & \cdots\cdots & U^*(x_\varrho,y_\varrho) \end{pmatrix}.$$

**Lemma 14.12.** *For $(x,y) \in \omega_j^\lambda \times \omega_i^\kappa$ let $r_\varrho(x,y)$ be the residual of the approximation $U^*_\varrho(x,y)$. Then there holds*

$$r_\varrho(x,y) = E_\varrho(x,y) - \sum_{k=1}^{\varrho} \frac{\det M_{\varrho,k}(x)}{\det M_\varrho} E_\varrho(x_k,y).$$

*Proof.* By using Lemma 14.9 and the matrix representation of $s_\varrho(x,y)$, see Lemma 14.11, we first have

$$\begin{aligned} r_\varrho(x,y) &= U^*(x,y) - s_\varrho(x,y) \\ &= U^*(x,y) - (U^*(x,y_k))^\top_{k=1,\varrho} M_\varrho^{-1} (U^*(x_k,y))_{k=1,\varrho} \\ &= U^*(x,y) - (U^*(x,y_k))^\top_{k=1,\varrho} (L_k(y))_{k=1,\varrho} \\ &\quad - (U^*(x,y_k))^\top_{k=1,\varrho} M_\varrho^{-1} \Big[ (U^*(x_k,y))_{k=1,\varrho} - M_\varrho (L_k(y))_{k=1,\varrho} \Big]. \end{aligned}$$

Due to

$$\begin{aligned} (U^*(x_k,y))_{k=1,\varrho} - M_\varrho (L_k(y))_{k=1,\varrho} &= \left( U^*(x_k,y) - \sum_{\ell=1}^{\varrho} U^*(x_k,y_\ell) L_\ell(y) \right)_{k=1,\varrho} \\ &= (E_\varrho(x_k,y))_{k=1,\varrho} \end{aligned}$$

we then conclude

$$\begin{aligned} r_\varrho(x,y) &= E_\varrho(x,y) - (U^*(x,y_k))^\top_{k=1,\varrho} M_\varrho^{-1} (E_\varrho(x_k,y))_{k=1,\varrho} \\ &= E_\varrho(x,y) - \left( M_\varrho^{-\top} (U^*(x,y_\ell))_{\ell=1,\varrho} \right)^\top (E_\varrho(x_k,y))_{k=1,\varrho}\,. \end{aligned}$$

Let $\underline{e}^\ell \in \mathbb{R}^\varrho$ be the unit vectors satisfying $e_i^\ell = \delta_{i\ell}$. Then,

$$M_\varrho^\top \underline{e}^i = (U^*(x_i, y_k))_{k=1,\varrho}$$

and thus

$$\underline{e}^i = M_\varrho^{-\top} (U^*(x_i, y_k))_{k=1,\varrho} .$$

Hence we conclude

$$\begin{aligned} M_\varrho^{-\top} M_{\varrho,\ell}^\top(x) &= \\ &= M_\varrho^{-\top} \Big( (U^*(x_i, y_k))_{k=1,\varrho;i=1,\ell-1}, (U^*(x, y_k))_{k=1,\varrho}, (U^*(x_i, y_k))_{k=1,\varrho;i=\ell+1,\varrho} \Big) \\ &= \Big( \underline{e}^1, \ldots, \underline{e}^{\ell-1}, M_\varrho^{-\top} (U^*(x, y_k))_{k=1,\varrho}, \underline{e}^{\ell+1}, \ldots, \underline{e}^\varrho \Big), \end{aligned}$$

and with

$$\Big( M_\varrho^{-\top} (U^*(x, y_\ell)_{\ell=1,\varrho} \Big)_k = \det \left( M_\varrho^{-\top} M_{\varrho,k}^\top(x) \right) = \frac{\det M_{\varrho,k}(x)}{\det M_\varrho}$$

we finally get the assertion. □

**Corollary 14.13.** *Let the interpolation nodes $(x_k, y_k) \in \omega_j^\lambda \times \omega_i^\lambda$ be chosen such that*

$$|\det M_{\varrho,\ell}(x)| \leq |\det M_\varrho| \tag{14.32}$$

*is satisfied for all $\ell = 1, \ldots, \varrho$ and for all $x \in \omega_j^\lambda$. The residual $r_\varrho(x,y)$ then satisfies the estimate*

$$|r_\varrho(x,y)| \leq (1+\varrho) \sup_{x \in \omega_j^\lambda} |E_\varrho(x,y)| .$$

The criteria (14.32) to define the interpolation nodes $(x_k, y_k) \in \omega_j^\lambda \in \omega_i^\lambda$ seems not be very suitable for a practical realization. Hence we finally consider an alternative choice.

**Lemma 14.14.** *For $k = 1, \ldots, \varrho$ let the nodal pairs $(x_k, y_k) \in \omega_j^\lambda \in \omega_i^\lambda$ be chosen such that*

$$|r_{k-1}(x_k, y_k)| \geq |r_{k-1}(x, y_k)| \quad \textit{for all } x \in \omega_j^\lambda \tag{14.33}$$

*is satisfied. Then there holds*

$$\sup_{x \in \omega_j^\lambda} \frac{|\det M_{k,\ell}(x)|}{|\det M_k|} \leq 2^{k-\ell} .$$

*Proof.* As in the proof of Lemma 14.10 we have for $\det M_{k,\ell}(x)$ and $1 \leq \ell < k$ the recursion

$$\det M_{k,\ell}(x) = r_{k-1}(x_k, y_k) \det M_{k-1,\ell}(x) - r_{k-1}(x, y_k) \det M_{k-1,\ell}(x_k),$$

or,

$$\det M_{1,1}(x) = r_0(x, y_1), \quad \det M_{k,k}(x) = r_{k-1}(x, y_k) \det M_{k-1}$$

for $k = 2, 3, \ldots, \varrho$, as well as

$$\det M_k = r_{k-1}(x_k, y_k) \det M_{k-1} .$$

Hence we have for $1 \leq \ell < k$ by using the assumption (14.33)

$$\frac{\det M_{k,\ell}(x)}{\det M_k} = \frac{\det M_{k-1,\ell}(x)}{\det M_{k-1}} - \frac{r_{k-1}(x, y_k)}{r_{k-1}(x_k, y_k)} \frac{\det M_{k-1,\ell}(x)}{\det M_{k-1}}$$

and therefore

$$\sup_{x \in \omega_j^\lambda} \frac{|\det M_{k,\ell}(x)|}{|\det M_k|} \leq 2 \sup_{x \in \omega_j^\lambda} \frac{|\det M_{k-1,\ell}(x)|}{|\det M_{k-1}|},$$

or,

$$\left| \frac{\det M_{k,k}(x)}{\det M_k} \right| = \left| \frac{r_{k-1}(x, y_k)}{r_{k-1}(x_k, y_k)} \right| \leq 1 . \qquad \square$$

By using Lemma 14.14 we obtain from Lemma 14.12 an upper bound of the residual $r_\varrho(x, y)$ by the Lagrange interpolation error $E_\varrho(x, y)$.

**Corollary 14.15.** *For $k = 1, \ldots, \varrho$ let the nodal pairs $(x_k, y_k) \in \omega_j^\lambda \times \omega_i^\lambda$ be chosen such that assumption* (14.33) *is satisfied. Then there holds*

$$|r_\varrho(x, y)| \leq 2^\varrho \sup_{x \in \omega_j^\lambda} |E_\varrho(x, y)| .$$

Contrary to the one–dimensional Lagrange interpolation the interpolation in more space dimensions is quite difficult. In particular the uniqueness of the interpolation polynomial depends on the choice of the interpolation nodes. Moreover, there is no explicit representation of the remainder $E_\varrho(x, y)$ known. Hence we skip a more detailed discussion at this point. By using results of [121] one can derive similar error estimates as in Lemma 14.6, see [13].

The adaptive cross approximation algorithm to approximate a scalar function as described in this subsection can be generalized in a straightforward way to define low rank approximations of a matrix, see, e.g. [12, 14], and [117] for a more detailed discussion.

## 14.3 Wavelets

In this subsection we introduce wavelets as hierarchical basis functions to be used in the Galerkin discretization of the single layer potential $V$. As in standard boundary element methods this leads to a dense stiffness matrix, but by neglecting small matrix entries one can define a sparse approximation.

This reduces both the amount of storage, and amount to realize a matrix by vector multiplication.

Without loss of generality we assume that the Lipschitz boundary $\Gamma = \partial\Omega$ of a bounded domain $\Omega \subset \mathbb{R}^2$ is given via a parametrization $\Gamma = \chi(\mathcal{Q})$ with respect to a parameter domain $\mathcal{Q} = [0,1]$ where we assume that the Jacobian $|\dot{\chi}(\xi)|$ is constant for all $\xi \in \mathcal{Q}$. Moreover, we extend the parametrization $\Gamma = \chi([0,1])$ periodically onto $\mathbb{R}$. Hence we can assume the estimates

$$c_1^\chi\,|\xi - \eta| \leq |\chi(\xi) - \chi(\eta)| \leq c_2^\chi\,|\xi - \eta| \quad \text{for all } \xi, \eta \in \mathbb{R} \tag{14.34}$$

with some positive constants $c_1^\chi$ and $c_2^\chi$. In the case of a piecewise smooth Lipschitz boundary $\Gamma$ all following considerations have to be transfered to the non–periodic parametrizations describing the parts $\Gamma_j$ satisfying $|\dot{\chi}_j| = c_j$.

For $j \in \mathbb{N}$ we consider a decomposition of the parameter domain $\mathcal{Q} = [0,1]$ into $N_j = 2^j$ finite elements $q_\ell^j$ of mesh size $|q_\ell^j| = 2^{-j}$, $\ell = 1, \ldots, N_j$,

$$\mathcal{Q}_j = \bigcup_{\ell=1}^{2^j} \overline{q}_\ell^j, \quad q_\ell^j := ((\ell-1)2^{-j}, \ell\, 2^{-j}) \quad \text{for } \ell = 1, \ldots, 2^j.$$

A decomposition $\mathcal{Q}_j$ implies an associated trial space of piecewise constant functions,

$$\mathcal{V}_j := S_j^0(\mathcal{Q}) = \text{span}\{\widetilde{\varphi}_\ell^j\}_{\ell=1}^{N_j} \subset L_2(\mathcal{Q}), \quad \dim \mathcal{V}_j = N_j = 2^j,$$

where the basis functions are given as

$$\widetilde{\varphi}_\ell^j(x) = \begin{cases} 1 & \text{for } \xi \in q_\ell^j, \\ 0 & \text{elsewhere.} \end{cases}$$

By construction we have the nested inclusions

$$\mathcal{V}_0 \subset \mathcal{V}_1 \subset \cdots \subset \mathcal{V}_L = S_L^0(\mathcal{Q}) \subset \mathcal{V}_{L+1} \subset \cdots \subset L_2(\mathcal{Q}).$$

For any $j > 0$ we now construct subspaces $\mathcal{W}_j$ as $L_2(\mathcal{Q})$–orthogonal complements of $\mathcal{V}_{j-1}$ in $\mathcal{V}_j$, i.e.

$$\mathcal{W}_0 := \mathcal{V}_0, \quad \mathcal{W}_j := \left\{\widetilde{\varphi}_\ell^j \in \mathcal{V}_j : \langle \widetilde{\varphi}_\ell^j, \widetilde{\varphi}_i^{j-1}\rangle_{L_2(\mathcal{Q})} = 0 \quad \text{for all } \widetilde{\varphi}_i^{j-1} \in \mathcal{V}_{j-1}\right\}$$

where

$$\dim\mathcal{W}_0 = 1, \quad \dim\mathcal{W}_j = \dim\mathcal{V}_j - \dim\mathcal{V}_{j-1} = 2^{j-1} \quad \text{for } j > 0.$$

Hence we obtain a multilevel decomposition of the trial space $\mathcal{V}_L = S_L^0(\mathcal{Q})$ as

$$\mathcal{V}_L = \mathcal{W}_0 \oplus \mathcal{W}_1 \oplus \cdots \oplus \mathcal{W}_L.$$

Due to $\mathcal{W}_0 = \mathcal{V}_0$ we have

$$\mathcal{W}_0 = \text{span}\{\widetilde{\psi}_1^0\} \quad \text{where } \widetilde{\psi}_1^0(\xi) = 1 \quad \text{for } \xi \in \mathcal{Q} = [0,1], \quad \|\widetilde{\psi}_1^0\|_{L_2(\mathcal{Q})} = 1.$$

It remains to construct a basis of

$$\mathcal{W}_j = \text{span}\{\widetilde{\psi}_\ell^j\}_{\ell=1}^{2^{j-1}} \quad \text{for } j = 1, \ldots, L.$$

By using $\dim \mathcal{W}_1 = 1$ we have to determine one basis function only. By setting

$$\widetilde{\psi}_1^1(\xi) = a_1\, \widetilde{\varphi}_1^1(\xi) + a_2\, \widetilde{\varphi}_2^1(\xi) \quad \text{for } \xi \in \mathcal{Q} = [0,1]$$

we obtain from the orthogonality condition

$$0 = \int_{\mathcal{Q}} \widetilde{\psi}_1^0(\xi)\widetilde{\psi}_1^1(\xi)d\xi = \frac{1}{2}(a_1 + a_2)$$

and therefore $a_2 = -a_1$. Hence we can define the basis function as

$$\widetilde{\psi}_1^1(\xi) = \begin{cases} 1 & \text{for } \xi \in q_1^1 = (0, \frac{1}{2}), \\ -1 & \text{for } \xi \in q_2^1 = (\frac{1}{2}, 1), \end{cases} \qquad \|\widetilde{\psi}_1^1\|_{L_2(\mathcal{Q})} = 1.$$

By applying this recursively we obtain the following representation of the basis functions,

$$\widehat{\psi}_\ell^j(\xi) = \begin{cases} 1 & \text{for } \xi \in ((2\ell-2)2^{-j}, (2\ell-1)2^{-j}), \\ -1 & \text{for } \xi \in ((2\ell-1)2^{-j}, 2\ell\, 2^{-j}), \\ 0 & \text{elsewhere} \end{cases}$$

where $\ell = 1, \ldots, 2^{j-1}$, $j = 2, 3, \ldots$. Moreover, we have

$$\|\widehat{\psi}_\ell^j\|_{L_2(\mathcal{Q})}^2 = \left|\text{supp}\, \widehat{\psi}_\ell^j\right| = 2^{1-j} \quad \text{for } j \geq 1.$$

Hence we can define normalized basis functions as

$$\widetilde{\psi}_\ell^j(\xi) = 2^{(j-1)/2}\, \widehat{\psi}_\ell^j(\xi) \quad \text{for } \ell = 1, \ldots, 2^{j-1},\ j \geq 1. \tag{14.35}$$

Due to the orthogonality relation

$$\int_{\mathcal{Q}} \widetilde{\psi}_\ell^j(\xi)\widetilde{\psi}_1^0(\xi)d\xi = 0 \quad \text{for all } \ell = 1, \ldots, 2^{j-1}, j \geq 1$$

we also obtain the moment condition

$$\int_{\mathcal{Q}} \widetilde{\psi}_\ell^j(\xi)d\xi = 0 \quad \text{for all } \ell = 1, \ldots, 2^{j-1}, j \geq 1 \tag{14.36}$$

which holds for piecewise constant wavelets $\widetilde{\psi}_\ell^j$. The basis functions (14.35) as constructed above are also denoted as Haar wavelets, see Fig. 14.4.

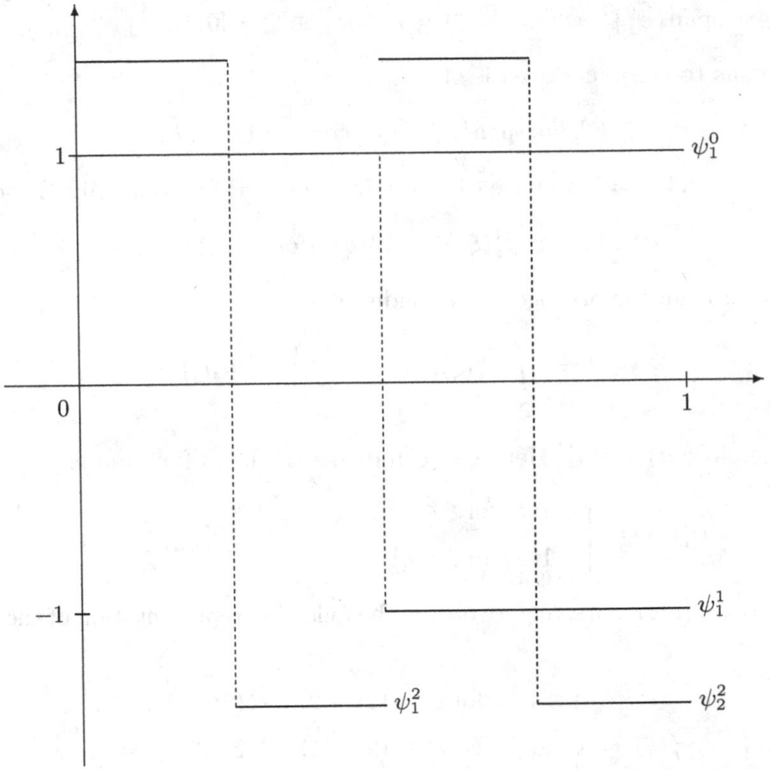


**Fig. 14.4.** Piecewise constant wavelets for $j = 0, 1, 2$.

Via the parametrization $\Gamma = \chi(\mathcal{Q})$ we can also define boundary elements $\tau_\ell^j = \chi(q_\ell^j)$ of the mesh size

$$h_\ell^j = \int_{\tau_\ell^j} ds_x = \int_{q_\ell^j} |\dot{\chi}(\xi)| d\xi = |\dot{\chi}|\, 2^{-j}.$$

The global mesh size of the boundary element mesh $\Gamma_{N_j} = \bigcup_{\ell=1}^{N_j} \overline{\tau}_\ell^j$ is then given as $h_j = |\dot{\chi}|\, 2^{-j}$. Moreover, we can lift both the piecewise constant basis functions $\widetilde{\varphi}_\ell^j \in \mathcal{V}_j$ as well as the wavelets $\widetilde{\psi}_\ell^j \in \mathcal{W}_j$ on the boundary $\Gamma = \partial\Omega$, for $x = \chi(\xi) \in \Gamma$ we have

$$\varphi_\ell^j(x) := \widetilde{\varphi}_\ell^j(\xi), \quad \psi_\ell^j(x) := \widetilde{\psi}_\ell^j(\xi) \quad \text{for } \xi \in \mathcal{Q} = [0, 1].$$

For $j > 0$ these basis functions define the trial spaces

$$V_j = \operatorname{span}\{\varphi_\ell^j\}_{\ell=1}^{2^j}, \quad W_j = \operatorname{span}\{\psi_\ell^j\}_{\ell=1}^{\max\{1,2^{j-1}\}},$$

and for $L \in \mathbb{N}$ we have

$$V_L = S^0_{h_L}(\Gamma) = \operatorname{span}\{\varphi^L_\ell\}_{\ell=1}^{N_L} = \operatorname{span}\{\psi^j_\ell\}_{\ell=1,\ldots,\max\{1,2^{j-1}\},j=1,\ldots,L},$$

i.e. any function $w_{h_L} \in V_L$ can be described as

$$w_{h_L} = \sum_{i=0}^{L} \sum_{k=1}^{\max\{1,2^{i-1}\}} w^i_k \psi^i_k \in V_L. \tag{14.37}$$

Due to

$$\langle \psi^i_k, \psi^j_\ell \rangle_{L_2(\Omega)} = \int_\Gamma \psi^i_k(x)\psi^j_\ell(x) ds_x$$
$$= |\dot{\chi}| \int_{\mathcal{Q}} \widetilde{\psi}^i_k(\xi)\widetilde{\psi}^j_\ell(\xi)d\xi = \begin{cases} |\dot{\chi}| & \text{for } i = j, k = \ell, \\ 0 & \text{elsewhere} \end{cases}$$

the orthogonality of the trial spaces $\mathcal{W}^i_k$ in the parameter domain is transfered to the trial spaces $W^i_k$ which are defined with respect to the boundary element mesh.

By using Remark 13.21 we can derive spectrally equivalent norm representations by means of the multilevel representation (14.37) of a given function $w_{h_L} \in V_L$.

**Lemma 14.16.** *Let $w_{h_L} \in V_L$ be given as in (14.37). Then,*

$$\|w_{h_L}\|^2_{L,s} = \sum_{i=0}^{L} 2^{2si} \sum_{k=1}^{\max\{1,2^{i-1}\}} |w^i_k|^2 .$$

*defines an equivalent norm in $H^s(\Gamma)$ for all $s \in (-\frac{1}{2}, \frac{1}{2})$.*

*Proof.* Since $w_{h_L} \in V_L = S^0_{h_L}(\Gamma)$ is a piecewise constant function, and by using Remark 13.21, the bilinear form of the multilevel operator $B^s$,

$$\langle B^s w_{h_L}, w_{h_L} \rangle_{L_2(\Gamma)} = \sum_{i=0}^{L} h_i^{-2s} \|(Q_i - Q_{i-1})w_{h_L}\|^2_{L_2(\Gamma)},$$

defines an equivalent norm in $H^s(\Gamma)$, $s \in (-\frac{1}{2}, \frac{1}{2})$. Thereby, the $L_2$ projection $Q_i w_{h_L} \in V_i = S^0_{h_i}(\Gamma)$ is the unique solution of the variational problem

$$\langle Q_i w_{h_L}, v_{h_i} \rangle_{L_2(\Gamma)} = \langle w_{h_L}, v_{h_i} \rangle_{L_2(\Gamma)} \quad \text{for all } v_{h_i} \in V_i.$$

By using the orthogonality of basis functions we conclude for the $L_2$ projection the representation

$$Q_i w_{h_L} = \frac{1}{|\dot{\chi}|} \sum_{j=0}^{i} \sum_{\ell=1}^{\max\{1,2^{j-1}\}} \langle w_{h_L}, \psi^j_\ell \rangle_{L_2(\Gamma)} \psi^j_\ell = \sum_{j=0}^{i} \sum_{\ell=1}^{\max\{1,2^{j-1}\}} w^j_\ell \psi^j_\ell .$$

When taking the difference of two succeeding $L_2$ projections this gives

$$(Q_i - Q_{i-1})w_{h_L} \; = \; \sum_{k=1}^{\max\{1,2^{i-1}\}} w_k^i \psi_k^i$$

and therefore

$$\begin{aligned}\langle B^s w_{h_L}, w_{h_L}\rangle_{L_2(\Gamma)} &= \sum_{i=0}^{L} h_i^{-2s} \left\| \sum_{k=1}^{\max\{1,2^{i-1}\}} w_k^i \psi_k^i \right\|_{L_2(\Gamma)}^2 \\ &= |\dot{\chi}| \sum_{i=0}^{L} h_i^{-2s} \sum_{k=1}^{\max\{1,2^{i-1}\}} |w_k^i|^2 .\end{aligned}$$

By inserting the mesh sizes $h_i = |\dot{\chi}|\, 2^{-i}$ this concludes the proof. □

By using the representation (14.37) the variational problem (14.1) is equivalent to

$$\sum_{i=0}^{L} \sum_{k=1}^{\max\{1,2^{i-1}\}} w_k^i \langle V\psi_k^i, \psi_\ell^j \rangle_\Gamma \; = \; \langle g, \psi_\ell^j \rangle_\Gamma \quad \text{for all } \psi_\ell^j \in V_L. \tag{14.38}$$

Hence we have to compute the entries of the stiffness matrix

$$V^L[(\ell,j),(k,i)] \; = \; -\frac{1}{2\pi} \int_\Gamma \psi_\ell^j(x) \int_\Gamma \log|x-y| \psi_k^i(y) ds_y ds_x$$

for $i = 1, \ldots, \max\{1, 2^{\ell-1}\}$, $j = 1, \ldots, \max\{1, 2^{k-1}\}$ and $k, \ell = 0, \ldots, L$. Now we can estimate the matrix entries $V^L[(\ell,j),(k,i)]$ when assuming a certain relation of the supports of the basis functions $\psi_k^i$ and $\psi_\ell^j$. For this we first define the support of $\psi_k^i$ as

$$S_k^i \; := \; \operatorname{supp}\left(\psi_k^i\right) \subset \Gamma,$$

and

$$d_{k\ell}^{ij} \; := \; \operatorname{dist}\left(S_k^i, S_\ell^j\right) \; = \; \min_{(x,y)\in S_k^i \times S_\ell^j} |x-y|$$

describes the distance between the supports of the basis functions $\psi_k^i$ and $\psi_\ell^j$, respectively.

**Lemma 14.17.** *Assume that for $i, j \geq 2$ the condition $d_{k\ell}^{ij} > 0$ is satisfied. Then there holds the estimate*

$$|V^L[(\ell,j),(k,i)]| \; \leq \; \frac{|\dot{\chi}|^4}{2\pi}\, 2^{-3(i+j)/2} \left(d_{k\ell}^{ij}\right)^{-2} .$$

*Proof.* By inserting the parametrization $\Gamma = \chi(\mathcal{Q})$ and by using the normalized basis functions (14.35) the entries of the stiffness matrix $V^L$ can be computed as

$$\begin{aligned} V^L[(\ell,j),(k,i)] &= -\frac{1}{2\pi}\int_\Gamma \psi_\ell^j(x)\int_\Gamma \log|x-y|\psi_k^i(y)ds_y ds_x \\ &= -\frac{1}{2\pi}\int_\mathcal{Q} \widetilde{\psi}_\ell^j(\xi)\int_\mathcal{Q} \log|\chi(\eta)-\chi(\xi)|\,\widetilde{\psi}_k^i(\eta)\,|\dot{\chi}(\eta)|\,d\eta\,|\dot{\chi}(\xi)|\,d\xi \\ &= -\frac{|\dot{\chi}|^2}{2\pi}2^{(i-1)/2}2^{(j-1)/2}\int_\mathcal{Q} \widehat{\psi}_\ell^j(\xi)\int_\mathcal{Q} \log|\chi(\eta)-\chi(\xi)|\,\widehat{\psi}_k^i(\eta)\,d\eta\,d\xi. \end{aligned}$$

With the substitutions

$$\eta = \eta(s) = (2k-2)2^{-i} + s\,2^{1-i}, \quad \xi = \xi(t) = (2\ell-2)2^{-j} + t\,2^{1-j}$$

for $s,t \in \mathcal{Q} = [0,1]$ this is equivalent to

$$V^L[(\ell,j),(k,i)] = -\frac{|\dot{\chi}|^2}{\pi}\,2^{-(i+j)/2}\int_\mathcal{Q} \widehat{\psi}_1^1(\xi(t))\int_\mathcal{Q} k(s,t)\widehat{\psi}_1^1(\eta(s))\,ds\,dt$$

where the kernel function is given by

$$k(s,t) = \log|\chi(\eta(s)) - \chi(\xi(t))|\,.$$

Due to the moment condition (14.36) we can replace the kernel function $k(s,t)$ by $r(s,t) := k(s,t) - P_1(s) - P_2(t)$ where $P_1(s)$ and $P_2(t)$ correspond to the first terms of the Taylor expansion of $k(s,t)$, i.e. $r(s,t)$ corresponds to the remainder of the Taylor expansion. The Taylor expansion of the kernel function $k(s,t)$ with respect to $s_0 = \frac{1}{2}$ gives

$$k(s,t) = k\left(\frac{1}{2},t\right) + \left(s-\frac{1}{2}\right)\left[\frac{\partial}{\partial s}k(s,t)\right]_{s=\bar{s}}$$

with a suitable $\bar{s} \in \mathcal{Q}$. Applying another Taylor expansion with respect to $t_0 = \frac{1}{2}$ we obtain

$$\left[\frac{\partial}{\partial s}k(s,t)\right]_{s=\bar{s}} = \left[\frac{\partial}{\partial s}k\left(s,\frac{1}{2}\right)\right]_{s=\bar{s}} + \left(t-\frac{1}{2}\right)\left[\frac{\partial^2}{\partial s\partial t}k(s,t)\right]_{(s,t)=(\bar{s},\bar{t})}$$

where $\bar{t} \in \mathcal{Q}$. Hence we have

$$\begin{aligned} k(s,t) - k\left(\frac{1}{2},t\right) - \left(s-\frac{1}{2}\right)\left[\frac{\partial}{\partial s}k\left(s,\frac{1}{2}\right)\right]_{s=\bar{s}} \\ = \left(s-\frac{1}{2}\right)\left(t-\frac{1}{2}\right)\left[\frac{\partial^2}{\partial s\partial t}k(s,t)\right]_{(s,t)=(\bar{s},\bar{t})}. \end{aligned}$$

Due to the moment condition (14.36) we conclude

$$\int_{\mathcal{Q}} \widehat{\psi}_1^1(\xi(t)) \left(s - \frac{1}{2}\right) \left[\frac{\partial}{\partial s} k\left(s, \frac{1}{2}\right)\right]_{s=\bar{s}} dt \;=\; \int_{\mathcal{Q}} \widehat{\psi}_1^1(\eta(s)) k\left(\frac{1}{2}, t\right) ds \;=\; 0\,.$$

Hence we obtain

$$\begin{aligned} V^L[(\ell,j),(k,i)] &= -\frac{|\dot{\chi}|^2}{\pi}\, 2^{-(i+j)/2} \int_{\mathcal{Q}} \widehat{\psi}_1^1(\xi(t)) \int_{\mathcal{Q}} k(s,t)\widehat{\psi}_1^1(\eta(s))\, ds\, dt \\ &= -\frac{|\dot{\chi}|^2}{\pi}\, 2^{-(i+j)/2} \int_{\mathcal{Q}} \widehat{\psi}_1^1(\xi(t)) \int_{\mathcal{Q}} \left(s - \frac{1}{2}\right)\left(t - \frac{1}{2}\right) \cdot \\ &\qquad \cdot \left[\frac{\partial^2}{\partial s \partial t} k(s,t)\right]_{(s,t)=(\bar{s},\bar{t})} \widehat{\psi}_1^1(\eta(s))\, ds\, dt \end{aligned}$$

and therefore

$$\left|V^L[(\ell,j),(k,i)]\right| \;\leq\; \frac{|\dot{\chi}|^2}{\pi}\, 2^{-(i+j)/2}\, \frac{1}{16} \max_{(s,t)\in\mathcal{Q}\times\mathcal{Q}} \left|\frac{\partial^2}{\partial s\partial t} k(s,t)\right|\,.$$

By applying the chain rule we further have

$$\frac{\partial^2}{\partial s \partial t} k(s,t) \;=\; 2^{1-i} 2^{1-j} \frac{\partial^2}{\partial \eta \partial \xi} \log|\chi(\eta) - \chi(\xi)|\,.$$

Moreover,

$$\frac{\partial}{\partial \eta} \log|\chi(\eta) - \chi(\xi)| \;=\; \sum_{i=1}^{2} \frac{\partial}{\partial y_i} \log|y - x(\xi)|_{|y=\chi(\eta)} \frac{\partial}{\partial \eta}\chi_j(\eta),$$

as well as

$$\frac{\partial^2}{\partial \eta\partial\xi} \log|\chi(\eta) - \chi(\xi)| \;=\; \sum_{i,j=1}^{2} \frac{\partial^2}{\partial y_i \partial x_j} \log|y - x|_{|y=\chi(\eta),x=\chi(\xi)} \frac{\partial}{\partial\eta}\chi_i(\eta) \frac{\partial}{\partial \xi}\chi_i(\xi)\,.$$

Applying the Cauchy–Schwarz inequality twice this gives

$$\left|\frac{\partial^2}{\partial\eta\partial\xi} \log|\chi(\eta) - \chi(\xi)|\right| \;\leq\; |\dot{\chi}|^2 \left(\sum_{i,j=1}^{2} \left[\frac{\partial^2}{\partial y_i \partial x_j} \log|y - x|_{|y=\chi(\eta),x=\chi(\xi)}\right]^2\right)^{1/2}.$$

By using

$$\begin{aligned} \frac{\partial^2}{\partial x_i \partial y_i} \log|x - y| &= -\frac{1}{|x-y|^2} + 2\frac{(x_i - y_i)^2}{|x-y|^4}, \\ \frac{\partial^2}{\partial x_1 \partial y_2} \log|x - y| &= 2\frac{(x_1 - y_1)(x_2 - y_2)}{|x-y|^4} \end{aligned}$$

we finally obtain

$$\max_{(s,t)\in\mathcal{Q}\times\mathcal{Q}}\left|\frac{\partial^2}{\partial s\partial t}k(s,t)\right| \le 2^{2-(i+j)}\,|\dot{\chi}|^2 \max_{(x,y)\in S_k^i\times S_\ell^j}\frac{2}{|x-y|^2}. \qquad\square$$

The estimate of Lemma 14.17 describes the decay of the matrix entries $V^L[(\ell,j),(k,i)]$ when considering wavelets $\psi_k^i$ and $\psi_\ell^j$ with supports $S_k^i$ and $S_\ell^j$ which are far to each other. By defining an appropriate compression parameter we therefore can characterize matrix entries $V^L[(\ell,j),(k,i)]$ which can be neglected when computing the stiffness matrix $V^L$. For real valued parameter $\alpha,\kappa\ge 1$ we first define a symmetric parameter matrix by

$$\tau_{ij} := \alpha\,2^{\kappa L-i-j}.$$

This enables the definition of a symmetric approximation $\widetilde{V}^L$ of the stiffness matrix $V^L$ by

$$\widetilde{V}^L[(\ell,j),(k,i)] := \begin{cases} V^L[(\ell,j),(k,i)] & \text{if } d_{k\ell}^{ij}\le\tau_{ij}, \\ 0 & \text{elsewhere.}\end{cases} \tag{14.39}$$

For the following considerations, in particular to estimate the number of nonzero elements of the matrix $\widetilde{V}^L$ as well as for the related stability and error analysis we define for fixed $i,j\ge 2$ block matrices

$$V_{ij}^L := \left(V^L[(\ell,j),(k,i)]\right)_{k=1,\ldots,2^{i-1},\ell=1,\ldots,2^{j-1}},$$

and the corresponding approximation $\widetilde{V}_{ij}^L$, respectively.

**Lemma 14.18.** *The number of nonzero elements of the approximated stiffness matrix $\widetilde{V}^L$ as defined in (14.39) is $\mathcal{O}(N^\kappa(\log_2 N)^2)$.*

*Proof.* For $i=0,1$ and $j=0,1$ the number of nonzero elements is $4(N-1)$. For $i,j\ge 2$ we estimate the number of nonzero elements of the approximate block matrix $\widetilde{V}_{ij}^L$ as follows.

By using the parametrization $\Gamma=\chi(\mathcal{Q})$ we can identify the basis functions $\psi_k^i$ and $\psi_\ell^j$ with basis functions $\widetilde{\psi}_k^i$ and $\widetilde{\psi}_\ell^j$ which are defined in the parameter domain $\mathcal{Q}$. For the support of the basis functions $\widetilde{\psi}_k^i$ and $\widetilde{\psi}_\ell^j$ we then obtain

$$\widetilde{S}_k^i = ((2(k-1)2^{-i}, 2k\,2^{-i}), \quad \widetilde{S}_\ell^j = (2(\ell-1)2^{-j}, 2\ell\,2^{-j}).$$

For an arbitrary but fixed $\ell=1,\ldots,2^{j-1}$ we first determine all basis functions $\widetilde{\psi}_k^i$ where the supports $\widetilde{S}_k^i$ and $\widetilde{S}_\ell^j$ do not overlap, i.e.

$$2k\,2^{-i}\le 2(\ell-1)2^{-j}, \quad 2\ell\,2^{-j}\le 2(k-1)2^{-i}.$$

Then it follows that

$$1 \le k \le (\ell-1)2^{i-j}, \quad 1+\ell\, 2^{i-j} \le k \le 2^{i-1}.$$

Next we choose from the above those basis functions $\widetilde{\psi}_i^k$ which in addition satisfy the distance condition

$$d_{k\ell}^{ij} \le \tau_{ij} = \alpha\, 2^{\kappa L-i-j}.$$

Due to assumption (14.34) we have $d_{k\ell}^{ij} \le c_2^X \operatorname{dist}(\widetilde{S}_k^i, \widetilde{S}_\ell^j)$. By requesting

$$c_2^X \operatorname{dist}(\widetilde{S}_k^i, \widetilde{S}_\ell^j) \le \alpha\, 2^{\kappa L-i-j}$$

we can find an upper bound for the number of related basis functions. By using

$$0 < \operatorname{dist}(\widetilde{S}_k^i, \widetilde{S}_\ell^j) = 2(k-1)2^{-i} - 2\ell\, 2^{-j} \le \frac{\alpha}{c_2^X}\, 2^{\kappa L-i-j}$$

for $\widetilde{S}_\ell^j \ni \eta < \xi \in \widetilde{S}_k^i$ we obtain the estimate

$$k \le 1+\ell 2^{i-j} + \frac{\alpha}{2c_2^X}\, 2^{\kappa L-j}$$

and therefore

$$1+\ell\, 2^{i-j} \le k \le \min\left\{2^{i-1}, 1+\ell 2^{i-j} + \frac{\alpha}{2c_2^X} 2^{\kappa L-j}\right\}.$$

Hence we can estimate the number of related nonzero elements as

$$\frac{\alpha}{2c_2^X}\, 2^{\kappa L-j}.$$

This results follows analogously in the case $\widetilde{S}_k^i \ni \xi < \eta \in \widetilde{S}_\ell^j$. Thus, for a fixed $\ell = 1, \ldots, 2^{j-1}$ there exist maximal

$$\frac{\alpha}{c_2^X}\, 2^{\kappa L-j}$$

nonzero elements. The number of nonzero elements of the approximate block matrix $\widetilde{V}_{ij}^L$ is therefore bounded by

$$2^{j-1} \frac{\alpha}{c_2^X}\, 2^{\kappa L-j} = \frac{\alpha}{2c_2^X}\, 2^{\kappa L}.$$

By taking the sum over all block matrices $\widetilde{V}_{ij}^L$ where $k, \ell = 2, \ldots, L$ we can estimate, by taking into account the special situation for $i, j = 0, 1$, the number of nonzero elements of $\widetilde{V}^L$ by

$$4(N-1) + (L-1)^2 \frac{\alpha}{2c_2^X}\, 2^{\kappa L}.$$

A similar estimate follows when considering basis functions with overlapping supports. When inserting $N = 2^L$ or $L = \log_2 N$ this concludes the proof. □

To estimate the approximation error $\|V^L - \widetilde{V}^L\|$ of the stiffness matrix $V^L$ we first consider the approximation errors $\|V_{ij}^L - \widetilde{V}_{ij}^L\|$ of the block matrices $V_{ij}^L$.

**Lemma 14.19.** *For $i, j = 2, \ldots, L$ let $V_{ij}^L$ be the exact Galerkin stiffness matrix of the single layer potential $V$ with respect to the trial spaces $W_i$ and $W_j$. Let $\widetilde{V}_{ij}^L$ be the approximation as defined in (14.39). Then there hold the error estimates*

$$\|V_{ij}^L - \widetilde{V}_{ij}^L\|_\infty \leq c_1\, 2^{-(i+j)/2}\, 2^{-j}\, (\tau_{ij})^{-1},$$

$$\|V_{ij}^L - \widetilde{V}_{ij}^L\|_1 \leq c_2\, 2^{-(i+j)/2}\, 2^{-i}\, (\tau_{ij})^{-1}.$$

*Proof.* By using Lemma 14.17 we first have

$$\begin{aligned}\|V_{ij}^L - \widetilde{V}_{ij}^L\|_\infty &= \max_{\ell=1,\ldots,2^{j-1}} \sum_{k=1}^{2^{i-1}} \left| V^L[(\ell,j),(k,i)] - \widetilde{V}^L[(\ell,j),(k,i)] \right| \\ &= \max_{\ell=1,\ldots,2^{j-1}} \sum_{\substack{k=1 \\ d_{k\ell}^{ij} > \tau_{ij}}}^{2^{i-1}} \left| V^L[(\ell,j),(k,i)] \right| \\ &\leq c\, 2^{-3(i+j)/2} \max_{\ell=1,\ldots,2^{j-1}} \sum_{\substack{k=1 \\ d_{k\ell}^{ij} > \tau_{ij}}}^{2^{i-1}} (d_{k\ell}^{ij})^{-2}.\end{aligned}$$

Since we assume (14.34) to be satisfied for the parametrization of the boundary $\Gamma$ we then conclude

$$\|V_{ij}^L - \widetilde{V}_{ij}^L\|_\infty \leq \widetilde{c}\, 2^{-3(i+j)/2} \max_{\ell=1,\ldots,2^{j-1}} \sum_{\substack{k=1 \\ c_1^\chi \mathrm{dist}(\widetilde{S}_k^i, \widetilde{S}_\ell^j) > \tau_{ij}}}^{2^{i-1}} (\mathrm{dist}(\widetilde{S}_k^i, \widetilde{S}_\ell^j))^{-2}.$$

For an arbitrary but fixed $\ell = 1, \ldots, 2^{j-1}$ the sum can be further estimated by

$$\begin{aligned}\sum_{\substack{k=1 \\ c_1^\chi \mathrm{dist}(\widetilde{S}_k^i, \widetilde{S}_\ell^j) > \tau_{ij}}}^{2^{i-1}} (\mathrm{dist}(\widetilde{S}_k^i, \widetilde{S}_\ell^j))^{-2} &\leq 2 \sum_{k > 1 + \frac{\alpha}{2c_1^\chi} 2^{\kappa L - j} + \ell 2^{i-j}} \left(2(k-1)2^{-i} - 2\ell 2^{-j}\right)^{-2} \\ &= 2^{2i-1} \sum_{n > \frac{\alpha}{2c_1^\chi} 2^{\kappa L - j}} \frac{1}{n^2}.\end{aligned}$$

Let $n_1 \in \mathbb{N}$ be the smallest number satisfying $n_1 \geq \frac{\alpha}{2c_1^\chi} 2^{\kappa L - j}$. Then we have

$$\sum_{n=n_1}^\infty \frac{1}{n^2} = \frac{1}{n_1^2} + \sum_{n=n_1+1}^\infty \frac{1}{n^2} \leq \frac{1}{n_1^2} + \int_{x=n_1}^\infty \frac{1}{x^2} dx = \frac{1}{n_1^2} + \frac{1}{n_1} \leq \frac{2}{n_1} \leq \frac{4c_1^\chi}{\alpha} 2^{j-\kappa L}$$

which gives immediately the first estimate. The second estimate follows in the same way. □

By applying Lemma 13.17 (Schur Lemma) we can now estimate the error of the approximation $\widetilde{V}^L$.

**Theorem 14.20.** *For $w_h, v_h \in V_L = S^0_{h_L}(\Gamma) \leftrightarrow \underline{w}, \underline{v} \in \mathbb{R}^N$ there holds the error estimate*

$$\left|((V^L - \widetilde{V}^L)\underline{w}, \underline{v})\right| \le c\gamma(h_L, \sigma_1, \sigma_2)\, \|w_h\|_{H^{\sigma_1}(\Gamma)} \|v_h\|_{H^{\sigma_2}(\Gamma)}$$

*where*

$$\gamma(h_L, \sigma_1, \sigma_2) := \begin{cases} h_L^{\kappa+\sigma_1+\sigma_2} & \text{for } \sigma_1, \sigma_2 \in (-\frac{1}{2}, 0), \\ h_L^{\kappa}\, |\ln h_L| & \text{for } \sigma_1 = \sigma_2 = 0, \\ h_L^{\kappa} & \text{for } \sigma_1, \sigma_2 \in (0, \frac{1}{2}), \\ h_L^{\kappa+\sigma_2} & \text{for } \sigma_1 \in (0, \frac{1}{2}), \sigma_2 \in (-\frac{1}{2}, 0). \end{cases}$$

*Proof.* First we note that for $i = 0, 1$ and $j = 0, 1$ there is no approximation of the matrix entries of $V^L$. Then, by using the Cauchy–Schwarz inequality and Lemma 14.16 we obtain

$$\begin{aligned} &\left|((V^L - \widetilde{V}^L)\underline{w}, \underline{v})\right| \\ &= \left| \sum_{i=2}^{L} \sum_{j=2}^{L} \sum_{k=1}^{2^{i-1}} \sum_{\ell=1}^{2^{j-1}} w_k^i v_\ell^j \left[ V^L[(\ell, j) - (k, i)] - \widetilde{V}^L[(\ell, j), (k, i)] \right] \right| \\ &= \left| \sum_{i=2}^{L} \sum_{j=2}^{L} \sum_{k=1}^{2^{i-1}} \sum_{\ell=1}^{2^{j-1}} 2^{\sigma_1 i} w_k^i 2^{\sigma_2 j} v_\ell^j 2^{-\sigma_1 i - \sigma_2 j} \right. . \\ &\qquad \left. \cdot \left[ V^L[(\ell, j) - (k, i)] - \widetilde{V}^L[(\ell, j), (k, i)] \right] \right| \\ &\le \|A\|_2 \left( \sum_{i=2}^{L} \sum_{k=1}^{2^{i-1}} 2^{2\sigma_1 i} \left| w_k^i \right|^2 \right)^{1/2} \left( \sum_{j=2}^{L} \sum_{\ell=1}^{2^{j-1}} 2^{2\sigma_2 j} \left| v_\ell^j \right|^2 \right)^{1/2} \\ &\le \|A\|_2 \|w_h\|_{L,\sigma_1} \|v_h\|_{L,\sigma_2} \\ &\le \|A\|_2 \|w_h\|_{H^{\sigma_1}(\Gamma)} \|v_h\|_{H^{\sigma_2}(\Gamma)} \end{aligned}$$

where the matrix $A$ is defined by

$$A[(\ell, j), (k, i)] = 2^{-\sigma_1 i - \sigma_2 j} \left[ V^L[(\ell, j) - (k, i)] - \widetilde{V}^L[(\ell, j), (k, i)] \right].$$

By using (13.17) we now can estimate the spectral norm $\|A\|_2$ for an arbitrary $s$ as

$$\|A\|_2^2 \leq \sup_{\substack{j=2,\ldots,L\\ \ell=1,\ldots,2^{j-1}}} \sum_{i=2}^{L} \sum_{k=1}^{2^{i-1}} |A[(\ell,j),(k,i)]| 2^{s(i-j)} \,.$$

$$\cdot \sup_{\substack{i=2,\ldots,L\\ k=1,\ldots,2^{i-1}}} \sum_{j=2}^{L} \sum_{\ell=1}^{2^{j-1}} |A[(\ell,j),(k,i)]| 2^{s(j-i)}.$$

By using Lemma 14.19 we can bound the first term by

$$\begin{aligned}
A_1 &= \sup_{\substack{j=2,\ldots,L\\ \ell=1,\ldots,2^{j-1}}} \sum_{i=2}^{L} \sum_{k=1}^{2^{i-1}} \left| V^L[(\ell,j),(k,i)] - \widetilde{V}^L[(\ell,j),(k,i)] \right| 2^{s(i-j)} 2^{-\sigma_1 i - \sigma_2 j} \\
&= \sup_{j=2,\ldots,L} \sum_{i=2}^{L} 2^{s(i-j)} 2^{-\sigma_1 i - \sigma_2 j} \sup_{\ell=1,\ldots,2^{j-1}} \sum_{k=1}^{2^{i-1}} \left| V^L[(\ell,j),(k,i)] - \widetilde{V}^L[(\ell,j),(k,i)] \right| \\
&= \sup_{j=2,\ldots,L} \sum_{i=2}^{L} 2^{s(i-j)} 2^{-\sigma_1 i - \sigma_2 j} \|V_{ij}^L - \widetilde{V}_{ij}^L\|_\infty \\
&\leq c \sup_{j=2,\ldots,L} \sum_{i=2}^{L} 2^{s(i-j)} \, 2^{-\sigma_1 i - \sigma_2 j} \, 2^{-(i+j)/2} \, 2^{-j} \, (\tau_{ij})^{-1} \\
&= c \sup_{j=2,\ldots,L} \sum_{i=2}^{L} 2^{s(i-j)} \, 2^{-\sigma_1 i - \sigma_2 j} \, 2^{-(i+j)/2} \, 2^{-j} \frac{1}{\alpha} 2^{i+j-\kappa L} \\
&= \widetilde{c} 2^{-\kappa L} \sup_{j=2,\ldots,L} 2^{j(-s-\frac{1}{2}-1+1-\sigma_2)} \sum_{i=2}^{L} 2^{i(s-\frac{1}{2}+1-\sigma_1)} \\
&= \widetilde{c} 2^{-\kappa L} \sup_{j=2,\ldots,L} 2^{-\sigma_2 j} \sum_{i=2}^{L} 2^{-\sigma_1 i}
\end{aligned}$$

where $s = -\frac{1}{2}$. Note that

$$\sum_{i=2}^{L} 2^{-\sigma_1 i} \leq c \begin{cases} 2^{-\sigma_1 L} & \text{for } \sigma_1 \in (-\frac{1}{2}, 0), \\ L & \text{for } \sigma_1 = 0, \\ 1 & \text{for } \sigma_1 \in (0, \frac{1}{2}) \end{cases}$$

and

$$\sup_{j=2,\ldots,L} 2^{-\sigma_2 j} \leq c \begin{cases} 2^{-\sigma_2 L} & \text{for } \sigma_2 \in (-\frac{1}{2}, 0), \\ 1 & \text{for } \sigma_2 \in [0, \frac{1}{2}). \end{cases}$$

To estimate the second term we proceed in an analogous way, and inserting $h_L = 2^{-L}$ finally gives the announced error estimate. □

By using Theorem 8.3 (Strang Lemma) and Theorem 14.20 we can now derive the stability and error analysis of the approximated stiffness matrix $\widetilde{V}^L$.

For this we first need to establish the positive definiteness of the approximated stiffness matrix $\widetilde{V}^L$.

**Theorem 14.21.** *For $\kappa > 1$ and for a sufficient small global boundary element mesh size $h_L$ the approximated stiffness matrix $\widetilde{V}^L$ is positive definite, i.e.*

$$(\widetilde{V}^L \underline{w}, \underline{w}) \geq \frac{1}{2} c_1^V \, \|w_{h_L}\|^2_{H^{-1/2}(\Gamma)}$$

*is satisfied for all $w_{h_L} \in V_L \leftrightarrow \underline{w} \in \mathbb{R}^N$.*

*Proof.* For $\sigma \in (-\frac{1}{2}, 0)$ we have by using Theorem 14.20, the $H^{-1/2}(\Gamma)$–ellipticity of the single layer potential $V$, and by applying the inverse inequality in $V_L$

$$\begin{aligned}
(\widetilde{V}^L \underline{w}, \underline{w}) &= (V^L \underline{w}, \underline{w}) + ((\widetilde{V}^L - V^L)\underline{w}, \underline{w}) \\
&\geq \langle V w_{h_L}, w_{h_L} \rangle_\Gamma - |((\widetilde{V}^L - V^L)\underline{w}, \underline{w})| \\
&\geq c_1^V \, \|w_{h_L}\|^2_{H^{-1/2}(\Gamma)} - c \, h_L^{\kappa+2\sigma} \, \|w_{h_L}\|^2_{H^\sigma(\Gamma)} \\
&\geq c_1^V \, \|w_{h_L}\|^2_{H^{-1/2}(\Gamma)} - c \, h_L^{\kappa+2\sigma} \, c_I \, h_L^{-1-2\sigma} \, \|w_{h_L}\|^2_{H^{-1/2}(\Gamma)} \\
&= \left[c_1^V - \widetilde{c} \, h_L^{\kappa-1}\right] \|w_h\|^2_{H^{-1/2}(\Gamma)}.
\end{aligned}$$

Now, if $\widetilde{c} h_L^{\kappa-1} \leq \frac{1}{2} c_1^V$ is satisfied the assertion follows. □

Instead of the linear system $V^L \underline{w} = \underline{f}$ which corresponds to the variational problem (14.38) we now have to solve the perturbed linear system $\widetilde{V}^L \underline{\widetilde{w}} = \widetilde{f}$ where $\underline{\widetilde{w}} \in \mathbb{R}^N \leftrightarrow \widetilde{w}_{h_L} \in V_L$ defines the associated approximate solution.

**Theorem 14.22.** *Let $w \in H^1_{pw}(\Gamma)$ be the unique solution of the boundary integral equation $Vw = g$. For the approximate solution $\widetilde{w}_{h_L} \in V_L \leftrightarrow \underline{\widetilde{w}} \in \mathbb{R}^N$ of the perturbed linear system $\widetilde{V}^L \underline{\widetilde{w}} = \underline{f}$ there holds the error estimate*

$$\|w - \widetilde{w}_{h_L}\|_{H^{-1/2}(\Gamma)} \leq c_1 \, h_L^{3/2} \, \|w\|_{H^1_{pw}(\Gamma)} + c_2 \, h_L^{\kappa-1/2} \, \|w\|_{H^{1/2}(\Gamma)}.$$

*Proof.* The solutions $\underline{w}, \underline{\widetilde{w}} \in \mathbb{R}^N$ of the linear systems $V^L \underline{w} = \underline{f}$ and $\widetilde{V}^L \underline{\widetilde{w}} = \underline{f}$ satisfy the orthogonality relation

$$(V^L \underline{w} - \widetilde{V}^L \underline{\widetilde{w}}, \underline{v}) = 0 \quad \text{for all } \underline{v} \in \mathbb{R}^N.$$

By using the positive definiteness of the approximated stiffness matrix $\widetilde{V}^L$ we then obtain

$$\begin{aligned}
\frac{1}{2} c_1^V \, \|w_{h_L} - \widetilde{w}_{h_L}\|^2_{H^{-1/2}(\Gamma)} &\leq (\widetilde{V}^L(\underline{w} - \underline{\widetilde{w}}), \underline{w} - \underline{\widetilde{w}}) \\
&= ((\widetilde{V}^L - V^L)\underline{w}, \underline{w} - \underline{\widetilde{w}}).
\end{aligned}$$

By applying Theorem 14.20 we conclude for $\sigma_1 \in (0, \frac{1}{2})$ and $\sigma_2 \in (-\frac{1}{2}, 0)$

$$\frac{1}{2}c_1^V \, \|w_{h_L} - \widetilde{w}_{h_L}\|_{H^{-1/2}(\Gamma)} \leq c\, h_L^{\kappa+\sigma_2} \, \|w_{h_L}\|_{H^{\sigma_1}(\Gamma)} \|w_{h_L} - \widetilde{w}_{h_L}\|_{H^{\sigma_2}(\Gamma)}.$$

Further we have as in Lemma 12.2

$$\|w_{h_L}\|_{H^{\sigma_1}(\Gamma)} \leq \|w\|_{H^{\sigma_1}(\Gamma)} + \|w - w_{h_L}\|_{H^{\sigma_1}(\Gamma)} \leq c\,\|w\|_{H^{\sigma_1}(\Gamma)} \leq c\,\|w\|_{H^{1/2}(\Gamma)}.$$

On the other hand, by using the inverse inequality this gives

$$\|w_{h_L} - \widetilde{w}_{h_L}\|_{H^{\sigma_2}(\Gamma)} \leq c_I \, h_L^{-\frac{1}{2}-\sigma_2} \, \|w_{h_L} - \widetilde{w}_{h_L}\|_{H^{-1/2}(\Gamma)}.$$

Hence we have

$$\frac{1}{2}c_1^V \, \|w_{h_L} - \widetilde{w}_{h_L}\|_{H^{-1/2}(\Gamma)} \leq c\, h_L^{\kappa-1/2} \, \|w\|_{H^{1/2}(\Gamma)}.$$

Now the assertion follows from applying the triangle inequality

$$\|w - \widetilde{w}_{h_L}\|_{H^{-1/2}(\Gamma)} \leq \|w - w_{h_L}\|_{H^{-1/2}(\Gamma)} + \|w_{h_L} - \widetilde{w}_{h_L}\|_{H^{-1/2}(\Gamma)}$$

and by using the error estimate (14.3). □

*Remark 14.23.* The error estimate of Theorem 14.22 is not optimal with respect to the regularity of the solution $w \in H^1_{\text{pw}}(\Gamma)$. Since Lemma 14.16 is only valid for $s \in (-\frac{1}{2}, \frac{1}{2})$ the higher regularity $w \in H^1_{\text{pw}}(\Gamma)$ is not recognized in the error estimate. Formally, this yields the error estimate

$$\|w - \widetilde{w}_{h_L}\|_{H^{-1/2}(\Gamma)} \leq c \left[ h_L^{3/2} + h_L^{\kappa} \right] \|w\|_{H^1_{\text{pw}}(\Gamma)}.$$

When summarizing the results of Lemma 14.18 on the numerical amount of work and the error estimate of Remark 14.23 we have to notice that it is not possible to choose the compression parameter $\kappa \geq 1$ in (14.39) in an optimal way. In particular for $\kappa = \frac{3}{2}$ we obtain in Remark 14.23 the same asymptotic accuracy as in the error estimate (14.3) of the standard Galerkin boundary element method, but the number of nonzero elements of the stiffness matrix $\widetilde{V}^L$ is $\mathcal{O}(N_L^{3/2}(\log_2 N)^2)$ and therefore not optimal. On the other hand, by choosing $\kappa = 1$ we would obtain $\mathcal{O}(N_L(\log_2 N)^2)$ nonzero elements, but for a sufficient regular solution $w \in H^1_{\text{pw}}(\Gamma)$ we will lose accuracy. The theoretical background of this behavior is given in the proof of Lemma 14.17. There, the moment condition (14.36) is used for piecewise constant wavelets, i.e. they are orthogonal on constant functions. To obtain a higher order of approximation we therefore have to require higher order moment conditions, e.g. orthogonality with respect to linear functions. This can be ensured when using piecewise linear wavelets [76] but their construction is a quite challenging task.

## 14.4 Exercises

**14.1** Consider the finite element stiffness matrix of Exercise 11.1 for $h = 1/9$,

$$K_h = 9 \begin{pmatrix} 2 & -1 & & & & & & \\ -1 & 2 & -1 & & & & & \\ & -1 & 2 & -1 & & & & \\ & & -1 & 2 & -1 & & & \\ & & & -1 & 2 & -1 & & \\ & & & & -1 & 2 & -1 & \\ & & & & & -1 & 2 & -1 \\ & & & & & & -1 & 2 \end{pmatrix}.$$

Write $K_h$ as a hierarchical matrix and compute the inverse $K_h^{-1}$ as a hierarchical matrix.

**14.2.** The solution of the Dirichlet boundary value problem

$$-u''(x) = f(x) \quad \text{for } x \in (0,1), \quad u(0) = u(1) = 0$$

is given by

$$u(x) = (Nf)(x) = \int_0^1 G(x,y)f(y)dy \quad \text{for } x \in (0,1)$$

where $G(x,y)$ is the associated Green function, cf. Exercise 5.2. Discuss the Galerkin discretization of $Nf$ when using piecewise linear continuous basis functions with respect to a uniform decomposition of $(0,1)$.

# 15

# Domain Decomposition Methods

Domain decomposition methods are a modern numerical tool to handle partial differential equations with jumping coefficients, and to couple different discretization methods such as finite and boundary element methods [43]. Moreover, domain decomposition methods allow the derivation and parallelization of efficient solution strategies [95] in a natural setting. For a more detailed study of domain decomposition methods we refer, for example, to [18, 68, 114, 139, 149].

As a model problem we consider the potential equation

$$\begin{aligned} -\operatorname{div}[\alpha(x)\nabla u(x)] &= 0 && \text{for } x \in \Omega, \\ \gamma_0^{\text{int}} u(x) &= g(x) && \text{for } x \in \Gamma = \partial\Omega \end{aligned} \tag{15.1}$$

where $\Omega \subset \mathbb{R}^d$ is a bounded Lipschitz domain for which a non–overlapping domain decomposition is given, see Fig. 15.1,

$$\overline{\Omega} = \bigcup_{i=1}^{p} \overline{\Omega}_i, \quad \Omega_i \cap \Omega_j = \emptyset \quad \text{for } i \neq j. \tag{15.2}$$

The subdomains $\Omega_i$ are assumed to be Lipschitz with boundaries $\Gamma_i = \partial\Omega_i$. By

$$\Gamma_S := \bigcup_{i=1}^{p} \Gamma_i$$

we denote the skeleton of the domain decomposition.

We assume that in (15.1) the coefficient $\alpha(x)$ is piecewise constant, i.e.

$$\alpha(x) = \alpha_i \quad \text{for } x \in \Omega_i, \quad i = 1, \ldots, p. \tag{15.3}$$

For an approximate solution of the boundary value problem (15.1) we will use a boundary element method within the subdomains $\Omega_1, \ldots, \Omega_q$ while for the remaining subdomains $\Omega_{q+1}, \ldots, \Omega_p$ a finite element method will be applied.

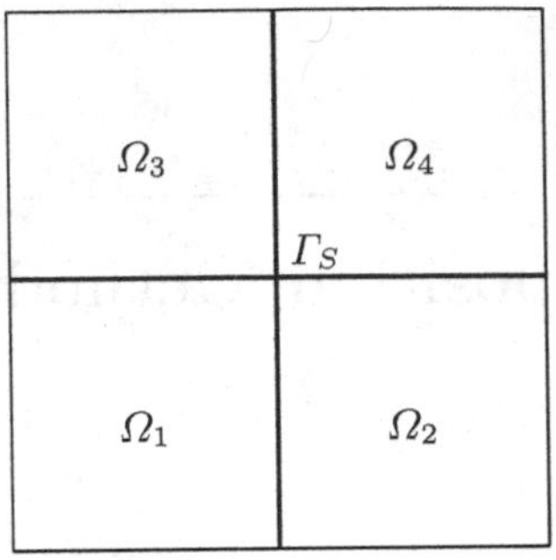


**Fig. 15.1.** Domain decomposition with four subdomains.

By using the results of Chapter 4 the variational formulation of the Dirichlet boundary value problem (15.1) is to find $u \in H^1(\Omega)$ with $\gamma_0^{\text{int}} u = g$ such that

$$\int_\Omega \alpha(x) \nabla u(x) \nabla v(x) dx = 0 \qquad (15.4)$$

is satisfied for all $v \in H_0^1(\Omega)$.

Due to the non–overlapping domain decomposition (15.2) and by using assumption (15.3) the variational formulation (15.4) is equivalent to

$$\sum_{i=1}^{p} \alpha_i \int_{\Omega_i} \nabla u(x) \nabla v(x) dx = 0 \quad \text{for all } v \in H_0^1(\Omega).$$

The application of Green's first formula (1.5) with respect to the subdomains $\Omega_i$ for $i = 1, \ldots, q \leq p$ results in a variational problem to find $u \in H^1(\Omega)$ with $\gamma_0^{\text{int}} u = g$ such that

$$\sum_{i=1}^{q} \alpha_i \int_{\partial\Omega_i} \gamma_{1,i}^{\text{int}} u(x) \gamma_{0,i}^{\text{int}} v(x) ds_x + \sum_{i=q+1}^{p} \alpha_i \int_{\Omega_i} \nabla u(x) \nabla v(x) dx = 0 \qquad (15.5)$$

is satisfied for all $v \in H_0^1(\Omega)$.

The Cauchy data $\gamma_{0,i}^{\text{int}} u$ and $\gamma_{1,i}^{\text{int}} u$ of the solution $u$ are solutions of the boundary integral equations (6.22) on $\Gamma_i = \partial\Omega_i$, $i = 1, \ldots, q$, i.e.

$$\begin{pmatrix} \gamma_{0,i}^{\text{int}} u \\ \gamma_{1,i}^{\text{int}} u \end{pmatrix} = \begin{pmatrix} \frac{1}{2} I - K_i & V_i \\ D_i & \frac{1}{2} I + K_i' \end{pmatrix} \begin{pmatrix} \gamma_{0,i}^{\text{int}} u \\ \gamma_{1,i}^{\text{int}} u \end{pmatrix}. \qquad (15.6)$$

Inserting the second equation of (15.6) into the variational formulation (15.5) this results in the variational problem to find $u \in H^1(\Omega)$ with $\gamma_0^{\text{int}} u = g$ and $\gamma_{1,i}^{\text{int}} u \in H^{-1/2}(\Gamma_i)$ for $i = 1, \ldots, q$ such that

$$\sum_{i=1}^{q} \alpha_i \langle D_i \gamma_{0,i}^{\text{int}} u + (\frac{1}{2}I + K_i')\gamma_{1,i}^{\text{int}} u, \gamma_{0,i}^{\text{int}} v\rangle_{\Gamma_i} + \sum_{i=q+1}^{p} \alpha_i \int_{\Omega_i} \nabla u(x) \nabla v(x) dx = 0$$

$$\alpha_i \left[ \langle V_i \gamma_{1,i}^{\text{int}} u - (\frac{1}{2}I + K_i)\gamma_{0,i}^{\text{int}} u, \tau_i \rangle_{\Gamma_i} \right] = 0$$

is satisfied for all $v \in H_0^1(\Omega)$ and$\tau_i \in H^{-1/2}(\Gamma_i)$, $i = 1, \ldots, q$.

By introducing the bilinear form

$$\begin{aligned} a(u, \gamma_{1,1}^{\text{int}} u, \ldots, \gamma_{1,q}^{\text{int}} u; v, \tau_1, \ldots, \tau_q) \\ := \sum_{i=1}^{q} \alpha_i \langle D_i \gamma_{0,i}^{\text{int}} u + (\frac{1}{2}I + K_i')\gamma_{1,i}^{\text{int}} u, \gamma_{0,i}^{\text{int}} v\rangle_{\Gamma_i} \\ + \sum_{i=1}^{q} \alpha_i \left[ \langle V_i \gamma_{1,i}^{\text{int}} u - (\frac{1}{2}I + K_i)\gamma_{0,i}^{\text{int}} u, \tau_i \rangle_{\Gamma_i} \right] \\ + \sum_{i=q+1}^{p} \alpha_i \int_{\Omega_i} \nabla u(x) \nabla v(x) dx \end{aligned}$$

we finally obtain a variational formulation to find $u \in H^1(\Omega)$ with $\gamma_0^{\text{int}} u = g$ and $\gamma_{1,i}^{\text{int}} u \in H^{-1/2}(\Gamma_i)$, $i = 1, \ldots, q$, such that

$$a(u, \underline{\gamma}_1^{\text{int}} u; v, \underline{\tau}) = 0 \tag{15.7}$$

is satisfied for all $v \in H_0^1(\Omega)$ and $\tau_i \in H^{-1/2}(\Gamma_i)$, $i = 1, \ldots, q$.

**Theorem 15.1.** *There exists a unique solution of the variational problem* (15.7).

*Proof.* It is sufficient to prove all assumptions of Theorem 3.8. For this we define

$$X := H_0^1(\Omega) \times H^{-1/2}(\Gamma_1) \times \cdots \times H^{-1/2}(\Gamma_q)$$

where the norm is given by

$$\|(u, \underline{t})\|_X^2 := \sum_{i=1}^{q} \left[ \|\gamma_{0,i}^{\text{int}} u\|_{H^{1/2}(\Gamma_i)}^2 + \|t_i\|_{H^{-1/2}(\Gamma_i)}^2 \right] + \sum_{i=q+1}^{p} \|u\|_{H^1(\Omega_i)}^2 .$$

The boundedness of the bilinear form $a(\cdot, \cdot)$ follows from the boundedness of all local boundary integral operators, and from the boundedness of the local Dirichlet forms.

For arbitrary $(v, \underline{\tau}) \in X$ we have

$$a(v,\underline{\tau};v,\underline{\tau}) = \sum_{i=1}^q \alpha_i \Big[ \langle V_i \tau_i, \tau_i\rangle_{\Gamma_i} + \langle D_i \gamma_{0,i}^{\text{int}} v, \gamma_{0,i}^{\text{int}} v\rangle_{\Gamma_i}\Big] + \sum_{i=q+1}^p \alpha_i \, \|\nabla v\|^2_{L_2(\Omega_i)}$$

$$\geq \min_{i=1,p} \left\{ \alpha_i c_{1,i}^V, \alpha_i c_{1,i}^D, \alpha_i \right\} \left\{ \sum_{i=1}^q \Big[ \|\tau_i\|^2_{H^{-1/2}(\Gamma_i)} |\gamma_{0,i}^{\text{int}} u|^2_{H^{1/2}(\Gamma_i)} \Big] + \sum_{i=q+1}^p \|\nabla v\|^2_{L_2(\Omega_i)} \right\}.$$

Due to $v \in H_0^1(\Omega)$ we therefore conclude the $X$–ellipticity of the bilinear form $a(\cdot,\cdot)$ and thus the unique solvability of the variational problem (15.7). □

Let

$$X_h := S_h^1(\Omega) \times S_h^0(\Gamma_1) \times \cdots \times S_h^0(\Gamma_q) \subset X$$

be a conforming trial space of piecewise linear basis functions to approximate the potential $u \in H_0^1(\Omega)$ and of piecewise constant basis functions to approximate the local Neumann data $\gamma_{1,i}^{\text{int}} u \in H^{-1/2}(\Gamma_i)$, $i = 1, \ldots, q$. All degrees of freedom of the trial space $S_h^1(\Omega) \subset H_0^1(\Omega)$ are depicted in Fig. 15.2.

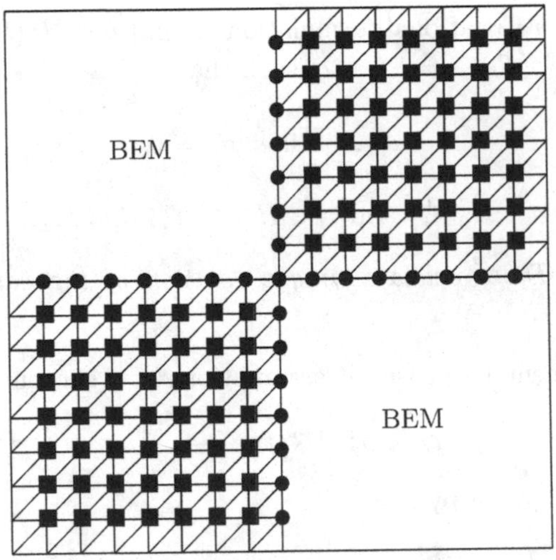


**Fig. 15.2.** Degrees of freedom of the trial space $S_h^1(\Omega) \subset H_0^1(\Omega)$.

The global trial space $S_h^1(\Omega) \subset H_0^1(\Omega)$ is decomposed into local trial spaces

$$S_h^1(\Omega_i) := S_h^1(\Omega)_{|\Omega_i} \cap H_0^1(\Omega_i) = \text{span}\{\varphi_{i,k}^1\}_{k=1}^{M_i}, \quad i = q+1, \ldots, p,$$

and into a global one

$$S_h^1(\Gamma_S) = \text{span}\{\varphi_{S,k}^1\}_{k=1}^{M_S},$$

which is defined with respect to the skeleton $\Gamma_S$. All global degrees of freedom are characterized in Fig. 15.2 by • while all local degrees of freedom correspond to ■ . From the decomposition

$$S_h^1(\Omega) = S_h^1(\Gamma_S) \cup \bigcup_{i=q+1}^{p} S_h^1(\Omega_i)$$

it follows that a function $u_h \in S_h^1(\Omega) \cap H_0^1(\Omega)$ allows the representation

$$u_h(x) = \sum_{k=1}^{M_S} u_{S,k}\varphi_{S,k}^1(x) + \sum_{i=q+1}^{p}\sum_{k=1}^{M_i} u_{i,k}\varphi_{i,k}^1(x).$$

Accordingly, the coefficient vector $\underline{u} \in \mathbb{R}^M$ can be written as

$$\underline{u} = \begin{pmatrix} \underline{u}_S \\ \underline{u}_{q+1} \\ \vdots \\ \underline{u}_p \end{pmatrix} = \begin{pmatrix} \underline{u}_S \\ \underline{u}_L \end{pmatrix}.$$

Finally we introduce the trial space

$$S_h^0(\Gamma_i) = \text{span}\{\varphi_{i,k}^0\}_{k=1}^{N_i} \subset H^{-1/2}(\Gamma_i), \quad i = 1, \ldots, q,$$

of piecewise constant basis functions.

Let $u_g \in H^1(\Omega)$ be a bounded extension of the given Dirichlet datum $g \in H^{1/2}(\Gamma)$. Then the Galerkin variational formulation of (15.7) is to find $u_{0,h} \in S_h^1(\Omega) \cap H_0^1(\Omega)$ and $t_{i,h} \in S_h^0(\Gamma_i)$, $i = 1, \ldots, q$, such that

$$a(u_{0,h} + u_g, \underline{t}_h; v_h, \underline{\tau}_h) = 0 \tag{15.8}$$

is satisfied for all $v_h \in S_h^1(\Omega) \cap H_0^1(\Omega)$ and $\tau_{i,h} \in S_h^0(\Gamma_i)$, $i = 1, \ldots, q$.

By applying Theorem 8.1 (Cea's Lemma) the Galerkin variational formulation (15.8) has a unique solution which satisfies the a priori error estimate

$$\|(u_0 - u_{0,h}, \underline{\gamma}_1^{\text{int}} u - \underline{t}_h)\|_X \leq c \inf_{(v_h, \underline{\tau}_h) \in X_h} \|(u_0 - v_h, \underline{\gamma}_1^{\text{int}} u - \underline{\tau}_h)\|_X.$$

Hence, convergence for $h \to 0$ will follow from the approximation properties of the trial spaces $S_h^1(\Omega)$ and $S_h^0(\Gamma_i)$.

The Galerkin variational formulation (15.8) is equivalent to an algebraic system of linear equations,

$$\begin{pmatrix} V_h & -\frac{1}{2}M_h - K_h & \\ \frac{1}{2}M_h^\top + K_h^\top & D_h + A_{SS} & A_{LS} \\ & A_{SL} & A_{LL} \end{pmatrix} \begin{pmatrix} \underline{t} \\ \underline{u}_S \\ \underline{u}_L \end{pmatrix} = \begin{pmatrix} \underline{f}_B \\ \underline{f}_S \\ \underline{f}_L \end{pmatrix} \tag{15.9}$$

where the global stiffness matrices are given by

$$D_h[\ell,k] = \sum_{i=1}^{q} \alpha_i \langle D_i \gamma_{0,i}^{\text{int}} \varphi_{S,k}^1, \gamma_{0,i}^{\text{int}} \varphi_{S,\ell}^1 \rangle_{\Gamma_i},$$

$$A_{SS}[\ell,k] = \sum_{i=q+1}^{p} \alpha_i \int_{\Omega_i} \nabla \varphi_{S,k}^1(x) \nabla_{S,\ell}^1(x) dx$$

for $k,\ell = 1,\ldots,M_S$, and where the local stiffness matrices are

$$V_h = \operatorname{diag} V_{h,i}, \quad A_{LL} = \operatorname{diag} A_{h,i} \tag{15.10}$$

with

$$V_{h,i} = \alpha_i \langle V_i \varphi_{i,k}^0, \varphi_{i,\ell}^0 \rangle_{\Gamma_i} \quad \text{for } k,\ell = 1,\ldots,N_i, i = 1,\ldots,q,$$

and

$$A_{h,i} = \alpha_i \int_{\Omega_i} \nabla \varphi_{i,k}^1(x) \nabla \varphi_{i,\ell}^1(x) dx \quad \text{for } k,\ell = 1,\ldots,M_i, i = q+1,\ldots,p.$$

In addition, for $k = 1,\ldots,M_S$ and $i = 1,\ldots,q$ we have the block matrices

$$M_{h,i}[\ell,k] = \alpha_i \langle \gamma_{0,i}^{\text{int}} \varphi_{S,k}^1, \varphi_{i,\ell}^0 \rangle_{\Gamma_i},$$
$$K_{h,i}[\ell,k] = \alpha_i \langle K_i \gamma_{0,i}^{\text{int}} \varphi_{S,k}^1, \varphi_{i,\ell}^0 \rangle_{\Gamma_i}$$

for $\ell = 1,\ldots,N_1$, while for $i = q+1,\ldots,p$

$$A_{SL,i}[\ell,k] = \alpha_i \int_{\Omega_i} \nabla \varphi_{S,k}^1(x) \nabla \varphi_{i,\ell}^1(x) dx$$

for $\ell = 1,\ldots,M_i$ and $A_{LS} = A_{SL}^\top$.

The global stiffness matrix of the linear system (15.9) results from an assembling of local stiffness matrices, which stem either from a local boundary element or from a local finite element discretization. The vector of the right hand side in (15.9) correspondingly results from an evaluation of $a(u_g, 0; \cdot, \cdot)$,

$$f_{B,i,\ell} = \alpha_i \langle (\frac{1}{2} + K_i) \gamma_{0,i}^{\text{int}} u_g, \psi_{i,\ell} \rangle_{\Gamma_i}, \quad \ell = 1,\ldots,N_i, i = 1,\ldots,q,$$

$$f_{S,\ell} = -\sum_{i=1}^{q} \alpha_i \langle D_i \gamma_{0,i}^{\text{int}} u_g, \gamma_{0,i}^{\text{int}} \varphi_{S,\ell}^1 \rangle_{\Gamma_i}$$
$$- \sum_{i=q+1}^{p} \alpha_i \int_{\Omega_i} \nabla u_g(x) \nabla \varphi_{S,\ell}^1(x) dx, \quad \ell = 1,\ldots,M_S,$$

$$f_{L,i,\ell} = -\alpha_i \int_{\Omega_i} \nabla u_g(x) \nabla \varphi_{i,\ell}^1(x) dx, \quad \ell = 1,\ldots,M_i, i = q+1,\ldots,p.$$

The linear system (15.9) corresponds to the general system (13.20), hence we can apply all iterative methods of Chapter 13.3 to solve (15.9). In particular, when eliminating the local degrees of freedom $\underline{t}$ and $\underline{u}_L$ we obtain the Schur complement system

$$\left[D_h + (\frac{1}{2}M_h^\top + K_h^\top)V_h^{-1}(\frac{1}{2}M_h + K_h) + A_{SS} - A_{LS}A_{LL}^{-1}A_{SL}\right]\underline{u}_S = \underline{f} \tag{15.11}$$

where the modified right hand side is given by

$$\underline{f} := \underline{f}_S - (\frac{1}{2}M_h^\top + K_h^\top)V_h^{-1}\underline{f}_B - A_{LS}A_{LL}^{-1}\underline{f}_L\,.$$

Due to (15.10) the inversion of the local stiffness matrices $V_h$ and $A_{LL}$ can be done in parallel. This corresponds to the solution of local Dirichlet boundary value problems. In general we have to use local preconditioners for the local stiffness matrices $V_{h,i}$ and $A_{h,i}$. For the solution of the global Schur complement system (15.11) where the system matrix $S_h$ is symmetric and positive definite, we can use a preconditioned conjugate gradient scheme. The definition of an appropriate preconditioning matrix is then based on spectral equivalence inequalities of the corresponding Schur complement matrices,

$$S_h^{\text{BEM}} := D_h + (\frac{1}{2}M_h^\top + K_h^\top)V_h^{-1}(\frac{1}{2}M_h + K_h),$$

and

$$S_h^{\text{FEM}} := A_{SS} - A_{SL}A_{LL}^{-1}A_{LS},$$

and with the Galerkin discretization $D_h$ of the hypersingular boundary integral operator, see, e.g., [36, 139].

# References


1. Adams, R. A.: Sobolev Spaces. Academic Press, New York, London, 1975.
2. Ainsworth, M., Oden, J. T.: A Posteriori Error Estimation in Finite Element Analysis. John Wiley & Sons, New York, 2000.
3. Atkinson, K. E.: The Numerical Solution of Integral Equations of the Second Kind. Cambridge University Press, 1997.
4. Axelsson, O.: Iterative Solution Methods. Cambridge University Press, Cambridge, 1994.
5. Axelsson, O., Barker, V. A.: Finite Element Solution of Boundary Value Problems: Theory and Computation. Academic Press, Orlando, 1984.
6. Aziz, A., Babuška, I.: On the angle condition in the finite element method. SIAM J. Numer. Anal. 13 (1976) 214–226.
7. Babuška, I.: The finite element method with Lagrangian multipliers. Numer. Math. 20 (1973) 179–192.
8. Babuška, I., Strouboulis, T.: The Finite Element Method and its Reliability. The Clarendon Press, New York, 2001.
9. Banerjee, P. K.: The Boundary Element Methods in Engineering. McGraw, London, 1994.
10. Bangerth, W., Rannacher, R.: Adaptive Finite Element Methods for Differential Equations. Birkhäuser, Basel, 2003.
11. Barrett, R. et al.: Templates for the Solution of Linear Systems: Building Blocks for Iterative Methods. SIAM, Philadelphia, 1993.
12. Bebendorf, M.: Effiziente numerische Lösung von Randintegralgleichungen unter Verwendung von Niedrigrang–Matrizen. Dissertation, Universität des Saarlandes, Saarbrücken, 2000.
13. Bebendorf, M.: Approximation of boundary element matrices. Numer. Math. 86 (2000) 565–589.
14. Bebendorf, M., Rjasanow, S.: Adaptive low–rank approximation of collocation matrices. Computing 70 (2003) 1–24.
15. Beer, G.: Programming the Boundary Element Method. John Wiley & Sons, Chichester, 2001.
16. Bergh, J., Löfström, J.: Interpolation Spaces. An Introduction. Springer, Berlin, New York, 1976.
17. Bespalov, A., Heuer, N.: The *p*–version of the boundary element method for hypersingular operators on piecewise plane open surfaces. Numer. Math. 100 (2005) 185–209.

18. Bjørstad, P., Gropp, W., Smith, B.: Domain Decomposition. Parallel Multilevel Methods for Elliptic Partial Differential Equations. Cambridge University Press, 1996.
19. Bonnet, M.: Boundary Integral Equation Methods for Solids and Fluids. John Wiley & Sons, Chichester, 1999.
20. Bowman, F.: Introduction to Bessel Functions. Dover, New York, 1958.
21. Braess, D.: Finite Elemente. Springer, Berlin, 1991.
22. Braess, D.: Finite Elements: Theory, Fast Solvers and Applications in Solid Mechanics. Cambridge University Press, 1997.
23. Brakhage, H., Werner, P.: Über das Dirichletsche Aussenraumproblem für die Helmholtzsche Schwingungsgleichung. Arch. Math. 16 (1965) 325–329.
24. Bramble, J. H.: The Lagrange multiplier method for Dirichlet's problem. Math. Comp. 37 (1981) 1–11.
25. Bramble, J. H.: Multigrid Methods. Pitman Research Notes in Mathematics Series, vol. 294, Longman, Harlow, 1993.
26. Bramble, J. H., Pasciak, J. E.: A preconditioning technique for indefinite systems resulting from mixed approximations of elliptic problems. Math. Comp. 50 (1988) 1–17.
27. Bramble, J. H., Pasciak, J. E., Steinbach, O.: On the stability of the $L_2$ projection in $H^1(\Omega)$. Math. Comp. 71 (2002) 147–156.
28. Bramble, J. H., Pasciak, J. E., Xu, J.: Parallel multilevel preconditioners. Math. Comp. 55 (1990) 1–22.
29. Bramble, J. H., Zlamal, M.: Triangular elements in the finite element method. Math. Comp. 24 (1970) 809–820.
30. Brebbia, C. A., Telles, J. C. F., Wrobel, L. C.: Boundary Element Techniques: Theory and Applications in Engineering. Springer, Berlin, 1984.
31. Brenner, S., Scott, R. L.: The Mathematical Theory of Finite Element Methods. Springer, New York, 1994.
32. Breuer, J.: Wavelet–Approximation der symmetrischen Variationsformulierung von Randintegralgleichungen. Diplomarbeit, Mathematisches Institut A, Universität Stuttgart, 2001.
33. Buffa, A., Hiptmair, R.: Regularized combined field integral equations. Numer. Math. 100 (2005) 1–19.
34. Buffa, A., Sauter, S.: Stabilisation of the acoustic single layer potential on non–smooth domains. SIAM J. Sci. Comput. 28 (2003) 1974–1999.
35. Carstensen, C.: A unifying theory of a posteriori finite element error control. Numer. Math. 100 (2005) 617–637.
36. Carstensen, C., Kuhn, M., Langer, U.: Fast parallel solvers for symmetric boundary element domain decomposition methods. Numer. Math. 79 (1998) 321–347.
37. Carstensen, C., Stephan, E. P.: A posteriori error estimates for boundary element methods. Math. Comp. 64 (1995) 483–500.
38. Carstensen, C., Stephan, E. P.: Adaptive boundary element methods for some first kind integral equations. SIAM J. Numer. Anal. 33 (1996) 2166–2183.
39. Chen, G., Zhou, J.: Boundary Element Methods. Academic Press, New York, 1992.
40. Cheng, A. H.–D., Cheng, D. T.: Heritage and early history of the boundary element method. Engrg. Anal. Boundary Elements 29 (2005) 268–302.
41. Ciarlet, P. G.: The Finite Element Method for Elliptic Problems. North–Holland, 1978.

42. Clement, P.: Approximation by finite element functions using local regularization. RAIRO Anal. Numer. R–2 (1975) 77–84.
43. Costabel, M.: Symmetric methods for the coupling of finite elements and boundary elements. In: Boundary Elements IX (C. A. Brebbia, G. Kuhn, W. L. Wendland eds.), Springer, Berlin, pp. 411–420, 1987.
44. Costabel, M.: Boundary integral operators on Lipschitz domains: Elementary results. SIAM J. Math. Anal. 19 (1988) 613–626.
45. Costabel, M., Stephan, E. P.: Boundary integral equations for mixed boundary value problems in polygonal domains and Galerkin approximations. In: Mathematical Models and Methods in Mechanics. Banach Centre Publ. 15, PWN, Warschau, pp. 175–251,1985.
46. Dahmen, W., Prössdorf, S., Schneider, R.: Wavelet approximation methods for pseudodifferential equations I: Stability and convergence. Math. Z. 215 (1994) 583–620.
47. Dahmen, W., Prössdorf, S., Schneider, R.: Wavelet approximation methods for pseudodifferential equations II: Matrix compression and fast solution. Adv. Comput. Math. 1 (1993) 259–335.
48. Darve, E.: The Fast Multipole Method: Numerical implementation. J. Comp. Phys. 160 (2000) 195–240.
49. Dauge, M.: Elliptic boundary value problems on corner domains. Smoothness and asymptotics of solutions. Lecture Notes in Mathematics, vol. 1341, Springer, Berlin, 1988.
50. Dautray, R., Lions, J. L.: Mathematical Analysis and Numerical Methods for Science and Technology. Volume 4: Integral Equations and Numerical Methods. Springer, Berlin, 1990.
51. Demkowicz, L.: Computing with *hp* Adaptive Finite Elements. Vol. 1. Chapman & Hall/CRC, Boca Raton, 2007.
52. Douglas, C. C., Haase, G., Langer, U.: A Tutorial on Elliptic PDE Solvers and their Parallelization. SIAM, Philadelphia, 2003.
53. Duvaut, G., Lions, J. L.: Inequalities in Mechanics and Physics. Springer, Berlin, 1976.
54. Engleder, S., Steinbach, O.: Modified Boundary Integral Formulations for the Helmholtz Equation. J. Math. Anal. Appl. 331 (2007) 396–407.
55. Erichsen, S., Sauter, S. A.: Efficient automatic quadrature in $3-d$ Galerkin BEM. Comp. Meth. Appl. Mech. Eng. 157 (1998) 215–224.
56. Faermann, B.: Local a posteriori error indicators for the Galerkin discretization of boundary integral equations. Numer. Math. 79 (1998) 43–76.
57. Fix, G. J., Strang, G.: An Analysis of the Finite Element Method. Prentice Hall Inc., Englewood Cliffs, 1973.
58. Fortin, M.: An analysis of the convergence of mixed finite element methods. R.A.I.R.O. Anal. Numer. 11 (1977) 341–354.
59. Fox, L., Huskey, H. D., Wilkinson, J. H.: Notes on the solution of algebraic linear simultaneous equations. Quart. J. Mech. Appl. Math. 1 (1948) 149–173.
60. Gatica, G. N., Hsiao, G. C.: Boundary–Field Equation Methods for a Class of Nonlinear Problems. Pitman Research Notes in Mathematics Series, vol. 331. Longman, Harlow, 1995.
61. Gaul, L., Kögl, M., Wagner, M.: Boundary Element Methods for Engineers and Scientists. Springer, Berlin, 2003.

62. Giebermann, K.: Schnelle Summationsverfahren zur numerischen Lösung von Integralgleichungen für Streuprobleme im $\mathbb{R}^3$. Dissertation, Universität Karlsruhe, 1997.
63. Gradshteyn, I. S., Ryzhik, I. M.: Table of Integrals, Series, and Products. Academic Press, New York, 1980.
64. Greengard, L.: The Rapid Evaluation of Potential Fields in Particle Systems. The MIT Press, Cambridge, MA, 1987.
65. Greengard, L., Rokhlin, V.: A fast algorithm for particle simulations. J. Comput. Phys. 73 (1987) 325–348.
66. Grisvard, P.: Elliptic Problems in Nonsmooth Domains. Pitman, Boston, 1985.
67. Guiggiani, G., Gigante, A.: A general algorithm for multidimensional Cauchy principal value integrals in the boundary element method. ASME J. Appl. Mech. 57 (1990) 906–915.
68. Haase, G.: Parallelisierung numerischer Algorithmen für partielle Differentialgleichungen. B. G. Teubner, Stuttgart, Leipzig, 1999.
69. Hackbusch, W.: Multi–Grid Methods and Applications. Springer, Berlin, 1985.
70. Hackbusch, W.: Iterative Lösung grosser schwachbesetzter Gleichungssysteme. B. G. Teubner, Stuttgart, 1993.
71. Hackbusch, W.: Theorie und Numerik elliptischer Differentialgleichungen. B. G. Teubner, Stuttgart, 1996.
72. Hackbusch, W.: A sparse matrix arithmetic based on $\mathcal{H}$–matrices. I. Introduction to $\mathcal{H}$–matrices. Computing 62 (1999) 89–108.
73. Hackbusch, W., Nowak, Z. P.: On the fast matrix multiplication in the boundary element method by panel clustering. Numer. Math. 54, 463–491 (1989).
74. Hackbusch, W., Wittum, G. (eds.): Boundary Elements: Implementation and Analysis of Advanced Algorithms. Notes on Numerical Fluid Mechanics 54, Vieweg, Braunschweig, 1996.
75. Han, H.: The boundary integro–differential equations of three–dimensional Neumann problem in linear elasticity. Numer. Math. 68 (1994) 269–281.
76. Harbrecht, H.: Wavelet Galerkin schemes for the boundary element method in three dimensions. Doctoral Thesis, TU Chemnitz, 2001.
77. Hartmann, F.: Introduction to Boundary Elements. Springer, Berlin, 1989.
78. Hestenes, M., Stiefel, E.: Methods of conjugate gradients for solving linear systems. J. Res. Nat. Bur. Stand 49 (1952) 409–436.
79. Hörmander, L.: The Analysis of Linear Partial Differential Operators I, Springer, Berlin, 1983.
80. Hsiao, G. C., Stephan, E. P., Wendland, W. L.: On the integral equation method for the plane mixed boundary value problem of the Laplacian. Math. Meth. Appl. Sci. 1 (1979) 265–321.
81. Hsiao, G. C., Wendland, W. L.: A finite element method for some integral equations of the first kind. J. Math. Anal. Appl. 58 (1977) 449–481.
82. Hsiao, G. C., Wendland, W. L.: The Aubin–Nitsche lemma for integral equations. J. Int. Equat. 3 (1981) 299–315.
83. Hsiao, G. C., Wendland, W. L.: Integral Equation Methods for Boundary Value Problems. Springer, Heidelberg, to appear.
84. Jaswon, M. A., Symm, G. T.: Integral Equation Methods in Potential Theory and Elastostatics. Academic Press, London, 1977.
85. Jung, M., Langer, U.: Methode der finiten Elemente für Ingenieure. B. G. Teubner, Stuttgart, Leipzig, Wiesbaden, 2001.

86. Jung, M., Steinbach, O.: A finite element–boundary element algorithm for inhomogeneous boundary value problems. Computing 68 (2002) 1–17.
87. Kieser, R., Schwab, C., Wendland, W. L.: Numerical evaluation of singular and finite–part integrals on curved surfaces using symbolic manipulation. Computing 49 (1992) 279–301.
88. Kress, R.: Linear Integral Equations. Springer, Heidelberg, 1999.
89. Kupradze, V. D.: Three–dimensional problems of the mathematical theory of elasticity and thermoelasticity. North–Holland, Amsterdam, 1979.
90. Kythe, P. K.: Fundamental Solutions for Differential Operators and Applications. Birkhäuser, Boston, 1996.
91. Ladyzenskaja, O. A.: Funktionalanalytische Untersuchungen der Navier–Stokesschen Gleichungen. Akademie–Verlag, Berlin, 1965.
92. Ladyzenskaja, O. A., Ural'ceva, N. N.: Linear and quasilinear elliptic equations. Academic Press, New York, 1968.
93. Lage, C., Sauter, S. A.: Transformation of hypersingular integrals and black–box cubature. Math. Comp. 70 (2001) 223–250.
94. Lage, C., Schwab, C.: Wavelet Galerkin algorithms for boundary integral equations. SIAM J. Sci. Comput. 20 (1999) 2195–2222.
95. Langer, U.: Parallel iterative solution of symmetric coupled fe/be equations via domain decomposition. Contemp. Math. 157 (1994) 335–344.
96. Langer, U., Pusch, D.: Data–sparse algebraic multigrid methods for large scale boundary element equations. Appl. Numer. Math. 54 (2005) 406–424.
97. Langer, U., Steinbach, O.: Coupled boundary and finite element tearing and interconnecting methods. In: Domain Decomposition Methods in Science and Engineering (R. Kornhuber et. al. eds.), Lecture Notes in Computational Science and Engineering, vol. 40, Springer, Heidelberg, pp. 83–97, 2004.
98. Langer, U., Steinbach, O., Wendland, W. L.: Computing and Visualization in Science. Special Issue on Fast Boundary Element Methods in Industrial Applications. Volume 8, Numbers 3–4, 2005.
99. Maischak, M.: *hp*–Methoden für Randintegralgleichungen bei 3D–Problemen. Theorie und Implementierung. Doctoral Thesis, Universität Hannover, 1995.
100. Maischak, M., Stephan, E. P.: The *hp*–version of the boundary element method in $\mathbb{R}^3$. The basic approximation results. Math. Meth. Appl. Sci. 20 (1997) 461–476.
101. Maue, A. W.: Zur Formulierung eines allgemeinen Beugungsproblems durch eine Integralgleichung. Z. f. Physik 126 (1949) 601–618.
102. Mazya, V. G.: Boundary integral equations. In: Analysis IV (V. G. Mazya, S. M. Nikolskii eds.), Encyclopaedia of Mathematical Sciences, vol. 27, Springer, Heidelberg, pp. 127–233, 1991.
103. McLean, W.: Strongly Elliptic Systems and Boundary Integral Equations. Cambridge University Press, 2000.
104. McLean, W., Steinbach, O.: Boundary element preconditioners for a hypersingular boundary integral equation on an intervall. Adv. Comput. Math. 11 (1999) 271–286.
105. McLean, W., Tran, T.: A preconditioning strategy for boundary element Galerkin methods. Numer. Meth. Part. Diff. Eq. 13 (1997) 283–301.
106. Nečas, J.: Les Methodes Directes en Theorie des Equations Elliptiques. Masson, Paris und Academia, Prag, 1967.
107. Nedelec, J. C.: Integral equations with non integrable kernels. Int. Eq. Operator Th. 5 (1982) 562–572.

108. Of, G., Steinbach, O.: A fast multipole boundary element method for a modified hypersingular boundary integral equation. In: Analysis and Simulation of Multifield Problems (W. L. Wendland, M. Efendiev eds.), Lecture Notes in Applied and Computational Mechanics 12, Springer, Heidelberg, 2003, pp. 163–169.
109. Of, G., Steinbach, O., Wendland, W. L.: Applications of a fast multipole Galerkin boundary element method in linear elastostatics. Comput. Vis. Sci. 8 (2005) 201–209.
110. Of, G., Steinbach, O., Wendland, W. L.: The fast multipole method for the symmetric boundary integral formulation. IMA J. Numer. Anal. 26 (2006) 272–296.
111. Petersdorff, T. von, Stephan, E. P.: On the convergence of the multigrid method for a hypersingular integral equation of the first kind. Numer. Math. 57 (1990) 379–391.
112. Plemelj, J.: Potentialtheoretische Untersuchungen. Teubner, Leipzig, 1911.
113. Prössdorf, S., Silbermann, B.: Numerical Analysis for Integral and Related Operator Equations. Birkhäuser, Basel, 1991.
114. Quarteroni, A., Valli, A.: Domain Decomposition Methods for Partial Differential Equations. Oxford Science Publications, 1999.
115. Rathsfeld, A.: Quadrature methods for 2D and 3D problems. J. Comput. Appl. Math. 125 (2000) 439–460.
116. Reidinger, B., Steinbach, O.: A symmetric boundary element method for the Stokes problem in multiple connected domains. Math. Meth. Appl. Sci. 26 (2003) 77–93.
117. Rjasanow, S., Steinbach, O.: The Fast Solution of Boundary Integral Equations. Mathematical and Analytical Techniques with Applications to Engineering. Springer, New York, 2007.
118. Rudin, W.: Functional Analysis. McGraw–Hill, New York, 1973.
119. Ruotsalainen, K., Wendland, W. L.: On the boundary element method for some nonlinear boundary value problems. Numer. Math. 53 (1988) 299–314.
120. Saad, Y., Schultz, M. H.: A generalized minimal residual algorithm for solving nonsymmetric linear systems. SIAM J. Sci. Stat. Comput. 7 (1985) 856–869.
121. Sauer, T., Xu, Y.: On multivariate Lagrange interpolation. Math. Comp. 64 (1995) 1147–1170.
122. Sauter, S. A.: Variable order panel clustering. Computing 64 (2000) 223–261.
123. Sauter, S. A., Schwab, C.: Quadrature of *hp*–Galerkin BEM in $\mathbb{R}^3$. Numer. Math. 78 (1997) 211–258.
124. Sauter, S. A., Schwab, C.: Randelementmethoden. Analyse, Numerik und Implementierung schneller Algorithmen. B. G. Teubner, Stuttgart, Leipzig, Wiesbaden, 2004.
125. Schanz, M., Steinbach, O. (eds.): Boundary Element Analysis: Mathematical Aspects and Applications. Lecture Notes in Applied and Computational Mechanics, vol. 29, Springer, Heidelberg, 2007.
126. Schatz, A. H., Thomée, V., Wendland, W. L.: Mathematical Theory of Finite and Boundary Element Methods. Birkhäuser, Basel, 1990.
127. Schneider, R.: Multiskalen– und Wavelet–Matrixkompression: Analysisbasierte Methoden zur effizienten Lösung grosser vollbesetzter Gleichungssysteme. Advances in Numerical Mathematics. B. G. Teubner, Stuttgart, 1998.
128. Schulz, H., Steinbach, O.: A new a posteriori error estimator in adaptive direct boundary element methods. Calcolo 37 (2000) 79–96.

129. Schulz, H., Wendland, W. L.: Local a posteriori error estimates for boundary element methods. In: ENUMATH 2007, World Sci. Publ., River edge, pp. 564–571, 1998.
130. Schwab, C.: p– and hp–Finite Element Methods. Theory and Applications in Solid and Fluid Mechanics. Clarendon Press, Oxford, 1998.
131. Schwab, C., Suri, M.: The optimal $p$–version approximation of singularities on polyhedra in the boundary element method. SIAM J. Numer. Anal. 33 (1996) 729–759.
132. Schwab, C., Wendland, W. L.: On numerical cubatures of singular surface integrals in boundary element methods. Numer. Math. 62 (1992) 343–369.
133. Scott, L. R., Zhang, S.: Finite element interpolation of nonsmooth functions satisfying boundary conditions. Math. Comp. 54 (1990) 483–493.
134. Sirtori, S.: General stress analysis method by means of integral equations and boundary elements. Meccanica 14 (1979) 210–218.
135. Sloan, I. H.: Error analysis of boundary integral methods. Acta Numerica 92 (1992) 287–339.
136. Steinbach, O.: Fast evaluation of Newton potentials in boundary element methods. East–West J. Numer. Math. 7 (1999) 211–222.
137. Steinbach, O.: On the stability of the $L_2$ projection in fractional Sobolev spaces. Numer. Math. 88 (2001) 367–379.
138. Steinbach, O.: On a generalized $L_2$ projection and some related stability estimates in Sobolev spaces. Numer. Math. 90 (2002) 775–786.
139. Steinbach, O.: Stability estimates for hybrid coupled domain decomposition methods. Lecture Notes in Mathematics 1809, Springer, Heidelberg, 2003.
140. Steinbach, O.: Numerische Näherungsverfahren für elliptische Randwertprobleme. Finite Elemente und Randelemente. B. G. Teubner, Stuttgart, Leipzig, Wiesbaden, 2003.
141. Steinbach, O.: A robust boundary element method for nearly incompressible elasticity. Numer. Math. 95 (2003) 553–562.
142. Steinbach, O.: A note on the ellipticity of the single layer potential in two–dimensional linear elastostatics. J. Math. Anal. Appl. 294 (2004) 1–6.
143. Steinbach, O.: Lösungsverfahren für lineare Gleichungssysteme. Algorithmen und Anwendungen. B. G. Teubner, Stuttgart, Leipzig, Wiesbaden, 2005.
144. Steinbach, O., Wendland, W L.: The construction of some efficient preconditioners in the boundary element method. Adv. Comput. Math. 9 (1998) 191–216.
145. Steinbach, O., Wendland, W. L.: On C. Neumann's method for second order elliptic systems in domains with non–smooth boundaries. J. Math. Anal. Appl. 262 (2001) 733–748.
146. Stephan, E. P.: The h–p version of the boundary element method for solving 2– and 3–dimensional problems. Comp. Meth. Appl. Mech. Eng. 133 (1996) 183–208.
147. Stephan, E. P.: Multilevel methods for the $h$–, $p$–, and $hp$–versions of the boundary element method. J. Comput. Appl. Math. 125 (2000) 503–519.
148. Stroud, A. H.: Approximate Calculations of Multiple Integrals. Prentice Hall, Englewood Cliffs, 1973.
149. Toselli, A., Widlund, O.: Domain Decomposition Methods–Algorithms and Theory. Springer, Berlin, 2005.
150. Triebel, H.: Höhere Analysis. Verlag Harri Deutsch, Frankfurt/M., 1980.
151. Tyrtyshnikov, E. E.: Mosaic–skeleton approximations. Calcolo 33 (1996) 47–57.

152. Verchota, G.: Layer potentials and regularity for the Dirichlet problem for Laplace's equation in Lipschitz domains. J. Funct. Anal. 59 (1984) 572–611.
153. Verfürth, R.: A Review of A Posteriori Error Estimation and Adaptive Mesh–Refinement. John Wiley & Sons, Chichester, 1996.
154. Vladimirov, V. S.: Equations of Mathematical Physics. Marcel Dekker, New York, 1971.
155. van der Vorst, H. A.: Bi–CGSTAB: A fast and smoothly converging variant of Bi–CG for the solution of nonsymmetric linear systems. SIAM J. Sci. Stat. Comput. 13 (1992) 631–644.
156. Walter, W.: Einführung in die Theorie der Distributionen. BI Wissenschaftsverlag, Mannheim, 1994.
157. Wendland, W. L.: Elliptic Systems in the Plane. Pitman, London, 1979.
158. Wendland, W. L. (ed.): Boundary Element Topics. Springer, Heidelberg, 1997.
159. Wendland, W. L., Zhu, J.: The boundary element method for three–dimensional Stokes flow exterior to an open surface. Mathematical and Computer Modelling 15 (1991) 19–42.
160. Wloka, J.: Funktionalanalysis und Anwendungen. Walter de Gruyter, Berlin, 1971.
161. Wloka, J.: Partielle Differentialgleichungen. B. G. Teubner, Stuttgart, 1982.
162. Xu, J.: An introduction to multilevel methods. In: Wavelets, multilevel methods and elliptic PDEs. Numer. Math. Sci. Comput., Oxford University Press, New York, 1997, pp. 213–302.
163. Yosida, K.: Functional Analysis. Springer, Berlin, Heidelberg, 1980.

# Index